The Application of Viruses to Biotechnology

The Application of Viruses to Biotechnology

Editors

Carla Varanda
Patrick Materatski

MDPI • Basel • Beijing • Wuhan • Barcelona • Belgrade • Manchester • Tokyo • Cluj • Tianjin

Editors
Carla Varanda
Mediterranean Institute for
Agriculture, Environment and
Development
University of Evora
Evora
Portugal

Patrick Materatski
Mediterranean Institute for
Agriculture, Environment and
Development
University of Evora
Evora
Portugal

Editorial Office
MDPI
St. Alban-Anlage 66
4052 Basel, Switzerland

This is a reprint of articles from the Special Issue published online in the open access journal *Viruses* (ISSN 1999-4915) (available at: www.mdpi.com/journal/viruses/special_issues/viruses_biotechnology).

For citation purposes, cite each article independently as indicated on the article page online and as indicated below:

LastName, A.A.; LastName, B.B.; LastName, C.C. Article Title. *Journal Name* **Year**, *Volume Number*, Page Range.

ISBN 978-3-0365-2539-6 (Hbk)
ISBN 978-3-0365-2538-9 (PDF)

Contents

About the Editors vii

Carla Varanda, Maria do Rosário Félix, Maria Doroteia Campos and Patrick Materatski
An Overview of the Application of Viruses to Biotechnology
Reprinted from: *Viruses* **2021**, *13*, 2073, doi:10.3390/v13102073 1

Darrick L. Yu, Natalie S. M. Chow, Byram W. Bridle and Sarah K. Wootton
Macrophage Depletion via Clodronate Pretreatment Reduces Transgene Expression from AAV
Vectors In Vivo
Reprinted from: *Viruses* **2021**, *13*, 2002, doi:10.3390/v13102002 7

Srividhya Venkataraman and Kathleen Hefferon
Application of Plant Viruses in Biotechnology, Medicine, and Human Health
Reprinted from: *Viruses* **2021**, *13*, 1697, doi:10.3390/v13091697 21

Jacquelyn J. Bower, Liujiang Song, Prabhakar Bastola and Matthew L. Hirsch
Harnessing the Natural Biology of Adeno-Associated Virus to Enhance the Efficacy of Cancer
Gene Therapy
Reprinted from: *Viruses* **2021**, *13*, 1205, doi:10.3390/v13071205 43

Hasan Arsın, Andrius Jasilionis, Håkon Dahle, Ruth-Anne Sandaa, Runar Stokke, Eva Nordberg Karlsson and Ida Helene Steen
Exploring Codon Adjustment Strategies towards *Escherichia coli*-Based Production of Viral
Proteins Encoded by HTH1, a Novel Prophage of the Marine Bacterium *Hypnocyclicus thermotrophus*
Reprinted from: *Viruses* **2021**, *13*, 1215, doi:10.3390/v13071215 59

Carla M. R. Varanda, Maria do Rosário Félix, Maria Doroteia Campos, Mariana Patanita and Patrick Materatski
Plant Viruses: From Targets to Tools for CRISPR
Reprinted from: *Viruses* **2021**, *13*, 141, doi:10.3390/v13010141 81

Gorben P. Pijlman, Carissa Grose, Tessy A. H. Hick, Herman E. Breukink, Robin van den Braak, Sandra R. Abbo, Corinne Geertsema, Monique M. van Oers, Dirk E. Martens and Dominic Esposito
Relocation of the attTn7 Transgene Insertion Site in Bacmid DNA Enhances Baculovirus
Genome Stability and Recombinant Protein Expression in Insect Cells
Reprinted from: *Viruses* **2020**, *12*, 1448, doi:10.3390/v12121448 101

Zhihao Wang, Jielan Mi, Yulong Wang, Tingting Wang, Xiaole Qi, Kai Li, Qing Pan, Yulong Gao, Li Gao, Changjun Liu, Yanping Zhang, Xiaomei Wang and Hongyu Cui
Recombinant Lactococcus Expressing a Novel Variant of Infectious Bursal Disease Virus VP2
Protein Can Induce Unique Specific Neutralizing Antibodies in Chickens and Provide Complete
Protection
Reprinted from: *Viruses* **2020**, *12*, 1350, doi:10.3390/v12121350 119

Mo Wang, Shilei Gao, Wenzhi Zeng, Yongqing Yang, Junfei Ma and Ying Wang
Plant Virology Delivers Diverse Toolsets for Biotechnology
Reprinted from: *Viruses* **2020**, *12*, 1338, doi:10.3390/v12111338 133

Kenneth Lundstrom
Application of Viral Vectors for Vaccine Development with a Special Emphasis on COVID-19
Reprinted from: *Viruses* **2020**, *12*, 1324, doi:10.3390/v12111324 . . . . **149**

Darrick L. Yu, Natalie Chow and Sarah K. Wootton
JSRV Intragenic Enhancer Element Increases Expression from a Heterologous Promoter and
Promotes High Level AAV-Mediated Transgene Expression in the Lung and Liver of Mice
Reprinted from: *Viruses* **2020**, *12*, 1266, doi:10.3390/v12111266 . . . . **177**

**Laura Fernandez-Garcia, Olga Pacios, Mónica González-Bardanca, Lucia Blasco, Inés Bleriot,
Antón Ambroa, María López, German Bou and Maria Tomás**
Viral Related Tools against SARS-CoV-2
Reprinted from: *Viruses* **2020**, *12*, 1172, doi:10.3390/v12101172 . . . . **191**

Chin-Wei Hsu, Ming-Hao Chang, Hui-Wen Chang, Tzong-Yuan Wu and Yen-Chen Chang
Parenterally Administered Porcine Epidemic Diarrhea Virus-Like Particle-Based Vaccine
Formulated with CCL25/28 Chemokines Induces Systemic and Mucosal Immune Protectivity
in Pigs
Reprinted from: *Viruses* **2020**, *12*, 1122, doi:10.3390/v12101122 . . . . **211**

**Muhammad Bashir Bello, Khatijah Yusoff, Aini Ideris, Mohd Hair-Bejo, Abdurrahman
Hassan Jibril, Ben P. H. Peeters and Abdul Rahman Omar**
Exploring the Prospects of Engineered Newcastle Disease Virus in Modern Vaccinology
Reprinted from: *Viruses* **2020**, *12*, 451, doi:10.3390/v12040451 . . . . **233**

**Kiyoaki Maeda, Teppei Kikuchi, Ichiro Kasajima, Chungjiang Li, Noriko Yamagishi,
Hiroyuki Yamashita and Nobuyuki Yoshikawa**
Virus-Induced Flowering by Apple Latent Spherical Virus Vector: Effective Use to Accelerate
Breeding of Grapevine
Reprinted from: *Viruses* **2020**, *12*, 70, doi:10.3390/v12010070 . . . . **257**

**Karen Fong, Denise M. Tremblay, Pascal Delaquis, Lawrence Goodridge, Roger C. Levesque,
Sylvain Moineau, Curtis A. Suttle and Siyun Wang**
Diversity and Host Specificity Revealed by Biological Characterization and Whole Genome
Sequencing of Bacteriophages Infecting *Salmonella enterica*
Reprinted from: *Viruses* **2019**, *11*, 854, doi:10.3390/v11090854 . . . . **277**

About the Editors

Carla Varanda

Carla Varanda graduated in Agricultural Engineering in University of Évora (UE), Portugal, in 2002, obtained the Master's Degree in Biology of Pests and Diseases of Plants in the UE in 2005 and the PhD in Agricultural Sciences in the UE in 2011. She is a researcher at Mediterranean Institute for Agriculture, Environment and Development (MED), University of Évora. She has dedicated her research to 'Pant Protection', namely to 'Plant Pathology', where she has worked on the development and optimization of methods for diagnosis of plant pathogens and on the development of innovative plant protective methods against several pathogens, by using modified viruses.

Patrick Materatski

Patrick Materatski obtained his Master's Degree in Ecology at University of Coimbra, Portugal, in 2010 and conducted his PhD at the University of Évora, Portugal (2015). He is a researcher at Mediterranean Institute for Agriculture, Environment and Development (MED), University of Évora. His research areas are focused on Plant pathology, more specifically on two main domains. The first concerns the study of plant viruses, either as pathogens and/or as biotechnological tools for plant protection. The second concerns the study of endophytic and pathogenic fungi in important crops, focused on the understanding of plant-pathogen interactions for the development of methods for biological control.

Editorial

An Overview of the Application of Viruses to Biotechnology

Carla Varanda [1,*], **Maria do Rosário Félix** [2], **Maria Doroteia Campos** [1] **and Patrick Materatski** [1,*]

[1] MED–Mediterranean Institute for Agriculture, Environment and Development, Instituto de Investigação e Formação Avançada, Universidade de Évora, Pólo da Mitra, Ap. 94, 7006-554 Évora, Portugal; mdcc@uevora.pt

[2] MED–Mediterranean Institute for Agriculture, Environment and Development & Departamento de Fitotecnia, Escola de Ciências e Tecnologia, Universidade de Évora, Pólo da Mitra, Ap. 94, 7006-554 Évora, Portugal; mrff@uevora.pt

* Correspondence: carlavaranda@uevora.pt (C.V.); pmateratski@uevora.pt (P.M.)

Abstract: Viruses may cause devastating diseases in several organisms; however, they are simple systems that can be manipulated to be beneficial and useful for many purposes in different areas. In medicine, viruses have been used for a long time in vaccines and are now being used as vectors to carry materials for the treatment of diseases, such as cancer, being able to target specific cells. In agriculture, viruses are being studied to introduce desirable characteristics in plants or render resistance to biotic and abiotic stresses. Viruses have been exploited in nanotechnology for the deposition of specific metals and have been shown to be of great benefit to nanomaterial production. They can also be used for different applications in pharmacology, cosmetics, electronics, and other industries. Thus, viruses are no longer only seen as enemies. They have shown enormous potential, covering several important areas in our lives, and they are making our lives easier and better. Although viruses have already proven their potential, there is still a long road ahead. This prompt us to propose this theme in the Special Issue "The application of viruses to biotechnology". We believe that the articles gathered here highlight recent significant advances in the use of viruses in several fields, contributing to the current knowledge on virus applications.

Keywords: gene expression; gene therapy; nanotechnology; vaccines; viral vectors

Citation: Varanda, C.; Félix, M.d.R.; Campos, M.D.; Materatski, P. An Overview of the Application of Viruses to Biotechnology. *Viruses* **2021**, *13*, 2073. https://doi.org/10.3390/v13102073

Received: 6 October 2021
Accepted: 9 October 2021
Published: 14 October 2021

1. Introduction

Viruses are microscopic agents that exist worldwide and are present in humans, animals, plants, and other living organisms in which they can cause devastating diseases. In agriculture, viral diseases may cause losses in all crop yield or be responsible for a drastic decrease in product quality, threatening not only world population nourishment but also the production of fibers, ornamental plants, and medicinal products essential for mankind [1]. In medicine, viruses, such as the ones causing smallpox, influenza, and AIDS, have left their mark on human history, as will also be the case of COVID-19. Therefore, it is not a surprise that viruses do not have the best public image and are fairly seen as our enemies.

The advances of biotechnology and next-generation sequencing technologies have accelerated novel virus discovery, identification, sequencing, and manipulation, showing that they present unique characteristics that place them as valuable tools for a wide variety of biotechnological applications [2]. Viruses possess geometrically sophisticated architectures that make them attractive for materials science and nanotechnology. In addition, they present an efficient machinery and a comprehensive genome structure, which make them easy to manipulate.

Despite this recent technological developments, we can go back to the 18th century to find the first reports of the use of viruses as beneficial, with the first vaccine used against smallpox [3]. By then, it was verified that milkmaids who, due to their close work with cows, contracted cowpox, a cow disease that caused only mild symptoms in humans, and were immune to the human form of the disease, smallpox. After these observations, in 1796, Edward Jenner developed an experiment consisting of scratching a cowpox pustule into the arm of a child who was then exposed to human smallpox virus and revealed to be immune to human smallpox. In the following years, thousands of people were vaccinated by the same way. Of course, not every human disease has an animal analog that confers immunity, and research evolved by developing alternative methods that included the use of disarmed viruses to activate immune system in a putative later virus infection. This was the case of the vaccine against rabies developed by the end of the 19th century by Louis Pasteur as well as vaccines developed later for measles, rubeola, influenza, and polio, to name a few. With the advances of science, new vaccines have been developed with no need to use the entire viral particles and using specific viral proteins instead, overcoming the possible undesirable effect of viral vaccines to cause the disease by themselves. At present, there are extraordinary methods for the production of vaccines. Viral genomes have become easily decodable, and researchers have developed vaccines that rely on the injection of viral RNA into humans so that humans produce viral proteins and activate the immune system. This is the case of the latest vaccines developed for COVID-19 [4]. It is expected, as novel viruses continue to emerge, novel technologies and strategies also continue to arise.

However, vaccines are only one of the many examples of how viruses can be used as beneficial agents. Viruses are involved in many biological processes that have revolutionized some areas, namely gene editing, whose impact is patent through the recent Nobel prize in Chemistry in 2020, attributed to Emmanuelle Carpentier and Jennifer Doudna, for the discovery of tracrRNA as part of the bacteria's ancient immune system, CRISPR/Cas, that disarms viruses by cleaving their DNA and that has already been applied in several areas as a methodology for highly precise changes in genes [5].

Among viruses, the ones that infect bacteria, the bacteriophages, have also been investigated for their potential to be used in therapies, as they are able to target, infect, and destroy specific bacteria and are harmless to humans [6]. Bacteriophages present a very interesting potential, especially to be used against highly antibiotic-resistant bacteria, one of the biggest public health challenges. In addition, bacteriophages have contributed to biotechnology with many important proteins.

Viruses can also be used as vectors by essentially removing their pathogenic parts while retaining their gene-delivery capacities, making them incredibly versatile tools to carry and deliver genetic material. Viral vectors have been used in gene therapy, i.e., for the introduction of functioning genes into human cells. There are several types of viral vectors used in mammalian cells, including lentivirus, adenovirus, and, the most used virus vectors in gene therapy, the adeno-associated virus (AAV). One example of their use is Luxturna, a gene therapy product, approved in the EU since 2018, that uses AAV to deliver a functional copy of a genetically mutated gene into retinal cells and restore vision of patients with progressive vision loss due to that specific gene mutation. Another example of the use of viruses for gene therapy is to treat cancer, as viruses are able to target and specifically infect cancer cells without harming healthy cells and making tumors more visible to immune system.

In addition to all these biotechnological applications focusing medicine and pharmaceutical industry, viruses have also been used in other fields, such as the agriculture and materials industry. Many plant viruses have been developed as vectors for either the expression and production of a specific protein or for silencing by down regulating the expression of a homologous gene leading to a loss of function, also called virus-induced gene silencing (VIGS) [7,8]. Many applications of viruses have been used for agricultural purposes, namely concerning plant breeding and plant protection. Nevertheless, it is interesting to mention that plants have also many advantages to be used in vaccine production, such as the low cost and low risks they entail, showing once more the versatility of the use of viruses in biotechnology.

Although it will obviously never be ignored that viruses are responsible for devastating diseases, it is clear that the more they are studied, the more possibilities they offer to us. They are now on the front line of the most revolutionizing techniques in several fields, providing advances that would not be possible without their existence.

2. Special Issue Overview

This Special Issue of Viruses, "The Application of Viruses to Biotechnology", contains eight original articles and seven reviews that demonstrate the current work developed using viruses in biotechnology. These articles were brought by experts that focus on the development and applications of many viruses in several fields, such as agriculture, the pharmaceutical industry, and medicine.

Bacteriophages, the natural predators of bacteria, besides being a promising alternative to antibiotics and especially so for many antibiotic resistant bacteria, are among the many viruses used in biotechnology, mostly due to their easy manipulation and to the many valuable enzymes they possess, such as DNA polymerases and ligases, as well as lytic enzymes that degrade bacterial cell walls and with high potential for phage therapy [9,10]. The study presented by Arsin et al. (2021) allowed the identification and description of a novel marine prophage found in the gram-negative bacterium *Hypnocyclicus thermotrophus*, with similarities to phages infecting gram-positive bacteria [11]. Authors identified a set of nine genes with putative functions, such as cell and nucleotide lysis and replication, interesting for potential biotechnology applications. In addition, the study by Fong et al. (2019) also allowed the isolation and characterization of several phages of the human pathogen *Salmonella enterica*, showing the high phage diversity and phage–host interactions, providing valuable results for the advance on the developing phage-based applications in sectors such as biocontrol [12].

Adeno-associated virus (AAV) vectors are the top platform for gene delivery for the treatment of many human diseases. The review presented by Bower et al. (2021) summarizes recombinant AAV approaches for the treatment of many types of cancer, including interactions with the cellular host machinery to contribute to the enhancement of current strategies to treat cancer. The research articles by Yu et al. (2021) and Yu et al. (2020) presented in this special issue give great contributions for the development of new gene therapy protocols using AAV vectors and provide significant advances for high-level transgene expression from AAV vectors. Yu et al. (2021) showed that, in opposition to what happens with transgene expression using adenoviral vectors, pretreatment with clodronate liposomes does not appear to improve AAV-mediated gene delivery; in fact, it seems to have the opposite effect [13]. In a different study, Yu et al. (2020) demonstrated that an intragenic transcriptional enhancer element within the 3' end of the env gene of Jaagsiekte sheep retrovirus (JSRV) may be used as a promoter capable of directing transgene expression from AAV vectors in a variety of tissues, particularly liver and lung [14]. These authors also report that these promoter cassettes are small in size, which is suitable for genome constraints of the AAV vector systems.

Baculovirus expression vectors have been used for the commercial production of complex glycoproteins in eukaryotic cells. Genome engineering of single-copy baculovirus infectious clones (bacmids) has been valuable for the study of baculovirus biology. However, despite their potential, bacmids are not yet widely applied as expression vectors mainly due to the easy loss of gene-of-interest (GOI) expression. The study by Pijlman et al. (2020) revealed that the relocation of the attTn7 transgene insertion site away from the mini-F replicon in single-copy baculovirus infectious clones prevents deletion of the gene of interest, thereby resulting in higher and prolonged recombinant protein expression levels [15]. In this work, the authors were able to use a novel bacmid to produce chikungunya virus-like particles for industrial vaccines. This study shows great advances on the use of bacmids as expression vectors, whose limitation is the rapid loss of GOI expression. Hsu et al. (2020) optimized a polycistronic baculovirus expression vector to express virus-like particles (VLPs) containing several portions of *Porcine epidemic diarrhea virus* (PEDV) to elicit immunity against PEDV in pigs [16]. These authors also verified that pigs immunized with VLPs together with a mucosal adjuvant showed a higher protection against PEDV, which is of great interest for the development of other enteric viral vaccines.

The lactic acid bacterium *Lactococcus lactis* is widely used in dairy fermentation and is considered safe to be used as host for biopharmaceutical development [17]. In the study presented in this special issue by Wang et al. (2020), authors constructed a recombinant *Lactococcus lactis* expressing a protein of the novel variant of the infectious bursal disease virus (IBDV), against which the current conventional IBDV vaccine cannot completely protect [18]. With this new vaccine, they were able to immunize chickens, which produced unique, neutralizing antibodies and provide results showing the potential of *L. lactis* in vaccine development.

Among the many viruses that have been used as vaccine vectors is the Newcastle disease virus (NDV), in the review presented by Bello et al. (2020), the molecular biology and approaches to engineer NDV into an efficient vaccine vector is discussed, focusing on the prospects of the virus as a vehicle of vaccines against cancer and infectious diseases in humans and animals [19].

The study by Maeda et al. (2020) gives important contributions on the use of viruses to accelerate plant breeding. Authors used *Apple latent spherical virus* (ALSV) to induce flowering in grapevine (virus-induced flowering, VIF) by expression of the Arabidopsis flowering locus T gene; this study shows the potential of ALSV vectors as VIF to shorten the generation time of grapevine seedlings [20].

This special issue also covered aspects concerning the use of viruses against the new disease that changed the world in the recent years, COVID-19, demonstrating how viruses can be rapidly redirected to fight against a new threat. In the review by Lundstrom (2020), the advantages of using viral particles and RNA replicons and DNA replicon vectors of RNA viruses for vaccine development are presented, with a special emphasis on COVID-19 viral-based vaccines [21]. The review by Fernandez-Garcia et al. (2020) summarizes the recent research on viruses for therapy and diagnosis of COVID-19, namely viral-vector vaccines, bacteriophages to find SARS-CoV-2 antibodies, and their use as a treatment [22].

The latest progress on the development of viruses as vectors in biotechnology and their many applications, such as molecular breeding, functional genomic studies, and vaccines, are reviewed by Wang et al. (2020). These authors summarize available viral vectors for economically important crops [23].

Virus-like particles (VLPs) and virus nanoparticles (VNPs) are increasingly being used for a variety of applications in biotechnology. In the review by Venkataraman and Hefferon (2021), the use of *Tobacco mosaic virus* (TMV), *Potato virus X* (PVX), *Cowpea mosaic virus* (CPMV), and geminiviruses for biotechnological purposes is discussed with a great focus on the major achievements of these viruses as expression vectors in medicine and human health [24].

We also had the privilege to contribute to this special issue by providing a review where we focused on the advances on the CRISPR technology to target viruses and achieve plant viral resistance but also, and in line with this special issue, the use of viruses as vectors for CRISPR technology, discussing the advantages and disadvantages of their use as alternatives to other platforms. It is interesting to mention that during the preparation of the review, an increasing number of studies concerning the use of different and new viruses were constantly being published, and a great effort was made to keep the review as up to date as possible, showing the continuous, growing applications of the use of viruses in biotechnology [25].

Author Contributions: Conceptualization, C.V. and P.M.; writing—original draft preparation, C.V. and P.M.; writing—review and editing, C.V., P.M., M.d.R.F. and M.D.C.; funding acquisition, C.V. and P.M. All authors have read and agreed to the published version of the manuscript.

Funding: This work was funded by the project "Control of olive anthracnose through gene silencing and gene expression using a plant virus vector", with the references ALT20-03-0145-FEDER-028263 and PTDC/ASP-PLA/28263/2017, and the project "Development of a new virus-based vector to control TSWV in tomato plants", with the references ALT20-03-0145-FEDER-028266 and PTDC/ASP-PLA/28266/2017, co-financed by the European Union through the European Regional Development Fund, under the ALENTEJO 2020 (Regional Operational Program of the Alentejo), ALGARVE 2020 (Regional Operational Program of the Algarve) and through the Foundation for Science and Technology, in its national component.

Institutional Review Board Statement: Not applicable.

Informed Consent Statement: Not applicable.

Acknowledgments: We would like to thank all the authors that contributed to this Special Issue. We would also like to thank Gloria Gao, Assistant Editor, for all the support provided that contributed to the success of this SI.

Conflicts of Interest: The authors declare no conflict of interest.

References

1. Jones, R.A.C. Global plant virus disease pandemics and epidemics. *Plants* **2021**, *10*, 233. [CrossRef]
2. Materatski, P.; Jones, S.; Patanita, M.; Campos, M.D.; Dias, A.B.; Felix, M.R.; Varanda, C.M.R. A bipartite geminivirus with a highly divergent genomic organization identified in olive trees may represent a novel evolutionary direction in the family Geminiviridae. *Viruses* **2021**, *13*, 2035. [CrossRef]
3. Fermin, G.; Tennant, P. Chapter 1—Introduction: A Short History of Virology. In *Viruses: Molecular Biology, Host Interactions, and Applications to Biotechnology*, 1st ed.; Tennant, P., Fermin, G., Foster, J.E.B.T.-V., Eds.; Academic Press: New York, NY, USA, 2018; pp. 1–16, ISBN 978-0-12-811257-1.
4. Turner, J.S.; O'Halloran, J.A.; Kalaidina, E.; Kim, W.; Schmitz, A.J.; Zhou, J.Q.; Lei, T.; Thapa, M.; Chen, R.E.; Case, J.B.; et al. SARS-CoV-2 mRNA vaccines induce persistent human germinal centre responses. *Nature* **2021**, *596*, 109–113. [CrossRef]
5. Uyhazi, K.E.; Bennett, J. A CRISPR view of the 2020 nobel prize in chemistry. *J. Clin. Investig.* **2021**, *131*, 1–3. [CrossRef]
6. Ul Haq, I.; Chaudhry, W.N.; Akhtar, M.N.; Andleeb, S.; Qadri, I. Bacteriophages and their implications on future biotechnology: A review. *Virol. J.* **2012**, *9*, 9. [CrossRef]
7. Zaidi, S.S.E.A.; Mansoor, S. Viral vectors for plant genome engineering. *Front. Plant Sci.* **2017**, *8*, 539. [CrossRef] [PubMed]
8. Varanda, C.M.R.; Materatski, P.; Campos, M.D.; Clara, M.I.E.; Nolasco, G.; Félix, M.D.R. Olive mild mosaic virus coat protein and P6 are suppressors of RNA silencing, and their silencing confers resistance against OMMV. *Viruses* **2018**, *10*, 416. [CrossRef]
9. Dale, R.M.K.; McClure, B.A.; Houchins, J.P. A rapid single-stranded cloning strategy for producing a sequential series of overlapping clones for use in DNA sequencing: Application to sequencing the corn mitochondrial 18 S rDNA. *Plasmid* **1985**, *13*, 31–40. [CrossRef]
10. Doherty, A.J.; Ashford, S.R.; Subramanya, H.S.; Wigley, D.B. Bacteriophage T7 DNA Ligase: Overexpression, purification, crystallization, and characterization (*). *J. Biol. Chem.* **1996**, *271*, 11083–11089. [CrossRef] [PubMed]
11. Arsin, H.; Jasilionis, A.; Dahle, H.; Sandaa, R.A.; Stokke, R.; Nordberg Karlsson, E.; Steen, I.H. Exploring codon adjustment strategies towards escherichia coli-based production of viral proteins encoded by hth1, a novel prophage of the marine bacterium hypnocyclicus thermotrophus. *Viruses* **2021**, *13*, 1215. [CrossRef] [PubMed]
12. Fong, K.; Tremblay, D.M.; Delaquis, P.; Goodridge, L.; Levesque, R.C.; Moineau, S.; Suttle, C.A.; Wang, S. Diversity and Host Specificity Revealed by Biological Characterization and Whole Genome Sequencing of Bacteriophages Infecting Salmonella enterica. *Viruses* **2019**, *11*, 854. [CrossRef] [PubMed]
13. Yu, D.L.; Chow, N.S.M.; Bridle, B.W.; Wootton, S.K. Macrophage Depletion via Clodronate Pretreatment Reduces Transgene Expression from AAV Vectors in Vivo. *Viruses* **2021**, *13*, 2002. [CrossRef]
14. Yu, D.L.; Chow, N.; Wootton, S.K. Jsrv intragenic enhancer element increases expression from a heterologous promoter and promotes high level aav-mediated transgene expression in the lung and liver of mice. *Viruses* **2020**, *12*, 1266. [CrossRef]
15. Pijlman, G.P.; Grose, C.; Hick, T.A.H.; Breukink, H.E.; van den Braak, R.; Abbo, S.R.; Geertsema, C.; van Oers, M.M.; Martens, D.E.; Esposito, D. Relocation of the atttn7 transgene insertion site in bacmid dna enhances baculovirus genome stability and recombinant protein expression in insect cells. *Viruses* **2020**, *12*, 1448. [CrossRef]
16. Hsu, C.W.; Chang, M.H.; Chang, H.W.; Wu, T.Y.; Chang, Y.C. Parenterally Administered Porcine Epidemic Diarrhea. *Viruses* **2020**, *12*, 1122. [CrossRef]
17. Jørgensen, C.M.; Vrang, A.; Madsen, S.M. Recombinant protein expression in Lactococcus lactis using the P170 expression system. *FEMS Microbiol. Lett.* **2014**, *351*, 170–178. [CrossRef]
18. Wang, Z.; Mi, J.; Wang, Y.; Wang, T.; Qi, X.; Li, K.; Pan, Q.; Gao, Y.; Gao, L.; Liu, C.; et al. Recombinant lactococcus expressing a novel variant of infectious bursal disease virus vp2 protein can induce unique specific neutralizing antibodies in chickens and provide complete protection. *Viruses* **2020**, *12*, 1350. [CrossRef] [PubMed]
19. Bashir Bello, M.; Yusoff, K.; Ideris, A.; Hair-Bejo, M.; Hassan Jibril, A.; Peeters, B.P.H.; Rahman Omar, A. Exploring the prospects of engineered Newcastle disease virus in modern vaccinology. *Viruses* **2020**, *12*, 451. [CrossRef]

20. Maeda, K.; Kikuchi, T.; Kasajima, I.; Li, C.; Yamagishi, N.; Yamashita, H.; Yoshikawa, N. Virus-induced flowering by apple latent spherical virus vector: Effective use to accelerate breeding of grapevine. *Viruses* **2020**, *12*, 70. [CrossRef]
21. Lundstrom, K. Application of viral vectors for vaccine development with a special emphasis on COVID-19. *Viruses* **2020**, *12*, 1324. [CrossRef] [PubMed]
22. Fernandez-Garcia, L.; Pacios, O.; González-Bardanca, M.; Blasco, L.; Bleriot, I.; Ambroa, A.; López, M.; Bou, G.; Tomás, M. Viral related tools against SARS-CoV-2. *Viruses* **2020**, *12*, 1172. [CrossRef] [PubMed]
23. Wang, M.; Gao, S.; Zeng, W.; Yang, Y.; Ma, J.; Wang, Y. Plant virology delivers diverse toolsets for biotechnology. *Viruses* **2020**, *12*, 1338. [CrossRef] [PubMed]
24. Venkataraman, S.; Hefferon, K. Application of Plant Viruses in Biotechnology, Medicine, and Human Health. *Viruses* **2021**, *13*, 1697. [CrossRef]
25. Varanda, C.M.R.; Félix, M.D.R.; Campos, M.D.; Patanita, M.; Materatski, P. Plant viruses: From targets to tools for crispr. *Viruses* **2021**, *13*, 141. [CrossRef] [PubMed]

Article

Macrophage Depletion via Clodronate Pretreatment Reduces Transgene Expression from AAV Vectors In Vivo

Darrick L. Yu, Natalie S. M. Chow, Byram W. Bridle and Sarah K. Wootton *

Department of Pathobiology, University of Guelph, Guelph, ON N1G 2W1, Canada;
darrickyu@gmail.com (D.L.Y.); nataliesmchow@gmail.com (N.S.M.C.); bbridle@uoguelph.ca (B.W.B.)
* Correspondence: kwootton@uoguelph.ca; Tel.: +1-519-824-4120 (ext. 54729)

Abstract: Adeno-associated virus is a popular gene delivery vehicle for gene therapy studies. A potential roadblock to widespread clinical adoption is the high vector doses required for efficient transduction in vivo, and the potential for subsequent immune responses that may limit prolonged transgene expression. We hypothesized that the depletion of macrophages via systemic delivery of liposome-encapsulated clodronate would improve transgene expression if given prior to systemic AAV vector administration, as has been shown to be the case with adenoviral vectors. Contrary to our expectations, clodronate liposome pretreatment resulted in significantly reduced transgene expression in the liver and heart, but permitted moderate transduction of the white pulp of the spleen. There was a remarkable localization of transgene expression from the red pulp to the center of the white pulp in clodronate-treated mice compared to untreated mice. Similarly, a greater proportion of transgene expression could be observed in the medulla located in the center of the lymph node in mice treated with clodronate-containing liposomes as compared to untreated mice where transgene expression was localized primarily to the cortex. These results underscore the highly significant role that the immune system plays in influencing the distribution and relative numbers of transduced cells in the context of AAV-mediated gene delivery.

Keywords: adeno-associated virus; clodronate; macrophage; gene therapy

check for **updates**

Citation: Yu, D.L.; Chow, N.S.M.; Bridle, B.W.; Wootton, S.K. Macrophage Depletion via Clodronate Pretreatment Reduces Transgene Expression from AAV Vectors In Vivo. *Viruses* **2021**, *13*, 2002. https://doi.org/10.3390/v13102002

Academic Editors: Carla Varanda and Patrick Materatski

Received: 24 August 2021
Accepted: 30 September 2021
Published: 6 October 2021

1. Introduction

Adeno-associated virus (AAV)-mediated gene therapy is approaching a level previously not achieved by any other gene therapy vector. The recent regulatory approval of Luxturna, an AAV gene therapy vector designed to treat patients with a rare form of inherited vision loss [1], and Zolgensma, an AAV gene therapy for type 1 spinal muscular atrophy (SMA) [2], highlights the increasing success such gene therapy strategies are experiencing in the clinic [3]. For continued clinical success, however, there is a need to establish improved vector administration protocols so that maximum transduction of target cells can be achieved while immune responses against the vector itself as well as the transgene product are minimized. In early AAV clinical trials investigating the delivery of factor IX to the liver of hemophilia B patients, it was discovered that although therapeutic levels of factor IX could be achieved at the highest dose of AAV-factor IX, immune responses against the vector capsid limited prolonged expression of the transgene beyond 8 weeks [4]. Previous work has demonstrated that AAV capsids are able to inadvertently package DNA sequences that are entirely devoid of the cis-acting AAV packaging signals, the inverted terminal repeats (ITRs), resulting in the potential for unintended encapsidation of *rep* and *cap* DNA sequences into recombinant vectors [5]. De novo synthesis of capsid proteins after the transduction of host cells might result in the induction of immune responses against the AAV vector and subsequent elimination by the immune system. This severe limitation may be alleviated by using a specialized capsid expression cassette that has an oversized intron, rendering the capsid DNA too large for packaging into an AAV virion [5]. Such strategies may improve the successful transduction of target sites that lie outside of immunoprivileged sites such as the eye. Indeed, AAV trials targeting the eye (subretinal space) have seen more success than tissues more heavily surveyed by the immune system [6]. In a phase 3 trial utilizing an AAV2 vector encoding the hRPE65 required for isomerohydrolase activity of the retinal pigment epithelium, patients demonstrated sustained improvement in both subjective and objective measurements of vision, including improved light sensitivity, visual fields, and functional vision under dim lighting conditions [7], with such benefits lasting for at least three years [8].

In addition to adaptive immunity, innate immune mechanisms might play a role in eliminating much of the administered vector prior to transduction. Marginal zone macrophages play an essential role in eliminating blood-borne pathogens present in circulation as blood filters through the spleen [9,10]. The bacteria *Streptococcus pneumoniae*, *Haemophilus influenzae*, and *Neisseria meningitidis* and viruses adenovirus serotype 5 and cowpox virus may be cleared in this manner [11]. Marginal zone macrophages of the spleen have been shown to accumulate adenovirus serotype 5 in vivo following IV injection [12].

In contrast to the relatively low immunogenicity associated with AAV administration, adenoviral vectors exhibit a much higher degree of adversity when facing the innate and adaptive arms of the immune system [13,14]. AAV can transiently induce the expression of chemokines TNF-alpha, RANTES, IP-10, MIP-1B, MCP-1, and MIP-2 following intravenous administration in mice; however, levels decline to baseline levels 6 h post vector administration [15]. In contrast, adenoviral vectors induced the expression of TNF-alpha and chemokines for over 24 h post administration [15] and induced the expression of inflammatory cytokines in vitro, whereas AAV vectors did not [15]. Several attempts to tame host roadblocks to successful transduction have incorporated the pre-administration of liposome-encapsulated bisphosphonate clodronate in order to deplete phagocytic cells from experimental animals prior to the administration of adenoviral vectors [16–18]. Transient depletion of Kupffer cells through intravenous administration of chlodronate liposomes resulted in improved delivery of adenoviral DNA to the liver, prolonged as well as increased transgene expression, and delayed the clearance of vector DNA [18].

Based on the favorable results observed from the studies utilizing adenovirus, we hypothesized that transduction studies with AAV vectors might also benefit with pretreatment using clodronate-containing liposomes. Mice were given clodronate liposomes, and 48 h later, adeno-associated virus vectors expressing a human placental alkaline phosphatase reporter gene. Contrary to previous reports utilizing adenoviral vectors, clodronate liposome pretreatment resulted in a significant decrease in transgene expression in heart and liver tissue ($p < 0.05$), and a possible trend towards a decrease in the spleen.

2. Materials and Methods

2.1. Cell Culture

Human embryonic kidney (HEK) 293 cells were maintained in Dulbecco's modified Eagle's medium supplemented with 10% fetal bovine serum (GIBCO, Invitrogen, Carlsbad, CA, USA), 100 units/mL penicillin, 100 µg/mL streptomycin, and 2 mM L-glutamine in a humidified 5% CO_2 atmosphere at 37 °C.

2.2. AAV Vector Plasmid Construction

Vectors encoding human alkaline phosphatase reporter gene (hPLAP) were derived from an AAV vector plasmid, $A_{EE}E_1AP$, described previously [19]. The vector was modified to replace the Enzootic-Nasal Tumor Virus-1 (ENTV-1) enhancer/promoter component with enhancer elements derived from Jaagsiekte Sheep Retrovirus (JSRV), acting on the chicken beta actin promoter to drive expression of a human alkaline phosphatase reporter gene (hPLAP). Splicing of the hPLAP encoding transcript was promoted by the presence of a murine leukemia virus polymerase intron found between the enhancer/promoter and the hPLAP gene. Following the hPLAP gene was the SV40 polyA tail for the polyadenylation of transcripts. This enhancer/promoter combination has proven to be effective in driving constitutive expression in a variety of tissues, but is especially active in the lung and liver.

2.3. AAV Vector Production and Quantification

The AAV vector was produced as described previously [19]. Briefly, the plasmid containing the AAV genome was co-transfected into cells along with pDGM6, a packaging plasmid that expresses AAV6 capsid. AAV vector titers were determined by quantitative polymerase chain reaction analysis as previously described [20].

2.4. AAV Vector Delivery

Mouse experiments were performed in accordance with the guidelines set forth by the Canadian Council on Animal Care (CCAC) and Animal Utilization Protocol #3827 (approved 22 September 2017). Eight-week-old C57BL6/J mice were obtained from Charles River Laboratories (Saint-Constant, QC, Canada). Systemic delivery of AAV vectors was performed by intravenous delivery of a phosphate-buffered saline (PBS) solution containing 1×10^{11} vector genomes of AAV vectors injected in a 200 μL volume into the tail vein. Mice were euthanized 1 month post vector administration, and lungs were perfused through the heart with 20 mL of PBS and then separated into individual lobes. For consistency, the same lobe from each mouse was either flash frozen in liquid nitrogen or fixed in 2% paraformaldehyde–PBS for 2.5 h at 22 °C. Half of other major organs, including the liver, spleen, pancreas, lymph node, heart, and kidney, were fixed for 24 h at 22 °C, with the other half placed into liquid nitrogen for subsequent enzymatic assay of hPLAP activity.

2.5. Depletion of Macrophages Using Clodronate Liposomes

Clodronate liposomes were prepared as described previously [21]. Forty-eight hours prior to administration of AAV vectors, mice were given a 33 μL injection of clodronate liposomes through the tail vein [22,23]. At this volume, each mouse received 165 μg of liposome-encapsulated clodronate. A 200 μL injection was also given to determine if the effect could be ramped up in a dose-dependent manner.

2.6. Quantification of Total Splenocytes in Mice

Splenocytes were enumerated in the following manner: spleens were cut in half, with both halves pressed between the frosted ends of microscope slides to make a single-cell suspension, followed by manual counting using an improved Neubauer hemocytometer. Mean and standard deviation were calculated for PBS and the 33 μL treatment group.

2.7. Quantification of Mouse Splenic Macrophages

The method of Rose et al. was used to quantify numbers of marginal zone macrophages [24]. Briefly, spleens were harvested and a cell suspension was made after dicing spleens and treatment with DNase I and collagenase D. 1×10^6 to 2×10^6 cells were pelleted at 1500 rpm for 3 min and supernatant discarded. Cell suspension was incubated with 0.5 μg Fc Block (BD Biosciences, San Jose, USA) for 10 min at RT, followed by a wash with FACS buffer (0.5% BSA in saline). Supernatant was discarded once again and the following antibodies were added to each well: CD45R (B220) clone RA3-6B2 PE-Texas Red (1:400, BD Biosciences), Ly6C clone AL-21 APC-Cy7 (1:500, BD Biosciences), Ly6G clone 1A8 PE (1:400, BD Biosciences), NK1.1 clone PK136 APC (1:300, BD Biosciences), CD11b clone M1/70 FITC (1:160, eBioscience, San Diego, CA, USA), CD11c clone HL3 PE-Cy7 (1:125, BD Biosciences) with 20% 7-AAD in FACS buffer. The stain and cell mixture was incubated for 20 min on ice in the dark, and washed twice in FACS buffer. Lastly, samples were resuspended in FACS buffer and kept on ice, in the dark, until being run on a FACS Aria flow cytometer (BD Biosciences). Mean and standard deviation were calculated for PBS and the 33 μL treatment group, and Student's t-test was used to determine if differences between these groups was significant.

2.8. hPLAP Staining

Tissues were stained for vector-encoded heat-stable hPLAP expression as described previously [19]. Gross pictures of stained tissues were taken using a Zeiss dissecting scope (Zeiss Canada, Toronto, Canada). After fixation tissues were embedded in paraffin wax using an automatic tissue processor, tissue sections (5 μm) were cut and placed on positively charged slides. Since the hPLAP staining intensity is reduced after tissue processing, tissue sections are re-stained to detect hPLAP expression. Tissue sections were deparaffinized and equilibrated in hPLAP buffer (100 mM Tris-HCl pH 8.5, 100 mM NaCl, 50 mM $MgCl_2$) for 5 min. Slides were incubated overnight in hPLAP stain (100 mM Tris-HCl pH 8.5, 100 mM NaCl, 50 mM $MgCl_2$, 0.34 mg/mL nitroblue tetrazolium salt, 0.17 mg/mL X-phos) on rotator in the dark. After washing twice for 2 min in PBS, tissues were counterstained for 1 min with nuclear fast red solution (Millipore Sigma, Oakville, ON, Canada).

2.9. Determination of hPLAP Enzymatic Activity

Mouse tissues were harvested 28 days after AAV vector administration, snap-frozen in liquid nitrogen, and stored at $-80\ ^\circ$C until assayed. Tissues were homogenized in TMNC lysis buffer (50 mM Tris HCl pH 7.5, 5 mM $MgCl_2$, 100 mM NaCl, 4% (*wt/vol*) CHAPS) using a Precellys 24 homogenizer (Bertin Technologies, Montigny-le-Bretonneux, France) with ~200 μL of TMNC buffer in a FastPrep™ Lysing Matrix A tube (MP Bio, Santa Ana, CA, USA). Tissue homogenates were placed in a water bath at 65 $^\circ$C for 1 h to inactivate endogenous heat-labile AP activity and subsequently clarified by centrifugation at 17,900× g for 15 min at 4°C to remove cell debris. The protein content of each sample was determined by the method of Bradford, and the AP activity in tissue lysates was determined by a fluorometric assay using the 4-methylumbelliferyl phosphate (MUP) (Sigma, St. Louis, MO, USA) substrate, as described previously [19]. The mean and standard deviation were calculated for AP activity for treated and non-treated groups and in each of the organs: lung, liver, heart, and spleen. Student's *t*-test (unpaired, two-tailed) was used to determine if differences between clodronate-treated and untreated groups were significant.

3. Results

3.1. Absolute Splenocyte Quantification by Manual Counting and Marginal Zone Macrophage Quantification by Flow Cytometry Following Clodronate Liposome Pretreatment

A decrease in absolute splenocyte numbers was observed for mice given two different doses (33 and 200 μL) of clodronate liposomes (Figure 1A). A trend towards decreasing numbers of total splenocytes was observed in mice given clodronate compared to control mice. Flow cytometric analysis revealed a significant decrease in marginal zone macrophage number was observed following clodronate pretreatment ($p < 0.007$) (Figure 1B). In particular, there was an observed 65% decline in the numbers of marginal zone macrophages following a 33 μL intravenous administration of clodronate liposomes through the tail vein. Increasing the dose to 200 μL resulted in an 80% reduction in marginal zone macrophages; however, increasing the dose to this level may have also affected other cell types as well.

3.2. Gross Analysis of Human Placental Alkaline Phosphatase (hPLAP) Transgene Expression in Various Tissues

hPLAP reporter gene expression, as evidenced by purple staining, was observed in a variety of tissues 28 days following intravenous delivery of 1×10^{11} vg of an AAV6 vector expressing hPLAP with and without prior clodronate treatment. Most notably, expression appeared to be lower in clodronate-treated tissues than in non-treated controls (Figure 2). This was especially evident in the tissues of the heart (Figure 2A) and liver (Figure 2B). Following intravenous AAV administration, the expression of the reporter gene was restricted to particular patches in the lungs of both clodronate-treated and non-treated mice, likely owing to the architecture of the lungs and the fact that intravenously administered vector could access the lungs only at pulmonary or bronchial blood vessels

(Figure 3). However, in terms of improved expression, overall, the trend appeared to favor untreated animals.

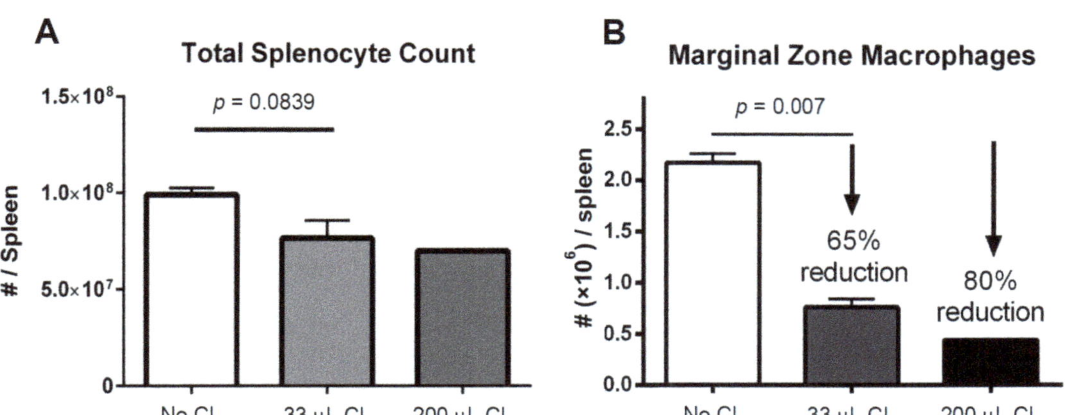

Figure 1. (**A**) Manual count of total numbers of splenocytes following treatment of mice with clodronate liposome treatment (CL) at two different volumes (n = 4 per group). There appeared to be a greater decline in total splenocyte numbers with the higher dose of clodronate liposomes, but this was not statistically significant. (**B**) Marginal zone macrophage count as assessed by flow cytometry following clodronate liposome treatment. Treatment with 33 µL of CL yielded a 65% reduction in numbers of marginal zone macrophages. This difference was statistically significant ($p < 0.05$). With a higher dose of 200 µL, marginal zone macrophage numbers could be reduced even further (80%).

3.3. Quantification of hPLAP Expression in Various Tissues

Enzymatic assay quantification of hPLAP was conducted after the homogenization of treated and untreated mouse tissues, as described previously [19]. Mean and standard deviation were calculated for treated and untreated groups for lung, liver, heart, and spleen, and Student's *t*-test (unpaired, two-tailed) was used to determine if differences were significant ($p < 0.05$). Pooled data from two independent experiments (n = 3 and n = 4 for the first and second experiment, respectively) demonstrated a significant decrease in expression for clodronate-treated mice in the liver and heart compared to untreated mice ($p < 0.005$ and $p < 0.05$, respectively). No significant trend was observed in the lungs, probably due to the low level of expression in this organ. A possible statistical trend towards a decrease in transgene expression was observed in the spleen ($p < 0.1$) (Figure 4).

3.4. Histological Analysis of hPLAP Transgene Expression in Clodronate-Treated Mice and PBS-Treated Controls

Tissues of untreated mice and mice treated with clodronate were paraffin embedded, sectioned, and viewed under a light microscope. Differences were observed in the pattern of expression of hPLAP within the spleens of treated and untreated mice (Figure 5). A greater proportion of hPLAP expression was observed within the center of white pulp in clodronate-treated mice compared to untreated mice. Due to the decline in marginal zone macrophage numbers because of clodronate treatment, a greater proportion of the vector may have been able to penetrate into the white pulp and persist long enough to transduce cells within the white pulp as compared to the non-treated mice. The white pulp of the spleen is a lymph node-like organ composed of lymphoid sheaths with both B and T cell compartments surrounding a central arteriole [25]. Interestingly, marginal zone B cells can become potent antigen presenting cells after uptake of soluble antigen [25]. Dendritic cells can also migrate into the white pulp following activation [25]. The dark alkaline phosphatase staining observed in the white pulp of clodronate-treated mice may have been antigen presenting cells that had been transduced by the vector and subsequently migrated into the white pulp following activation. A similar pattern of expression was observed for the lymph node, where a greater proportion of expression could be observed in the medulla located in the center of the lymph node.

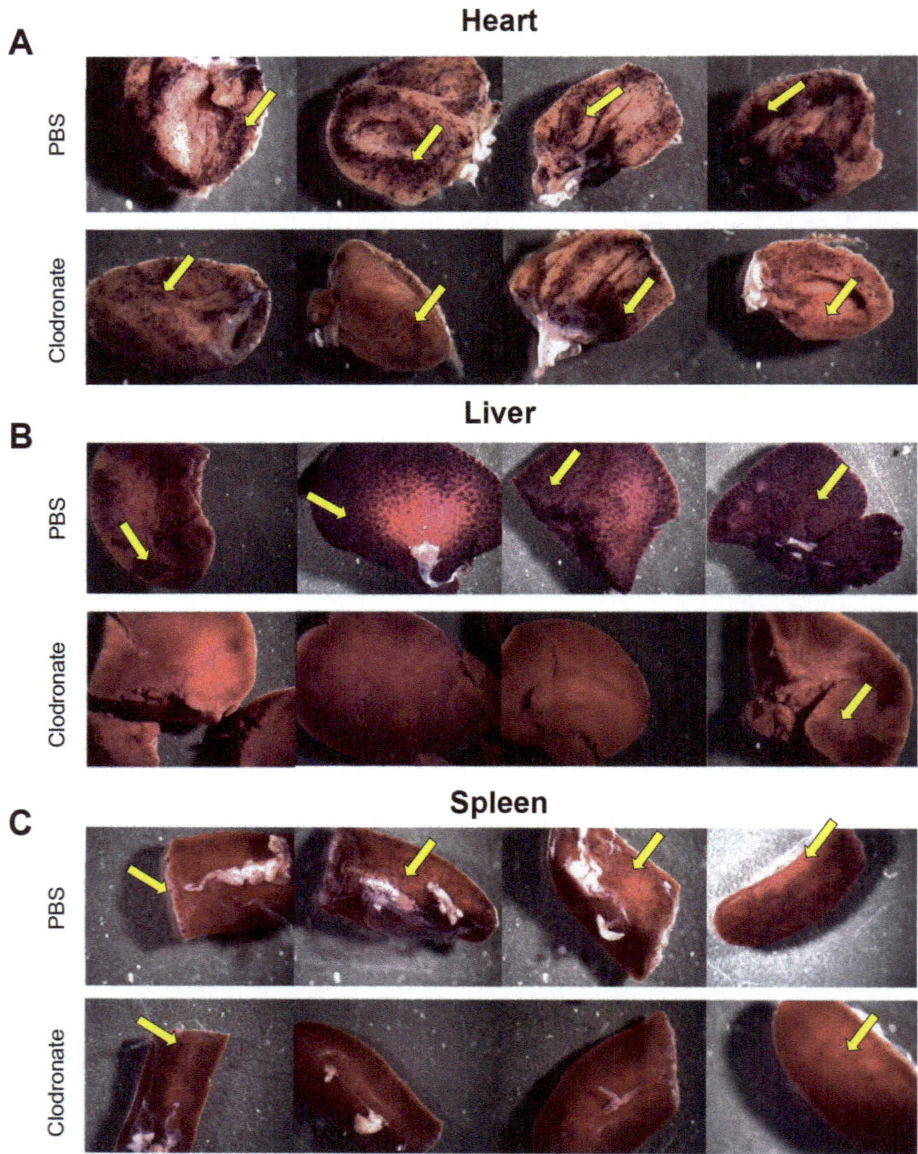

Figure 2. Gross comparison of mice (n = 4) transduced systemically (via tail vein) with 1×10^{11} vg of AAV6 vector expressing human placental alkaline phosphatase (hPLAP) after being given an intravenous injection of clodronate liposomes or PBS, 48 h prior. Mice were euthanized 28 days post-AAV administration and tissues fixed and stained for hPLAP expression. There appeared to be a trend towards more expression in clodronate-untreated mice, particularly in the heart (**A**) and liver (**B**), but less so in the spleen (**C**). Images pictured are from individual mice. Yellow arrows point to regions of purple punctate staining representative of hPLAP expression.

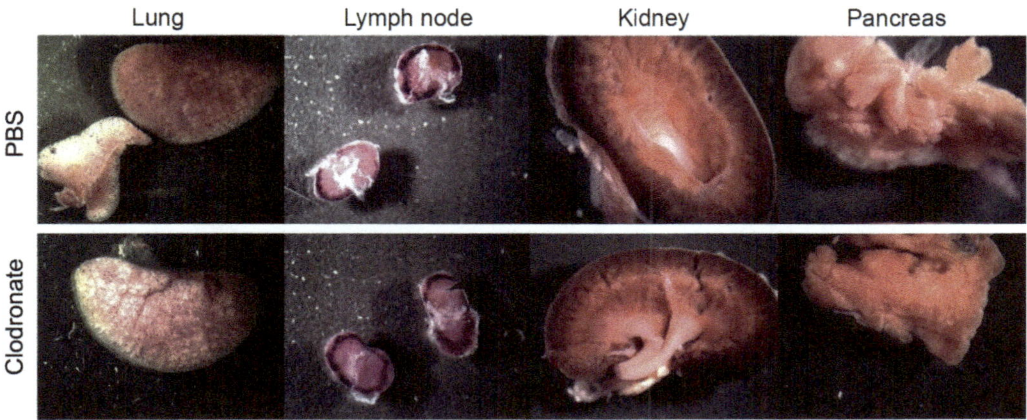

Figure 3. Representative gross images of lung, lymph node, spleen, and kidney from clodronate-treated mice and control mice. Both groups ($n = 4$) were administered 1×10^{11} vg of AAV vector expressing human placental alkaline phosphatase (hPLAP) via tail vein injection and euthanized 28 days later. Tissues were fixed and stained for hPLAP expression. There did not appear to be grossly visible differences between these two groups of mice in these particular organs. Expression of hPLAP in these organs was relatively low.

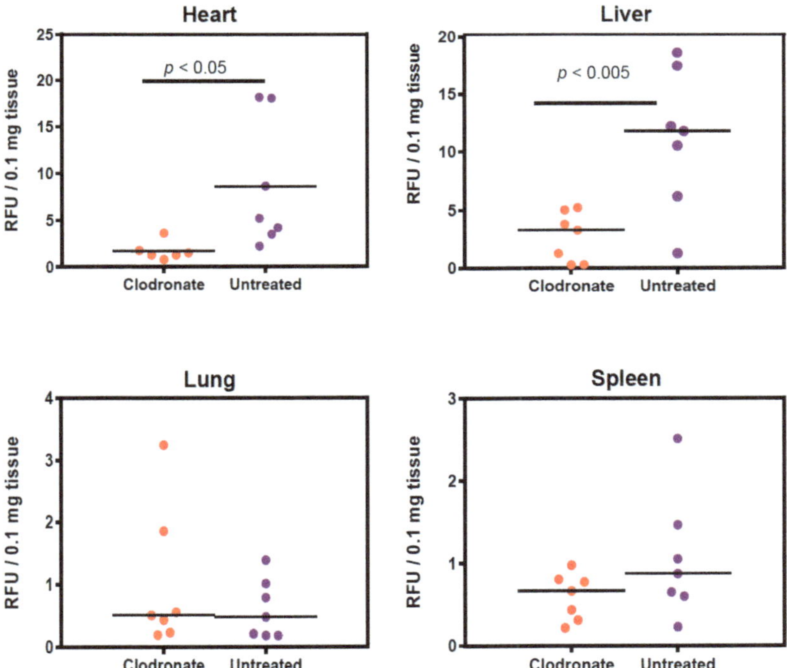

Figure 4. Quantification of alkaline phosphatase activity following systemic administration of 1×10^{11} vg of an AAV6 vector expressing human placental alkaline phosphatase (hPLAP) with and without prior intravenous administration of clodronate. In the heart ($p < 0.05$) and liver ($p < 0.005$), a statistically significant difference could be observed in clodronate-treated and untreated mice, with untreated mice possessing on average ~4 fold more alkaline phosphatase activity. No significant difference could be observed in the levels of alkaline phosphatase activity in mouse lung and spleen.

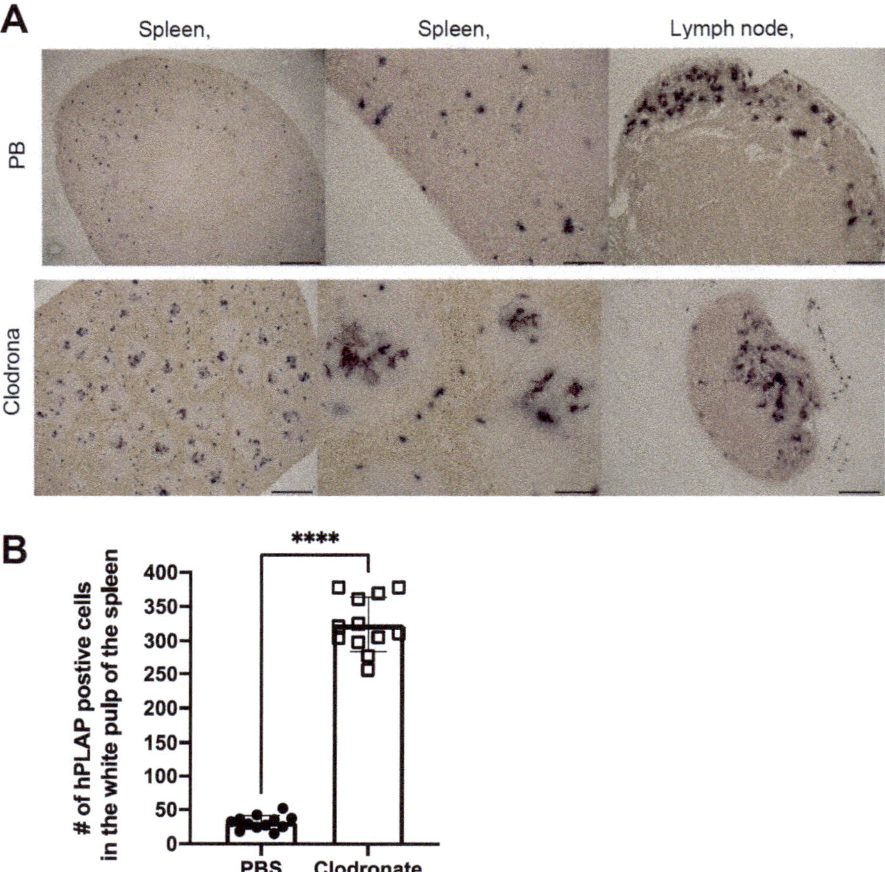

Figure 5. Histochemical staining of sections from spleen and lymph node in clodronate-treated and control mice. (**A**) Representative tissue sections of spleen and lymph node from clodronate-treated mice and control mice. Both groups were administered 1×10^{11} vg of AAV vector expressing human placental alkaline phosphatase (hPLAP) via tail vein injection and euthanized 28 days later. Tissues were formalin fixed, paraffin embedded, and stained for hPLAP expression. Administration of clodronate appeared to allow improved AAV transduction of white pulp relative to control mice. hPLAP staining in the center of the white pulp could be observed for clodronate-treated mice, whereas with control mice, no such staining could be observed. The presence of marginal zone macrophages likely prevented efficient transduction of white pulp in control mice. Within the lymph node, staining seemed to be restricted to the periphery of the lymph node for control mice, whereas in clodronate-treated mice, staining could be observed in the center. (**B**) Quantification of hPLAP positive foci in the white pulp of spleens from clodronate-treated and control mice. Foci of hPLAP stained cells in the white pulp of the spleen from three tissue sections (4×) per mouse per group were counted. 4× scale bar = 200 μM, 10× scale bar = 100 μM, 20× scale bar = 50 μM. Data were analyzed using an unpaired *t*-test, **** = $p < 0.0001$.

As was observed grossly, there was a marked reduction in transgene expression in tissues sections from the heart and liver of clodronate-treated mice, whereas there was no obvious difference in hPLAP staining in lung sections from clodronate-treated mice compared to PBS controls (Figure 6). Little to no transgene expression was observed in tissue sections from the kidney and pancreas in either treatment group. In tissues where transgene expression was detected, there did not appear to be a difference in the cell types that were transduced.

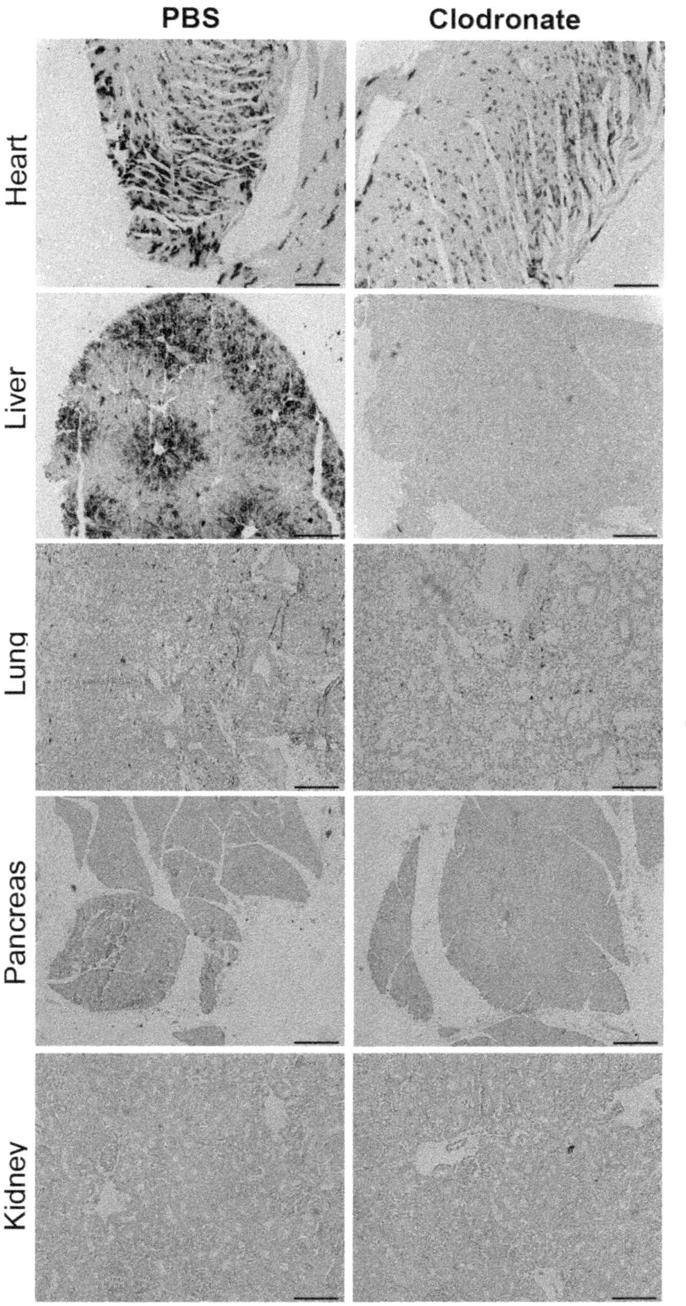

Figure 6. Representative tissue sections of the heart, liver, lung, and pancreas from clodronate-treated and control mice (PBS). Both groups were administered 1×10^{11} vg of AAV vector expressing human placental alkaline phosphatase (hPLAP) via tail vein injection and euthanized 28 days later. Tissues were formalin fixed, paraffin embedded, and stained for hPLAP expression. Images were taken at 10× magnification for all tissues except the pancreas, which was imaged at 4×. 4× scale bar = 200 μM, 10× scale bar = 100 μM.

4. Discussion

In contrast to the improvements in transgene expression observed with adenoviral vectors, pretreatment with clodronate liposomes did not appear to improve AAV-mediated gene delivery, and may have actually had the opposite effect. Significantly reduced expression was observed in mice that had been pretreated with clodronate, particularly in the heart and liver. Little or no difference was observed in transgene levels in the lungs between treated and non-treated groups. This might be attributed to the fact that the transduction of the lungs was poor due to the intravenous route of administration of the AAV vector, resulting in only minute patches of transduction where the vasculature is able to access lung tissue. However, in all other tissues assayed, transduction was lower in the clodronate-treated group. It is unclear why transduction might be lower in these tissues; however, several theories are postulated. Intravenous injection of clodronate liposomes ensures that all phagocytes having direct access to the blood can potentially engulf clodronate liposomes [26]. While clodronate-containing liposomes can deplete all professional phagocytes, including both macrophages and dendritic cells, the liver and spleen are so efficient at removing foreign particles from the bloodstream that the exposure of phagocytes outside of these organs to blood-borne clodronate liposomes would be limited [27]. Particular macrophage populations eliminated by clodronate treatment may have a role in tolerizing the host to components of the vector or the transgene itself. For example, the liver has been known to participate in the induction of tolerance to foreign antigens [28]. Breous and colleagues determined that resident liver macrophages, known as Kupffer cells, can work in conjunction with hepatic regulatory T cells to create a local immunosuppressive environment that inhibits the establishment of a cytotoxic T lymphocyte (CTL) response [29]. Specifically, they found that the elimination of the Kupffer cell population using clodronate was able to completely abolish the expression of the immunosuppressive cytokine IL-10. Furthermore, primary administration of an AAV vector encoding human alpha-1 anti-trypsin (hAAT) enabled systemic tolerance to a secondary administration of an adenoviral vector also encoding hAAT, resulting in greatly improved expression compared to the adenovirus vector alone. The induction of tolerance was shown to be specific to hAAT, as transduction with an Ad-LacZ vector after prior AAV-hAAT administration elicited a strong CTL response [29]. To the best of our knowledge, there are no reports in the literature of macrophages facilitating the distribution of AAV particles to other target tissues, although it remains a possibility. However, it is well known that tissue macrophages engulf AAV non-specifically by phagocytosis, process the capsid, and following migration to a draining lymph node, present capsid-derived peptides to effector lymphocytes [30].

The results demonstrated herein and in prior publications underscore the highly significant role that the immune system plays in sculpting the distribution and relative numbers of transduced cells in the context of AAV-mediated gene delivery. It appears that the phagocytic activity of macrophages is greatly outweighed by the contributions of adaptive immunity, at least for prolonged periods of transgene expression matching or exceeding four weeks. The inhibition of phagocytic uptake by macrophage depletion via clodronate liposomes may, however, be beneficial when a transient, limited period of expression is desired. Initial transduction may be greater under these circumstances due to decreased numbers of phagocytic cells capable of ingesting vector particles, but would likely decrease over time as an immune response against the transgene is mounted. Future experiments will evaluate AAV-mediated transgene expression in clodronate liposome-treated and untreated mice at earlier time points (e.g., 10–14 days post-transduction) so as to allow for the quantification of transgene expression prior to the onset of the host immune response to the therapy. Additionally, we would like to investigate the effect of administering clodronate liposomes via the intranasal route to specifically deplete alveolar macrophages [31], which might allow for greater lung specific transduction while leaving Kupffer cells intact.

In contrast to the results presented here, gene delivery studies by Wolff and colleagues employing adenovirus resulted in a marked increase in transgene expression when clodronate pretreatment occurred in advance of adenovirus transduction [18]. This difference may be attributed to differences in the biology of adenovirus and adeno-associated virus. Whereas adenovirus is highly immunogenic, inducing the expression of inflammatory cytokines for prolonged periods of time upon cell transduction, AAV is far less so [15]. Therefore, the innately high immunogenicity of adenovirus may make it more difficult to induce tolerance to these vectors, and therefore it might not occur as readily as with adeno-associated virus. Thus, the inhibition of phagocytosis might be a relatively less important process than removing phagocytic cells prior to administration, within the particular context of adenovirus vectors. In addition, differences in vector titers at the point of administration might account for differences in transgene expression between Ad and AAV vectors. Whereas in our study an AAV vector dose of 1×10^{11} vector genomes was delivered intravenously, Wolff et al. administered either 5×10^6 or 1×10^7 PFU [18]. Comparisons between genome copy number and plaque forming units might be fraught with difficulty; however, if taken at face value, the number of vector particles was greater in favor of AAV by a factor of 10,000. This number also does not account for the presence of empty AAV capsid particles, which capsids can vary from 10% to 90% [32]. Empty particles might therefore act in a "blocking" fashion, possibly overwhelming the phagocytic capacity of macrophages and thereby making the transient elimination of phagocytic cells unnecessary for AAV transduction [22]. In another study by Aalbers et al., intravenous administration of clodronate liposomes 48 h prior to intra-articular AAV5 vector administration improved transgene expression [22]. One reason for the discrepancy between their findings and ours may have to do with how AAV was administered. AAV was administered systemically in our study, whereas AAV was administered locally to the knee joint in the study by Aalbers et al., thereby limiting the exposure of AAV to synovial macrophages.

Future gene therapy protocols may seek to increase the number of Kupffer cells rather than eliminate them, or to promote the interaction of Kupffer cells with hepatic regulatory T cells with the aim of boosting tolerance [29]. Improved protocols might also incorporate targeted transduction of the liver in order to increase the likelihood of the induction of tolerance, even if the target tissue is distal to the liver. With increased immune tolerance, a reduction in the number of vector particles required for efficient transduction might be possible, leading to the decreased presentation of vector-associated antigens and opening the possibility of repeated vector administrations using the same capsid. A reduction in the number of vector particles required for successful transduction in large animals such as humans would likewise also be a welcome change, as the production of the vast quantities of AAV vectors needed for clinical use is difficult and costly.

Author Contributions: Conceptualization, D.L.Y. and S.K.W.; methodology, D.L.Y., N.S.M.C. and B.W.B.; formal analysis, D.L.Y.; writing—original draft preparation, D.L.Y.; writing—review and editing, S.K.W.; supervision, S.K.W.; funding acquisition, S.K.W. All authors have read and agreed to the published version of the manuscript.

Funding: This research was funded by NSERC Discovery Grant number RGPIN-2013-04737.

Institutional Review Board Statement: The study was conducted in accordance with the Canadian Council on Animal Care (CCAC) guidelines, and approved by the Animal Care Committee at the University of Guelph (animal utilization protocol 3827 approved 22 September 2017).

Informed Consent Statement: Not applicable.

Data Availability Statement: Raw data were generated at the University of Guelph. Derived data supporting the findings of this study are available from the corresponding author on request.

Acknowledgments: We would like to thank all those who were involved in the care of the animals for these studies.

Conflicts of Interest: The authors declare no conflict of interest. The funders had no role in the design of the study; in the collection, analyses, or interpretation of data; in the writing of the manuscript, or in the decision to publish the results.

References

1. Darrow, J.J. Luxturna: FDA documents reveal the value of a costly gene therapy. *Drug Discov. Today* **2019**, *24*, 949–954. [CrossRef]
2. Hoy, S.M. Onasemnogene Abeparvovec: First Global Approval. *Drugs* **2019**, *79*, 1255–1262. [CrossRef]
3. He, X.; Urip, B.A.; Zhang, Z.; Ngan, C.C.; Feng, B. Evolving AAV-delivered therapeutics towards ultimate cures. *J. Mol. Med.* **2021**. [CrossRef] [PubMed]
4. Manno, C.S.; Pierce, G.F.; Arruda, V.R.; Glader, B.; Ragni, M.; Rasko, J.J.; Ozelo, M.C.; Hoots, K.; Blatt, P.; Konkle, B.; et al. Successful transduction of liver in hemophilia by AAV-Factor IX and limitations imposed by the host immune response. *Nat. Med.* **2006**, *12*, 342–347. [CrossRef] [PubMed]
5. Halbert, C.L.; Metzger, M.J.; Lam, S.L.; Miller, A.D. Capsid-expressing DNA in AAV vectors and its elimination by use of an oversize capsid gene for vector production. *Gene Ther.* **2011**, *18*, 411–417. [CrossRef] [PubMed]
6. Tuohy, G.P.; Megaw, R. A Systematic Review and Meta-Analyses of Interventional Clinical Trial Studies for Gene Therapies for the Inherited Retinal Degenerations (IRDs). *Biomolecules* **2021**, *11*, 760. [CrossRef]
7. Russell, S.; Bennett, J.; Wellman, J.A.; Chung, C.D.; Zi-Fan, Y.; Tillman, A.; Wittes, J.; Pappas, J.; Elci, O.; McCague, S.; et al. Efficacy and safety of voretigene neparvovec (AAV2-hRPE65v2) in patients with RPE65-mediated inherited retinal dystrophy: A randomised, controlled, open-label, phase 3 trial. *Lancet* **2017**, *390*, 849–860. [CrossRef]
8. Bennett, J.; Wellman, J.; Marshall, A.K.; McCague, S.; Ashtari, M.; DiStefano-Pappas, J.; Elci, U.O.; Chung, C.D.; Sun, J.; Wright, J.F.; et al. Safety and durability of effect of contralateral-eye administration of AAV2 gene therapy in patients with childhood-onset blindness caused by RPE65 mutations: A follow-on phase 1 trial. *Lancet* **2016**, *388*, 661–672. [CrossRef]
9. A-Gonzalez, N.; Castrillo, A. Origin and specialization of splenic macrophages. *Cell Immunol.* **2018**, *330*, 151–158. [CrossRef]
10. Kashimura, M. The human spleen as the center of the blood defense system. *Int. J. Hematol.* **2020**, *112*, 147–158. [CrossRef] [PubMed]
11. Den Haan, J.M.; Kraal, G. Innate immune functions of macrophage subpopulations in the spleen. *J. Innate Immun.* **2021**, *4*, 437–445. [CrossRef]
12. Di Paolo, N.C.; Miao, E.; Iwakura, Y.; Murali-Krishna, K.; Aderem, A.; Flavell, R.A.; Papayannopoulou, T.; Shayakhmetov, D.M.; Miao, E.; Iwakura, Y.; et al. Virus Binding to a Plasma Membrane Receptor Triggers Interleukin-1 alphaα-Mediated Proinflammatory Macrophage Response In Vivo. *Immunity* **2009**, *31*, 110–121. [CrossRef]
13. Gilgenkrantz, H.; Duboc, D.; Juillard, V.; Couton, D.; Pavirani, A.; Guillet, G.J.; Briand, P.; Kahn, A. Transient expression of genes transferred in vivo into heart using first-generation adenoviral vectors: Role of the immune response. *Hum. Gene Ther.* **1995**, *6*, 1265–1274. [CrossRef]
14. Yang, Y.; Ertl, H.C.; Wilson, J.M. MHC class I-restricted cytotoxic T lymphocytes to viral antigens destroy hepatocytes in mice infected with E1-deleted recombinant adenoviruses. *Immunity* **1994**, *1*, 433–442. [CrossRef]
15. Zaiss, A.K.; Liu, Q.; Bowen, G.P.; Wong, N.C.W.; Bartlett, J.S.; Muruve, D.A.; Liu, Q.; Bowen, G.P.; Wong, N.C.W.; Bartlett, J.S.; et al. Differential Activation of Innate Immune Responses by Adenovirus and Adeno-Associated Virus Vectors. *J. Virol.* **2002**, *76*, 4580–4590. [CrossRef]
16. Wang, S.; Baum, B.J.; Kagami, H.; Zheng, C.; O'Connell, B.C.; Atkinson, J.C. Effect of clodronate on macrophage depletion and adenoviral-mediated transgene expression in salivary glands. *J. Oral Pathol. Med.* **1999**, *28*, 145–151. [CrossRef]
17. Alzuguren, P.; Hervas-Stubbs, S.; Gonzalez-Aseguinolaza, G.; Poutou, J.; Fortes, P.; Mancheño, U.; Buñuales, M.; Olagüe, C.; Razquin, N.; Van Rooijen, N.; et al. Transient depletion of specific immune cell populations to improve adenovirus-mediated transgene expression in the liver. *Liver Int.* **2014**, *35*, 1274–1289. [CrossRef]
18. Wolff, G.; Worgall, S.; van Rooijen, N.; Song, W.R.; Harvey, B.G.; Crystal, R.G. Enhancement of in vivo adenovirus-mediated gene transfer and expression by prior depletion of tissue macrophages in the target organ. *J. Virol.* **1997**, *71*, 624–629. [CrossRef]
19. Yu, D.L.; Linnerth-Petrik, N.M.; Halbert, C.L.; Walsh, S.R.; Miller, A.D.; Wootton, S.K. Jaagsiekte Sheep Retrovirus and Enzootic Nasal Tumor Virus Promoters Drive Gene Expression in All Airway Epithelial Cells of Mice but Only Induce Tumors in the Alveolar Region of the Lungs. *J. Virol.* **2011**, *85*, 7535–7545. [CrossRef]
20. Rghei, A.D.; Stevens, B.A.Y.; Thomas, S.P.; Yates, J.G.E.; McLeod, B.M.; Karimi, K.; Susta, L.; Bridle, B.W.; Wootton, S.K. Production of Adeno-Associated Virus Vectors in Cell Stacks for Preclinical Studies in Large Animal Models. *J. Vis. Exp.* **2021**, e62727. [CrossRef]
21. Van Rooijen, N.; van Kesteren-Hendrikx, E. "In vivo" depletion of macrophages by liposome-mediated "suicide". *Methods Enzymol.* **2003**, *373*, 3–16.
22. Aalbers, C.J.; Broekstra, N.; van Geldorp, M.; Kramer, E.; Ramiro, S.; Tak, P.P.; Vervoordeldonk, M.J.; Finn, J.D. Empty Capsids and Macrophage Inhibition/Depletion Increase rAAV Transgene Expression in Joints of Both Healthy and Arthritic Mice. *Hum. Gene Ther.* **2017**, *28*, 168–178. [CrossRef]
23. Van Rooijen, N.; Hendrikx, E. Liposomes for specific depletion of macrophages from organs and tissues. *Methods Mol. Biol.* **2010**, *605*, 189–203.

24. Rose, S.; Misharin, A.; Perlman, H. A novel Ly6C/Ly6G-based strategy to analyze the mouse splenic myeloid compartment. *Cytometry* **2012**, *81*, 343–350. [CrossRef]
25. Mebius, R.E.; Kraal, G. Structure and function of the spleen. *Nat. Rev. Immunol.* **2005**, *5*, 606–616. [CrossRef]
26. Patrakka, J.; Tryggvason, K. Molecular make-up of the glomerular filtration barrier. *Biochem. Biophys. Res. Commun.* **2010**, *396*, 164–169. [CrossRef] [PubMed]
27. Weisser, S.B.; van Rooijen, N.; Sly, L.M. Depletion and reconstitution of macrophages in mice. *J. Vis. Exp.* **2012**, *66*, 4105. [CrossRef]
28. Erhardt, A.; Biburger, M.; Papadopoulos, T.; Tiegs, G. IL-10, regulatory T cells, and Kupffer cells mediate tolerance in concanavalin A-induced liver injury in mice. *Hepatology* **2007**, *45*, 475–485. [CrossRef] [PubMed]
29. Breous, E.; Somanathan, S.; Vandenberghe, L.H.; Wilson, J.M. Hepatic regulatory T cells and Kupffer cells are crucial mediators of systemic T cell tolerance to antigens targeting murine liver. *Hepatology* **2009**, *50*, 612–621. [CrossRef] [PubMed]
30. Hamilton, B.A.; Wright, J.F. Challenges Posed by Immune Responses to AAV Vectors: Addressing Root Causes. *Front. Immunol.* **2021**, *12*, 675897. [CrossRef]
31. De Haan, A.; Groen, G.; Prop, J.; van Rooijen, N.; Wilschut, J. Mucosal immunoadjuvant activity of liposomes: Role of alveolar macrophages. *Immunology* **1996**, *89*, 488–493. [CrossRef]
32. Flotte, T.R. Empty Adeno-Associated Virus Capsids: Contaminant or Natural Decoy? *Hum. Gene Ther.* **2017**, *28*, 147–148. [CrossRef]

viruses

MDPI

Review

Application of Plant Viruses in Biotechnology, Medicine, and Human Health

Srividhya Venkataraman and Kathleen Hefferon *

Department of Cell and Systems Biology, University of Toronto, Toronto, ON M5S 3B2, Canada; byokem@hotmail.com
* Correspondence: kathleen.hefferon@alumni.utoronto.ca

Abstract: Plant-based nanotechnology programs using virus-like particles (VLPs) and virus nanoparticles (VNPs) are emerging platforms that are increasingly used for a variety of applications in biotechnology and medicine. Tobacco mosaic virus (TMV) and potato virus X (PVX), by virtue of having high aspect ratios, make ideal platforms for drug delivery. TMV and PVX both possess rod-shaped structures and single-stranded RNA genomes encapsidated by their respective capsid proteins and have shown great promise as drug delivery systems. Cowpea mosaic virus (CPMV) has an icosahedral structure, and thus brings unique benefits as a nanoparticle. The uses of these three plant viruses as either nanostructures or expression vectors for high value pharmaceutical proteins such as vaccines and antibodies are discussed extensively in the following review. In addition, the potential uses of geminiviruses in medical biotechnology are explored. The uses of these expression vectors in plant biotechnology applications are also discussed. Finally, in this review, we project future prospects for plant viruses in the fields of medicine, human health, prophylaxis, and therapy of human diseases.

Keywords: expression vectors; aspect ratio; VLPs; VNPs; TMV; PVX; CPMV; geminivirus; cancer; theranostics; CRISPR-cas9

check for
updates

Citation: Venkataraman, S.; Hefferon, K. Application of Plant Viruses in Biotechnology, Medicine, and Human Health. *Viruses* **2021**, *13*, 1697. https://doi.org/10.3390/v13091697

Academic Editors: Carla Varanda and Patrick Materatski

Received: 18 December 2020
Accepted: 12 July 2021
Published: 26 August 2021

Publisher's Note: MDPI stays neutral with regard to jurisdictional claims in published maps and institutional affiliations.

1. Introduction

Over the past few years, plant viruses have increased in visibility for a wide range of applications in biotechnology. Plant viruses are highly suitable for production of vaccines as they are recognized by the innate immune system through the pathogen associated molecular pattern (PAMP) receptors [1] while being non-pathogenic to mammals. Plant viruses can elicit both cell-mediated immunity [2,3] and a humoral immune response when delivered through mucosal [4] or parenteral [5] routes.

Plant virus genomes have been engineered to express heterologous open reading frames. For example, deconstructed virus vectors (Figure 1) were generated first using TMV [6,7] and PVX [8], of which the former was produced commercially by Icon Genetics as the magniCON vector [6]. Using this TMV magnifection technology, full immunoglobulin IgG was produced in under 2 weeks at high yields (4.8 g/kg fresh weight tissue) [9].

Plant viruses have also been developed as VLPs and VNPs in order to present them as epitope display systems for vaccine production and as scaffolds for the conjugation of drugs or molecules used in diagnostics (Figure 2). VNPs and VLPs based on plant viruses are favorable because they are non-pathogenic to humans, and hence preclude any unwanted side effects/contamination. VNPs are nanoparticle formulations based on viruses that can be employed as building blocks for novel nanomaterials exhibiting a variety of molecular characteristics [10]. VNPs are self-assembling highly symmetrical systems that are dynamic, polyvalent, and monodisperse. They are advantageous due to reasons such as their robustness and ability to be generated in short time periods while serving as programmable molecular scaffolds. Additionally, VNPs are superior to synthetic nanomaterials by virtue of being biocompatible and biodegradable. Several self-assembly mechanisms have been adopted to encapsulate ligands such as small chemical modifiers, peptides, proteins, or even additional nanoparticles into the VNPs for which a wide range of conjugation chemistries have been employed [11,12]. These include strategies such as encapsulation, mineralization, chemical bioconjugation, and genetic engineering.

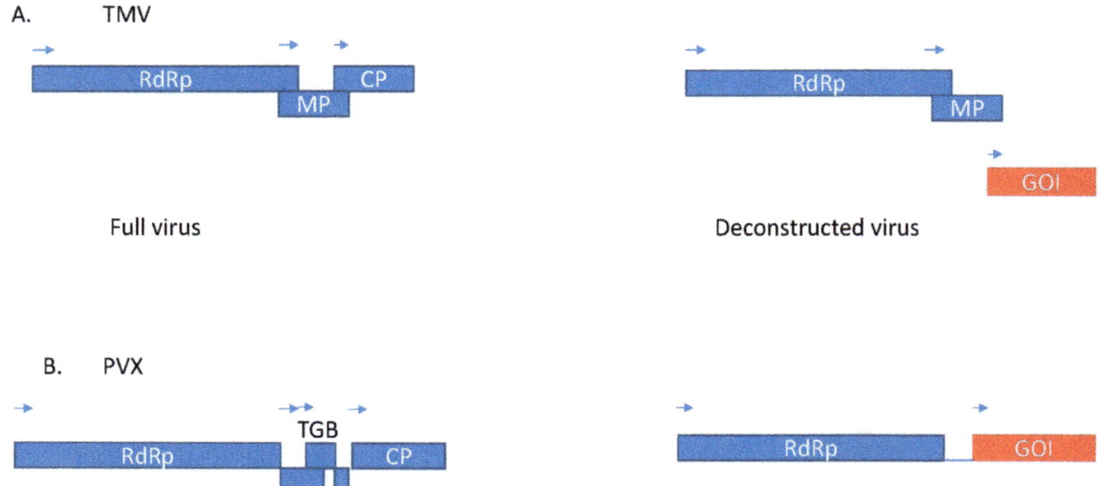

Figure 1. Schematic diagrams of full-length genome vs. deconstructed vectors of TMV (**A**) and PVX (**B**). RdRp, RNA-dependent RNA polymerase; MP, movement protein; CP, coat protein; GOI, gene of interest; TGB, triple gene block.

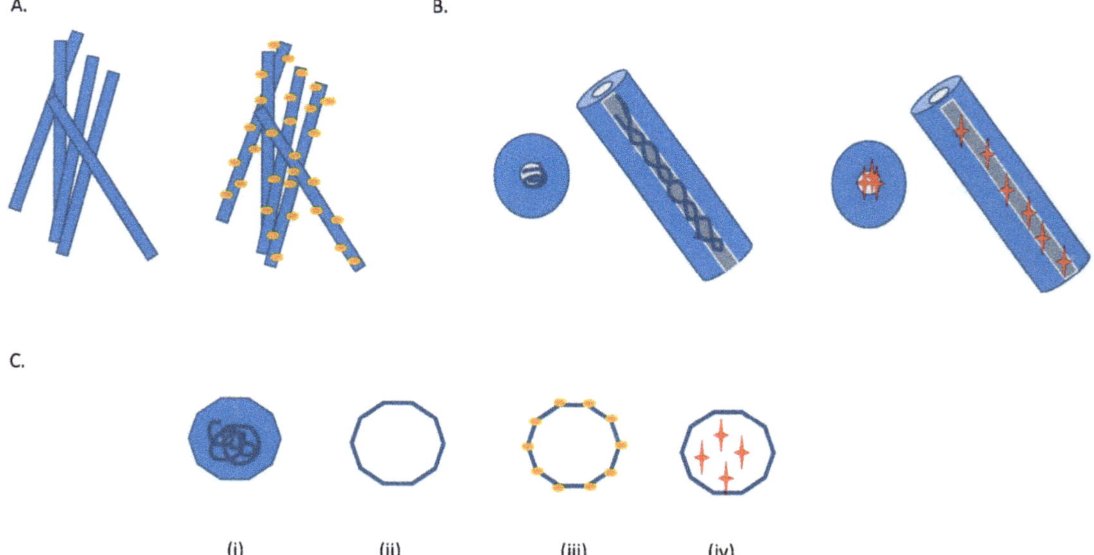

Figure 2. Schematic diagrams of TMV and CPMV wild-type virus vs. virus nanoparticles: (**A**) WT TMV (left hand side) and TMV nanoparticle (right hand side) displaying drug moieties conjugated to the surface of virus particle; (**B**) WT TMV (left hand side) containing viral RNA and TMV nanoparticle (right hand side), in this case RNA genome is replaced with drug moieties on the interior of the virus particle; (**C**) (**i**) CPMV intact virion, (**ii**) empty virus-like particle (eVLP), (**iii**) drug moieties conjugated to surface of eVLP, (**iv**) with drug moieties captured within eVLP.

VLPs are a subset of the VNPs but bereft of any nucleic acid genome, thus, making them noninfectious. VLPs are powerful vaccine candidates as they simulate the conformations of native viruses, utilizing their intrinsic immunogenicity while not compromising their safety. They accomplish this by having no viral genome, and therefore being unable to replicate [13]. Hence, VLPs have become popular as subunit vaccines while several plant viruses have been used to generate VNPs. VLPs evoke effectual immune responses as they are readily internalized by the antigen presenting cells (APCs) and are ideal platforms for antigen processing and epitope presentation to the immune system. Additionally, VLPs are increasingly used in cancer immunotherapy wherein their inherent ability to stimulate immune reactions can be employed to prime the tumor microenvironment towards launching antitumor immunity. VLPs occur as repetitive, multivalent molecular scaffolds by virtue of being composed of their capsid proteins in multiple copies that facilitate multivalent presentation of antigens. Therefore, VLP vaccines afford superior immunogenicity as compared with antigens in their soluble states. Additionally, plant viral VLPs and VNPs possess inherent adjuvant properties dispensing with the use of additional adjuvants to evoke strong immune responses.

Knowledge and insight into the molecular structure of TMV [14] and PVX [15], for example, have enabled their use for several applications as biocatalysts [16], fluorescent markers [17], nanoparticles for in vivo imaging [18], nanoparticles for biologics purification [19], vaccines [20–22], and assembly units for memory devices [23]. Plant viral VNPs serve toward a variety of applications such as immunotherapy [24], chemotherapy [25], vaccines [26], gene delivery [27], and plant virus-assisted sensors [28].

The following review describes many of the uses of plant viruses in biotechnology, with examples based on TMV, PVX, CPMV, and geminiviruses. In the last section, we conclude with a future projection of the significance of plant viruses in the fields of medicine and engineering.

2. Molecular Characteristics of TMV Advantageous for Biotechnological Use

TMV was initially characterized in the 19th century and has since become a paradigm for our current perspective on the morphogenesis of self-assembling viral particle structures [29]. TMV is the most well-studied plant virus, and it is also the most important plant virus both scientifically and economically [30,31]. In recent times, this knowledge has been translated toward the generation of novel compounds and structures that could be used in nanotechnology and medicine. TMV can be easily produced and purified in bulk amounts, and therefore has become of tremendous importance in molecular biology and virology [32]. TMV has been used to detect translational enhancers for the augmented expression of heterologous genes [33,34], and for the design of effective vectors for virus-induced gene silencing and transient expression in plant systems [35], as well as for creating virus-resistant plant lines [36,37].

TMV is also simple and well-characterized with respect to particle structure and genome organization. Thus, it is well suited as a highly amenable experimental system for different applications. The rod-shaped virus particle measures 300 nm in length and 18 nm in diameter, and contains a 6.7 kb viral RNA genome that is encapsidated by 2130 identical copies of the capsid protein assembled in a helical arrangement. The crystal structure of the 158 amino acid capsid protein has been determined [38]. The genomic RNA contains a stretch of 432 nucleotide bases that forms the origin-of-assembly sequence (OAS) sufficient for viral assembly [39]. At neutral pH and without its RNA, the coat protein (CP) assembles itself into an 18 nm double disk, a 20S aggregate or nano-ring containing two layers of 17 CP molecules which can serve as a nanoscale scaffold. The amino acid sequence of the CP has many accessible regions for chemical modifications both at the inner and outer surfaces [30]. TMV can also assemble into spherical nanoparticles of 100–800 nm, in the absence of its RNA genome, by thermal processing [40]. Moreover, the TMV RNA genome can self-assemble with its purified CP in vitro to generate infectious virus particles [41], in addition to its ability to self-assemble in vivo. Therefore, TMV has become a model system for RNA-protein recognition.

Different strategies can be used to modify TMV, such as the modification of the interior or exterior surface of the capsid through genetic engineering, chemical conjugation, or a combination of both processes. The interaction and transport of heterologous cargo within the virus inner cavity or generation of multivalent structures by particle integration have thus been adopted. The conformation of the TMV CP facilitates the insertion of foreign peptides at both its N- and C-termini. In addition to this, the loop formed from CP amino acids 59–66 can be used towards surface display of foreign peptides on intact virions or on CP assemblies [42].

3. The Use of Genetically Engineered TMV in Biochemistry, Nanotechnology, and Plant Biotechnology

Table 1 illustrates some examples of plant viral expression vectors derived from TMV [43–48], CPMV [49–51], PVX [52], and bean yellow dwarf virus (BeYDV) [53] for generation of foreign proteins.

Table 1. Examples of plant viruses used as expression vectors for foreign proteins.

Recombinant Protein or Vaccine or VLP	Viral Vector
Cholera toxin b subunit	TMV [43]
Human anti-non-Hodgkin's lymphoma single-chain Fv (scFv) immunoglobulins	Hybrid TMV and odontoglossum ringspot virus (ORSV) [44]
Rice a-amylase	Hybrid TMV and tomato mosaic virus (ToMV) [45]
Assembled full-size monoclonal antibody	Combination of non-competing viral vectors TMV and PVX [46]
Human growth hormone	Hybrid crucifer-infecting TMV (cr-TMV) and turnip vein-clearing virus (TVCV) [47]
Plant-produced VLP developed for drug delivery	TMV [48]
Plant-produced chimaeric virus vaccine for influenza virus	TMV [21]
Assembled full-size monoclonal antibody	CPMV [49]
Plant-produced chimaeric virus vaccine for human rhinovirus 14 and human immunodeficiency virus	CPMV [50]
Plant-produced VLP developed for encapsulation of metals	CPMV [51]
Plant-produced chimaeric virus vaccine for hepatitis C virus	PVX [52]
Hepatitis B core Norwalk virus capsid protein (NVCP)	BeYDV [53]

(Adapted from Ibrahim et al., 2019 [54]).

The location of C-terminus of the TMV CP on the exterior surface of assembled TMV virions makes it the most used site for insertion of foreign peptides. Table 2 presents some examples of the plant viruses (TMV [55–57], PVX [58], and CPMV [59,60]) used as drug delivery systems and the respective regions within their coat proteins that are amenable to genetic modifications.

Table 2. Examples of plant viruses used in drug delivery systems.

Virus	Symmetry	Family	Locations within the CP Amenable to Genetic Modification
TMV	Rod-like	Tombusviridae	Threonine 104/158, serine 123, N/C-terminal of coat protein [55–57]
PVX	Rod-like	Potexviridae	N-terminal of coat protein [58]
CPMV	Icosahedral	Comoviridae	βB-βC loop of the small subunit/βE-βF loop of the large subunit [59,60]

(Adapted from Sokullu et al., 2019 [61]).

TMV particles have been exploited for active enzyme display, with wide-ranging uses in biodetection, sensor development, medicine, and enzymatic conversion. Enzymes such as penicillinase [62,63], horseradish peroxidase [64], and glucose oxidase [65] have been expressed on the TMV surface, as TMV exhibits a strong stabilizing effect on these enzymes. TMV adapter rods have been incorporated on sensor surfaces, which have facilitated bioaffinity-derived presentation of streptavidin conjugates of the above enzymes at surface densities that are not attainable on supports free of TMV. Enhanced reusability and augmented target detection ranges of these high-performance TMV-based biosensors have been reported and present great promise for multiple applications.

TMV membranes have been engineered that could be recruited as tissue engineering frameworks by sequentially altered layering of two TMV variants with different charges. Recently, these TMV-based carrier templates have been used to prepare surfaces that promote cellular attachment and differentiation [66–68].

Some cells have been cultivated on TMV-covered culture supports and peptide ligands have been presented in a spatially defined manner over nanometric scales. Arginine–glycine–aspartic acid peptide associated TMV layers have been used for osteogenesis of stem cells from bone marrow [67,68]. TMV has been employed as a carrier for peptide motifs and is capable of cell-binding that simulates extracellular matrix proteins. TMV-derived nanorod fibers synthesized from complexation with electrospun composite polymers have been used to generate mats for better handling [69].

Transgenic plants expressing TMV CP were generated by Powell Abel et al. (1986) [37]. These plants showed resistance to TMV challenge and, as a result, initiated the theory of "capsid protein-induced resistance" [70]. TMV has also been used to engineer virus-induced gene silencing (VIGS) systems for *Colletotrichum acutatum*, a phytopathogenic fungus which proved to efficiently assemble virus particles inside hyphal cells [71].

4. The Use of TMV in Medicine, Cancer, Imaging, and Theranostics

TMV disks have a flat and round morphology that yields a high aspect ratio. TMV particles, by virtue of their flexuous rod-like structures, marginate toward blood vessel walls, enhancing the likelihood of invading diseased areas of the body, while accumulating inside tumor tissues [61,72]. In contrast to their spherical equivalents, the helical virus derived VLPs and VNPs transit more efficiently through tissues and membranes [73]. As compared with VLPs, VNPs are more effective because their RNA genome cargo functions as a ruler to define the length of the nucleoprotein–virus complex. In addition, the surface characteristics of these viruses can be altered by means of genetic or chemical approaches without compromising virus structural integrity. Consequentially, the positions of functional units such as drugs, contrast agents, or targeting ligands can be spatially controlled which enables the engineering of multifunctional systems that harbor different combinations of these moieties [74].

Molecular imaging is an emerging biomedical field which facilitates the visualization, identification, and evaluation of biological mechanisms in vivo. Some of these imaging technologies include magnetic resonance imaging (MRI), computed tomography (CT), positron emission tomography (PET), and optical imaging, which enable the monitoring of molecular and cellular processes in normal and diseased conditions in living subjects. Ideally, a given molecular imaging technique should readily afford optimal signal-to-noise ratios within the target site while minimizing toxicity [13].

VLPs are more beneficial for molecular imaging technologies than synthetic nanoparticles, due to their short half-life in circulation and their low retention times, which thus reduce probable side effects [10]. Furthermore, VLPs can be developed to carry a wide array of contrast agents and fluorescent labels, as they can be modified with antibodies, peptides, and aptamers to enable enhanced targeting to specific tissues and cells.

TMV has been successfully used for imaging, targeting atherosclerosis, and thrombosis [75]. Cargo mRNA encoding the green fluorescent protein (GFP) was encapsulated within TMV, which when administered into mice, elicited an immune response against

GFP. This provided a proof-of-concept that this technology can be utilized for vaccine development [76]. TMV has also been engineered to display the iLOV protein, which acts as a fluorescent probe [77]. TMV has also been used in theranostics (drugs and/or techniques combined to both diagnose and treat medical conditions), for enabling photoacoustic imaging and MRI capabilities to photothermal therapy (PTT) treatment.

Table 3a presents some examples of the use of the engineered TMV for treating diseases, while Table 3b shows a list of studies wherein TMV has been used for cancer treatment. Table 3c presents examples of studies using TMV in theranostic applications.

Table 3. Applications of TMV in biotechnology and medicine: (**a**) Applications of TMV in medicine; (**b**) applications of TMV in cancer treatment; (**c**) applications of TMV in theranostics.

(a) Applications of TMV in medicine		
Engineered Modifications	**Effects**	**Reference**
The extreme C-terminus of the TMV CP fused to the 11 amino acid epitope of the foot and mouth disease virus (FMDV) VP1 protein	This nanoparticle protected animals against FMDV challenge	[78]
Peptides from the coronavirus murine hepatitis virus spike protein displayed on the surface of TMV particles	Increased antibody titers and protected mice against murine hepatitis virus challenge	[79]
An epitope from Pseudomonas aeruginosa outer membrane protein F fused to the C-terminus of the TMV CP	Demonstrated immunity to Pseudomonas aeruginosa	[80]
The influenza virus M2e epitope displayed by fusion near the C-terminus of the TMV CP	Afforded protective anti-influenza immune response in mice	[21]
TMV conjugated to the thrombolytic tissue plasminogen activator tPA	Functioned efficiently equivalent to free tPA and enhanced safety profile as shown by diminished average bleeding times and therefore applicable for cardiovascular therapy	[81]
(b) Applications of TMV in cancer treatment		
Engineered Modifications	**Effects**	**Reference**
TMV employed to display a weakly immunogenic tumor-associated carbohydrate antigen, the Tn antigen (GalNAc-α-O-Ser/Thr)	Potent immune responses were observed when the Tn antigen was conjugated to Tyr 139 of TMV	[82]
TMV CP used as nanocarrier for a highly hydrophobic, insoluble peptide that binds to the neuropilin (NRP1) receptor transmembrane domain in cancer cells	Shown to be anti-angiogenic by reducing cancer cell growth and migration	[30]
Doxorubicin (DOX) loaded onto TMV disks	Increased rates of survival of mice bearing intracranial glioblastoma	[83]
DOX loaded onto TMV VNPs coated with albumin	Antitumor effects	[84]
Cisplatin and phenanthriplatin loaded into the cavity of TMV by formation of stable covalent adduct or by charge-based reaction	Enhanced absorption by cancer cells and improved cytotoxicity	[85,86]
TMV VNPs loaded with cisplatin modified using lactose and mannose moieties on their external surface	This construction assisted the VNP's recognition by the asialoglycoprotein receptor that is present on cell membranes and demonstrated augmented cytotoxicity in cancer cell lines	[87]
Modification of the TMV coat protein with a molecular fluorous ponytail incorporated at specific sites which resulted in self-assembly of the virus into spherical VNPs	These spherical VNP's conferred greater stability of for the cisplatin-VNP complexes formed via metal-ligated coordination	[88]

Table 3. Cont.

(b) Applications of TMV in cancer treatment

Engineered Modifications	Effects	Reference
Mitoxanthrone (MTO) loaded onto TMV VNPs by a charge-driven mechanism	Increased antitumor effects in mice	[89]
Antimitotic drug, valine-citrulline monomethyl auristatin E loaded onto external surface of TMV VNPs	Effective targeting and cytotoxicity in non-Hodgkin's lymphoma cell line, Karpas 299; internal entry of TMV VNPs into endolysosomal components accompanied by protease-encoded release of the drug	[90]
Transacting activation transduction (TAT) peptide fused to the external surface of TMV	The engineered TAT-tagged TMV was internalized; this delivered RNA silencing in nude mice hepatocellular carcinoma tumors upon intravenous and intratumoral delivery	[91]
Zn-EpPor (5-(4-ethynylphenyl)-10,15,20-tris(4-methylpyridin-4-ium-1-yl)porphyrin-zinc(II) triiodide), a photosensitizer drug loaded onto the interior of the TMV particles	Demonstrated high stability and shelf-life; drug was released into endolysosomes and showed augmented cell-killing efficiency	[61]
Zn-Por+3 loaded TMV conjugated to F3 peptide	Targeted the nucleolin shuttle protein overexpressed on Hela cells; drug accumulated on cell membranes along with increased cell-killing efficiency likely due to disruption of the cell membrane through light activation followed by drug release and cellular uptake	[92]

(c) Applications of TMV in theranostics

Engineered Modifications	Effects	Reference
A near infrared fluorescent (NIR) dye as well as a peptide targeting S100A9 (a myeloid-related protein 14 present in atherosclerotic lesions and a molecular marker for acute myocardial infarctions) were conjugated to TMV	These targeted TMV particles were able to identify atherosclerotic lesions in apolipoprotein E-deficient (ApoE-/-) mice upon intravenous injection, showing that TMV can be used as a platform to detect at-risk lesions	[75]
A TMV-MOF (metal-organic framework) hybrid nanoparticle engineered	Increased retention of the TMV VNPs observed in mice	[93]
A Cy5-encapsulated TMV coated with zeolitic imidazolate framework-8 (Cy5-TMV@ZIF)	Improved the fluorescence retention time by 2.5 times more than that of the Cy5-TMV alone; this TMV@ZIF was recalcitrant to harsh conditions and proved to be highly stable and non-toxic	[93]
Gd-dodecane tetraacetic acid (Gd-DOTA) loaded onto TMV particles altered to target the vascular cell adhesion molecule, VCAM-1	Facilitated the sensitive identification and depiction of atherosclerotic plaques in ApoE-/- mice, using low doses of the contrast agent wherein the augmented relaxivity and slower tumbling of the Gd-DOTA coupled with the TMV carrier improved the signal-to-noise ratio; also, this coupling afforded greater sensitivity of imaging, allowing $40\times$ decrease in Gd dose in comparison with the standard clinical doses	[94]
Packing of a dysprosium (Dy3+) complex within the interior cavity of TMV	Enhanced T2 relaxivity towards MRI; this enabled NIR fluorescent dye delivery, which facilitated dual optical-MR imaging. The exterior surface of TMV was labeled with an Asp-Gly-Glu-Ala peptide that enabled target specificity to integrin $\alpha2\beta1$ molecules on prostate cancer cells	[13]
A metal-free paramagnetic nitroxide organic radical contrast agent (ORCA) loaded onto TMV particles to generate electron paramagnetic resonance and MRI probes towards the detection of superoxide	This augmented in vitro r1 and r2 relaxivities and these probes worked as both T1 as well as T2 contrast agents, facilitating their suitability for preclinical and clinical MRI scanning	[95]

Table 3. *Cont.*

(c) Applications of TMV in theranostics		
Engineered Modifications	**Effects**	**Reference**
TMV conjugated to a derivative of the aminoxyl radical TEMPO (tetramethylpiperidin-1-oxyl, coined Compound 6) by means of a copper catalyzed azide-alkyne cyclo-addition reaction	Subsequent interaction with cucurbit [8] uril (CB [8]) generated an aminoxyl-based ORCA (semitroxane) that was silent for MRI; the r1 (relaxivity) values for TMV-6 emulated that of Gd-DOTA	[96]
TMV nanorods loaded with Gd and coated with polydopamine (PDA)	The PDA enhanced the MRI properties and provided PDA contrast, while simultaneously facilitating photothermal therapy (PTT); strong in vitro NIR absorption was observed along with increased photothermal conversion efficiency, compared to that of gold nanocages [97] and nanorods [98]; also, these VNPs demonstrated potent efficiency with lowered cytotoxicity in treating 4T1 breast and PC-3 prostate cancer cells in vitro	[99,100]

5. Molecular Characteristics of PVX Advantageous for Biotechnological Use

PVX is a single-stranded, positive-sense RNA virus with a flexuous rod-like morphology. The PVX genome is 6430 bases in length [101] and contains a 5′ cap structure and 3′ poly-A tail. There are five open reading frames (ORFs) encoding the ORF1 replicase protein for viral replication, the ORF 2, 3, and 4 triple gene block (TGB) proteins which mediate virus movement and the ORF5 capsid protein for encapsidation and cell-to-cell movement. Protein overexpression systems based on plant viruses are more economical and easier to implement as compared with stable transformation which is more laborious and could take protracted lengths of time [102], whereas infecting plants with genetically engineered viruses directly or through Agrobacterium-mediated infiltration enables easy, rapid, highly efficient transient expression of heterologous proteins. Particularly, the sequence between the TGB and the CP can be modified to clone and express foreign genes [8,103,104].

6. PVX as an Expression Vector and Repurposing PVX for Use in Medicine, Cancer, and Theranostics

PVX has been widely explored as an expression vector for several biopharmaceutical applications such as for antigenic epitopes displayed on the virus outer surface, as well as for expressing full-length and fusion proteins [105]. Virus-derived biocatalysts have been generated using filamentous PVX that was integrated with the enzyme lipase [16]. The major advantage of this scaffold is the ability of the PVX-lipase complex to self-replicate, unlike the equivalent synthetic systems. Such enzymes can be positioned in or on the virus capsid, thus, spatially combining several different enzymes into specific groups that can simulate metabolic cascades.

Of note is the engineering of PVX to serve various biomedical purposes. Uhde-Holzem et al. (2016) [106] reported genetically altered PVX which displayed *Staphylococcus aureus* protein A fragments on its surface, and proved to be easily functionalized with IgG to be used in biosensing plant viruses [107]. PVX has also been widely used in biotechnology, disease diagnostics, development of vaccines/antibodies against infectious diseases, as well as cancer research and treatment. The CP of PVX is not capable of forming VLPs on its own [108,109]. PVX nanoparticles have been shown to inhibit tumor growth in both cell lines and animal models [110]. They are increasingly being used for immunotherapy of tumor microenvironments.

PVX-based VLPs and VNPs are ideal tools in molecular imaging and unlike synthetic nanoparticles, they have limited half-lives in circulation as well as diminished retention times, thereby, decreasing the chances of unwanted side effects. Additional studies have reported that PVX has been conjugated to fluorescent reporters that could be applied towards theranostics, nanomedicine, and in vivo imaging [111]. The small fluorescent iLOV protein was expressed on PVX through genetic engineering, and the resultant engineered PVX served as a fluorescent probe which could be of potential use in vivo imaging. Shukla et al. (2018) [112] reportedly produced PVX VNPs that displayed mCherry or GFP on their N-termini in *N. benthamiana* plants. Significantly, fluorescent PVX could successfully be used for in vivo particle tracking in an HT-29 murine model, for in vitro imaging of HT-29 cells, and for tracing viral infection within plants.

In plant systems, PVX has been used in the identification of pathogenicity determinants of various viruses, fungi, and bacteria (Table 4a). Table 4b presents examples of studies using PVX for diagnosis, prophylaxis, and therapy of infectious diseases, while Table 4c shows instances where PVX has been successfully used in the treatment of cancer.

Table 4. Applications of PVX in biotechnology and medicine: (**a**) Applications of PVX in identifying pathogenicity determinants and in VIGS; (**b**) applications of PVX in the diagnosis, prophylaxis and therapy of infectious diseases; (**c**) applications of PVX in cancer.

(a) Applications of PVX in identifying pathogenicity determinants and in VIGS		
Engineered Modifications	**Effects**	**Reference**
PVX used as an expression vector for the production of V2, C1, and C4 proteins of a novel monopartite begomovirus, the Ageratum leaf curl Sichuan virus in N. benthamiana	Deletion and mutational analysis of the C4 protein using this PVX-derived vector showed that C4 is the major pathogenicity determinant which impacted symptom expression and virus accumulation	[113]
Phytophthora sojae virulence effector Avh148 expressed in plants using a PVX-based vector and a virus-induced virulence effector (VIVE) assay to detect putative effectors encoded by various plant pathogens	This PVX-Avh148 vector infected plants with strong viral symptoms and led to elevated levels of Avh148 effector and viral RNA accumulation; Avh148 was found to be essential for full pathogenic virulence; this VIVE assay could detect putative effectors encoded by various plant pathogens including even unculturable pathogens using this PVX-based expression vector	[114]
Grapevine leafroll-associated virus 2 (GLRaV-2) encodes a p24 polypeptide (a suppressor of RNA-silencing) that was expressed in a PVX-based vector	p24 causes systemic necrosis in N. benthamiana wherein a cytoplasmic Zn2+-binding protein, NbRAR1 is involved and the symptoms are characteristic of a hypersensitive response; the essential role of p24 in GLRaV-2 pathogenesis was elucidated using the PVX expression vector wherein both silencing suppression and p24 self-interaction are critical for the pathogenic activity of p24	[115]
Tomato torrado virus (ToTV) capsid protein subunits Vp23, Vp26, and Vp35 expressed transiently from a PVX-derived vector in Solanum lycopersicum	Of these, Vp26 protein was shown to be the necrosis and pathogenicity determinant responsible for severe systemic necrosis of the plants accompanied by increased ribonuclease and oxidative activities	[116]
PVX has been developed as a VIGS vector in potatoes wherein VIGS mediates silencing of endogenous plant genes, thus helping to investigate the functions of the silenced genes	This caused the silencing of the endogenous phytoene desaturase gene in potato plants which led to characteristic photobleaching symptoms in the leaves by interference of the carotenoid biosynthetic pathway	[117]
(b) Applications of PVX in the diagnosis, prophylaxis, and therapy of infectious diseases		
Engineered Modifications	**Effects**	**Reference**
The scFv-TM43-E10 and scFv-Fc-TM43-E10 antibody derivatives specific for the recognition of the Salmonella typhimurium Omp D protein expressed in a deconstructed PVX vector deficient for virus movement	These PVX vector-based antibodies exhibited similar antigen-binding specificities as that of their mammalian/microbial cell-generated counterparts and were able to successfully recognize the *S. typhimurium* Omp D antigen; therefore showed great promise as new diagnostic tools for the detection of *S. typhimurium* infection	[118]
The Severe Acute Respiratory Syndrome Coronavirus (SARS-CoV) N and M proteins expressed using PVX	The presence of antibodies specific to the SARS-CoV N protein could be detected in SARS-CoV patient sera using the plant-derived N protein	[119]

Table 4. *Cont.*

(b) Applications of PVX in the diagnosis, prophylaxis, and therapy of infectious diseases		
Engineered Modifications	**Effects**	**Reference**
The M2e peptide of H1N1 Influenza virus was fused to bacterial flagellin to augment immunogenicity and then expressed in a PVX vector	The yield of the fusion protein was as high as 30% of the total soluble protein and mice inoculated with the PVX-derived protein exhibited protection against Influenza virus infection	[120]
The hyper variable region 1 (HVR-1) epitope of Hepatitis C Virus (HCV) expressed in a PVX Vector and administered parenterally	This elicited IgG immune response and the PVX-HVR1 epitope reacted positively with the serum of chronic HCV patients	[53]
A second capsid protein promoter of PVX used to express a chimaeric protein derived from fusion of the HCV core antigen with the hepatitis B virus (HBV) surface antigen (HBsAg)	This PVX-based polytopic HCVpc-HBsAg construct could be a potential plant-derived HCV vaccine	[121]
(c).Applications of PVX in cancer		
Engineered Modifications	**Effects**	**Reference**
PVX used as an expression vector for Mambalgin-1, a peptide that functions as a potent analgesic by obstructing acid-sensing ion channels (ASIC) in nerve cells wherein the ASIC is involved in the growth and proliferation of cancer cells	This resulted in the production of Mambalgin-1 which exhibited cytotoxicity towards nervous (SH-SY5Y) cancer cells, inhibited ASIC channels and potentiated anticancer effects	[122]
Monoclonal antibodies of Herceptin or Trastuzumab loaded onto PVX nanofilaments	This successfully induced apoptosis in breast cancer cell lines	[123]
PVX used as an expression vector for a mutant form of the HPV16 E7 oncoprotein, by fusing it with lichenase	This elicited protection against tumor progression in mice by inducing robust cytotoxic T-cell response	[124]
The filamentous PVX used to deliver DOX	These DOX-loaded PVX VNPs greatly diminished the growth of tumors in athymic mice harboring breast cancer xenografts	[125]
PVX-DOX combination	Prolonged mouse survival and stimulated chemokine/cytokine levels in mouse intradermal melanoma models	[126]
PVX used to display tumor necrosis factor-related apoptosis inducing ligand (TRAIL)	Multivalent display of TRAIL enabled increased recruitment and stimulation of death receptors expressed on cancer cell lines and successfully suppressed tumor growth in mice breast cancer models	[127]
PVX conjugated to an idiotypic (Id) tumor-associated antigen (TAA) recombinant through a biotin/streptavidin linker	This elicited a 7 times higher anti-Id IgG response as compared with Id alone in a mouse B-cell lymphoma model; IFN-α and IL-12 were induced; also TLR7 was found to be essential for viral RNA recognition	[128]

7. Molecular Characteristics of CPMV Advantageous for Biotechnological Applications

CPMV is the type member of the genus *Comovirus*, composed of two separately encapsidated positive-strand RNAs. RNA-1 is capable of independent replication in plant cells; however, RNA-2 (encoding the viral movement and structural proteins) depends on RNA-1 for its replication. CPMV virions are icosahedral in shape and are comprised of 60 copies each of a large (L) and a small (S) coat protein [129].

8. Applications of Comoviruses CPMV and Cowpea Chlorotic Mottle Virus (CCMV) in Medical Biotechnology and Cancer

CPMV has been developed as an autonomously replicating virus vector for the expression of either peptides or polypeptides in plants (Table 5). Examples of CPMV used as an epitope presentation system include epitopes from the outer membrane (OM) protein F of *Pseudomonas aeruginosa* which were shown to protect mice against bacterial challenge, and an epitope expressing the 30 amino acid D2 domain of the fibronectin-binding protein (FnBP) from *Staphylococcus aureus*, which has been shown to be able to protect rats against endocarditis [130].

In addition to the use of CPMV to present peptides, replicating and non-replicating expression vectors based on CPMV have been developed [131]. The non-replicating expression system is based on a disabled version of RNA-2 of CPMV. A gene of interest is positioned between the 5' leader sequence and 3' untranslated region (UTR) of RNA-2, and the vector is introduced to the plant via Agrobacterium-mediated transient transformation [50]. By deleting an in-frame initiation codon located upstream of the main translation initiation site of RNA-2, a massive increase in foreign protein accumulation has been observed. This CPMV non-replicating system generated high quality purified anti-HIV-1 antibody in plants [132]. The vector has also been used to express influenza vaccine proteins.

Meshcheriakova et al. (2017) compared the differences between empty virus-like particles (eVLPs) of CPMV and intact virus containing its RNA genome, for their potential use as nanoparticles [133]. eVLPs are noninfectious and could be loaded with heterologous material, which has increased the number of possible applications for CPMV-based particles. In addition to this, they have distinct yet overlapping immunostimulatory effects resulting from virus RNA in wild-type particles, and therefore can be used for different immunotherapeutic strategies [134].

As described for TMV, CPMV has been explored for its potential to block cancer [135]. Steinmetz et al. (2011) found that CPMV nanoparticles could bind to vimentin, a protein found on the surface of most cells [136]. Vimentin is upregulated during tumor progression, making it an attractive target for cancer therapy. The fact that surface vimentin expression correlated with CPMV uptake in this study demonstrated the ability of CPMV to detect invasive cancer cells. Soon after this discovery, Lizotte et al. (2016) found that inhaled CPMV nanoparticles could be rapidly taken up by lung cancer cells in a mouse model and activated neutrophils in the tumor microenvironment to initiate an antitumor immune response [137]. CPMV nanoparticles also demonstrated antitumor immunity in ovarian, colon, and breast tumor models in mice.

Patel et al. (2018) used CPMV nanoparticles in conjunction with radiotherapy to delay ovarian tumor growth in a mouse model [138]. The treatment was able to result in an increase in tumor infiltrating lymphocytes (TILs), suggesting that this combined treatment could act as a future in situ tumor vaccine. Further studies by Wang and Steinmetz (2019) found that a protein known as CD47, which is widely expressed on tumor cells, prevents the action of T cells and phagocytic cells. The authors used a combination therapy of CD47-blocking antibodies and CPMV nanoparticles to act synergistically and elicit an antitumor immune response [139]. The same research group also used low doses of cyclophosphamide (CPA) and CPMV nanoparticles as a combination therapy to successfully reduce mouse tumors in vivo [140].

Recently, Albakri et al. (2019) explored how CPMV particles could activate human monocytes, dendritic cells (DCs), and macrophages [141]. Monocytes, upon incubation with CPMV in vitro, released the chemokines CXCL10, MIP-1α, and MIP-1β into cell culture supernatants. Dendritic cells and monocyte-derived macrophages also were activated after incubation with CPMV. The authors found that activation was part of SYK signaling. Shukla et al. (2020) were able to demonstrate that CPMV outperformed many other types of virus-like particles, and therefore was a particularly strong immune stimulant [142].

Plant VLPs based on CCMV have been employed to deliver mRNA. For example, CCMV was used to successfully deliver enhanced yellow fluorescent protein (EYFP) mRNA to mammalian BHK-21 cells, using transfection with lipofectamine. In this case, the mRNA was successfully delivered and released from the VLPs into the cytoplasm of the BHK-21 cells, facilitating EYFP expression [27]. Furthermore, CCMV can be used to deliver mRNA vaccines, and a proof of concept has been demonstrated with a variety of reporter genes [143].

There are other examples of how icosahedral VLPs can be utilized in medicine. For example, CCMV can be disassembled and reassembled to encapsulate CpG ODNs (oligodeoxynucleotides). CpG ODNs are ligands of the toll-like receptor 9 (TLR9). Upon activation, TLR9 has the capability to induce macrophages. The CpG loaded CCMV VLPs showed significantly enhanced uptake by tumor associated macrophages and inhibited the growth of solid CT26 colon cancer and B16F10 melanoma tumors in Balb/c mice via the macrophage activation [144].

As another example, encapsulated drug-activating enzymes within plant VLPs such as CCMV can be utilized for therapeutic purposes [145]. Cytochrome P450 family enzymes can convert chemotherapeutic prodrugs into an active format. Using plant VLPs to encapsulate these enzymes can reduce side effects while increasing retention and targeting to the tumor site [25,146]. CCMV has been used, for example, to encapsulate bacterial cytochrome, CYPBM3, to activate the prodrugs into activated forms of tamoxifen and resveratrol.

9. Molecular Features of Geminiviruses Advantageous for Biotechnological Use

Plant viruses with ssDNA genomes offer an exceptional alternative format for expression vector design. These plant viruses tend to have small genomes that can readily incorporate open reading frames of unrestricted sizes. They replicate using a rolling circle mechanism and can express genes of interest at extremely high levels; they also infect a broad range of different plant varieties. Geminivirus constructs, for example, require only the virus origin of replication, the gene of interest, and the replication-associated protein (Rep) gene provided in cis or trans format for potential expression in a wide range of plant families [147,148]. Geminiviruses are considered unique for their twinned capsid morphology. Although they are transmitted in the wild by insects, they are readily amenable to genetic engineering and can be introduced easily into plants in a laboratory setting.

10. The Use of Geminiviruses in Biotechnology and Medicine

Bean yellow dwarf virus (BeYDV) is a geminivirus frequently used for expression of pharmaceutical proteins. BeYDV has recently been used to produce norovirus, HIV, HPV, and hepatitis B virus subunit vaccines, monoclonal antibodies to West Nile virus and Ebola virus, as well as earthworm-derived Lumbrokinase (PI239), used to dissolve fibrin and blood clots [149,150]. Besides using higher plants such as tobacco as hosts, geminiviruses have also been used to express proteins in algae [151]. In this case, a microalgae-based system known as Algevir was utilized to produce Ebola virus vaccine protein as well as the highly immunogenic B subunit of the heat-labile *Escherichia coli* enterotoxin. The authors generated a yield of 1.25 mg/g fresh biomass (6 mg/L of culture), within 3 days after transformation.

More examples of the use of geminiviruses for pharmaceutical production include the expression of plant-made recombinant immune complex (RIC) vaccines [152,153]. In one instance, a bio-better vaccine toward Zika virus (ZIKV) was established. The antigen fusion site ZE3 on the RIC platform was altered to accommodate an N-terminal fusion to the IgG heavy chain (N-RIC) with an improvement of 40% in RIC expression. This construct produced a strong antibody titer that correlated with neutralization of the Zika virus. Moreover, when these RICs were co-delivered with plant-produced hepatitis B core (HBc) virus-like particles (VLP) displaying ZE3, there was a five-fold greater antibody titer (>1,000,000) that more strongly neutralized ZIKV than using either RICs or VLPs alone, in the absence of adjuvant and after only two doses [154].

In another recent study, a variety of plant-made human IgG1 fusion vaccine candidates were examined using Zika virus (ZIKV) envelope domain III (ZE3) as a model antigen. These fusion constructs were altered to make RICs and generated using geminivirus vectors in plants which had their glycosylation pathways altered to make the plant more humanized in its glycan profile. The results of this study were the generation of a vaccine candidate at 1.5 mg IgG fusion per g leaf fresh weight that generated high titers of antibodies specific for Zika virus [155].

Future directions for use of geminivirus expression vectors follow the blossoming new field of genome editing, with this expression vector carrying CRISPR/Cas9 machinery to enable precise gene editing through homologous recombination [156].

Table 5. Medical applications of comovirus and geminivirus vectors.

Virus	Application	References
Comovirus CPMV	Delays tumor growth using combination therapy	[137,138]
	CPMV and cyclosposphamide	[140]
	Activation of monocytes, dendritic cells, macrophages	[141]
Comovirus CCMV	mRNA vaccine delivery	[143]
	Encapsulate CpG oligonucleotides, activated macrophages and inhibit growth	[144]
	Encapsulate drug-activating enymes to reduce side effects, increase targeting to tumor site	[145,146]
Geminivirus BeYDV	Vaccines and monoclonal antibodies	[149]
	Monoclonal antibodies to West Nile Virus, Ebola Virus	[149]
	RIC vaccines to ZIKV	[154,155]

11. Viral Expression Vectors and the CRISPR/Cas9 Technology

The agricultural industry has been greatly burdened by infections due to plant viruses and several genetic engineering techniques have been applied to confront plant viral infections. Since the turn of the century, RNA interference has been used effectively for this purpose. In a reported pioneering investigation by Zhang et al. (2018), FnCas9 from Francisella novicida and its guide RNA were used to target the RNA genome of TMV and cucumber mosaic virus (CMV) to engineer virus resistance [157]. Three sites of the TMV genome were targeted which inhibited virus accumulation by 40–80%. Additionally, it was found that the FnCas9 bound the RNA genome, but did not cleave it, thus, limiting the chances of the emergence of viral escape mutants and facilitating durable resistance towards virus control in the long term.

Ariga et al. (2020) reported the use of a PVX vector expressing the cas9 gene and single-guide RNA for highly effective targeted mutagenesis in the model system, *N. benthamiana* [158]. The virus vector was introduced through Agrobacterium transformation that enabled transgene-free gene editing. On the one hand, this coupled with high level expression by amplification of the viral RNA, wherein the PVX can accommodate the large size of the Cas9 gene, is of great applicability in precise editing of the plant genome. On the other hand, other viruses such as the TMV, beet necrotic yellow vein virus, and the tobacco rattle virus cannot accommodate the Cas9 gene due to their size limitations and are known to work only with the Cas9 applied in trans. Deconstructed geminiviruses have been used to express Cas9 successfully, however, such deconstructed forms are not infectious.

One of the most significant antiviral mechanisms of plants is RNA silencing. This is executed through the essential function of the small RNA guided Argonaute proteins which act as agents of viral restriction. One of these proteins is AGO2 which has been proven to be involved in antiviral responses in the host Arabidopsis thaliana. In a study by Ludman et al. (2017), the role of AGO2 in conferring antiviral immunity was explored using *Nicotiana benthamiana* as the host plant [159]. In this investigation, the CRISPR/Cas9 technology was used to inactivate the AGO2 gene which plays an important role in the immune responses of the plant against PVX and other plant viruses.

12. Conclusions

During the 1980s, the brome mosaic virus (BMV) and the cauliflower mosaic virus (CaMV) were genetically engineered as the first RNA and DNA plant virus vectors, respectively, to express bacterial genes [160,161]. Since then, several vectors based on plant viruses have been designed as efficient tools for the expression of recombinant proteins and to advance genomic research. Thus far, many plant viruses have been recruited as delivery vectors for several purposes. These include viruses infecting dicotyledonous plants such as potexviruses [8,162–164], tobamoviruses [165,166], furovirus [167], potyvirus [168–172], geminiviruses [171], comoviruses [172,173], Necrovirus [174], and Caulimovirus [161]. Further, viruses such as Foxtail mosaic virus [175], barley stripe mosaic virus [176–178], wheat streak mosaic virus [179] and soil-borne wheat mosaic virus [167] capable of infecting monocotyledonous plants have been repurposed as expression vectors. Plant viral expression vectors are increasingly being used in basic and applied research requiring the expression of pharmaceutical peptides, antibodies, and other functional complex heterologous proteins. Furthermore, these vectors have been used in functional genomics applications such as virus-based miRNA expression, VIGS, identification of virulence effectors, and virus-mediated genome editing. This review discusses the use of the most popular plant viruses namely the TMV, PVX, CPMV, and geminiviruses for biotechnological purposes, medicine, and human health. While this is not by any means exhaustive considering the wealth of recent and older literature in this area, it addresses some of the major achievements in the use of these viruses as expression vectors.

In the current review, we highlight the use of plant virus based VLPs and VNPs as diagnostic and therapeutic agents for biotechnological and biomedical applications such as VIGs, identification of virulence effectors of plant pathogens, vaccines against cancer and infectious diseases, theranostics and nanocarriers for imaging modalities. VNPs and VLPs play a major role in the future of nanotechnology and nanomedicines. Viruses and VNPs are natural carriers of nucleic acid molecules which protect and transport their cargo, and this is the major property used for drug delivery. Through a combination of chemistries and by attachment of a wide range of functional groups, drug cargo can be encapsulated, infused, conjugated, or absorbed to the exterior and interior surfaces of their coat protein interfaces [180]. This affords molecular flexibility towards protection of cargo with proteinaceous matrices, reversible binding of active molecules, and specific targeting to the sites of action. VNPs are advantageous as natural delivery carriers because of their structural uniformity, water solubility, biocompatibility, ease of functionalization, and high uptake efficacy [181]. Nanosized cages afford ideal approaches for imaging and drug delivery while conferring high stability, cell-targeting, cell penetrability, and appropriate pharmacokinetics. In addition, VNPs do not show tissue tropisms, and therefore can be employed for targeting and binding cell surface receptors, crossing membranes and penetrating the nucleus [182].

The structures of several viruses are known at atomic resolution enabling modifications with spatial selectivity in a precise manner. By genetic engineering, the VLPs can be formulated to obtain new structures having predictable interactions with biological systems]. VLPs can be engineered to display on their surface functional groups such as ligands for targeting, epitopes, imaging dyes, and drug payloads. The VLPs by virtue of their size and shape facilitate vascular transport, active cellular uptake, and molecular interactions. VLPs can tolerate harsh environments while being biocompatible. In addition, high doses of VLPs are mostly well tolerated and the VLPs are completely and rapidly cleared by proteolytic degradation to enable diminished side effects. Moreover, the characteristic ability of the VLPs to self-assemble coupled with novel molecular design using chemical biology technologies enable the production of functionalized hybrid VLP nanomaterials. The field of VNP- and VLP-based technologies for drug delivery applications continues to evolve with several candidates in clinical trials that should lead to advanced therapeutics, in the near future. In the future, it is very much likely that more plant viruses would be genetically engineered and repurposed for further use in biotechnology and medicine.

Author Contributions: Both S.V. and K.H. contributed equally to writing this manuscript. All authors have read and agreed to the published version of the manuscript.

Funding: This research received no external funding.

Data Availability Statement: Not applicable.

Conflicts of Interest: The authors declare no conflict of interest.

References

1. Acosta-Ramírez, E.; Pérez-Flores, R.; Majeau, N.; Pastelin-Palacios, R.; Gil-Cruz, C.; Ramírez-Saldaña, M.; Manjarrez-Orduño, N.; Cervantes-Barragan, L.; Santos-Argumedo, L.; Flores-Romo, L.; et al. Translating innate response into long-lasting antibody response by the intrinsic antigen-adjuvant properties of papaya mosaic virus. *Immunology* **2008**, *124*, 186–197. [CrossRef] [PubMed]

2. Yusibov, V.; Mett, V.; Davidson, C.; Musiychuk, K.; Gilliam, S.; Farese, A.; MacVittie, T.; Mann, D. Peptide-based candidate vaccine against respiratory syncytial virus. *Vaccine* **2005**, *23*, 2261–2265. [CrossRef] [PubMed]

3. Kemnade, J.O.; Seethammagari, M.; Collinson-Pautz, M.; Kaur, H.; Spencer, D.M.; McCormick, A.A. Tobacco mosaic virus efficiently targets DC uptake, activation and antigen-specific T cell responses in vivo. *Vaccine* **2014**, *32*, 4228–4233. [CrossRef] [PubMed]

4. Brennan, F.R.; Bellaby, T.; Helliwell, S.M.; Jones, T.D.; Kamstrup, S.; Dalsgaard, K.; Flock, J.-I.; Hamilton, W.D.O. Chimeric plant virus particles administered nasally or orally induce systemic and mucosal immune responses in mice. *J. Virol.* **1999**, *73*, 930–938. [CrossRef] [PubMed]

5. Brennan, F.; Jones, T.; Longstaff, M.; Chapman, S.; Bellaby, T.; Smith, H.; Xu, F.; Hamilton, W.; Flock, J.-I. Immunogenicity of peptides derived from a fibronectin-binding protein of S. aureus expressed on two different plant viruses. *Vaccine* **1999**, *17*, 1846–1857. [CrossRef]

6. Gleba, Y.; Klimyuk, V.; Marillonnet, S. Magnifection—A new platform for expressing recombinant vaccines in plants. *Vaccine* **2005**, *23*, 2042–2048. [CrossRef]

7. Lindbo, J.A. TRBO: A High-Efficiency Tobacco Mosaic Virus RNA-Based Overexpression Vector. *Plant Physiol.* **2007**, *145*, 1232–1240. [CrossRef]

8. Chapman, S.; Kavanagh, T.; Baulcombe, D. Potato virus X as a vector for gene expression in plants. *Plant J.* **1992**, *2*, 549–557.

9. Bendandi, M.; Marillonnet, S.; Kandzia, R.; Thieme, F.; Nickstadt, A.; Herz, S.; Fröde, R.; Inogés, S.; de Cerio, A.L.-D.; Soria, E.; et al. Rapid, high-yield production in plants of individualized idiotype vaccines for non-Hodgkin's lymphoma. *Ann. Oncol.* **2010**, *21*, 2420–2427. [CrossRef] [PubMed]

10. Steinmetz, N.F. Viral nanoparticles as platforms for next-generation therapeutics and imaging devices. *Nanomedicine* **2010**, *6*, 634–641. [CrossRef]

11. Young, M.; Debbie, W.; Uchida, M.; Douglas, T. Plant Viruses as Biotemplates for Materials and Their Use in Nanotechnology. *Annu. Rev. Phytopathol.* **2008**, *46*, 361–384. [CrossRef]

12. Steinmetz, N.F.; Evans, D.J. Utilisation of plant viruses in bionanotechnology. *Org. Biomol. Chem.* **2007**, *5*, 2891–2902. [CrossRef]

13. Chung, Y.H.; Cai, H.; Steinmetz, N.F. Viral nanoparticles for drug delivery, imaging, immunotherapy, and theranostic applications. *Adv. Drug Deliv. Rev.* **2020**, *156*, 214–235. [CrossRef] [PubMed]

14. Namba, K.; Pattanayek, R.; Stubbs, G. Visualization of protein-nucleic acid interactions in a virus: Refined structure of intact tobacco mosaic virus at 2.9 Å resolution by X-ray fiber diffraction. *J. Mol. Biol.* **1989**, *208*, 307–325. [CrossRef]

15. Parker, L.; Kendall, A.; Stubbs, G. Surface features of potato virus X from fiber diffraction. *Virology* **2002**, *300*, 291–295. [CrossRef]

16. Carette, N.; Engelkamp, H.; Akpa, E.; Pierre, S.J.; Cameron, N.R.; Christianen, P.C.M.; Maan, J.C.; Thies, J.C.; Weberskirch, R.; Rowan, A.E.; et al. A virus-based biocatalyst. *Nat. Nanotechnol.* **2007**, *2*, 226–229. [CrossRef]

17. Yi, H.; Nisar, S.; Lee, S.-Y.; Powers, M.A.; Bentley, W.E.; Payne, G.F.; Ghodssi, R.; Rubloff, G.W.; Harris, M.T.; Culver, J.N. Patterned Assembly of Genetically Modified Viral Nanotemplates via Nucleic Acid Hybridization. *Nano Lett.* **2005**, *5*, 1931–1936. [CrossRef]

18. Niehl, A.; Appaix, F.; Boscá, S.; Van Der Sanden, B.; Nicoud, J.-F.; Bolze, F.; Heinlein, M. Fluorescent Tobacco mosaic virus-Derived Bio-Nanoparticles for Intravital Two-Photon Imaging. *Front. Plant Sci.* **2016**, *6*, 1244. [CrossRef]

19. Werner, S.; Marillonnet, S.; Hause, G.; Klimyuk, V.; Gleba, Y. Immunoabsorbent nanoparticles based on a tobamovirus displaying protein A. *Proc. Natl. Acad. Sci. USA* **2006**, *103*, 17678–17683. [CrossRef] [PubMed]

20. Smolenska, L.; Roberts, I.M.; Learmonth, D.; Porter, A.J.; Harris, W.J.; Wilson, T.; Cruz, S.S. Production of a functional single chain antibody attached to the surface of a plant virus. *FEBS Lett.* **1998**, *441*, 379–382. [CrossRef]

21. Petukhova, N.; Gasanova, T.; Stepanova, L.; Rusova, O.; Potapchuk, M.; Korotkov, A.; Skurat, E.; Tsybalova, L.; Kiselev, O.; Ivanov, P.; et al. Immunogenicity and Protective Efficacy of Candidate Universal Influenza A Nanovaccines Produced in Plants by Tobacco Mosaic Virus-based Vectors. *Curr. Pharm. Des.* **2013**, *19*, 5587–5600. [CrossRef]

22. Thérien, A.; Bédard, M.; Carignan, D.; Rioux, G.; Gauthier-Landry, L.; Laliberté-Gagné, M.-È.; Bolduc, M.; Savard, P.; Leclerc, D. A versatile papaya mosaic virus (PapMV) vaccine platform based on sortase-mediated antigen coupling. *J. Nanobiotechnol.* **2017**, *15*, 54. [CrossRef]

23. Tseng, R.J.; Tsai, C.; Ma, L.; Ouyang, J.; Ozkan, C.S.; Yang, Y. Digital memory device based on tobacco mosaic virus conjugated with nanoparticles. *Nat. Nanotechnol.* **2006**, *1*, 72–77. [CrossRef] [PubMed]

24. Venuti, A.; Curzio, G.; Mariani, L.; Paolini, F. Immunotherapy of HPV-associated cancer: DNA/plant-derived vaccines and new orthotopic mouse models. *Cancer Immunol. Immunother.* **2015**, *64*, 1329–1338. [CrossRef] [PubMed]
25. Sánchez-Sánchez, L.; Cadena-Nava, R.D.; Palomares, L.A.; Ruiz-Garcia, J.; Koay, M.S.; Cornelissen, J.J.; Vazquez-Duhalt, R. Chemotherapy pro-drug activation by biocatalytic virus-like nanoparticles containing cytochrome P450. *Enzym. Microb. Technol.* **2014**, *60*, 24–31. [CrossRef]
26. Phelps, J.P.; Dang, N.; Rasochova, L. Inactivation and purification of cowpea mosaic virus-like particles displaying peptide antigens from Bacillus anthracis. *J. Virol. Methods* **2007**, *141*, 146–153. [CrossRef] [PubMed]
27. Azizgolshani, O.; Garmann, R.F.; Cadena-Nava, R.; Knobler, C.M.; Gelbart, W.M. Reconstituted plant viral capsids can release genes to mammalian cells. *Virology* **2013**, *441*, 12–17. [CrossRef]
28. Eiben, S.; Koch, C.; Altintoprak, K.; Southan, A.; Tovar, G.; Laschat, S.; Weiss, I.M.; Wege, C. Plant virus-based materials for biomedical applications: Trends and prospects. *Adv. Drug Deliv. Rev.* **2019**, *145*, 96–118. [CrossRef]
29. Lomonossoff, G.P.; Wege, C. TMV Particles: The Journey from Fundamental Studies to Bionanotechnology Applications. *Adv. Virus Res.* **2018**, *102*, 149.
30. Gamper, C.; Spenlé, C.; Boscá, S.; Van Der Heyden, M.; Erhardt, M.; Orend, G.; Bagnard, D.; Heinlein, M. Functionalized Tobacco Mosaic Virus Coat Protein Monomers and Oligomers as Nanocarriers for Anti-Cancer Peptides. *Cancers* **2019**, *11*, 1609. [CrossRef]
31. Scholthof, K.-B.; Adkins, S.; Czosnek, H.; Palukaitis, P.; Jacquot, E.; Hohn, T.; Hohn, B.; Saunders, K.; Candresse, T.; Ahlquist, P.; et al. Top 10 plant viruses in molecular plant pathology. *Mol. Plant Pathol.* **2011**, *12*, 938–954. [CrossRef] [PubMed]
32. Lomonossoff, G.P. So what have plant viruses ever done for virology and molecular biology? *Adv. Virus Res.* **2018**, *100*, 145. [PubMed]
33. Gallie, D.R.; Sleat, D.E.; Watts, J.W.; Turner, P.C.; Wilson, T.M.A. The 5′-leader sequence of tobacco mosaic virus RNA enhances the expression of foreign gene transcripts in vitro and in vivo. *Nucleic Acids Res.* **1987**, *15*, 3257. [CrossRef] [PubMed]
34. Wilson, T.M.A. Plant viruses: A tool-box for genetic engineering and crop protection. *Bioessays* **1989**, *10*, 179. [PubMed]
35. Peyret, H.; Lomonossoff, G.P. When plant virology met Agrobacterium: The rise of the deconstructed clones. *Plant Biotechnol. J.* **2015**, *13*, 1121–1135. [CrossRef]
36. Golemboski, D.B.; Lomonossoff, G.P.; Zaitlin, M. Plants transformed with a tobacco mosaic virus nonstructural gene sequence are resistant to the virus. *Proc. Natl. Acad. Sci. USA* **1990**, *87*, 6311–6315. [CrossRef]
37. Abel, P.P.; Nelson, R.S.; De, B.; Hoffmann, N.; Rogers, S.G.; Fraley, R.T.; Beachy, R.N. Delay of disease development in transgenic plants that express the tobacco mosaic virus coat protein gene. *Science* **1986**, *232*, 738–743. [CrossRef] [PubMed]
38. Fromm, S.; Bharat, T.A.; Jakobi, A.J.; Hagen, W.; Sachse, C. Seeing tobacco mosaic virus through direct electron detectors. *J. Struct. Biol.* **2015**, *189*, 87–97. [CrossRef]
39. Sleat, D.; Turner, P.; Finch, J.; Butler, P.; Wilson, T. Packaging of recombinant RNA molecules into pseudovirus particles directed by the origin-of-assembly sequence from tobacco mosaic virus RNA. *Virology* **1986**, *155*, 299–308. [CrossRef]
40. Bruckman, M.; Hern, S.; Jiang, K.; Flask, C.A.; Yu, X.; Steinmetz, N.F. Tobacco mosaic virus rods and spheres as supramolecular high-relaxivity MRI contrast agents. *J. Mater. Chem. B* **2013**, *1*, 1482–1490. [CrossRef]
41. Fraenkel-Conrat, H.; Williams, R.C. Reconstitution of active tobacco mosaic virus from its inactive protein and nucleic acid components. *Proc. Natl. Acad. Sci. USA* **1955**, *41*, 690–698. [CrossRef] [PubMed]
42. Smith, M.L.; Fitzmaurice, W.P.; Turpen, T.H.; Palmer, K.E. Display of Peptides on the Surface of Tobacco Mosaic Virus Particles. *Curr. Top. Microbiol. Immunol.* **2009**, *332*, 13–31. [CrossRef]
43. Moore, L.; Hamorsky, K.; Matoba, N. Production of Recombinant Cholera Toxin B Subunit in Nicotiana benthamiana Using GENEWARE® Tobacco Mosaic Virus Vector. *Methods Protoc.* **2016**, *1385*, 129–137. [CrossRef]
44. McCormick, A.A.; Reinl, S.J.; Cameron, T.I.; Vojdani, F.; Fronefield, M.; Levy, R.; Tusé, D. Individualized human scFv vaccines produced in plants: Humoral anti-idiotype responses in vaccinated mice confirm relevance to the tumor Ig. *J. Immunol. Methods* **2003**, *278*, 95–104. [CrossRef]
45. Kumagai, M.H.; Donson, J.; Dellacioppa, G.R.; Grill, L.K. Rapid, high-level expression of glycosylated rice α-amylase in transfected plants by an RNA viral vector. *Gene* **2000**, *245*, 169–174. [CrossRef]
46. Giritch, A.; Marillonnet, S.; Engler, C.; van Eldik, G.; Botterman, J.; Klimyuk, V.; Gleba, Y. Rapid high-yield expression of full-size IgG antibodies in plants coinfected with noncompeting viral vectors. *Proc. Natl. Acad. Sci. USA* **2006**, *103*, 14701–14706. [CrossRef]
47. Gils, M.; Kandzia, R.; Marillonnet, S.; Klimyuk, V.; Gleba, Y. High-yield production of authentic human growth hormone using a plant virus-based expression system. *Plant Biotechnol. J.* **2005**, *3*, 613–620. [CrossRef] [PubMed]
48. Czapar, A.E.; Zheng, Y.; Riddell, I.A.; Shukla, S.; Awuah, S.G.; Lippard, S.J.; Steinmetz, N.F. Tobacco Mosaic Virus Delivery of Phenanthriplatin for Cancer therapy. *ACS Nano* **2016**, *10*, 4119–4126. [CrossRef]
49. Sainsbury, F.; Lomonossoff, G.P. Extremely High-Level and Rapid Transient Protein Production in Plants without the Use of Viral Replication. *Plant Physiol.* **2008**, *148*, 1212–1218. [CrossRef]
50. Porta, C.; Spall, V.E.; Loveland, J.; Johnson, J.E.; Barker, P.J.; Lomonossoff, G.P. Development of Cowpea Mosaic Virus as a High-Yielding System for the Presentation of Foreign Peptides. *Virology* **1994**, *202*, 949–955. [CrossRef]
51. Aljabali, A.A.A.; Sainsbury, F.; Lomonossoff, G.P.; Evans, D.J. Cowpea Mosaic Virus Unmodified Empty Viruslike Particles Loaded with Metal and Metal Oxide. *Small* **2010**, *6*, 818–821. [CrossRef]
52. Uhde-Holzem, K.; Schlösser, V.; Viazov, S.; Fischer, R.; Commandeur, U. Immunogenic properties of chimeric potato virus X particles displaying the hepatitis C virus hypervariable region I peptide R9. *J. Virol. Methods* **2010**, *166*, 12–20. [CrossRef]

53. Huang, Z.; Chen, Q.; Hjelm, B.; Arntzen, C.; Mason, H. A DNA replicon system for rapid high-level production of virus-like particles in plants. *Biotechnol. Bioeng.* **2009**, *103*, 706–714. [CrossRef] [PubMed]

54. Ibrahim, A.; Odon, V.; Kormelink, R. Plant Viruses in Plant Molecular Pharming: Toward the Use of Enveloped Viruses. *Front. Plant Sci.* **2019**, *10*, 803. [CrossRef] [PubMed]

55. Finbloom, J.A.; Han, K.; Aanei, I.L.; Hartman, E.C.; Finley, D.T.; Dedeo, M.T.; Fishman, M.; Downing, K.H.; Francis, M.B. Stable Disk Assemblies of a Tobacco Mosaic Virus Mutant as Nanoscale Scaffolds for Applications in Drug Delivery. *Bioconj. Chem.* **2016**, *27*, 2480–2485. [CrossRef] [PubMed]

56. Shukla, S.; Eber, F.J.; Nagarajan, A.S.; DiFranco, N.A.; Schmidt, N.; Wen, A.M.; Eiben, S.; Twyman, R.M.; Wege, C.; Steinmetz, N.F. The Impact of Aspect Ratio on the Biodistribution and Tumor Homing of Rigid Soft-Matter Nanorods. *Adv. Healthc. Mater.* **2015**, *4*, 874–882. [CrossRef]

57. Bazzini, A.A.; Hopp, H.E.; Beachy, R.N.; Asurmendi, S. Infection and coaccumulation of tobacco mosaic virus proteins alter microRNA levels, correlating with symptom and plant development. *Proc. Natl. Acad. Sci. USA* **2007**, *104*, 12157–12162. [CrossRef] [PubMed]

58. Shukla, S.; Dickmeis, C.; Nagarajan, A.S.; Fischer, R.; Commandeur, U.; Steinmetz, N.F. Molecular farming of fluorescent virus-based nanoparticles for optical imaging in plants, human cells and mouse models. *Biomater. Sci.* **2014**, *2*, 784–797. [CrossRef]

59. Wang, Q.; Kaltgrad, E.; Lin, T.; Johnson, J.; Finn, M. Natural Supramolecular Building Blocks: Wild-Type Cowpea Mosaic Virus. *Chem. Biol.* **2002**, *9*, 805–811. [CrossRef]

60. Huynh, N.T.; Hesketh, E.L.; Saxena, P.; Meshcheriakova, Y.; Ku, Y.-C.; Hoang, L.T.; Johnson, J.E.; Ranson, N.; Lomonossoff, G.P.; Reddy, V.S. Crystal Structure and Proteomics Analysis of Empty Virus-like Particles of Cowpea Mosaic Virus. *Structure* **2016**, *24*, 567–575. [CrossRef]

61. Sokullu, E.; Abyaneh, H.S.; Gauthier, M.A. Plant/Bacterial Virus-Based Drug Discovery, Drug Delivery, and Therapeutics. *Pharmaceutics* **2019**, *11*, 211. [CrossRef]

62. Koch, C.; Poghossian, A.; Schöning, M.J.; Wege, C. Penicillin Detection by Tobacco Mosaic Virus-Assisted Colorimetric Biosensors. *Nanotheranostics* **2018**, *2*, 184–196. [CrossRef]

63. Poghossian, A.; Jablonski, M.; Koch, C.; Bronder, T.S.; Rolka, D.; Wege, C.; Schoning, M.J. Field-effect biosensor using virus particles as scaffolds for enzyme immobilization. *Biosens. Bioelectron.* **2018**, *110*, 168. [CrossRef] [PubMed]

64. Koch, C.; Wabbel, K.; Eber, F.J.; Krolla-Sidenstein, P.; Azucena, C.; Gliemann, H.; Eiben, S.; Geiger, F.; Wege, C. Modified TMV particles as beneficial scaffolds to present sensor enzymes. *Front. Plant Sci.* **2015**, *6*, 1137. [CrossRef]

65. Bäcker, M.; Koch, C.; Eiben, S.; Geiger, F.; Eber, F.; Gliemann, H.; Poghossian, A.; Wege, C.; Schoening, M.J. Tobacco mosaic virus as enzyme nanocarrier for electrochemical biosensors. *Sens. Actuators B Chem.* **2017**, *238*, 716–722. [CrossRef]

66. Tiu, B.D.B.; Kernan, D.L.; Tiu, S.B.; Wen, A.M.; Zheng, Y.; Pokorski, J.K.; Advincula, R.C.; Steinmetz, N.F. Electrostatic layer-by-layer construction offibrous TMV biofilms. *Nanoscale* **2017**, *9*, 1580. [CrossRef] [PubMed]

67. Kaur, G.; Wang, C.; Sun, J.; Wang, Q. The synergistic effects of multivalent ligand display and nanotopography on osteogenic differentiation of rat bone marrow stem cells. *Biomaterials* **2010**, *31*, 5813–5824. [CrossRef] [PubMed]

68. Sitasuwan, P.; Lee, L.A.; Li, K.; Nguyen, H.G.; Wang, Q. RGD-conjugated rod-like viral nanoparticles on 2D scaffold improve bone differentiation of mesenchymal stem cells. *Front. Chem.* **2014**, *2*, 31. [CrossRef] [PubMed]

69. Wu, L.; Zang, J.; Lee, L.A.; Niu, Z.; Horvatha, G.C.; Braxtona, V.; Wibowo, A.C.; Bruckman, M.A.; Ghoshroy, S.; Loye, H.-C.Z.; et al. Electrospinning fabrication, structural and mechanical characterization of rod-like virus-based composite nanofibers. *J. Mater. Chem.* **2011**, *21*, 8550–8557. [CrossRef]

70. Beachy, R.N. Coat–protein–mediated resistance to tobacco mosaic virus: Discovery mechanisms and exploitation. *Philos. Trans. R. Soc. B Biol. Sci.* **1999**, *354*, 659–664. [CrossRef]

71. Mascia, T.; Nigro, F.; Abdallah, A.; Ferrara, M.; De Stradis, A.; Faedda, R.; Palukaitis, P.; Gallitelli, D. Gene silencing and gene expression in phytopathogenic fungi using a plant virus vector. *Proc. Natl. Acad. Sci. USA* **2014**, *111*, 4291–4296. [CrossRef]

72. Shukla, S.; Ablack, A.L.; Wen, A.M.; Lee, K.L.; Lewis, J.D.; Steinmetz, N.F. Increased Tumor Homing and Tissue Penetration of the Filamentous Plant Viral Nanoparticle Potato virus X. *Mol. Pharm.* **2013**, *10*, 33–42. [CrossRef]

73. Lee, K.L.; Hubbard, L.C.; Hern, S.; Yildiz, I.; Gratzl, M.; Steinmetz, N.F. Shape matters: The diffusion rates of TMV rods and CPMV icosahedrons in a spheroid model of extracellular matrix are distinct. *Biomater. Sci.* **2013**, *1*, 581–588. [CrossRef] [PubMed]

74. Rong, J.; Niu, Z.; Lee, L.A.; Wang, Q. Self-assembly of viral particles. *Curr. Opin. Colloid Interface Sci.* **2011**, *16*, 441–450. [CrossRef]

75. Park, J.; Gao, H.; Wang, Y.; Hu, H.; Simon, D.I.; Steinmetz, N.F. S100A9-targeted tobacco mosaic virus nanoparticles exhibit high specificity toward atherosclerotic lesions in ApoE−/− mice. *J. Mater. Chem. B* **2019**, *7*, 1842–1846. [CrossRef]

76. Grasso, S.; Santi, L. Viral nanoparticles as macromolecular devices for new therapeutic and pharmaceutical approaches. *Int. J. Physiol. Pathophysiol. Pharmacol.* **2010**, *2*, 161–178.

77. Chapman, S.; Faulkner, C.; Kaiserli, E.; Garcia-Mata, C.; Savenkov, E.I.; Roberts, A.G.; Oparka, K.J.; Christie, J.M. The photoreversible fluorescent protein iLOV outperforms GFP as a reporter of plant virus infection. *Proc. Natl. Acad. Sci. USA* **2008**, *105*, 20038–20043. [CrossRef]

78. Wu, L.; Jiang, L.; Zhou, Z.; Fan, J.; Zhang, Q.; Zhu, H.; Han, Q.; Xu, Z. Expression of foot-and-mouth disease virus epitopes in tobacco by a tobacco mosaic virus-based vector. *Vaccine* **2003**, *21*, 4390. [CrossRef]

79. Koo, M.; Bendahmane, M.; Lettieri, G.A.; Paoletti, A.D.; Lane, T.E.; Fitchen, J.H.; Buchmeier, M.J.; Beachy, R.N. Protective immunity against murine hepatitis virus (MHV) induced by intranasal or subcutaneous administration of hybrids of tobacco mosaic virus that carries an MHV epitope. *Proc. Natl. Acad. Sci. USA* **1999**, *96*, 7774–7779. [CrossRef] [PubMed]

80. Staczek, J.; Bendahmane, M.; Gilleland, L.B.; Beachy, R.N.; Gilleland, H. Immunization with a chimeric tobacco mosaic virus containing an epitope of outer membrane protein F of Pseudomonas aeruginosa provides protection against challenge with P. aeruginosa. *Vaccine* **2000**, *18*, 2266–2274. [CrossRef]

81. Pitek, A.S.; Park, J.; Wang, Y.; Gao, H.; Hu, H.; Simon, D.I.; Steinmetz, N.F. Delivery of thrombolytic therapy using rod-shaped plant viral nanoparticles decreases the risk of hemorrhage. *Nanoscale* **2018**, *10*, 16547–16555. [CrossRef]

82. Yin, Z.; Nguyen, H.G.; Chowdhury, S.; Bentley, P.; Bruckman, M.A.; Miermont, A.; Gildersleeve, J.C.; Wang, Q.; Huang, X. Tobacco mosaic virus as a new carrier for tumor associated carbohydrate antigens. *Bioconjug. Chem.* **2012**, *23*, 1694–1703. [CrossRef]

83. Finbloom, J.A.; Aanei, I.L.; Bernard, J.M.; Klass, S.H.; Elledge, S.K.; Han, K.; Ozawa, T.; Nicolaides, T.P.; Berger, M.S.; Francis, M.B. Evaluation of Three Morphologically Distinct Virus-Like Particles as Nanocarriers for Convection-Enhanced Drug Delivery to Glioblastoma. *Nanomaterials* **2018**, *8*, 1007. [CrossRef]

84. Pitek, A.S.; Hu, H.; Shukla, S.; Steinmetz, N.F. Cancer Theranostic Applications of Albumin-Coated Tobacco Mosaic Virus Nanoparticles. *ACS Appl. Mater. Interfaces* **2018**, *10*, 39468–39477. [CrossRef]

85. Franke, C.E.; Czapar, A.E.; Patel, R.; Steinmetz, N.F. Tobacco Mosaic Virus-Delivered Cisplatin Restores Efficacy in Platinum-Resistant Ovarian Cancer Cells. *Mol. Pharm.* **2017**, *15*, 2922–2931. [CrossRef]

86. Vernekar, A.; Berger, G.; Czapar, A.E.; Veliz, F.A.; Wang, D.I.; Steinmetz, N.F.; Lippard, S.J. Speciation of Phenanthriplatin and Its Analogs in the Core of Tobacco Mosaic Virus. *J. Am. Chem. Soc.* **2018**, *140*, 4279–4287. [CrossRef]

87. Liu, X.; Liu, B.; Gao, S.; Wang, Z.; Tian, Y.; Wu, M.; Jiang, S.; Niu, Z. Glyco-decorated tobacco mosaic virus as a vector for cisplatin delivery. *J. Mater. Chem. B* **2017**, *5*, 2078–2085. [CrossRef] [PubMed]

88. Gao, S.; Liu, X.; Wang, Z.; Jiang, S.; Wu, M.; Tian, Y.; Niu, Z. Fluorous interaction induced self-assembly of tobacco mosaic virus coat protein for cisplatin delivery. *Nanoscale* **2018**, *10*, 11732–11736. [CrossRef] [PubMed]

89. Lin, R.D.; Steinmetz, N.F. Tobacco mosaic virus delivery of mitoxantrone for cancer therapy. *Nanoscale* **2018**, *10*, 16307–16313. [CrossRef] [PubMed]

90. Kernan, D.L.; Wen, A.M.; Pitek, A.S.; Steinmetz, N.F. Featured Article: Delivery of chemotherapeutic vcMMAE using tobacco mosaic virus nanoparticles. *Exp. Biol. Med.* **2017**, *242*, 1405–1411. [CrossRef] [PubMed]

91. Tian, Y.; Zhou, M.; Shi, H.; Gao, S.; Xie, G.; Zhu, M.; Wu, M.; Chen, J.; Niu, Z. Integration of Cell-Penetrating Peptides with Rod-like Bionanoparticles: Virus-Inspired Gene-Silencing Technology. *Nano Lett.* **2018**, *18*, 5453–5460. [CrossRef] [PubMed]

92. Chariou, P.L.; Wang, L.; Desai, C.; Park, J.; Robbins, L.K.; von Recum, H.A.; Ghiladi, R.A.; Steinmetz, N.F. Let There Be Light: Targeted Photodynamic Therapy Using High Aspect Ratio Plant Viral Nanoparticles. *Macromol. Biosci.* **2019**, *19*, e1800407. [CrossRef] [PubMed]

93. Luzuriaga, M.A.; Welch, R.P.; Dharmarwardana, M.; Benjamin, C.E.; Li, S.; Shahrivarkevishahi, A.; Popal, S.; Tuong, L.H.; Creswell, C.T.; Gassensmith, J.J. Enhanced Stability and Controlled Delivery of MOF-Encapsulated Vaccines and Their Immunogenic Response In Vivo. *ACS Appl. Mater. Interfaces* **2019**, *11*, 9740–9746. [CrossRef] [PubMed]

94. Bruckman, M.; Jiang, K.; Simpson, E.J.; Randolph, L.N.; Luyt, L.; Yu, X.; Steinmetz, N.F. Dual-Modal Magnetic Resonance and Fluorescence Imaging of Atherosclerotic Plaques in Vivo Using VCAM-1 Targeted Tobacco Mosaic Virus. *Nano Lett.* **2014**, *14*, 1551–1558. [CrossRef]

95. Dharmarwardana, M.; Martins, A.F.; Chen, Z.; Palacios, P.M.; Nowak, C.M.; Welch, R.P.; Li, S.; Luzuriaga, M.A.; Bleris, L.; Pierce, B.S.; et al. Nitroxyl Modified Tobacco Mosaic Virus as a Metal-Free High-Relaxivity MRI and EPR Active Superoxide Sensor. *Mol. Pharm.* **2018**, *15*, 2973–2983. [CrossRef]

96. Lee, H.; Shahrivarkevishahi, A.; Lumata, J.L.; Luzuriaga, M.A.; Hagge, L.M.; Benjamin, C.E.; Brohlin, O.R.; Parish, C.R.; Firouzi, H.R.; Nielsen, S.O.; et al. Supramolecular and biomacromolecular enhancement of metal-free magnetic resonance imaging contrast agents. *Chem. Sci.* **2020**, *11*, 2045–2050. [CrossRef]

97. Wang, Y.; Black, K.C.L.; Luehmann, H.; Li, W.; Zhang, Y.S.; Cai, X.; Wan, D.; Liu, S.-Y.; Li, M.; Kim, P.; et al. Comparison Study of Gold Nanohexapods, Nanorods, and Nanocages for Photothermal Cancer Treatment. *ACS Nano* **2013**, *7*, 2068–2077. [CrossRef] [PubMed]

98. Vankayala, R.; Huang, Y.-K.; Kalluru, P.; Chiang, C.-S.; Hwang, K.C. First Demonstration of Gold Nanorods-Mediated Photodynamic Therapeutic Destruction of Tumors via Near Infra-Red Light Activation. *Small* **2014**, *10*, 1612–1622. [CrossRef] [PubMed]

99. Scholthof, K.-B.G. Tobaccomosaic Virus: A Model System for Plant Biology. *Annu. Rev. Phytopathol.* **2004**, *42*, 13–34. [CrossRef]

100. Hu, H.; Yang, Q.; Baroni, S.; Yang, H.; Aime, S.; Steinmetz, N.F. Polydopamine-decorated tobacco mosaic virus for photoacoustic/magnetic resonance bimodal imaging and photothermal cancer therapy. *Nanoscale* **2019**, *11*, 9760–9768. [CrossRef]

101. Yu, X.-Q.; Jia, J.-L.; Zhang, C.-L.; Li, X.-D.; Wang, Y.-J. Phylogenetic analyses of an isolate obtained from potato in 1985 revealed potato virus X was introduced to China via multiple events. *Virus Genes* **2010**, *40*, 447–451. [CrossRef]

102. Wang, Y.; Cong, Q.-Q.; Lan, Y.-F.; Geng, C.; Li, X.-D.; Liang, Y.-C.; Yang, Z.-Y.; Zhu, X.-P.; Li, X.-D. Development of new potato virus X-based vectors for gene over-expression and gene silencing assay. *Virus Res.* **2014**, *191*, 62–69. [CrossRef] [PubMed]

103. Lacomme, C.; Chapman, S. Use of potato virus X (PVX)-based vectors for geneexpression and virus-induced gene silencing (VIGS). *Curr. Protoc. Microbiol.* **2008**. [CrossRef] [PubMed]

104. Plchova, H.; Moravec, T.; Hoffmeisterova, H.; Folwarczna, J.; Čeřovská, N. Expression of Human papillomavirus 16 E7ggg oncoprotein on N- and C-terminus of Potato virus X coat protein in bacterial and plant cells. *Protein Expr. Purif.* **2011**, *77*, 146–152. [CrossRef] [PubMed]
105. Hefferon, K. Plant Virus Expression Vectors: A Powerhouse for Global Health. *Biomedicines* **2017**, *5*, 44. [CrossRef] [PubMed]
106. Uhde-Holzem, K.; McBurney, M.; Tiu, B.D.; Advincula, R.C.; Fischer, R.; Commandeur, U.; Steinmetz, N.F. Production of Immunoabsorbent Nanoparticles by Displaying Single-Domain Protein A on Potato Virus X. *Macromol. Biosci.* **2016**, *16*, 231–241. [CrossRef] [PubMed]
107. Yang, L.; Biswas, M.E.; Chen, P. Study of Binding between Protein A and Immunoglobulin G Using a Surface Tension Probe. *Biophys. J.* **2003**, *84*, 509–522. [CrossRef]
108. Arkhipenko, M.V.; Petrova, E.K.; Nikitin, N.A.; Protopopova, A.D.; Dubrovin, E.V.; Yaminskii, I.V.; Rodionova, N.P.; Karpova, O.; Atabekov, J.G.; Center, A.T. Characteristics of Artificial Virus-like Particles Assembled in vitro from Potato Virus X Coat Protein and Foreign Viral RNAs. *Acta Nat.* **2011**, *3*. [CrossRef]
109. Tyulkina, L.G.; Skurat, E.V.; Frolova, O.Y.; Komarova, T.V.; Karger, E.M.; Atabekov, I.G. New Viral Vector for Superproduction of Epitopes of Vaccine Proteins in Plants. *Acta Nat.* **2011**, *3*, 11. [CrossRef]
110. Shukla, S.; DiFranco, N.A.; Wen, A.M.; Commandeur, U.; Steinmetz, N.F. To Target or Not to Target: Active vs. Passive Tumor Homing of Filamentous Nanoparticles Based on Potato virus X. *Cell. Mol. Bioeng.* **2015**, *8*, 433–444. [CrossRef]
111. Röder, J.; Dickmeis, C.; Fischer, R.; Commandeur, U. Systemic infection of nicotiana benthamianawith potato virus X nanoparticles presenting a fluorescent iLOV polypeptide fused directly to the coat protein. *Biomed. Res. Int.* **2018**, e932867. [CrossRef]
112. Shukla, S.; Dickmeis, C.; Fischer, R.; Commandeur, U.; Steinmetz, N.F. In Planta Production of Fluorescent Filamentous Plant Virus-Based Nanoparticles. *Methods Mol. Biol.* **2018**, *1776*, 61–84. [CrossRef]
113. Li, P.; Jing, C.; Ren, H.; Jia, Z.; Ghanem, H.; Wu, G.; Li, M.; Qing, L.; Li, P.; Jing, C.; et al. Analysis of Pathogenicity and Virulence Factors of Ageratum leaf curl Sichuan virus. *Front. Plant Sci.* **2020**, *11*. [CrossRef]
114. Shi, J.; Zhu, Y.; Li, M.; Ma, Y.; Liu, H.; Zhang, P.; Fang, D.; Guo, Y.; Xu, P.; Qiao, Y. Establishment of a novel virus-induced virulence effector assay for the identification of virulence effectors of plant pathogens using a PVX-based expression vector. *Mol. Plant Pathol.* **2020**, *21*, 1654–1661. [CrossRef]
115. Wang, X.; Luo, C.; Xu, Y.; Zhang, C.; Bao, M.; Dou, J.; Wang, Q.; Cheng, Y. Expression of the p24 silencing suppressor of Grapevine leafroll-associated virus 2 from Potato virus X or Barley stripe mosaic virus vector elicits hypersensitive responses in Nicotiana benthamiana. *Plant Physiol Biochem.* **2019**, *142*, 34–42. [CrossRef]
116. Wieczorek, P.; Wrzesińska, B.; Frąckowiak, P.; Przybylska, A.; Obrępalska-Stęplowska, A. Contribution of Tomato torrado virus Vp26 coat protein subunit to systemic necrosis induction and virus infectivity in Solanum lycopersicum. *Virol. J.* **2019**, *16*, 9. [CrossRef] [PubMed]
117. Faivre-Rampant, O.; Gilroy, E.M.; Hrubikova, K.; Hein, I.; Millam, S.; Loake, G.J.; Birch, P.; Taylor, M.; Lacomme, C. Potato virus X-induced gene silencing in leavesand tubers of potato. *Plant Physiol.* **2004**, *134*, 1308–1316. [CrossRef] [PubMed]
118. Kopertekh, L.; Meyer, T.; Freyer, C.; Hust, M. Transient plant production of Salmonella Typhimurium diagnostic antibodies. *Biotechnol. Rep.* **2018**, *20*, e00314. [CrossRef]
119. Demurtas, O.C.; Massa, S.; Illiano, E.; De Martinis, D.; Chan, P.; Di Bonito, P.; Franconi, R. Antigen Production in Plant to Tackle Infectious Diseases Flare Up: The Case of SARS. *Front. Plant Sci.* **2016**, *7*, 54. [CrossRef]
120. Mardanova, E.S.; Kotlyarov, R.Y.; Kuprianov, V.V.; Stepanova, L.A.; Tsybalova, L.M.; Lomonosoff, G.P.; Ravin, N.V. Rapid high-yield expression of a candidate influenza vaccine based on the ectodomain of M2 protein linked to flagellin in plants using viral vectors. *BMC Biotechnol.* **2015**, *15*, 42. [CrossRef] [PubMed]
121. Mohammadzadeh, S.; Roohvand, F.; Memarnejadian, A.; Jafari, A.; Ajdary, S.; Salmanian, A.H.; Ehsani, P. Co-expression of hepatitis C virus polytope-HBsAg and p19-silencing suppressor protein in tobacco leaves. *Pharm. Biol.* **2016**, *54*, 465–473. [CrossRef]
122. Khezri, G.; Rouz, B.B.K.; Ofoghi, H.; Davarpanah, S.J. Heterologous expression of biologically active Mambalgin-1 peptide as a new potential anticancer, using a PVX-based viral vector in Nicotiana benthamiana. *Plant Cell Tissue Organ Cult.* **2020**, *142*, 1–11. [CrossRef] [PubMed]
123. Esfandiari, N.; Arzanani, M.K.; Soleimani, M.; Kohi-Habibi, M.; Svendsen, W.E. A new application of plant virus nanoparticles as drug delivery in breast cancer. *Tumor Biol.* **2016**, *37*, 1229–1236. [CrossRef] [PubMed]
124. Demurtas, O.C.; Massa, S.; Ferrante, P.; Venuti, A.; Franconi, R.; Giuliano, G. A Chlamydomonas-Derived Human Papillomavirus 16 E7 Vaccine Induces Specific Tumor Protection. *PLoS ONE* **2013**, *8*, e61473. [CrossRef]
125. Le, D.H.T.; Lee, K.L.; Shukla, S.; Commandeur, U.; Steinmetz, N.F. Potato virus X, a filamentous plant viral nanoparticle for doxorubicin delivery in cancer therapy. *Nanoscale* **2017**, *9*, 2348–2357. [CrossRef] [PubMed]
126. Lee, K.L.; Murray, A.A.; Le, D.H.T.; Sheen, M.R.; Shukla, S.; Commandeur, U.; Fiering, S.; Steinmetz, N.F. Combination of Plant Virus Nanoparticle-Based in Situ Vaccination with Chemotherapy Potentiates Antitumor Response. *Nano Lett.* **2017**, *17*, 4019–4028. [CrossRef] [PubMed]
127. Le, D.H.T.; Commandeur, U.; Steinmetz, N.F. Presentation and Delivery of Tumor Necrosis Factor-Related Apoptosis-Inducing Ligand via Elongated Plant Viral Nanoparticle Enhances Antitumor Efficacy. *ACS Nano* **2019**, *13*, 2501–2510. [CrossRef]

128. Jobsri, J.; Allen, A.; Rajagopal, D.; Shipton, M.; Kanyuka, K.; Lomonossoff, G.P.; Ottensmeier, C.; Diebold, S.S.; Stevenson, F.; Savelyeva, N. Plant Virus Particles Carrying Tumour Antigen Activate TLR7 and Induce High Levels of Protective Antibody. *PLoS ONE* **2015**, *10*, e0118096. [CrossRef] [PubMed]

129. Kruse, I.; Peyret, H.; Saxena, P.; Lomonossoff, G.P. Encapsidation of Viral RNA in Picornavirales: Studies on Cowpea Mosaic Virus Demonstrate Dependence on Viral Replication. *J. Virol.* **2019**, *93*, e01520-18. [CrossRef] [PubMed]

130. Liu, L.; Cañizares, M.; Monger, W.; Perrin, Y.; Tsakiris, E.; Porta, C.; Shariat, N.; Nicholson, L.; Lomonossoff, G.P. Cowpea mosaic virus-based systems for the production of antigens and antibodies in plants. *Vaccine* **2005**, *23*, 1788–1792. [CrossRef]

131. Sainsbury, F.; Cañizares, M.C.; Lomonossoff, G.P. Cowpea mosaicVirus: The Plant Virus–Based Biotechnology Workhorse. *Annu. Rev. Phytopathol.* **2010**, *48*, 437–455. [CrossRef]

132. Sainsbury, F.; Sack, M.; Stadlmann, J.; Quendler, H.; Fischer, R.; Lomonossoff, G.P. Rapid Transient Production in Plants by Replicating and Non-Replicating Vectors Yields High Quality Functional Anti-HIV Antibody. *PLoS ONE* **2010**, *5*, e13976. [CrossRef]

133. Meshcheriakova, Y.; Durrant, A.; Hesketh, E.L.; Ranson, N.; Lomonossoff, G.P. Combining high-resolution cryo-electron microscopy and mutagenesis to develop cowpea mosaic virus for bionanotechnology. *Biochem. Soc. Trans.* **2017**, *45*, 1263–1269. [CrossRef] [PubMed]

134. Wang, C.; Beiss, V.; Steinmetz, N.F. Cowpea Mosaic Virus Nanoparticles and Empty Virus-Like Particles Show Distinct but Overlapping Immunostimulatory Properties. *J. Virol.* **2019**, *93*. [CrossRef] [PubMed]

135. Dent, M.; Matoba, N. Cancer biologics made in plants. *Curr. Opin. Biotechnol.* **2020**, *61*, 82–88. [CrossRef] [PubMed]

136. Steinmetz, N.F.; Cho, C.-F.; Ablack, A.; Lewis, J.D.; Manchester, M. Cowpea mosaic virus nanoparticles target surface vimentin on cancer cells. *Nanomedicine* **2011**, *6*, 351–364. [CrossRef]

137. Lizotte, P.; Wen, A.M.; Sheen, M.R.; Fields, J.; Rojanasopondist, P.; Steinmetz, N.F.; Fiering, S. In situ vaccination with cowpea mosaic virus nanoparticles suppresses metastatic cancer. *Nat. Nanotechnol.* **2016**, *11*, 295–303. [CrossRef]

138. Patel, R.; Czapar, A.E.; Fiering, S.; Oleinick, N.L.; Steinmetz, N.F. Radiation Therapy Combined with Cowpea Mosaic Virus Nanoparticle in Situ Vaccination Initiates Immune-Mediated Tumor Regression. *ACS Omega* **2018**, *3*, 3702–3707. [CrossRef]

139. Wang, C.; Steinmetz, N.F. CD47 Blockade and Cowpea Mosaic Virus Nanoparticle In Situ Vaccination Triggers Phagocytosis and Tumor Killing. *Adv. Healthc. Mater.* **2019**, *8*, e1801288. [CrossRef] [PubMed]

140. Cai, H.; Wang, C.; Shukla, S.; Steinmetz, N.F. Cowpea Mosaic Virus Immunotherapy Combined with Cyclophosphamide Reduces Breast Cancer Tumor Burden and Inhibits Lung Metastasis. *Adv. Sci.* **2019**, *6*, 1802281. [CrossRef]

141. Albakri, M.; Veliz, F.A.; Fiering, S.N.; Steinmetz, N.F.; Sieg, S.F. Endosomal toll-like receptors play a key role in activation of primary human monocytes by cowpea mosaic virus. *Immunology* **2019**, *159*, 183–192. [CrossRef] [PubMed]

142. Shukla, S.; Wang, C.; Beiss, V.; Cai, H.; Washington, T., II; Murray, A.A.; Gong, X.; Zhao, Z.; Masarapu, H.; Zlotnick, A.; et al. The unique potency of Cowpea mosaic virus (CPMV) in situ cancer vaccine. *Biomater. Sci.* **2020**, *8*, 5489–5503. [CrossRef] [PubMed]

143. Biddlecome, A.; Habte, H.H.; McGrath, K.M.; Sambanthamoorthy, S.; Wurm, M.; Sykora, M.M.; Knobler, C.M.; Lorenz, I.C.; Lasaro, M.; Elbers, K.; et al. Delivery of self-amplifying RNA vaccines in in vitro reconstituted virus-like particles. *PLoS ONE* **2019**, *14*, e0215031. [CrossRef]

144. Cai, H.; Shukla, S.; Steinmetz, N.F. The Antitumor Efficacy of CpG Oligonucleotides is Improved by Encapsulation in Plant Virus-Like Particles. *Adv. Funct. Mater.* **2020**, *30*, 1908743. [CrossRef] [PubMed]

145. Koyani, R.D.; Pérez-Robles, J.; Cadena-Nava, R.D.; Vazquez-Duhalt, R. Biomaterial-based nanoreactors, an alternative for enzyme delivery. *Nanotechnol. Rev.* **2017**, *6*, 405–419. [CrossRef]

146. Strable, E.; Finn, M.G. Chemical Modification of Viruses and Virus-Like Particles. In *Viruses and Nanotechnology*; Manchester, M., Steinmetz, N.F., Eds.; Springer: Berlin/Heidelberg, Germany, 2009; pp. 1–21. [CrossRef]

147. Diamos, A.G.; Mason, H.S. High-level expression and enrichment of norovirus virus-like particles in plants using modified geminiviral vectors. *Protein Expr. Purif.* **2018**, *151*, 86–92. [CrossRef] [PubMed]

148. Rybicki, E.P.; Martin, D.P. Virus-Derived ssDNA Vectors for the Expression of Foreign Proteins in Plants. *Curr. Top. Microbiol. Immunol.* **2011**, *375*, 19–45. [CrossRef]

149. Hefferon, K.L. DNA Virus Vectors for Vaccine Production in Plants: Spotlight on Geminiviruses. *Vaccines* **2014**, *2*, 642–653. [CrossRef]

150. Dickey, A.; Wang, N.; Cooper, E.; Tull, L.; Breedlove, D.; Mason, H.; Liu, D.; Wang, K.Y. Transient Expression of Lumbrokinase (PI239) in Tobacco (Nicotiana tabacum) Using a Geminivirus-Based Single Replicon System Dissolves Fibrin and Blood Clots. *Evid. Based Complement. Altern. Med.* **2017**, *2017*, 1–9. [CrossRef]

151. Bañuelos-Hernández, B.; Monreal-Escalante, E.; Gonzalez-Ortega, O.; Angulo, C.; Rosales-Mendoza, S. Algevir: An Expression System for Microalgae Based on Viral Vectors. *Front. Microbiol.* **2017**, *8*, 1100. [CrossRef]

152. Lozano-Durán, R. Geminiviruses for biotechnology: The art of parasite taming. *New Phytol.* **2016**, *210*, 58–64. [CrossRef] [PubMed]

153. Yang, Q.-Y.; Ding, B.; Zhou, X.-P. Geminiviruses and their application in biotechnology. *J. Integr. Agric.* **2017**, *16*, 2761–2771. [CrossRef]

154. Diamos, A.G.; Pardhe, M.D.; Sun, H.; Hunter, J.G.L.; Kilbourne, J.; Chen, Q.; Mason, H.S. A Highly Expressing, Soluble, and Stable Plant-Made IgG Fusion Vaccine Strategy Enhances Antigen Immunogenicity in Mice Without Adjuvant. *Front. Immunol.* **2020**, *11*, 576012. [CrossRef] [PubMed]

155. Diamos, A.G.; Pardhe, M.D.; Sun, H.; Hunter, J.G.; Mor, T.; Meador, L.; Kilbourne, J.; Chen, Q.; Mason, H.S. Codelivery of improved immune complex and virus-like particle vaccines containing Zika virus envelope domain III synergistically enhances immunogenicity. *Vaccine* **2020**, *38*, 3455–3463. [CrossRef]
156. Čermák, T.; Curtin, S.J.; Gil-Humanes, J.; Čegan, R.; Kono, T.J.Y.; Konečná, E.; Belanto, J.J.; Starker, C.G.; Mathre, J.W.; Greenstein, R.L.; et al. A Multipurpose Toolkit to Enable Advanced Genome Engineering in Plants. *Plant Cell.* **2017**, *29*, 1196–1217. [CrossRef]
157. Zhang, T.; Zheng, Q.; Yi, X.; An, H.; Zhao, Y.; Ma, S.; Zhou, G. Establishing RNA virus resistance in plants by harnessing CRISPR immune system. *Plant Biotechnol. J.* **2018**, *16*, 1415–1423. [CrossRef]
158. Ariga, H.; Toki, S.; Ishibashi, K. Potato Virus X Vector-Mediated DNA-Free Genome Editing in Plants. *Plant Cell Physiol.* **2020**, *61*, 1946–1953. [CrossRef]
159. Ludman, M.; Burgyán, J.; Fátyol, K. Crispr/Cas9 Mediated Inactivation of Argonaute 2 Reveals its Differential Involvement in Antiviral Responses. *Sci. Rep.* **2017**, *7*, 1–12. [CrossRef]
160. Brisson, N.; Paszkowski, J.; Penswick, J.R.; Gronenborn, B.; Potrykus, I.; Hohn, T.M. Expression of a bacterial gene in plants by using a viral vector. *Nat. Cell Biol.* **1984**, *310*, 511–514. [CrossRef]
161. French, R.; Janda, M.; Ahlquist, P. Bacterial Gene Inserted in an Engineered RNA Virus: Efficient Expression in Monocotyledonous Plant Cells. *Science* **1986**, *231*, 1294–1297. [CrossRef]
162. Palmer, K.; Gleba, Y. Plant Viral Vectors. *Curr. Top. Microbiol. Immunol.* **2014**, *375*, 1–194. [CrossRef]
163. Baulcombe, D.; Chapman, S.; Cruz, S.S. Jellyfish green fluorescent protein as a reporter for virus infections. *Plant J.* **1995**, *7*, 1045–1053. [CrossRef]
164. Zhang, H.; Wang, L.; Hunter, D.; Voogd, C.; Joyce, N.; Davies, K. A Narcissus mosaic viral vector system for protein expression and flavonoid production. *Plant Methods* **2013**, *9*, 28. [CrossRef] [PubMed]
165. Takamatsu, N.; Ishikawa, M.; Meshi, T.; Okada, Y. Expression of bacterial chloramphenicol acetyltransferase gene in tobacco plants mediated by TMV-RNA. *EMBO J.* **1987**, *6*, 307–311. [CrossRef] [PubMed]
166. Yusibov, V.; Shivprasad, S.; Turpen, T.H.; Dawson, W.; Koprowski, H. Plant Viral Vectors Based on Tobamoviruses. *Plant Biotechnol.* **2000**, *240*, 81–94. [CrossRef]
167. Jarugula, S.; Charlesworth, S.R.; Qu, F.; Stewart, L.R. Soil-borne wheat mosaic virus infectious clone and manipulation for gene-carrying capacity. *Arch. Virol.* **2016**, *161*, 2291–2297. [CrossRef] [PubMed]
168. Lellis, A.D.; Kasschau, K.D.; Whitham, S.A.; Carrington, J.C. Loss-ofsusceptibility mutants of Arabidopsis thaliana reveal an essential role for eIF (iso)4E during potyvirus infection. *Curr. Biol.* **2002**, *12*, 1046–1051. [CrossRef]
169. Majer, E.; Navarro, J.-A.; Darós, J.-A. A potyvirus vector efficiently targets recombinant proteins to chloroplasts, mitochondria and nuclei in plant cells when expressed at the amino terminus of the polyprotein. *Biotechnol. J.* **2015**, *10*, 1792–1802. [CrossRef]
170. Seo, J.-K.; Choi, H.-S.; Kim, K.-H. Engineering of soybean mosaic virus as a versatile tool for studying protein–protein interactions in soybean. *Sci. Rep.* **2016**, *6*, 22436. [CrossRef]
171. Stanley, J. Geminiviruses: Plant viral vectors. *Curr. Opin. Genet. Dev.* **1993**, *3*, 91–96. [CrossRef]
172. Sainsbury, F.; Lavoie, P.O.; D'Aoust, M.A.; Vezina, L.P.; Lomonossoff, G.P. Expression of multiple proteins using full-length and deleted versions of Cowpea mosaic virus RNA-2. *Plant Biotechnol. J.* **2008**, *6*, 82–92. [CrossRef]
173. Zhang, C.; Bradshaw, J.D.; Whitham, S.A.; Hill, J.H. The Development of an Efficient Multipurpose Bean Pod Mottle Virus Viral Vector Set for Foreign Gene Expression and RNA Silencing. *Plant Physiol.* **2010**, *153*, 52–65. [CrossRef]
174. Zhou, B.; Zhang, Y.; Wang, X.-B.; Dong, J.; Wang, B.; Han, C.; Yu, J.; Li, D. Oral administration of plant-based rotavirus VP6 induces antigen-specific IgAs, IgGs and passive protection in mice. *Vaccine* **2010**, *28*, 6021–6027. [CrossRef]
175. Bouton, C.; King, R.C.; Chen, H.; Azhakanandam, K.; Bieri, S.; Hammond-Kosack, K.E.; Kanyuka, K. Foxtail mosaic virus: A Viral Vector for Protein Expression in Cereals. *Plant Physiol.* **2018**, *177*, 1352–1367. [CrossRef]
176. Cheuk, A.; Houde, M. A New Barley Stripe Mosaic Virus Allows Large Protein Overexpression for Rapid Function Analysis. *Plant Physiol.* **2018**, *176*, 1919–1931. [CrossRef] [PubMed]
177. Lee, W.-S.; Hammond-Kosack, K.; Kanyuka, K. Barley Stripe Mosaic Virus-Mediated Tools for Investigating Gene Function in Cereal Plants and Their Pathogens: Virus-Induced Gene Silencing, Host-Mediated Gene Silencing, and Virus-Mediated Overexpression of Heterologous Protein. *Plant Physiol.* **2012**, *160*, 582–590. [CrossRef] [PubMed]
178. Yuan, C.; Li, C.; Yan, L.; Jackson, A.O.; Liu, Z.; Han, C.; Yu, J.; Li, D. A High Throughput Barley Stripe Mosaic Virus Vector for Virus Induced Gene Silencing in Monocots and Dicots. *PLoS ONE* **2011**, *6*, e26468. [CrossRef] [PubMed]
179. Choi, I.-R.; Stenger, D.C.; Morris, T.J.; French, R. A plant virus vector for systemic expression of foreign genes in cereals. *Plant J.* **2000**, *23*, 547–555. [CrossRef] [PubMed]
180. Elzoghby, A.; Samy, W.; Elgindy, N.A. Protein-based nanocarriers as promising drug and gene delivery systems. *J. Control. Release* **2012**, *161*, 38–49. [CrossRef] [PubMed]
181. Ma, Y.; Nolte, R.; Cornelissen, J.J. Virus-based nanocarriers for drug delivery. *Adv. Drug Deliv. Rev.* **2012**, *64*, 811–825. [CrossRef]
182. Ferrer-Miralles, N.; Rodríguez-Carmona, E.; Corchero, J.L.; García-Fruitós, E.; Vázquez, E.; Villaverde, A. Engineering protein self-assembling in protein-based nanomedicines for drug delivery and gene therapy. *Crit. Rev. Biotechnol.* **2015**, *35*, 209–221. [CrossRef] [PubMed]

Review

Harnessing the Natural Biology of Adeno-Associated Virus to Enhance the Efficacy of Cancer Gene Therapy

Jacquelyn J. Bower [1,2,*], Liujiang Song [2,3], Prabhakar Bastola [2,3] and Matthew L. Hirsch [2,3,*]

1 Lineberger Comprehensive Cancer Center, University of North Carolina at Chapel Hill, Chapel Hill, NC 27514, USA
2 Department of Ophthalmology, University of North Carolina at Chapel Hill, Chapel Hill, NC 27599, USA; liujiang@email.unc.edu (L.S.); pbastola@email.unc.edu (P.B.)
3 Gene Therapy Center, University of North Carolina at Chapel Hill, Chapel Hill, NC 27599, USA
* Correspondence: jacquelyn_bower@med.unc.edu (J.J.B.); mhirsch@email.unc.edu (M.L.H.)

Abstract: Adeno-associated virus (AAV) was first characterized as small "defective" contaminant particles in a simian adenovirus preparation in 1965. Since then, a recombinant platform of AAV (rAAV) has become one of the leading candidates for gene therapy applications resulting in two FDA-approved treatments for rare monogenic diseases and many more currently in various phases of the pharmaceutical development pipeline. Herein, we summarize rAAV approaches for the treatment of diverse types of cancers and highlight the natural anti-oncogenic effects of wild-type AAV (wtAAV), including interactions with the cellular host machinery, that are of relevance to enhance current treatment strategies for cancer.

Keywords: adeno-associated virus; AAV; cancer gene therapy

Citation: Bower, J.J.; Song, L.; Bastola, P.; Hirsch, M.L. Harnessing the Natural Biology of Adeno-Associated Virus to Enhance the Efficacy of Cancer Gene Therapy. *Viruses* **2021**, *13*, 1205. https://doi.org/10.3390/v13071205

Academic Editors: Carla Varanda and Patrick Materatski

Received: 3 May 2021
Accepted: 10 June 2021
Published: 23 June 2021

Publisher's Note: MDPI stays neutral with regard to jurisdictional claims in published maps and institutional affiliations.

1. Adeno-Associated Virus: Discovery and Biology

Adeno-associated virus (AAV), a member of the *parvoviridae* family, was first discovered as a defective contaminant virus of an adenovirus preparation because it required a helper virus to replicate. It was largely thought to be non-pathogenic as no adverse events were noted upon infection in cell culture or in several mammalian species inoculated with the virus [1]. AAV harbors a single-stranded DNA genome of approximately 4.7 Kb that contains two primary open reading frames (ORFs) (Figure 1). Reports have shown that the first ORF, *rep*, encodes four versions of the replication protein (Rep78, Rep68, Rep52, and Rep40), and the large Rep proteins (78/68) maintain site-specific DNA binding and endonucleolytic activity conferred by approximately the first 200 N-terminal amino acids. The smaller Rep proteins maintain the helicase activity and are thought to assemble as a multimeric complex to elicit replication. The second ORF, *cap*, encodes three isoforms of the capsid protein (VP1, VP2, and VP3). The assembly activating protein (AAP) is encoded through an alternative ORF within the *cap* gene and is reported to aid in genome packaging of several capsid serotypes [2,3]. Two additional proteins encoded by *cap*, protein X and the membrane associated accessory protein (MAAP), have been described [4–6]. Although their precise functions in AAV biology are to date unclear, these proteins have been suggested to be involved in the enhancement of AAV replication and virion egress, respectively.

The AAV genome is terminated by inverted repeat DNA sequences (Inverted Terminal Repeats; ITRs) that are required for replication, capsid packaging, and long-term intracellular persistence [7]. To generate recombinant AAV (rAAV) for therapeutic applications, the AAV coding region is removed, and a transgenic sequence is positioned between the flanking ITRs. The transgenic sequence is then replicated and packaged into an AAV capsid via the expression of the wtAAV genome in trans and without ITRs, in the presence of AAV "helper" plasmids which influence wtAAV gene expression. The resultant particles are a protein capsid, with ITRs flanking the genetic sequence of interest (<5 kb), on a single strand of DNA of either polarity [7,8].

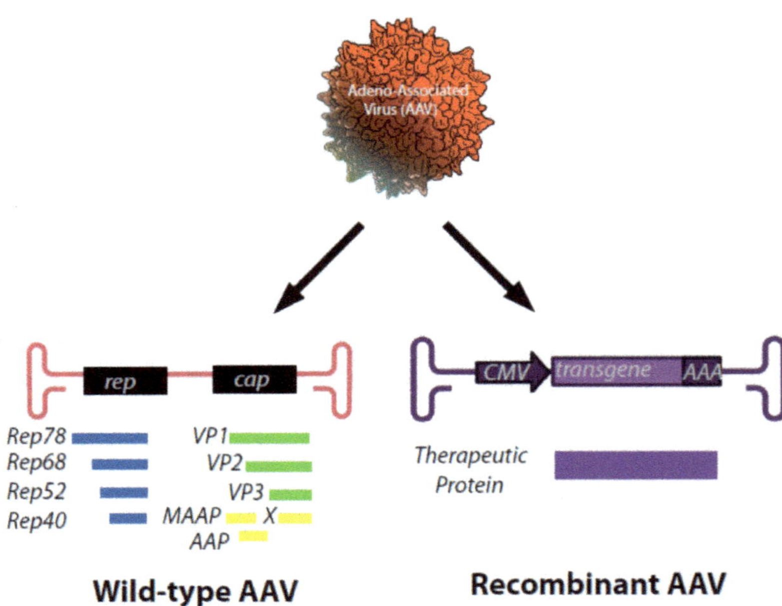

Figure 1. Schematic representation of genomes of a wild-type AAV virus (left) and a recombinant AAV particle (right). Proteins encoded by each ORF are listed below the appropriate gene.

AAV requires a helper virus for replication such as adenovirus or herpes simplex virus in a process that is not completely understood [1,9]. The large Rep proteins (68/78) are capable of binding DNA including a defined 16-nucleotide rep binding element (RBE) located within the AAV ITR. Rep binds to the RBE as a multimeric complex and is reported to elicit site- and strand-specific endonuclease, DNA/DNA helicase, RNA/DNA helicase, and ATPase activities [10]. In addition to these roles, Rep can also self-regulate the activity of its own p5 promoter in an orchestrated sequence related to external helper functions and to its own abundance [11,12]. Interestingly, Rep78 has also been shown to regulate non-native gene expression of other viruses and multiple human genes. For example, a purified Rep78-maltose binding protein (MBP) fusion was shown to bind and repress gene expression from the HIV-1 LTR [13]. Other studies have shown that Rep78 can bind to the human promoter sequences of multiple oncogenes, including *c-fos*, *c-myc*, and *H-ras*, and downregulate reporter gene activity [14,15]. Consistently, RBEs or RBE-like sequences have been identified in multiple locations of the human genome, other viral genomes, and other animal genomes: in fact, a bioinformatics search for the consensus 16-mer core Rep recognition sequence has also been found in or flanking multiple human genes, several of which play roles in DNA repair and cell cycle arrest including BRCA1, ERCC1, and GADD45 [16]. Although these binding sites were confirmed using a Rep68-MBP purified fusion protein in electrophoretic mobility shift assays, the endonuclease activity of Rep68 or effects on human gene expression were not reported. Thus, it is unclear whether Rep68 might mediate AAV genome integration and/or affect gene regulation at these loci.

The *cap* gene encodes three structural proteins: VP1, VP2, and VP3. These proteins assemble in a 1:1:10 ratio, respectively, to generate a relatively simple icosahedral capsid structure. The capsid proteins determine the serotype of the resulting AAV particle and influence its transduction efficiency at multiple discrete steps of the infection pathway. Once AAV enters the cell via receptor-mediated endocytosis, conformational changes of the capsid facilitate endosomal escape, nuclear trafficking, and entry, where it is thought to uncoat partially and/or completely, releasing the ssDNA genome [3,17]. The host cellular DNA replication machinery subsequently synthesizes the complementary sequence presumably via leading strand synthesis initiated by the 3′ ITR, termed "second-strand synthesis", and creating a single self-complementary DNA strand capable of duplex formation via self-annealing [18,19]. Depending on the abundance of nuclear ssDNA genomes, opposite polarity strand annealing also generates double-stranded AAV (dsAAV) DNA. Once the AAV genome assumes a double-stranded form, the genes encoded are competent for transcription regardless of whether the virus contained a wild-type genome or a transgenic sequence. Often, the AAV genome will undergo intra- or inter-molecular circularization, a process facilitated by the recombinogenic nature of the ITRs that is postulated to form larger concatemers as its persistent episomal form.

In addition to the Rep and Cap proteins, wtAAV relies on other exogenous proteins to complete its life cycle. Although wtAAV was discovered in the presence of an adenovirus, several other viruses can provide helper function for wtAAV replication, such as Herpes Simplex Virus and Human Papillomavirus [20,21]. It has even been suggested that wtAAV is not strictly a "defective" virus, as autonomous replication in particular contexts, such as in skin cells, has been reported [21]. Other biological effects of AAV, such as transduction efficiency, can also be enhanced by the presence of a helper virus (such as adenovirus) or by exposure to small molecule inhibitors [18,22]. Interestingly, Nicolson et al. showed that five different categories of small molecules could enhance rAAV transduction 2–200 fold including topoisomerase II poisons, DNA damaging agents, epigenetic modifiers such as HDAC inhibitors, DNA intercalators, and proteosome inhibitors [22]. These findings are relevant for all rAAV therapeutic applications; however, many of these drugs are currently used as chemotherapeutics for multiple types of cancer, suggesting that a well-designed combination of chemotherapy and AAV-based gene therapy approaches could produce a synergistic effect to enhance cancer cell death.

2. Overview of Cancer Gene Therapy Approaches with rAAV

The rAAV vectors, unlike wtAAV, contain only the flanking ITR regions of the viral sequence. To prepare a rAAV vector for gene therapy applications, portions of a helper virus are provided on a plasmid that facilitate replication of the rAAV, but cannot produce helper virus to prevent contamination of the therapeutic rAAV with other viral particles. The use of rAAV vectors for gene therapy boasts several advantages, including, but not limited to: (1) no known pathogenesis associated with AAV infection, (2) the ability to deliver genetic material to dividing and non-dividing cells, (3) conferring long-term transgene expression, and (4) a favorable safety profile. The FDA has approved AAV-based gene therapies for the treatment of the rare genetic diseases Leber's congenital amaurosis (Luxturna®) and spinal muscular atrophy (Zolgensma®), solidifying the role of rAAV therapeutic approaches for monogenetic diseases. Thus, it is not surprising that similar or related therapeutic strategies have been evaluated for multigenic diseases such as cancer. In fact, a multitude of rAAV approaches for cancer gene therapy are under examination in pre-clinical models. Curiously, rAAV clinical trials for cancer gene therapy applications to date have been relatively sparse when compared to the number of adenovirus-based cancer gene therapy trials (23 trials for rAAV vs. 436 trials for adenovirus) and are heavily focused on the use of rAAV to express the GM-CSF cytokine to induce an immune response against prostate cancer cells (Table 1) [23].

Due to the heterogeneous nature of cancer cells both among and within individual tumors, pre-clinical rAAV cancer gene therapy approaches have widely varied in an effort to take advantage of an assortment of cancer-driving cell signaling networks. For example, several groups have attempted to overexpress tumor suppressors and/or DNA repair genes, such as p53, to enhance apoptosis in a variety of cell line and xenograft models including breast cancer and cervical cancer models [24,25]. Another group has expressed the c-terminal portion of hTERT in AAV vectors to induce telomere dysfunction in a xenograft mouse model of glioblastoma [26]. Others have overexpressed endogenous anti-angiogenic factors such as endostatin or angiostatin in rAAV to reduce new blood vessel formation in tumors including melanoma, lung, and pancreatic cancer models in the hopes of reducing metastatic spread [27,28]. rAAV vectors expressing "suicide" genes such as HSV-TK in combination with ganciclovir have also been used to enhance death in breast cancer cell lines [29,30]. Finally, rAAV vectors encoding monoclonal antibodies such as bevacizumab to target VEGF and prevent angiogenesis have also been encoded in AAV vectors and reduce tumor burden in a xenograft mouse model of ovarian cancer [31]. Unfortunately, such gene therapy approaches targeting individual signaling pathways may be difficult to implement from a translational/commercial perspective, due to the inherent diversity of tumors. For example, there are at least five independent subtypes of breast cancer that differ in their presentation, molecular pathology, and response to chemotherapy [32,33]. Although a portion of each tumor subtype contain known cell signaling defects in a single pathway, such as the BRCA1 pathway, it is not feasible to identify, test, develop, and produce a rAAV vector targeting each individual component of the mutant BRCA1 signaling cascade. In addition, such an approach would require the development of high-level clinical diagnostics to determine the defective molecular pathway for each patient. Furthermore, many of the proteins involved in these DNA repair pathways are quite large; thus, not all components of these pathways are targetable due to the 4.8-Kb packaging limitation for rAAV.

Table 1. Clinical Trials Employing Viral-based Vectors for Cancer Gene Therapy Treatment [23]. Database Accessed on 6 March 2021.

Journal of Gene Medicine Database Trial ID	Start Date	Disease	Payload	Phase
CH-0025	2001	Malignant Melanoma	GM-CSF B7.2	1
CN-0020	2008	Malignant Solid Tumors	Tumor Antigen	1
CN-0028	2012	Gastric Cancer	Carcinoembryonic antigen (CEA)	1
ES-0021	2012	Pancreatic Cancer	Hyaluronidase	1
JP-0014	N/A	Hormone refractory metastatic prostate cancer	HSV-TK	1
NL-0012	2004	Hormone refractory prostate cancer	GM-CSF	1
NL-0013	2006	Metastatic Prostate Cancer	GM-CSF	3
NL-0014	2006	Prostate Cancer	GM-CSF	3
NL-0015	2006	Prostate Cancer	GM-CSF	3
NL-0016	N/A	Prostate Cancer	GM-CSF	1
NL-0021	2005	Prostate Cancer	IL-12	1
UK-0133	2005	Prostate Cancer	GM-CSF	3
UK-0134	2005	Prostate Cancer	GM-CSF	3
US-0459	2001	Hormone-Refractory Prostate Cancer	GM-CSF	1
US-0493	2001	Hormone Refractory Prostate Cancer	GM-CSF	1/2
US-0653	2004	Hormone-Refractory Prostate Cancer	GM-CSF	3
US-0675	2004	Prostate Cancer	GM-CSF	1/2
US-0708	2005	Prostate Cancer	GM-CSF	3
US-0903	2008	Prostate Cancer	GM-CSF	2
US-1165	2012	Prostate Cancer	GM-CSF	1/2
US-1748	2018	Non-Hodgkin's Lymphoma/B-cell Acute Lymphoblastic Leukemia	CD19, CD8a, N6 and TCRζ	1
US-1800	2018	Multiple Myeloma	CAR2-α-BCMA, CD28/CD3ζ	1

Other notable approaches using rAAV include the delivery of sequences that can alter/regulate gene expression specifically in cancer cells. These approaches include the expression of shRNA(s) that specifically target oncogenic driver genes such as FHL2 in a colorectal cancer model [34]. Another pre-clinical study using a cervical cancer xenograft mouse model was treated with rAAV2 expressing an shRNA to deplete the HPV-E6 protein; this approach reduced the tumor burden to non-detectable levels [35]. Alternate approaches expressing microRNAs (miRs) such as miR-26a specifically induced apoptosis in liver tumor cells in a genetically engineered mouse model [36]. Promoter restriction has also been used to specifically target cancer cells, such as the use of the CXCR4 promoter, which is overexpressed in breast cancer, and could be combined with the aforementioned "suicide gene therapy" approach [37]. From a feasibility standpoint, such approaches bypass some of the translational concerns including the size limitation of rAAV packaging as the genetic components are much smaller than full-length proteins and the need for an individual rAAV for each aberrant signaling protein, because it can be engineered to harbor multiple shRNAs/miRs. Furthermore, shRNA and/or miRs expressed in rAAV may be an excellent alternative to currently employed small molecule inhibitors, which often initially decrease the tumor burden of patients, but almost always ultimately result in acquired resistance. rAAV's ability to provide long-term genetic expression of multiple shRNA/miR cassettes may be able to overcome this challenge by expressing multiple cassettes and potentially decreasing targeted therapy resistance. However, they would still require advanced molecular diagnostic assays for each individual patient, which could impede clinical applications.

Another approach to rAAV cancer gene therapy has been to modify capsid proteins via rational design/directed evolution to enhance/restrict rAAV transduction of cancer cells. Several groups have attempted to rationally design capsid proteins fused to a ligand that binds a specific receptor on cancer cells, such as the designed ankyrin repeat proteins [38]. Others have endeavored to deliver rAAV specifically to increase transduction of dendritic cells to induce a T cell-mediated immune response to the cancer [39]. Additional studies have generated "locked" versions of AAV capsids that contain matrix metalloproteinase (MMP) cleavage sites to promote capsid uncoating and, therefore, transgene expression specifically in cancer cells that overexpress MMPs, which are thought to contribute to their metastatic behavior [40]. Directed evolution approaches that enhance transduction of glioma cells when compared to natural AAV capsid serotypes have also shown promise for previously difficult to transduce brain tumor cells [41]. Such approaches are expected to provide enhanced delivery of transgenes, and in combination with the aforementioned molecular strategies, have the potential to increase rAAV's efficacy through improved cancer cell specificity.

One of the most popular approaches for cancer gene therapy is to express modulators of the immune system from rAAV to generate an immune response to a particular cancer. For example, several groups have attempted to deliver immune stimulating cytokines via rAAV to induce a tumor-specific host immune response. Ma et al. have shown in a pre-clinical model that rAAV expressing soluble TRAIL can induce apoptosis in liver metastases via hepatic portal vein injection in a mouse xenograft model of lymphoma, suggesting that rAAV–TRAIL may be able to treat liver metastases arising from hematogenous metastasis [42]. Another immune based approach is to stimulate antigen presenting cells (APCs) with rAAV-encoded antigens such as carcinoembryonic antigen (CEA) as a vaccine for colon cancer; indeed, this approach was shown to reduce the development of colon tumors in mice [43]. Many of these studies demonstrated significant or complete tumor regression in multiple in vivo mouse model systems; however, clinical translation of cancer gene therapies remains on the horizon. Typically, clinical translation of cancer therapies from mice have been historically difficult; thus, it remains to be seen whether any of these approaches will have efficacy in the human context.

3. WT AAV Induces Cancer Cell Death

Although most approaches to cancer gene therapy have utilized rAAV strategies, wtAAV itself is reported to have inherent anti-tumorigenic properties that specifically disrupt cancer cell physiology. Much of the early work in this area was generated in the last century and focused on the ability of AAV to prevent viral transformation of mammalian cells upon co-infection with oncogenic viruses such as herpesvirus, bovine papillomavirus, and human papillomavirus [44–47]. In 1981, De la Maza and Carter showed that AAV2 particles harboring sub-genomic pieces of AAV2 DNA (known as defective interfering AAV particles or DI particles), presumably containing the ITR, suppressed Adenovirus-12-induced tumorigenesis in a newborn Syrian golden hamster model [48]. Subsequently, Khleif et al. used an adenovirus derived E1a/ras oncogene plasmid to induce transformation in mouse NIH/3T3 cells [49]. Co-transfection of a non-viral wild-type AAV2 plasmid almost completely suppressed transformation. After examining plasmids with either a defective *rep* or *cap* gene, it was determined that the *rep* mutant plasmid failed to suppress E1a/ras transformation. In contrast, a *cap* mutant plasmid or a plasmid that contained no viral origin of replication suppressed transformation as well as wt AAV2. *rep* mutant plasmids that expressed only Reps 52/40, spliced Reps 68/40, or a K446H mutant located in the purine binding site showed transformation suppression of varying levels. A plasmid containing ITRs, but no *rep* or *cap* sequences, did not suppress transformation, suggesting that the rep protein was required for suppressing the transformation of mouse NIH/3T3 fibroblasts by the adenovirus E1a and ras proteins, at least in a plasmid context [49]. Although the mechanism was not explored in this study, it was subsequently shown that Rep78 binds to p53 and prevents its degradation by Adenovirus, suggesting that p53 may play a role in wtAAV's capacity to reduce cellular transformation induced by helper viruses [50].

Later reports have demonstrated that wtAAV can directly induce cell death in multiple types of cancer cell lines. Furthermore, the toxicity of wtAAV appears to be specific to cancer cells, as it does not seem to induce cell death in most primary or diploid cell lines. For example, more recent work has suggested that wtAAV2 can induce death in HPV-infected cells, while HPV negative keratinocytes remain unaffected [51]. AAV2 appeared to induce apoptosis that correlated with the expression of several of the Rep isoforms (78, 68, and 40), whereas Rep expression was not observed in the HPV negative keratinocytes. AAV replication was detected in both HPV-infected cervical cancer cells and keratinocytes, albeit at slightly weaker levels in the HPV negative keratinocytes. AAV2 infection of the cervical cancer cells increased the number of cells in S phase, the number of cells with subG1 DNA content, CDK1 kinase activity, and active pRB, when compared to non-infected cells [51]. Furthermore, a reduction of p21/p16/p27 proteins normally associated with a G1 arrest was also reported. Taken together, these data suggest that the HPV-infected cells bypassed the G1 cell cycle checkpoint and facilitate entry into S phase, perhaps linking the observed cell death to Rep and/or wtAAV genome replication in the context of HPV-transformed cells.

wtAAV's tumoricidal effects are not limited to HPV-transformed cells. In fact, similar results were seen in multiple types of breast cancer cell lines including hormone receptor positive and triple negative breast cancer cell lines, suggesting that wtAAV exhibits a broader anti-tumorigenic effect that is not specific to a single cell line or class of cell lines [52]. Again, cell death correlated with Rep expression, AAV replication, and a higher percentage of cancer cells in S phase; however, the cell death occurred through both caspase-dependent and caspase-independent mechanisms among the cell lines, suggesting that AAV is capable of activating multiple cell death signaling mechanisms that can override survival signals inherent to tumor cells [52]. Finally, wtAAV2 infection was concurrent with Rep 78/52 expression and viral replication in a breast cancer cell line xenograft model, and tumor growth was retarded [53]. Because the Rep78 protein is reported to induce apoptosis via caspase 3 activation independently of p53, wtAAV may exhibit a broader range of tumoricidal activity for multiple types of cancers, regardless of p53 status [54].

Although the above studies demonstrate that Rep protein levels and viral replication are correlated with the tumoricidal effects of wtAAV, the mechanism(s) is/are not well defined. The Rep78 protein is thought to induce a prolonged S phase arrest via multiple mechanisms [55]. Rep78 binds to the Cdc25A phosphatase and prevents it from interacting with the cyclin-dependent kinases, Cdk2 and Cdk1, which are required for S phase entry and mitotic entry, respectively [55]. Both large Rep proteins (78/68) can also nick the chromatin and induce an ATM-dependent arrest leading to G1 and G2 arrest [55]. Additional requirements for the sustained S phase arrest appear to include hypophosphorylated pRB, the zinc finger domain of Rep78, and Rep78 endonuclease activity [55,56]. Taken together, these data suggest that wtAAV can stall cells in S phase, presumably for the purposes of enhancing its ability to undergo second-strand synthesis [57]. Because most cancer cells exhibit functional defects in one or more cell cycle checkpoints and wtAAV is considered non-pathogenic, it may be possible to enhance the tumoricidal activity of wt and/or rAAV by exploiting its natural life cycle to deliver transgenes that exhibit increased toxicity in the S phase of the cell cycle. Upon a wtAAV-induced DNA damage response, cancer cells are perhaps likely to be overwhelmed by the high levels of induced "DNA damage" signaling elicited upon AAV infection and forced to undergo apoptosis and/or necrosis. As non-tumorigenic cells generally have a higher capacity to repair damaged DNA and can maintain a cell cycle arrest until that damage is repaired, they are predicted to be less susceptible to cell death upon wtAAV infection.

In addition to the anti-tumor effects of the Rep proteins, it has also been suggested that the AAV ITRs are capable of inducing death in tumor cells lacking p53 both in vitro and in vivo. Raj et al. have demonstrated that p53-negative tumor incidence and tumor growth were decreased in response to UV-irradiated AAV infection in a mouse xenograft model, suggesting that Rep expression is not required for the tumoricidal activity of wtAAV in cells deficient for p53 [58]. Further experimentation showed that the short hairpin ITR sequences alone were sufficient to induce death in these p53-deficient cell lines, suggesting that a second genetic feature of wtAAV may enhance its tumoricidal effects. It is likely that the inherent sequence/structure of the hairpin ITRs, which elicit a cellular DNA damage response, are difficult for cancer cells to process and thus enhance the activation of additional apoptotic and necrotic cell death mechanisms, perhaps following a replicative catastrophe [59].

Consistently, several epidemiological studies have suggested that natural AAV infection measured via seropositivity is inversely associated with cervical cancer development [60–62]. Other studies have shown that 50–80% of all cervical tissue samples collected for routine Pap smears in women without cervical cancer contained AAV, offering incidental human data that AAV infection may be preventative for cervical carcinogenesis [63,64]. However, it should be noted that wtAAV is capable of integrating into the host genome, particularly at the AAVS1 site on human chromosome 19 [65]. Recently, whole genome sequencing data were mined from approximately 1400 liver biopsy samples, and although the authors observed wtAAV integration events in liver tissues, they were less frequent in malignant or benign tumors (8%) than in non-tumor liver tissue (18%), with a few instances of wtAAV insertions flanking or inserted into potential oncogenes (~1.2%) [66]. AAV integration events were also observed in a long-term study of a canine cohort receiving rAAV-based gene therapy for hemophilia; however, none of the dogs showed signs of liver disease including tumors [67]. Additional studies in humans revealed no difference between the presence of AAV in multiple tumor tissues when compared to the adjacent non-tumor tissue in more than 400 cancer patients [68], and functional p53 has been shown to reduce AAV integration events [69]. Taken together, these data suggest that although integration of wtAAV is possible, it likely is not a cause of tumorigenesis; rather, its integration and amplification are a consequence of the altered biology of cancer cells. Although the mechanism is not yet clear, potential factors may include alterations in chromatin condensation, increased cell proliferation, and loss of tumor suppressor genes.

Forthcoming data from ongoing human trials will shed further light on the relationship between rAAV gene therapy and cancer.

4. Combination Therapy—rAAV and Chemotherapy

In addition to using rAAV vectors to treat multiple types of cancer, an orthogonal approach has been to combine rAAV gene therapy with currently available chemotherapeutics. Cytotoxic chemotherapies have been the clinical standard-of-care across a plethora of cancer types for decades. Although these cytotoxic chemotherapies can display effective responses, they are often associated with severe side-effects [70,71]. Furthermore, cytotoxic chemotherapies are not universally effective for all cancer types, and a subset of patients will experience the recurrence of drug-resistant tumors, at which point subsequent chemotherapies can be less effective [72–74]. Combining additional modes of therapies (also known as adjuvant therapies) with standard-of-care chemotherapy regimens could help minimize the side-effects associated with these regimens by lowering the required effective dose. In addition, enhanced tumor cell killing achieved with effective combinations could help to eliminate residual disease, thereby reducing the recurrence of drug-resistant tumors. rAAV-mediated transgene expression has been investigated as a potential avenue to selectively trigger cell death in tumor cells (as discussed extensively in the "Overview of Cancer Gene Therapy Approaches with rAAV" section), and in conjunction with cytotoxic chemotherapies, represents the potential for more effective combination therapy regimens.

Several pre-clinical studies have demonstrated that rAAV gene therapy in combination with cytotoxic chemotherapies enhances tumor cell killing in multiple tumor types including ovarian, gastric, colon, hepatocellular, and head and neck cancers (Table 2). In an orthotopic ovarian cancer mouse model, the combination of carboplatin, a standard-of-care chemotherapy in ovarian cancer, and rAAV-mediated delivery of the anti-angiogenic mutant endostatin (rAAV-P125A-endostatin) resulted in a significant decrease in tumor burden and an increase in survival compared to the single agents alone or the untreated group [75]. Although the mechanism for such an enhanced effect was not investigated, the authors noted that carboplatin treatment could sensitize the endothelial cells to P125A-endostatin, resulting in a further reduction of angiogenesis in ovarian tumors. Similarly, another study using ovarian cancer mouse models showed that treatment with the chemotherapy agents topotecan and paclitaxel resulted in increased survival of mice following the rAAV-mediated expression of bevacizumab, a monoclonal antibody directed towards VEGF, resulting in decreased angiogenesis [31]. In gastric cancer models, rAAV-mediated expression of the dominant negative survivin mutant Thr34Ala (rAAV-Sur-Mut(T34A)) resulted in decreased cell proliferation and increased apoptosis. Survivin, a member of the inhibitor of apoptosis (IAP) gene family, has been shown to be overexpressed in gastric cancer cells resulting in the inhibition of apoptosis. Overexpression of the dominant-negative survivin (T34A) abolished the anti-apoptotic effect exerted by the wild-type survivin. Importantly, combining 5-Fluorouracil (5-FU), the first-line chemotherapy drug for gastric cancer, resulted in enhanced tumor cell killing compared to single agents alone or the untreated group [76]. 5-FU treatment has been shown to induce survivin levels; therefore, the authors postulated that overexpression of the dominant-negative survivin with 5-FU treatment resulted in enhanced cytotoxicity [76]. Furthermore, enhanced tumor cell killing was demonstrated following the rAAV-mediated expression of survivin mutant T34A (rAAV-Sur-Mut(T34A)) and oxaliplatin, used for the treatment of advanced colorectal cancer that is resistant to 5-FU, in colon cancer models in vivo [77]. Yet another study showed that the combination of 5-FU and rAAV-mediated overexpression of shRNA targeting FHL2 (Four and a half LIM-only protein 2) resulted in increased colon cancer cell death in vivo [34]. Lastly, cisplatin, another chemotherapeutic used in clinics against multiple solid tumors, has been combined to elicit enhanced cell death with rAAV–TRAIL in hepatocellular carcinoma and head and neck squamous cell carcinoma in vivo [78,79]. TRAIL, a member of TNF-super family, has been previously shown to selectively trigger apoptosis in transformed cells.

Table 2. Preclinical studies conducted using rAAV and chemotherapy combinations.

Cancer Type	Chemotherapy	Transgene	rAAV Capsid, Promoter	References
Ovarian	Carboplatin	Endostatin (P125A)	Unknown, CGA	[75]
Ovarian	Topotecan and Paclitaxel	Bevacizumab Ab	rh.10, CGA	[31]
Gastric	5-FU	Survivin (T34A)	Unknown, CGA	[76]
Colorectal	Oxaliplatin	Survivin (T34A)	Unknown, CGA	[77]
Colorectal	5-FU	shRNA FHL2	AAV2, U6	[34]
Hepatocellular	Cisplatin	TRAIL	AAV2, hTERT	[78]
Head and Neck	Cisplatin	TRAIL	Unknown, CGA	[79]

Notes: CGA = CMV enhancer, chicken beta-Actin promoter; hTERT = human reverse transcriptase component; 5-FU = 5-Fluorouracil.

The above studies clearly demonstrate that rAAV-mediated gene therapy could enhance clinical efficacy when combined with the standard-of-care chemotherapy; however, the mechanism associated with such enhanced efficacy following the combinations remains to be elucidated. It is well established that chemotherapy induces genotoxic and cytotoxic stress in tumor cells. Although each of these effective combinations could be acting through disparate mechanisms as discussed above, increased rAAV-mediated transduction following the increased stress response with chemotherapy may contribute to the enhanced cytotoxicity [22]. Interestingly, apart from paclitaxel, all of the aforementioned chemotherapies used in combination with rAAV vectors interfere with cellular processes occurring in S phase. Therefore, it is likely that an exacerbated stress response, which has been shown to enhance rAAV-mediated transgene expression, coupled with the chemotherapeutic drug's interference in S phase progression, refs [80–83] could result in enhanced tumor cell death. This would allow for the combination of other chemotherapies and newer targeted therapies that induce a tumor-mediated stress response with rAAV-mediated gene therapy in multiple tumor types.

These studies should be pursued cautiously since similar mechanisms could also increase rAAV off-target tissue transduction resulting in increased rAAV-mediated toxicity; thus, approaches to limit rAAV-mediated transgene expression to tumor cells through optimizing the promoter, capsid, and/or delivery route should be explored. As previously noted, many of the synergistic effects of rAAV occurred with cytotoxic chemotherapies that target one or more cellular processes occurring in S phase; thus, elucidation of rAAV's mechanistic interactions with these drugs in a cellular host context would likely lead to improved combination strategies for clinical applications. In summary, rAAV-mediated gene therapy could be explored as an avenue to enhance the efficacy and safety of the chemotherapeutic agents currently used in clinical settings.

5. Targeting Cancer Stem Cells with AAV

Cancer stem cells (CSCs) are a subpopulation of cancer cells residing within the solid tumor bulk (such as breast cancer or lung cancer) or hematological tumors (such as leukemia) [84] that have the potential for self-renewal and contribute to tumorigenesis and/or chemoresistance [85–87]. CSCs often participate in the epithelial–mesenchymal transition via the induction of an embryonic genetic program that re-expresses the genes SOX2, OCT4, NANOG, and DNMT1, and display stem cell-like profiles that contribute to tumor initiation, progression towards invasive phenotypes, therapeutic failure, and/or tumor recurrence [88]. Thus, it has been suggested that efficient tumor treatment requires eradication of the CSC population [89], which has generated considerable enthusiasm for the investigation of CSC-targeted therapies.

Gene therapy has long been proposed as a promising approach for cancer treatment [90], as illustrated in over 2144 clinical trials, accounting for 67.4% of all gene therapy clinical trials to date [90,91]. This technique could be applied to a wide range of tumor types, and a variety of genes and vectors are being used in clinical trials with successful outcomes [7]. Among these trials, 23 are currently underway to determine the efficacy of rAAV vectors for cancer therapeutics (Table 1). However, as summarized above, conventional strategies, such as enhancing the immune system to recognize the cancer cells, enhancing cancer cell apoptosis, and inducing anti-angiogenesis effects via regulation of the VEGF signaling pathway, are the major designs used in these cancer gene therapy clinical trials. Thus, alternative therapeutic approaches using AAV vectors to target CSCs would be a unique avenue of exploration.

One obvious strategy for CSC-targeted therapy is to target the CSC stemness features; this would require the use of AAV to mitigate the overexpression of the genes or perhaps miRs which play vital roles in maintaining the self-renewal capacity of CSCs. For example, one critical factor would be to identify an AAV capsid that could transduce the CSCs with high tropism and efficiency to confer CSC transduction specificity. Although the transduction efficiency of rAAV in different stem cells including embryonic stem cells (ESCs), hematopoietic stem cells (HSCs), and mesenchymal stem cells (MSCs) has been widely studied [92], to the best of our knowledge, there is no comparative transduction study of CSCs in vitro or in vivo. Although AAV6 capsids with site-specific modifications and CD34+ HSC-derived AAV variants are reported to have higher tropism to blood stem cells and support stable and efficient gene transfer [93], most other AAV serotypes may not productively transduce stem cells efficiently [94,95]. Unfortunately, cancer or CSC-derived AAV capsids have not yet been reported to date, consistent with reports of low to no germline transmission of AAV and an inverse relationship of wtAAV integrants in cancerous versus non-transformed cells. Indeed, Gao's group [68] recently published a large-scale molecular epidemiological analysis of AAV in a cancer patient population, and found no significant difference in AAV prevalence, abundance, and variation between cancer and normal tissues. In addition, no specific AAV sequences predominated in tumor samples, and no clonality was observed [68]. This indicates a minimal chance that a natural CSC-derived AAV capsid exists and highlights an opportunity for directed evolution studies to isolate novel AAV capsid variants for increased CSC targeting.

In addition, previous reports have suggested that AAV vectors induce toxicity in some types of stem-like cells [59,96,97]. Johnston et al. reported that AAV ablates neurogenesis, and the ITR sequence alone is sufficient to induce cell death, suggesting AAV-linked toxicity in adult neural progenitor cells in mice [96,98]. Similarly, Hirsch et al. found that ESCs are particularly sensitive to AAV-induced cell death via a p53-dependent apoptotic response that is elicited by a telomeric sequence within the AAV ITR [59,92]. Reports of toxicity to other types of stem-like cells have shown mixed results, alluding to potentially differential effects of AAV toxicity depending on the species, type of stem cell, and AAV serotype and have been thoroughly discussed in a previous review [92,95,99–102]. When toxicity is observed, it is mostly thought to be induced through TLR activation via DNA sensors, especially the palindromic ITR hairpin DNA sequences [103]. These reports further indicate that optimization of the ITR sequences [2,104] (such as CpG depletion in the ITR structure) may be a useful strategy to counteract the potential toxicity, thus ensuring safer applications of rAAV that protect adult stem cell populations.

Notably, it is reported that ESCs and CSCs share some common biomarkers, gene signatures, signaling pathways, and epigenetic regulators [88]. This highlights the possibility that CSCs may also demonstrate toxicity in response to infection by wt or rAAV and illustrates the vast potential of AAV vectors (especially the ITR sequences) as a possible CSC-targeted therapy to supplement current cancer therapies and reduce tumor recurrence. However, before this exciting concept can progress to clinical translation in humans, attempts to better understand the CSCs in the context of AAV biology are needed. These approaches could include: (i) determining the wtAAV persistence in CSCs and the clinical consequences of its potential replication on CSCs and cancer cells, (ii) exploiting the ITR toxicity in CSCs, (iii) examining the potential interaction between the p53 transcriptional binding sites on AAV ITR sequences and CSC stemness, and (iv) the potential inhibitory effects of wtAAV on CSCs. Thus, a growing understanding of CSCs together with the favorable outcomes obtained from rAAV clinical trials [105] provides a common conceptual and research framework for basic and applied cancer research. Exploring AAV as vectors for CSC-targeted therapies would complement and expand the existing repertoire of therapeutic strategies for the treatment of cancer.

6. Closing Remarks

In summary, AAV may be an excellent choice as an adjuvant therapy for multiple types of cancer. The genetic features of AAV combined with the cellular host response to AAV infection provide a unique opportunity to enhance the tumoricidal activity of cancer gene therapy vectors (Figure 2). Both the Rep protein(s) and the AAV ITRs exhibit independent tumoricidal activity in multiple contexts. Therefore, in combination with a transgene or other genetic components that specifically interfere with oncogenic signaling defects, AAV exhibits the potential to induce tumor regression and/or sustained remission, particularly when administered as an adjuvant therapy in conjunction with standard cytotoxic chemotherapies. Further mechanistic studies of AAV's interactions with the cellular host machinery and the exploration of alternative transgenes/genetic cassettes would provide a basic scaffold for the development of novel approaches to treat cancer and lead to promising new avenues of therapeutic inquiry.

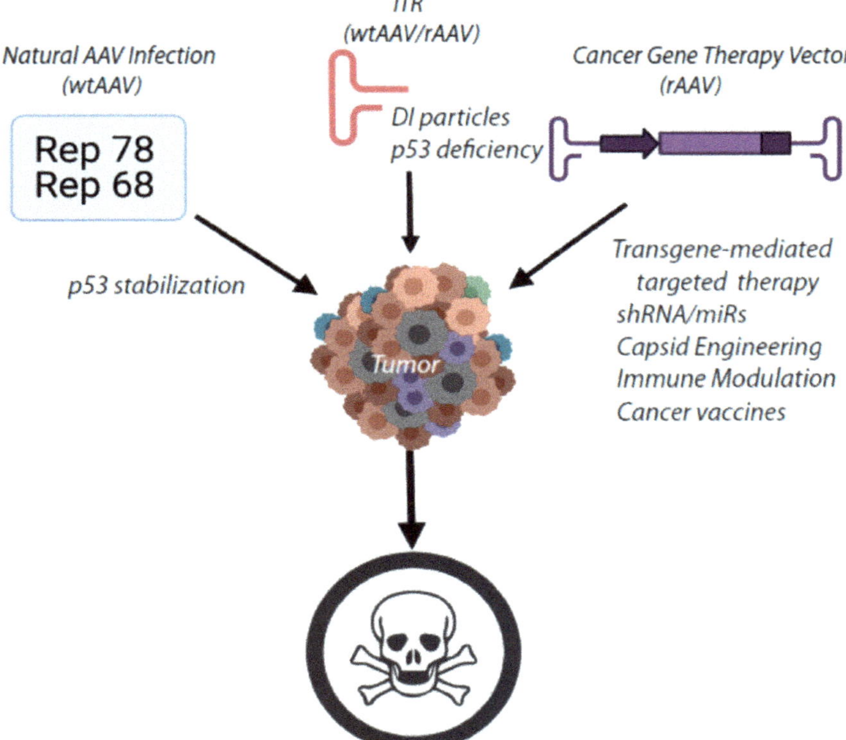

Figure 2. Summary of the potential attributes of AAV for cancer gene therapy.

Author Contributions: Conceptualization, J.J.B. and M.L.H.; writing—original draft preparation, J.J.B., L.S., and P.B.; writing—review and editing, J.J.B., L.S., P.B., and M.L.H. All authors have read and agreed to the published version of the manuscript.

Funding: This work was supported in part by the National Center for Advancing Translational Sciences (NCATS) Grant Award Number UL1TR002489 #550KR262105, a National Cancer Institute (NCI) Grant # R21CA231847, and the UNC Department of Ophthalmology. The content is solely the responsibility of the authors and does not necessarily represent the official vies of the NIH.

Conflicts of Interest: M.L.H. is a cofounder of Bedrock Therapeutics and RainBIO, Inc. M.L.H. has other unrelated technology licensed to Asklepios BioPharmaceutical for which he has received royalties. The remaining authors declare no conflict of interest.

References

1. Atchison, R.W.; Casto, B.C.; Hammon, W.M. Adenovirus-Associated Defective Virus Particles. *Science* **1965**, *149*, 754–755. [CrossRef] [PubMed]
2. Earley, L.F.; Conatser, M.L.; Lue, V.; Dobbins, M.A.L.; Li, C.; Hirsch, M.L.; Samulski, R.J. Adeno-Associated Virus Serotype-Specific Inverted Terminal Repeat Sequence Role in Vector Transgene Expression. *Hum. Gene Ther.* **2020**, *31*, 151–162. [CrossRef]
3. Sonntag, F.; Bleker, S.; Leuchs, B.; Fischer, R.; Kleinschmidt, J.A. Adeno-Associated Virus Type 2 Capsids with Externalized VP1/VP2 Trafficking Domains Are Generated prior to Passage through the Cytoplasm and Are Maintained until Uncoating Occurs in the Nucleus. *J. Virol.* **2006**, *80*, 11040–11054. [CrossRef]
4. Cao, M.; You, H.; Hermonat, P.L. The X Gene of Adeno-Associated Virus 2 (AAV2) Is Involved in Viral DNA Replication. *PLoS ONE* **2014**, *9*, e104596. [CrossRef]
5. Ogden, P.J.; Kelsic, E.D.; Sinai, S.; Church, G.M. Comprehensive AAV capsid fitness landscape reveals a viral gene and enables machine-guided design. *Science* **2019**, *366*, 1139–1143. [CrossRef]
6. Hermonat, P.L.; Santin, A.D.; De Greve, J.; De Rijcke, M.; Bishop, B.M.; Han, L.; Mane, M.; Kokorina, N. Chromosomal latency and expression at map unit 96 of a wild-type plus adeno-associated virus (AAV)/Neo vector and identification of p81, a new AAV transcriptional promoter. *J. Hum. Virol.* **2000**, *2*, 359–368.
7. Li, C.; Samulski, R.J. Engineering adeno-associated virus vectors for gene therapy. *Nat. Rev. Genet.* **2020**, *21*, 255–272. [CrossRef] [PubMed]
8. Grieger, J.C.; Choi, V.W.; Samulski, R.J. Production and characterization of adeno-associated viral vectors. *Nat. Protoc.* **2006**, *1*, 1412–1428. [CrossRef]
9. Buller, R.M.L.; Janik, J.E.; Sebring, E.D.; Rose, J.A. Herpes simplex virus types 1 and 2 completely help adenovirus-associated virus replication. *J. Virol.* **1981**, *40*, 241–247. [CrossRef]
10. Im, D.-S.; Muzyczka, N. The AAV origin binding protein Rep68 is an ATP-dependent site-specific endonuclease with DNA helicase activity. *Cell* **1990**, *61*, 447–457. [CrossRef]
11. Murphy, M.; Gomos-Klein, J.; Stankic, M.; Falck-Pedersen, E. Adeno-Associated Virus Type 2 p5 Promoter: A Rep-Regulated DNA Switch Element Functioning in Transcription, Replication, and Site-Specific Integration. *J. Virol.* **2007**, *81*, 3721–3730. [CrossRef] [PubMed]
12. Kyöstiö, S.R.; Wonderling, R.S.; Owens, R.A. Negative regulation of the adeno-associated virus (AAV) P5 promoter involves both the P5 rep binding site and the consensus ATP-binding motif of the AAV Rep68 protein. *J. Virol.* **1995**, *69*, 6787–6796. [CrossRef]
13. Kokorina, N.A.; Santin, A.D.; Li, C.; Hermonat, P.L. Involvement of protein-DNA interaction in adeno-associated virus Rep78-mediated inhibition of HIV-1. *J. Hum. Virol.* **1999**, *1*, 441–450.
14. Hermonat, P.L. Inhibition of H-ras expression by the adeno-associated virus Rep78 transformation suppressor gene product. *Cancer Res.* **1991**, *51*, 3373–3377.
15. Hermonat, P. Down-regulation of the human c-fos and c-myc proto-oncogene promoters by adeno-associated virus Rep78. *Cancer Lett.* **1994**, *81*, 129–136. [CrossRef]
16. Wonderling, R.S.; Owens, R.A. Binding sites for adeno-associated virus Rep proteins within the human genome. *J. Virol.* **1997**, *71*, 2528–2534. [CrossRef] [PubMed]
17. Bartlett, J.S.; Wilcher, R.; Samulski, R.J. Infectious Entry Pathway of Adeno-Associated Virus and Adeno-Associated Virus Vectors. *J. Virol.* **2000**, *74*, 2777–2785. [CrossRef]
18. Ferrari, F.K.; Samulski, T.; Shenk, T.; Samulski, R.J. Second-strand synthesis is a rate-limiting step for efficient transduction by recombinant adeno-associated virus vectors. *J. Virol.* **1996**, *70*, 3227–3234. [CrossRef] [PubMed]
19. Fisher, K.J.; Gao, G.P.; Weitzman, M.D.; DeMatteo, R.; Burda, J.F.; Wilson, J.M. Transduction with recombinant adeno-associated virus for gene therapy is limited by leading-strand synthesis. *J. Virol.* **1996**, *70*, 520–532. [CrossRef]
20. Boucher, D.W.; Melnick, J.L.; Mayor, H.D. Nonencapsidated Infectious DNA of Adeno-Satellite Virus in Cells Coinfected with Herpesvirus. *Science* **1971**, *173*, 1243–1245. [CrossRef]
21. Meyers, C.; Alam, S.; Mane, M.; Hermonat, P.L. Altered Biology of Adeno-associated Virus Type 2 and Human Papillomavirus during Dual Infection of Natural Host Tissue. *Virology* **2001**, *287*, 30–39. [CrossRef] [PubMed]
22. Nicolson, S.C.; Li, C.; Hirsch, M.L.; Setola, V.; Samulski, R.J. Identification and Validation of Small Molecules That Enhance Recombinant Adeno-associated Virus Transduction following High-Throughput Screens. *J. Virol.* **2016**, *90*, 7019–7031. [CrossRef] [PubMed]
23. Ginn, S.L.; Amaya, A.K.; Alexander, I.E.; Edelstein, M.; Abedi, M.R. Gene therapy clinical trials worldwide to 2017: An update. *J. Gene Med.* **2018**, *20*, e3015. [CrossRef] [PubMed]

24. Zhang, H.; Wang, Y.; Bai, Y.; Shao, Y.; Bai, J.; Ma, Z.; Liu, Q.; Wu, S. Recombinant adeno-associated virus expressing a p53-derived apoptotic peptide (37AA) inhibits HCC cells growth in vitro and in vivo. *Oncotarget* **2017**, *8*, 16801–16810. [CrossRef] [PubMed]

25. Qazilbash, M.; Xiao, X.; Seth, P.; Cowan, K.; Walsh, C. Cancer gene therapy using a novel adeno-associated virus vector expressing human wild-type p53. *Gene Ther.* **1997**, *4*, 675–682. [CrossRef]

26. Ng, S.S.M.; Gao, Y.; Chau, D.H.W.; Li, G.H.Y.; Lai, L.H.; Huang, P.T.; Huang, C.F.; Huang, J.J.; Chen, Y.; Kung, H.F.; et al. A novel glioblastoma cancer gene therapy using AAV-mediated long-term expression of human TERT C-terminal polypeptide. *Cancer Gene Ther.* **2007**, *14*, 561–572. [CrossRef]

27. Noro, T.; Miyake, K.; Suzuki-Miyake, N.; Igarashi, T.; Uchida, E.; Misawa, T.; Yamazaki, Y.; Shimada, T. Adeno-Associated Viral Vector-Mediated Expression of Endostatin Inhibits Tumor Growth and Metastasis in an Orthotropic Pancreatic Cancer Model in Hamsters. *Cancer Res.* **2004**, *64*, 7486–7490. [CrossRef] [PubMed]

28. Lalani, A.S.; Chang, B.; Lin, J.; Case, S.S.; Luan, B.; Wu-Prior, W.-W.; VanRoey, M.; Jooss, K. Anti-Tumor Efficacy of Human Angiostatin Using Liver-Mediated Adeno-Associated Virus Gene Therapy. *Mol. Ther.* **2004**, *9*, 56–66. [CrossRef]

29. Zi-Bo, L.I.; Zhao-Jun, Z.E.N.G.; Qian, C.H.E.N.; Sai-Qun, L.U.O.; Wei-Xin, H.U. Recombinant AAV-mediated HSVtk gene transfer with direct intratumoral injections and Tet-On regulation for implanted human breast cancer. *BMC Cancer* **2006**, *6*, 66. [CrossRef]

30. Zeng, Z.-J.; Xiang, S.-G.; Xue, W.-W.; Li, H.-D.; Ma, N.; Ren, Z.-J.; Xu, Z.-J.; Jiao, C.-H.; Wang, C.-Y.; Hu, W.-X. The cell death and DNA damages caused by the Tet-On regulating HSV-tk/GCV suicide gene system in MCF-7 cells. *Biomed. Pharmacother.* **2014**, *68*, 887–892. [CrossRef] [PubMed]

31. Xie, Y.; Hicks, M.; Kaminsky, S.; Moore, M.A.; Crystal, R.G.; Rafii, A. AAV-mediated persistent bevacizumab therapy suppresses tumor growth of ovarian cancer. *Gynecol. Oncol.* **2014**, *135*, 325–332. [CrossRef] [PubMed]

32. Perou, C.M.; Sørlie, T.; Eisen, M.B.; Van De Rijn, M.; Jeffrey, S.S.; Rees, C.A.; Pollack, J.R.; Ross, D.T.; Johnsen, H.; Akslen, L.A.; et al. Molecular portraits of human breast tumors. *Nature* **2000**, *406*, 747–752. [CrossRef]

33. Carey, L.A.; Dees, E.C.; Sawyer, L.; Gatti, L.; Moore, D.T.; Collichio, F.; Ollila, D.W.; Sartor, C.I.; Graham, M.L.; Perou, C.M. The Triple Negative Paradox: Primary Tumor Chemosensitivity of Breast Cancer Subtypes. *Clin. Cancer Res.* **2007**, *13*, 2329–2334. [CrossRef]

34. Wu, Y.; Guo, Z.; Zhang, D.; Zhang, W.; Yan, Q.; Shi, X.; Zhang, M.; Zhao, Y.; Zhang, Y.; Jiang, B.; et al. A novel colon cancer gene therapy using rAAV-mediated expression of human shRNA-FHL2. *Int. J. Oncol.* **2013**, *43*, 1618–1626. [CrossRef] [PubMed]

35. Sato, N.; Saga, Y.; Uchibori, R.; Tsukahara, T.; Urabe, M.; Kume, A.; Fujiwara, H.; Suzuki, M.; Ozawa, K.; Mizukami, H. Eradication of cervical cancer in vivo by an AAV vector that encodes shRNA targeting human papillomavirus type 16 E6/E7. *Int. J. Oncol.* **2018**, *52*, 687–696. [CrossRef]

36. Kota, J.; Chivukula, R.R.; O'Donnell, K.A.; Wentzel, E.A.; Montgomery, C.L.; Hwang, H.-W.; Chang, T.-C.; Vivekanandan, P.; Torbenson, M.; Clark, K.R.; et al. Therapeutic microRNA Delivery Suppresses Tumorigenesis in a Murine Liver Cancer Model. *Cell* **2009**, *137*, 1005–1017. [CrossRef] [PubMed]

37. Rajendran, S.; Collins, S.; Van Pijkeren, J.P.; O'Hanlon, D.; O'Sullivan, G.C.; Tangney, M. Targeting of breast metastases using a viral gene vector with tumour-selective transcription. *Anticancer. Res.* **2011**, *31*, 1627–1635.

38. Münch, R.C.; Janicki, H.; Völker, I.; Rasbach, A.; Hallek, M.; Büning, H.; Buchholz, C.J. Displaying High-affinity Ligands on Adeno-associated Viral Vectors Enables Tumor Cell-specific and Safe Gene Transfer. *Mol. Ther.* **2013**, *21*, 109–118. [CrossRef] [PubMed]

39. Pandya, J.; Ortiz, L.; Ling, C.; Rivers, A.E.; Aslanidi, G. Rationally designed capsid and transgene cassette of AAV6 vectors for dendritic cell-based cancer immunotherapy. *Immunol. Cell Biol.* **2013**, *92*, 116–123. [CrossRef] [PubMed]

40. Judd, J.; Ho, M.L.; Tiwari, A.; Gomez, E.J.; Dempsey, C.; Van Vliet, K.; Igoshin, O.A.; Silberg, J.J.; Agbandje-McKenna, M.; Suh, J. Tunable Protease-Activatable Virus Nanonodes. *ACS Nano* **2014**, *8*, 4740–4746. [CrossRef]

41. Maguire, C.A.; Gianni, D.; Meijer, D.H.; Shaket, L.A.; Wakimoto, H.; Rabkin, S.D.; Gao, G.; Sena-Esteves, M. Directed evolution of adeno-associated virus for glioma cell transduction. *J. Neuro Oncol.* **2010**, *96*, 337–347. [CrossRef]

42. Ma, H.; Liu, Y.; Liu, S.; Kung, H.F.; Sun, X.; Zheng, D.; Xu, R. Recombinant adeno-associated virus-mediated TRAIL gene therapy suppresses liver metastatic tumors. *Int. J. Cancer* **2005**, *116*, 314–321. [CrossRef]

43. Hensel, J.A.; Khattar, V.; Ashton, R.; Ponnazhagan, S. Recombinant AAV-CEA Tumor Vaccine in Combination with an Immune Adjuvant Breaks Tolerance and Provides Protective Immunity. *Mol. Ther. Oncolytics* **2019**, *12*, 41–48. [CrossRef] [PubMed]

44. Cukor, G.; Blacklow, N.R.; Kibrick, S.; Swan, I.C.; Sidney, K. Effect of Adeno-Associated Virus on Cancer Expression by Herpesvirus-Transformed Hamster Cells 2. *J. Natl. Cancer Inst.* **1975**, *55*, 957–959. [CrossRef] [PubMed]

45. Hermonat, P.L. The adeno-Associated virus Rep78 gene inhibits cellular transformation induced by bovine papillomavirus. *Virology* **1989**, *172*, 253–261. [CrossRef]

46. Hermonat, P.L. Adeno-associated virus inhibits human papillomavirus type 16: A viral interaction implicated in cervical cancer. *Cancer Res.* **1994**, *54*, 2278–2281. [PubMed]

47. Hermonat, P.L. Inhibition of bovine papillomavirus plasmid DNA replication by adeno-associated virus. *Virology* **1992**, *189*, 329–333. [CrossRef]

48. De La Maza, L.M.; Carter, B.J. Inhibition of adenovirus oncogenicity in hamsters by adeno-associated virus DNA. *J. Natl. Cancer Inst.* **1981**, *67*, 1323–1326.

49. Khleif, S.N.; Myersz, T.; Carter, B.J.; Trempe, J.P. Inhibition of Cellular transformation by the adeno-associated virus rep gene. *Virology* **1991**, *181*, 738–741. [CrossRef]

50. Batchu, R.B.; Shammas, M.A.; Wang, J.Y.; Munshi, N.C. Interaction of adeno-associated virus Rep78 with p53: Implications in growth inhibition. *Cancer Res.* **1999**, *59*, 3592–3595.
51. Alam, S.; Meyers, C. Adeno-associated virus type 2 induces apoptosis in human papillomavirus-infected cell lines but not in normal keratinocytes. *J. Virol.* **2009**, *83*, 10286–91022. [CrossRef]
52. Alam, S.; Bowser, B.S.; Conway, M.J.; Israr, M.; Tandon, A.; Meyers, C. Adeno-associated virus type 2 infection activates caspase dependent and independent apoptosis in multiple breast cancer lines but not in normal mammary epithelial cells. *Mol. Cancer* **2011**, *10*, 97. [CrossRef] [PubMed]
53. Alam, S.; Bowser, B.S.; Israr, M.; Conway, M.J.; Meyers, C. Adeno-associated virus type 2 infection of nude mouse human breast cancer xenograft induces necrotic death and inhibits tumor growth. *Cancer Biol. Ther.* **2014**, *15*, 1013–1028. [CrossRef] [PubMed]
54. Schmidt, M.; Afione, S.; Kotin, R.M. Adeno-Associated Virus Type 2 Rep78 Induces Apoptosis through Caspase Activation Independently of p53. *J. Virol.* **2000**, *74*, 9441–9450. [CrossRef] [PubMed]
55. Berthet, C.; Raj, K.; Saudan, P.; Beard, P. How adeno-associated virus Rep78 protein arrests cells completely in S phase. *Proc. Natl. Acad. Sci. USA* **2005**, *102*, 13634–13639. [CrossRef]
56. Saudan, P.; Vlach, J.; Beard, P. Inhibition of S-phase progression by adeno-associated virus Rep78 protein is mediated by hypophosphorylated pRb. *EMBO J.* **2000**, *19*, 4351–4361. [CrossRef]
57. Russell, D.W.; Miller, A.D.; Alexander, I.E. Adeno-associated virus vectors preferentially transduce cells in S phase. *Proc. Natl. Acad. Sci. USA* **1994**, *91*, 8915–8919. [CrossRef]
58. Raj, K.; Ogston, P.; Beard, P. Virus-mediated killing of cells that lack p53 activity. *Nat. Cell Biol.* **2001**, *412*, 914–917. [CrossRef]
59. Hirsch, M.L.; Fagan, B.M.; Dumitru, R.; Bower, J.J.; Yadav, S.; Porteus, M.H.; Pevny, L.H.; Samulski, R.J. Viral Single-Strand DNA Induces p53-Dependent Apoptosis in Human Embryonic Stem Cells. *PLoS ONE* **2011**, *6*, e27520. [CrossRef]
60. Coker, A.L.; Russell, R.B.; Bond, S.M.; Pirisi, L.; Liu, Y.; Mane, M.; Kokorina, N.; Gerasimova, T.; Hermonat, P.L. Adeno-Associated Virus Is Associated with a Lower Risk of High-Grade Cervical Neoplasia. *Exp. Mol. Pathol.* **2001**, *70*, 83–89. [CrossRef]
61. Mayor, H.D.; Drake, S.; Stahmann, J.; Mumford, D.M. Antibodies to adeno-associated satellite virus and herpes simplex in sera from cancer patients and normal adults. *Am. J. Obstet. Gynecol.* **1976**, *126*, 100–104. [CrossRef]
62. Georg-Fries, B.; Biederlack, S.; Wolf, J.; Hausen, H.Z. Analysis of proteins, helper dependence, and seroepidemiology of a new human parvovirus. *Virology* **1984**, *134*, 64–71. [CrossRef]
63. Han, L.; Parmley, T.H.; Keith, S.; Kozlowski, K.J.; Smith, L.J.; Hermonat, P.L. High prevalence of adeno-associated virus (AAV) type 2 rep DNA in cervical materials: AAV may be sexually transmitted. *Virus Genes* **1996**, *12*, 47–52. [CrossRef]
64. Tobiasch, E.; Rabreau, M.; Geletneky, K.; Laruë-Charlus, S.; Severin, F.; Becker, N.; Schlehofer, J.R. Detection of adeno-associated virus DNA in human genital tissue and in material from spontaneous abortion. *J. Med Virol.* **1994**, *44*, 215–222. [CrossRef]
65. Kotin, R.M.; Siniscalco, M.; Samulski, R.J.; Zhu, X.D.; Hunter, L.; Laughlin, C.A.; McLaughlin, S.; Muzyczka, N.; Rocchi, M.; Berns, K.I. Site-specific integration by adeno-associated virus. *Proc. Natl. Acad. Sci. USA* **1990**, *87*, 2211–2215. [CrossRef] [PubMed]
66. La Bella, T.; Imbeaud, S.; Peneau, C.; Mami, I.; Datta, S.; Bayard, Q.; Caruso, S.; Hirsch, T.Z.; Calderaro, J.; Morcrette, G.; et al. Adeno-associated virus in the liver: Natural history and consequences in tumour development. *Gut* **2020**, *69*, 737–747. [CrossRef] [PubMed]
67. Nguyen, G.N.; Everett, J.K.; Kafle, S.; Roche, A.M.; Raymond, H.E.; Leiby, J.; Wood, C.; Assenmacher, C.-A.; Merricks, E.P.; Long, C.T.; et al. A long-term study of AAV gene therapy in dogs with hemophilia A identifies clonal expansions of transduced liver cells. *Nat. Biotechnol.* **2021**, *39*, 47–55. [CrossRef]
68. Qin, W.; Xu, G.; Tai, P.W.L.; Wang, C.; Luo, L.; Li, C.; Hu, X.; Xue, J.; Lu, Y.; Zhou, Q.; et al. Large-scale molecular epidemiological analysis of AAV in a cancer patient population. *Oncogene* **2021**, *40*, 3060–3071. [CrossRef]
69. Zacharias, J.; Romanova, L.G.; Menk, J.; Philpott, N.J. p53 inhibits adeno-associated viral vector integration. *Hum. Gene. Ther.* **2011**, *22*, 1445–1451. [CrossRef]
70. Nurgali, K.; Jagoe, R.T.; Abalo, R. Editorial: Adverse Effects of Cancer Chemotherapy: Anything New to Improve Tolerance and Reduce Sequelae? *Front. Pharmacol.* **2018**, *9*, 245. [CrossRef] [PubMed]
71. Schirrmacher, V. From chemotherapy to biological therapy: A review of novel concepts to reduce the side effects of systemic cancer treatment (Review). *Int. J. Oncol.* **2019**, *54*, 407–419.
72. Chien, J.; Kuang, R.; Landen, C.; Shridhar, V. Platinum-Sensitive Recurrence in Ovarian Cancer: The Role of Tumor Microenvironment. *Front. Oncol.* **2013**, *3*, 251. [CrossRef]
73. Bastola, P.; Neums, L.; Schoenen, F.; Chien, J. VCP inhibitors induce endoplasmic reticulum stress, cause cell cycle arrest, trigger caspase-mediated cell death and synergistically kill ovarian cancer cells in combination with Salubrinal. *Mol. Oncol.* **2016**, *10*, 1559–1574. [CrossRef]
74. Nedeljković, M.; Damjanović, A. Mechanisms of Chemotherapy Resistance in Triple-Negative Breast Cancer—How We Can Rise to the Challenge. *Cells* **2019**, *8*, 957. [CrossRef]
75. Subramanian, I.V. Adeno-Associated Virus-Mediated Delivery of a Mutant Endostatin in Combination with Carboplatin Treatment Inhibits Orthotopic Growth of Ovarian Cancer and Improves Long-term Survival. *Cancer Res.* **2006**, *66*, 4319–4328. [CrossRef] [PubMed]
76. Dang, S.C.; Feng, S.; Wang, P.J.; Cui, L.; Qu, J.G.; Zhang, J.X. Overexpression of Survivin mutant Thr34Ala induces apoptosis and inhibits gastric cancer growth. *Neoplasma* **2015**, *62*, 81–87. [CrossRef]

77. Tu, S.P.; Xue, Z.; Sun, P.H.; Zhu, L.M.; Jiang, S.H.; Qiao, M.M.; Chi, A.L. Adeno-associated virus-mediated survivin mutant Thr34Ala cooperates with oxaliplatin to inhibit tumor growth and angiogenesis in colon cancer. *Oncol. Rep.* **2011**, *25*, 1039–1046. [CrossRef] [PubMed]

78. Wang, Y.; Huang, F.; Cai, H.; Wu, Y.; He, G.; Tan, W.-S. The efficacy of combination therapy using adeno-associated virus-TRAIL targeting to telomerase activity and cisplatin in a mice model of hepatocellular carcinoma. *J. Cancer Res. Clin. Oncol.* **2010**, *136*, 1827–1837. [CrossRef]

79. Jiang, M.; Liu, Z.; Xiang, Y.; Ma, H.; Liu, S.; Liu, Y.; Zheng, D. Synergistic antitumor effect of AAV-mediated TRAIL expression combined with cisplatin on head and neck squamous cell carcinoma. *BMC Cancer* **2011**, *11*, 54. [CrossRef]

80. Yan, Z.; Zak, R.; Zhang, Y.; Ding, W.; Godwin, S.; Munson, K.; Peluso, R.; Engelhardt, J.F. Distinct Classes of Proteasome-Modulating Agents Cooperatively Augment Recombinant Adeno-Associated Virus Type 2 and Type 5-Mediated Transduction from the Apical Surfaces of Human Airway Epithelia. *J. Virol.* **2004**, *78*, 2863–2874. [CrossRef]

81. Nathwani, A.C.; Cochrane, M.; McIntosh, J.; Ng, C.Y.; Zhou, J.; Gray, J.T.; Davidoff, A.M. Enhancing transduction of the liver by adeno-associated viral vectors. *Gene Ther.* **2008**, *16*, 60–69. [CrossRef]

82. Mitchell, A.M.; Li, C.; Samulski, R.J. Arsenic Trioxide Stabilizes Accumulations of Adeno-Associated Virus Virions at the Perinuclear Region, Increasing Transduction In Vitro and In Vivo. *J. Virol.* **2013**, *87*, 4571–4583. [CrossRef]

83. Zhong, L.; Qing, K.; Si, Y.; Chen, L.; Tan, M.; Srivastava, A. Heat-shock Treatment-mediated Increase in Transduction by Recombinant Adeno-associated Virus 2 Vectors Is Independent of the Cellular Heat-shock Protein 90. *J. Biol. Chem.* **2004**, *279*, 12714–12723. [CrossRef]

84. Bonnet, D.; Dick, J.E. Human acute myeloid leukemia is organized as a hierarchy that originates from a primitive hematopoietic cell. *Nat. Med.* **1997**, *3*, 730–737. [CrossRef]

85. Song, L.; Tao, X.; Lin, L.; Chen, C.; Yao, H.; He, G.; Zou, G.; Cao, Z.; Yan, S.; Lu, L.; et al. Cerasomal Lovastatin Nanohybrids for Efficient Inhibition of Triple-Negative Breast Cancer Stem Cells To Improve Therapeutic Efficacy. *ACS Appl. Mater. Interfaces* **2018**, *10*, 7022–7030. [CrossRef]

86. Carvalho, J. Cell Reversal from a Differentiated to a Stem-Like State at Cancer Initiation. *Front. Oncol.* **2020**, *10*, 541. [CrossRef]

87. Reya, T.; Morrison, S.J.; Clarke, M.F.; Weissman, I.L. Stem cells, cancer, and cancer stem cells. *Nature* **2001**, *414*, 105–111. [CrossRef]

88. Hadjimichael, C.; Chanoumidou, K.; Papadopoulou, N.; Arampatzi, P.; Papamatheakis, J.; Kretsovali, A. Common stemness regulators of embryonic and cancer stem cells. *World J. Stem Cells* **2015**, *7*, 1150–1184. [CrossRef] [PubMed]

89. Shibata, M.; Hoque, M.O. Targeting Cancer Stem Cells: A Strategy for Effective Eradication of Cancer. *Cancers* **2019**, *11*, 732. [CrossRef]

90. Cross, D.; Burmester, J.K. Gene Therapy for Cancer Treatment: Past, Present and Future. *Clin. Med. Res.* **2006**, *4*, 218–227. [CrossRef]

91. Ajith, T.A. Strategies used in the clinical trials of gene therapy for cancer. *J. Exp. Ther. Oncol.* **2015**, *11*, 33–39.

92. Brown, N.; Song, L.; Kollu, N.R.; Hirsch, M.L. Adeno-Associated Virus Vectors and Stem Cells: Friends or Foes? *Hum. Gene Ther.* **2017**, *28*, 450–463. [CrossRef] [PubMed]

93. Smith, L.J.; Ul-Hasan, T.; Carvaines, S.K.; Van Vliet, K.; Yang, E.; Wong, K.K.; Agbandje-McKenna, M.; Chatterjee, S. Gene Transfer Properties and Structural Modeling of Human Stem Cell-derived AAV. *Mol. Ther.* **2014**, *22*, 1625–1634. [CrossRef]

94. Song, L.; Kauss, M.A.; Kopin, E.; Chandra, M.; Ul-Hasan, T.; Miller, E.; Jayandharan, G.R.; Rivers, A.E.; Aslanidi, G.V.; Ling, C.; et al. Optimizing the transduction efficiency of capsid-modified AAV6 serotype vectors in primary human hematopoietic stem cells in vitro and in a xenograft mouse model in vivo. *Cytotherapy* **2013**, *15*, 986–998. [CrossRef] [PubMed]

95. Song, L.; Song, Z.; Fry, N.J.; Conatser, L.; Llanga, T.; Mei, H.; Kafri, T.; Hirsch, M.L. Gene Delivery to Human Limbal Stem Cells Using Viral Vectors. *Hum. Gene Ther.* **2019**, *30*, 1336–1348. [CrossRef] [PubMed]

96. Johnston, S.T.; Parylak, S.L.; Kim, S.; Mac, N.; Lim, C.K.; Gallina, I.S.; Bloyd, C.W.; Newberry, A.; Saavedra, C.D.; Novák, O.; et al. AAV Ablates Neurogenesis in the Adult Murine Hippocampus. *bioRxiv* **2020**. [CrossRef]

97. Rapti, K.; Stillitano, F.; Karakikes, I.; Nonnenmacher, M.; Weber, T.; Hulot, J.-S.; Hajjar, R.J. Effectiveness of gene delivery systems for pluripotent and differentiated cells. *Mol. Ther. Methods Clin. Dev.* **2015**, *2*, 14067. [CrossRef] [PubMed]

98. Kirschen, G.W.; Kery, R.; Liu, H.; Ahamad, A.; Chen, L.; Akmentin, W.; Kumar, R.; Levine, J.; Xiong, Q.; Ge, S. Genetic dissection of the neuro-glio-vascular machinery in the adult brain. *Mol. Brain* **2018**, *11*, 2. [CrossRef]

99. Yang, H.; Qing, K.; Keeler, G.D.; Yin, L.; Mietzsch, M.; Ling, C.; Hoffman, B.E.; Agbandje-McKenna, M.; Tan, M.; Wang, W.; et al. Enhanced Transduction of Human Hematopoietic Stem Cells by AAV6 Vectors: Implications in Gene Therapy and Genome Editing. *Mol. Ther. Nucleic Acids* **2020**, *20*, 451–458. [CrossRef]

100. Cromer, M.K.; Camarena, J.; Martin, R.M.; Lesch, B.J.; Vakulskas, C.A.; Bode, N.M.; Kurgan, G.; Collingwood, M.A.; Rettig, G.R.; Behlke, M.A.; et al. Gene replacement of α-globin with β-globin restores hemoglobin balance in β-thalassemia-derived hematopoietic stem and progenitor cells. *Nat. Med.* **2021**, *27*, 677–687. [CrossRef]

101. Cromer, M.K.; Vaidyanathan, S.; Ryan, D.E.; Curry, B.; Lucas, A.B.; Camarena, J.; Kaushik, M.; Hay, S.; Martin, R.M.; Steinfeld, I.; et al. Global Transcriptional Response to CRISPR/Cas9-AAV6-Based Genome Editing in CD34+ Hematopoietic Stem and Progenitor Cells. *Mol. Ther.* **2018**, *26*, 2431–2442. [CrossRef]

102. Basche, M.; Kampik, D.; Kawasaki, S.; Branch, M.J.; Robinson, M.; Larkin, D.F.; Smith, A.J.; Ali, R.R.; Larkin, F. Sustained and Widespread Gene Delivery to the Corneal Epithelium via In Situ Transduction of Limbal Epithelial Stem Cells, Using Lentiviral and Adeno-Associated Viral Vectors. *Hum. Gene Ther.* **2018**, *29*, 1140–1152. [CrossRef]

103. Chan, Y.K.; Wang, S.K.; Chu, C.J.; Copland, D.A.; Letizia, A.J.; Verdera, H.C.; Chiang, J.J.; Sethi, M.; Wang, M.K.; Neidermyer, W.J., Jr.; et al. Engineering adeno-associated viral vectors to evade innate immune and inflammatory responses. *Sci. Transl. Med.* **2021**, *13*, eabd3438. [CrossRef]
104. Espín-Palazón, R.; Stachura, D.L.; Campbell, C.A.; García-Moreno, D.; Del Cid, N.; Kim, A.D.; Candel, S.; Meseguer, J.; Mulero, V.; Traver, D. Proinflammatory Signaling Regulates Hematopoietic Stem Cell Emergence. *Cell* **2014**, *159*, 1070–1085. [CrossRef]
105. Berns, K.I. The Unusual Properties of the AAV Inverted Terminal Repeat. *Hum. Gene Ther.* **2020**, *31*, 518–523. [CrossRef]

viruses

Article

Exploring Codon Adjustment Strategies towards *Escherichia coli*-Based Production of Viral Proteins Encoded by HTH1, a Novel Prophage of the Marine Bacterium *Hypnocyclicus thermotrophus*

Hasan Arsın [1,2,*], Andrius Jasilionis [3], Håkon Dahle [2,4], Ruth-Anne Sandaa [1], Runar Stokke [1,2], Eva Nordberg Karlsson [3] and Ida Helene Steen [1,2,*]

1   Department of Biological Sciences, University of Bergen, N-5020 Bergen, Norway; Ruth.Sandaa@uib.no (R.-A.S.); Runar.Stokke@uib.no (R.S.)
2   Centre for Deep Sea Research, University of Bergen, N-5020 Bergen, Norway; Hakon.Dahle@uib.no
3   Division of Biotechnology, Lund University, P.O. Box 124, SE-221 00 Lund, Sweden; Andrius.jasilionis@biotek.lu.se (A.J.); Eva.nordberg_karlsson@biotek.lu.se (E.N.K.)
4   Computational Biology Unit, University of Bergen, N-5020 Bergen, Norway
*   Correspondence: Hasan.arsin@uib.no (H.A.); Ida.steen@uib.no (I.H.S.); Tel.: +47-555-88-375 (I.H.S.)

Citation: Arsın, H.; Jasilionis, A.; Dahle, H.; Sandaa, R.-A.; Stokke, R.; Nordberg Karlsson, E.; Steen, I.H. Exploring Codon Adjustment Strategies towards *Escherichia coli*-Based Production of Viral Proteins Encoded by HTH1, a Novel Prophage of the Marine Bacterium *Hypnocyclicus thermotrophus*. *Viruses* **2021**, *13*, 1215. https://doi.org/10.3390/v13071215

Academic Editors: Carla Varanda and Patrick Materatski

Received: 21 May 2021
Accepted: 18 June 2021
Published: 23 June 2021

Publisher's Note: MDPI stays neutral with regard to jurisdictional claims in published maps and institutional affiliations.

Abstract: Marine viral sequence space is immense and presents a promising resource for the discovery of new enzymes interesting for research and biotechnology. However, bottlenecks in the functional annotation of viral genes and soluble heterologous production of proteins hinder access to downstream characterization, subsequently impeding the discovery process. While commonly utilized for the heterologous expression of prokaryotic genes, codon adjustment approaches have not been fully explored for viral genes. Herein, the sequence-based identification of a putative prophage is reported from within the genome of *Hypnocyclicus thermotrophus*, a Gram-negative, moderately thermophilic bacterium isolated from the Seven Sisters hydrothermal vent field. A prophage-associated gene cluster, consisting of 46 protein coding genes, was identified and given the proposed name *Hypnocyclicus thermotrophus* phage H1 (HTH1). HTH1 was taxonomically assigned to the viral family *Siphoviridae*, by lowest common ancestor analysis of its genome and phylogeny analyses based on proteins predicted as holin and DNA polymerase. The gene neighbourhood around the HTH1 lytic cassette was found most similar to viruses infecting Gram-positive bacteria. In the HTH1 lytic cassette, an N-acetylmuramoyl-L-alanine amidase (Amidase_2) with a peptidoglycan binding motif (LysM) was identified. A total of nine genes coding for enzymes putatively related to lysis, nucleic acid modification and of unknown function were subjected to heterologous expression in *Escherichia coli*. Codon optimization and codon harmonization approaches were applied in parallel to compare their effects on produced proteins. Comparison of protein yields and thermostability demonstrated that codon optimization yielded higher levels of soluble protein, but codon harmonization led to proteins with higher thermostability, implying a higher folding quality. Altogether, our study suggests that both codon optimization and codon harmonization are valuable approaches for successful heterologous expression of viral genes in *E. coli*, but codon harmonization may be preferable in obtaining recombinant viral proteins of higher folding quality.

Keywords: prophage; hydrothermal vent; *Hypnocyclicus thermotrophus*; lytic cassette; *Escherichia coli*; heterologous expression; codon optimization; codon harmonization

1. Introduction

Hydrothermal vents host some of the most diverse microbial communities in marine environments. Diverse (hyper)thermophilic bacteria and archaea grow within the steep chemical and temperature gradients formed by rapid mixing of high temperature (up to above 300 °C) reduced vent fluids and cold seawater [1,2]. The discovery of the hydrothermal vent ecosystem remains one of the biggest breakthroughs in our understanding of how life can be sustained in extreme conditions, marked by the first vent observation on the Galápagos Rift, in the eastern Pacific [3] and the discovery of the first black smoker vents [4]. Today, hydrothermal vents are well-known as attractive sites for bioprospecting of biotechnologically interesting enzymes [5–8] and other valuable biomolecules with potential industrial applications [6,9,10]. As with other marine biomes [11–13], hydrothermal vent environments are observed to be abundant with viruses, especially tailed dsDNA bacteriophages of order *Caudovirales* [14,15]. These viruses remain a largely unexplored space of genetic diversity and, therefore, an under-utilized source for enzyme bioprospecting efforts [16,17].

The unique biology of host-reliant viral replication makes viruses remarkably interesting entities for biotechnology, where lytic enzymes can be found associated with their strategy of host infection [18,19]. While lytic phages reproduce by host cell lysis, lysogenic or temperate phages can remain dormant until induction, either as so-called "prophages" integrated into the host genome, or as extrachromosomal elements [20,21]. Temperate phages have been reported as particularly present in the microbial communities associated with vent fields [15,22], likely related to challenging environmental factors such as lower host abundances, limiting nutrient availability, and the fringe physical and chemical conditions present at these sites. In addition, the set of viral genes made available to the host via lysogeny may also produce fitness-enhancing phenotypes, increasing the host resilience in these environments [23–25].

The currently studied minority of bacteriophages have yielded numerous biotechnologically important enzymes. Some significant examples include enzymes acting on nucleic acids, such as DNA polymerases, DNA ligases from bacteriophages T4 [26,27] and T7 [28–30], and exonuclease from the bacteriophage T5 [31]. Furthermore, lytic enzymes such as endolysins, naturally arming the phages for the degradation of bacterial cell walls, are of increasing interest as bactericidal agents [32–35] and have been subjected to trials as phage therapy [36,37]. Many of the above viruses were studied from isolates and provide a glimpse into similar discoveries possible from within the vast viral sequence space in marine environments [13,16].

To be able to study discovered viral enzymes of potential biotechnological interest, molecular cloning and heterologous expression approaches are required to produce the enzymes in amounts needed for characterization experiments. Study of the heterologous expression of viral genes from marine metagenomes, however, has been extremely limited [38]. Extending the knowledge in this field has subsequently been a major task in the project Virus-X (Viral Metagenomics for Innovation Value) aiming to identify and characterize novel enzymes and other proteins from bacteriophages and archaeal viruses. To date, only a few examples of studies describing the expression of viral genes from environmental marine resources are reported [39,40]. For the heterologous production of most proteins, *Escherichia coli* remains a desirable host due to its ease of use, quick generation times and a wide genetic toolkit regarding cloning and expression vectors [41]. However, *E. coli* does present certain well-documented challenges in soluble protein production when expressing genes from genetically less-related sources [41,42]. Furthermore, the distinct codon usage bias of *E. coli* often presents a difference in the availability of tRNAs between the native organism and itself, adversely affecting protein expression efficacy [43,44].

Numerous approaches exist to increase soluble protein yields of recombinant genes in *E. coli*. The use of various fusion protein tags has been a popular and effective way to improve soluble yields for many years [45–48]. The use of transcription-level adjustments to improve soluble protein expression has been described in recent years, initially as "codon optimization" [49] and later as "codon harmonization" [50]. Both of these approaches rely on the modification of codons in the DNA sequence of the target prior to expression, to code for the same eventual polypeptide, but with a set of tRNAs tailored for the machinery of the expression host. The difference among these approaches can be summarized as such:

codon optimization substitutes rare codons in the native gene sequence with those that are most abundant in the heterologous host, potentially allowing a high-speed protein production, whereas codon harmonization aims to replicate the cadence of native gene expression in the host, potentially allowing for correct protein folding during expression. While codon optimization has been widely demonstrated to have some degree of success in expressing genes from a diverse range of native hosts [48,51], including viruses [52–54], codon harmonization is a more recent approach and, to our knowledge, has not yet been explored towards the expression of viral genes in *E. coli*.

In this work, we report the first study of a temperate phage infecting *H. thermotrophus*: a free-living, Gram-negative, moderately thermophilic bacterium isolated from a microbial mat collected from the Seven Sisters hydrothermal vent field located on the Arctic Mid-Ocean Ridge [55,56]. Within the phylum *Fusobacteria*, *Hypnocyclicus thermotrophus* IR-2^T (=DSM 100055 =JCM 30901) is listed as the current type strain of the genus *Hypnocyclicus*. In addition to describing the identification, gene organization and taxonomic analysis of the prophage via in silico methods, we also report on our efforts to identify and recombinantly express genes with potential links to various lytic and nucleic acid modifying enzymatic activities. In an effort to facilitate the soluble heterologous production of proteins in *E. coli*, we implemented the codon optimization and harmonization approaches in parallel for a set of nine diverse enzyme candidates. The comparison of proteins produced via these approaches revealed notable differences in their soluble yields and thermostability. Altogether, the combined strategy used herein presents a cohesive application of both bioinformatics and molecular biology to improve access to the viral genetic diversity present in marine environments.

2. Materials and Methods

2.1. Identification and Annotation of Prophage Genes

The annotated genome assembly of the bacterium *H. thermotrophus* was downloaded from NCBI GenBank (RefSeq GCF_004365575.1). Manual analysis of the genome indicated presence of prophage genes. To further assess these putative prophage genes, the GenBank file of the assembly was uploaded to the PHASTER (https://phaster.ca/, accessed on 10 December 2019) [57,58] online tool and compared against the PHASTER prophage/virus database (last updated in August 2019). The analysis output described the genome region(s) containing the prophage genes, along with putative functional annotations. In addition to annotations provided by NCBI and PHASTER, the HHpred server (https://toolkit.tuebingen.mpg.de/tools/hhpred, accessed on 1 December 2020) [59–61], and the eggNOG-Mapper (http://eggnog-mapper.embl.de accessed on 12 December 2019) [62,63] online services were also used for the functional annotation of the prophage genes using corresponding amino acid sequences.

When using the HHpred server for the pairwise comparison of profile hidden Markov models (HMMs), the databases queried were PDB_mmCIF70_29_Nov, Pfam-A_v33.1, COG_KOG_v1.0 and NCBI_Conserved_Domains(CDs)_v3.18.

2.2. Taxonomic Analysis of HTH1

To taxonomically characterize HTH1, the genes identified as phage-related using the PHASTER tool were subjected to a translated nucleotide to protein BLAST (blastx, accessed on 2 April 2020) search. The following parameters: organism = viruses (txid:10239), number of alignments = 100, word size = 6 were used. The resulting hits were then parsed and taxonomically assigned by lowest common ancestor (LCA) analysis [64] in MEGAN software (version 6.18.6) (Tübingen, Germany) [65]. The following parameters were used: minimum support = 2, minimum score = 70, top percent = 10. Megan Mapping Database file version October 2019 was used.

With a reported success rate of 93% when assigning tailed and unclassified phages to their defined head–neck–tail-based categories, the "Remote Homology Detection of Viral

Protein Families—Virfam" [66] (http://biodev.cea.fr/virfam, accessed on 3 April 2020) server was also used to further analyse the taxonomy of HTH1.

2.3. Analysis of Prophage Host Range

In order to analyse the currently documented host range of similar phages, DNA sequence of HTH1 was used to perform a translated nucleotide–protein BLAST (blastx) search as described above, except using the NCBI non-redundant (nr) nucleotide database. The species names of the top 5000 hits were parsed and uploaded to phyloT (https://phylot.biobyte.de/, accessed on 5 April 2020) (version 2) [67] online tool to visualize the taxonomic distribution by generating a phylogeny of the cumulative NCBI taxonomy lineages of each species on the list (Supplementary Materials Figure S1).

2.4. Phylogeny Analyses

The amino acid sequence of the holin (GenBank WP_134112787.1) identified in HTH1 was used as a basis for phylogeny analyses and relationship of the prophage to viruses in the NCBI (nr) database. A protein–protein BLAST (blastp) search was performed via NCBI BLAST [68] with the following parameters: organism = viruses (txid:10239), word size = 6. A list of 94 proteins exported from the BLAST search (including the holin from HTH1) was aligned using MAFFT (version 7.453) [69]. Gap regions were trimmed with trimAl (version 1.2 rev59) [70] using the *'gappyout'* command to automatically trim sequences based on gaps in the alignment. The resulting trimmed alignment comprising 106 amino acid positions was manually analysed and used as a basis to infer maximum likelihood (ML) phylogeny using IQ-TREE (version 1.6.12) [71] tool. The best-fitting model was automatically determined by ModelFinder [72], and ultrafast bootstrapping was performed with 1000 replicates [73]. The best-fitting model was identified as LG+I+G4 (general matrix with invariable site plus discrete gamma model [74,75]). The resulting tree was then annotated using the online Interactive Tree of Life (iTOL) (https://itol.embl.de/, accessed on 5 April 2020) (version 5.5.1) [76] software. Branches with less than 50% bootstrap support were collapsed (Figure 1).

The amino acid sequence of HTH1 holin was further analysed using a protein–protein BLAST (blastp) against the Integrated Microbial Genomics/Virus (IMG/VR) (https://img.jgi.doe.gov/vr/, accessed on 10 April 2020) [77] and Ocean Gene Atlas (http://tara-oceans.mio.osupytheas.fr/ocean-gene-atlas/, accessed on 10 April 2020) [78] databases to compare the prophage to viral genes from environmental samples, metagenomic datasets and other non-isolated virus genes. The top 100 hits with the highest percent identity from the IMG/VR search and all the hits (18) from the Ocean Gene Atlas were extracted in addition to the 94 sequences from NCBI as described above. After automatic and manual curation to remove duplicates or non-holin hits, a total of 211 holin-related sequences were aligned, trimmed and visualized as described above, with the best-fitting ML model for this group of sequences identified as LG+F+I+G4 (general matrix with invariable site plus discrete gamma model [74,75] with empirical codon frequencies counted from the data) (Supplementary Materials Figure S2).

The putative DNA polymerase (HTP4385) (GenBank WP_134112782.1) was also subjected to phylogeny analysis, using the same parameters as described above for the holin-based tree. In this analysis, a list of 101 protein entries was used to create a 625 amino acid long alignment for the construction of the tree shown in Supplementary Materials Figure S3.

2.5. Gene Neighbourhood Analysis

Gene neighbourhoods between genes of HTH1 and three highly similar viral gene clusters was compared. The similar viral gene clusters were selected based on the closest alignments to the HTH1 holin in the extended tree shown in Supplementary Materials Figure S2. Alongside HTH1, marine anoxygenic phototropic community R3 (MAPCR3) (IMG scaffold ID: Ga0071011_100294), *Streptococcus* phage Javan630 (SPJ630) (NCBI:txid2548289) and the *Erysipelothrix* phage phi1605 (EP1605) (NCBI:txid2006938) were inspected using Gene-Graphics (https://katlabs.cc/genegraphics/app, accessed on 20 April 2020) [79] (Figure 2) by uploading the relevant genome regions with their annotations for each entry in NCBI GenBank format to the online tool.

2.6. Selection of Genes for Expression Trials

In addition to the genes constituting the lytic cassette, genes with various putative functions on either side of the HTH1 lytic cassette were analysed. After inspection, nine genes were selected for expression trials for their putative activities related to lysis and DNA replication, including three genes with hypothetical function or conserved domains of unknown function (DUF). The selected genes were labelled with the prefix HTP (*H. thermotrophus* phage) followed by the last four digits of their corresponding locus tag in the NCBI GenBank annotation (such as HTP4435). The selected genes and their annotated domain structures predicted by the HMMER web service (https://www.ebi.ac.uk/Tools/hmmer/search/phmmer, accessed on 20 April 2020) [80–82] were visualized in Figure 3.

2.7. Preparation of Sequences for Protein Expression of Selected Genes

Codon optimization [49] and codon harmonization approaches [83,84] were used in parallel to evaluate their effectivity in obtaining properly folded, soluble protein from each of the selected genes tailored for heterologous expression in *E. coli*. Codon-optimized gene sequences were generated via GenSmart Codon Optimization (GenScript, Piscataway, NJ, USA) online tool following default codon optimization parameters. Codon Harmonizer developed by Claassens et al. [51] online tool was used to harmonize codon usage frequencies between the prophage host *H. thermotrophus* NCBI GenBank (RefSeq GCF_004365575.1) and the heterologous expression host *E. coli* BL21(DE3) NCBI GenBank (GenBank GCA_000022665.2) (accessed in April 2019). Codon Adaptation Index (CAI) and Codon Harmonization Index (CHI) values were calculated for each sequence. Both the codon-optimized and codon-harmonized target protein gene sequences (Supplementary Materials File S1) were ordered to be synthesized and delivered pre-cloned in pET-21b(+) (Merck, Darmstadt, Germany) [85] vector (GenScript, Leiden, the Netherlands), featuring a C-terminal hexa-histidine tag [86] to facilitate purification using affinity chromatography.

2.8. Protein Production in E. coli

All expression constructs were transformed into *E. coli* BL21(DE3) (Merck, Darmstadt, Germany) cells using the heat-shock protocol provided by the manufacturer, using 30 ng of plasmid per 15 μL of bacteria suspension. Single colonies were picked from Lysogeny Broth (LB)-agar plates containing 100 μg/mL ampicillin after plating and overnight growth at 37 °C, and 10 mL pre-cultures in LB were subsequently inoculated and incubated overnight at 37 °C with 220 rpm shaking. Expression cultures in Tryptic Soy Broth (Merck, Darmstadt, Germany) (adjusted to pH 7.4/RT) at 100 mL scale were inoculated with 5% (*v/v*) of each pre-culture and were grown at 37 °C and 220 rpm until an optical density at 600 nm of 0.5–0.6 was reached. The incubation temperature was then reduced to 28 °C and allowed to equilibrate for 30 min. Expression was induced with 0.5 mM isopropyl β-D-1-thiogalactopyranoside, at 28 °C for 5 h. Following the expression, cells were harvested by centrifugation at $5000 \times g$ at 4 °C for 10 min. Collected cells were re-suspended in 10 mL of lysis buffer containing 50 mM Tris-HCl pH 7.4/RT, 60 mM imidazole, 500 mM NaCl and 5% (*v/v*) glycerol and were lysed using ultrasonication performed at 4 °C using 5×30 s bursts at 15 s intervals, with 25% amplitude. An aliquot representing the total protein fraction was taken and stored at 4 °C from each crude lysate before clarification of lysates by centrifugation at $12{,}000 \times g$ at 4 °C for 3 min. After clarification, aliquots were taken from all samples representing the soluble protein fraction and stored at 4 °C.

2.9. Protein Solubility Assessment and Yield Estimation

Aliquots taken from lysed cell pellets, representing the total protein (crude lysate) and soluble protein (clear lysate) fractions were run on a gradient (8–16%) SDS-PAGE gel (GenScript, Piscataway, NJ, USA) to assess expression levels. Precision Plus Dual Color (Bio-Rad, Hercules, CA, USA) protein ladder was used for protein molecular mass determination. Equivalent volumes of protein samples were loaded onto the electrophoresis gels seeking to fractionate equal protein amounts. The gel was run at 200 V, and subsequently stained using InstantBlue (Expedeon, Cambridge, UK) using a staining protocol provided by the manufacturer. After staining was complete, unbound dye was washed off the gel using distilled water on a benchtop shaker to reveal protein bands. The gels were photographed using MiniBIS Pro system processing images with GelCapture (version 7.0.15) suite (DNR Bio-Imaging Systems, Neve Yamin, Israel).

Densitometry calculations to determine relative abundance of target proteins in the soluble lysate fractions were performed using GelQuantum Pro (version 12.2) suite (DNR Bio-Imaging Systems, Neve Yamin, Israel). Total protein concentration was measured with a NanoDrop 1000 spectrophotometer (operating software version 3.7; Thermo Fisher Scientific, Waltham, MA, USA), assuming A_{280} 1 = 1 mg/mL. Target protein soluble yields were estimated by combining the results of densitometry and total soluble protein quantification.

2.10. Protein Purification

HTH1 proteins obtained in soluble form were purified to near homogeneity from clear lysate fractions by nickel affinity chromatography. Soluble protein fraction in lysis buffer was loaded 1 mL/min into a HisTrap HP 1 mL (7 mm × 25 mm) column (Cytiva, Uppsala, Sweden) equilibrated with lysis buffer. Target proteins were eluted (2 column volumes (CV)) with elution buffer containing 50 mM Tris-HCl pH 7.4/RT, 500 mM imidazole, 500 mM NaCl and 5% (*v/v*) glycerol at 1 mL/min after extensive washing (5–8 CV) of unbound proteins with lysis buffer. The purified proteins were stored in elution buffer at 4 °C after filtering twice through regenerated cellulose 0.2 μm pore size syringe filters (GE Healthcare, Uppsala, Sweden).

Protein integrity and purity were assessed via SDS-PAGE. Protein concentrations were measured spectrophotometrically, considering calculated absorption coefficients for pure proteins. Purification yields were calculated comparing the target protein amount in the soluble protein fractions with the target protein amount obtained after the purification and filtration steps.

2.11. Protein Thermal Unfolding Assay

Nanoscale differential scanning fluorometry based on internal tryptophane as well as tyrosine content was performed to determine the melting temperatures (T_m, °C) of purified HTH1 proteins. These measurements were carried out on a Prometheus NT.48 system using standard grade capillaries (NanoTemper Technologies, Munich, Germany). The purified protein samples were diafiltrated into assay buffer containing 50 mM Tris-HCl pH 7.4/RT and 2% (*v/v*) glycerol using Amicon Ultra-0.5 mL (3 Kda) centrifugal filters (Merck, Darmstadt, Germany). Protein concentrations were adjusted to 0.2 mg/mL with assay buffer after diafiltration. Thermal unfolding assays were performed at adjusted 40% excitation power, with a temperature gradient between 20–95 °C and at a ramp rate of 1 °C/min. Finally, analysis of the recorded emission intensities, emission ratio (350 nm/330 nm) and first derivative calculations were processed using the PR.ThermControl software (version 2.0.4) (NanoTemper Technologies, Munich, Germany).

3. Results

3.1. Functional Annotation and Taxonomy Analysis of HTH1

Three regions of putative viral origin were identified within the *H. thermotrophus* using the PHASTER tool [58]. Region 1 (Supplementary Materials Table S1) was reported as an incomplete prophage region (PHASTER score: 10), consisting of eight conserved domains

(CDs) from locus tags EV215_RS03310 to EV215_RS03345 in the sense (+) strand. Region 2 was also predicted as incomplete (PHASTER score: 50), consisting of 33 CDs from locus tags EV215_RS04355 to EV215_RS04515. However, attachment sites attL and attR (nucleotide sequence TTACCATCTTA) were found between locus tags EV215_RS04470-EV215_RS04475 and EV215_RS04435-EV215_RS04440, respectively, within region 2, indicating that this region was likely associated with viral interaction and virus integration on to the host genome. Region 3 was predicted to be an intact prophage region (PHASTER score: 100) and contained 29 CDs from locus tags EV215_RS04440 to EV215_RS04580. There was an 11,971 bp overlap between regions 2 and 3, representing 16 CDs, with both regions found on the complementary (−) strand of the genome. Furthermore, regions 2 and 3 showed highly similar average G + C contents, 37.7% and 38.5%, respectively. In comparison, the average G + C contents of region 1 and the host genome were 27.9% and 24.8%, respectively. Due to their overlap, and coherent composition, regions 2 and 3 were considered as the "complete" prophage genome, totalling 46 CDs and a genome size of 41,571 bp. This region was subsequently designated with the proposed name *Hypnocyclicus thermotrophus* phage H1 (HTH1). With the combined use of various pipelines, functional annotations could be suggested for 34 HTH1 genes. The remaining 12 were noted as hypothetical, or to contain unknown elements as listed in Supplementary Materials Tables S1 and S2.

Taxonomic analysis based on the LCA algorithm in MEGAN suggested affiliation of HTH1 with the family *Siphoviridae* and the order *Caudovirales*. Consistently, the Virfam analysis (resulting identities provided in Supplementary Materials Table S3) identified the prophage head–neck–tail modules as being part of "Neck Type 1—Cluster 2" type of phages, noted to be associated with siphoviruses. Holin genes have previously been suggested as a phage-specific signature gene for siphoviruses [87]. Phylogeny analyses based on the HTH1 holin (Figure 1) as well as DNA polymerase (Supplementary Materials Figure S3) amino acid sequences revealed the closest affiliations to known phages from the Javan group of *Streptococci* phages [88] and to the *Erysipelothrix* phage phi1605 (NCBI:txid2006938). The sequence identity between the HTH1 holin and the holins from *Streptococcus* phage Javan630 (SPJ630) and *Erysipelothrix* phage phi1605 (EP1605) was found to be 75.7% and 75.0%, respectively.

The closest identified holin homologue from another phage infecting Gram-negative bacteria was that of the phage Funu2 (NCBI:txid1640978) (Figure 1), which is reported to infect *Fusobacterium nucleatum* [89] (sequence identity of 38.6%). This is an interesting hit, as to date, studies of viruses and viral genes associated with *Fusobacteria* remain limited, with only a small number of phages characterized thus far [37,90–92].

When the HTH1 holin was compared against environmental sequences from IMG/VR, an even closer hit at 99% sequence identity was observed against a metagenome-derived holin from a marine anoxygenic phototrophic community R3 (MAPCR3) sample (IMG genome ID 3300004816) originating from a shallow salt marsh pool in Falmouth, MA, USA (Supplementary Materials Figure S2 and Table S4). When the gene neighbourhood surrounding the lytic cassette of HTH1 was compared with those of MAPCR3, SPJ630 and the EP1605 (Figure 2), a remarkably close similarity was identified between the HTH1 and MAPCR3 lytic cassettes, particularly over the four genes corresponding to HTP4425 to HTP4410 in HTH1 (Supplementary Materials Table S4). The similarity was less significant when comparing to cassettes of SPJ630 and EP1605. Furthermore, the lytic cassette amidase (HTP4410) was observed to be replaced by a second glycosyl hydrolase (CAZy GH25) in SPJ630 and EP1605 when the gene annotation and protein domain structures were reviewed using a HMMER search [82] (Figure 3).

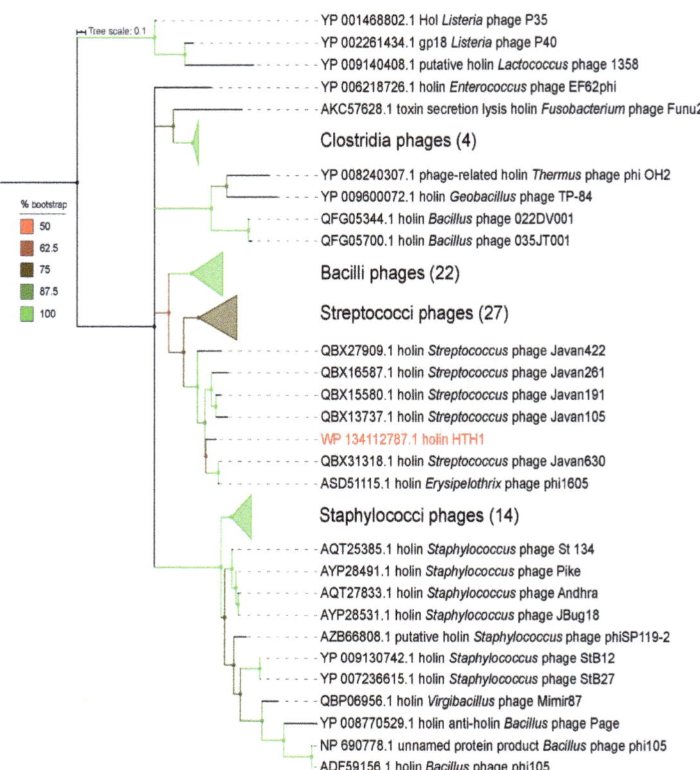

Figure 1. Phylogeny analysis of the prophage based on the alignment of 106 amino acid long region of holin proteins from 94 phages, using maximum likelihood, with 1000 bootstrap replicates. The tree is centre-rooted, and the scale bar represents the average number of amino acid substitutions per site. Numbers next to collapsed clades represent the number of leaves covered by each illustration. The HTH1 holin is highlighted in red.

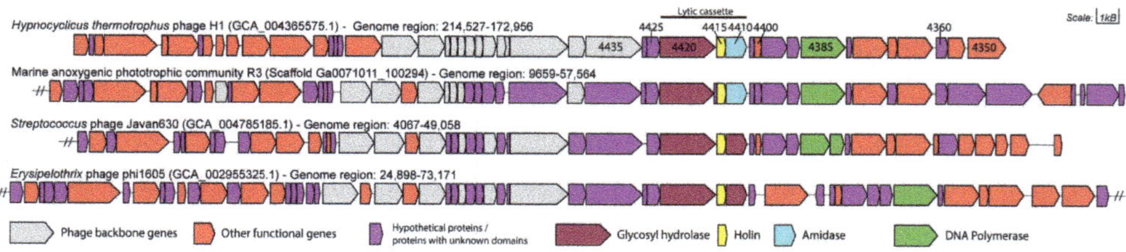

Figure 2. Gene neighbourhood map of HTH1 and the comparable regions of three closely related phage gene clusters aligned around the holin in their respective lytic cassettes. Displayed genes are drawn to scale, as shown on the top right. Respective organism or sample names, related accession numbers (in parentheses) and genome regions displayed (in bp ranges) are provided above each graphic. Genes chosen for expression of proteins from HTH1 are also labelled with their identifier numbers. Double dashes (/ /) indicate the presence of genes further up or downstream the gene regions displayed in this figure.

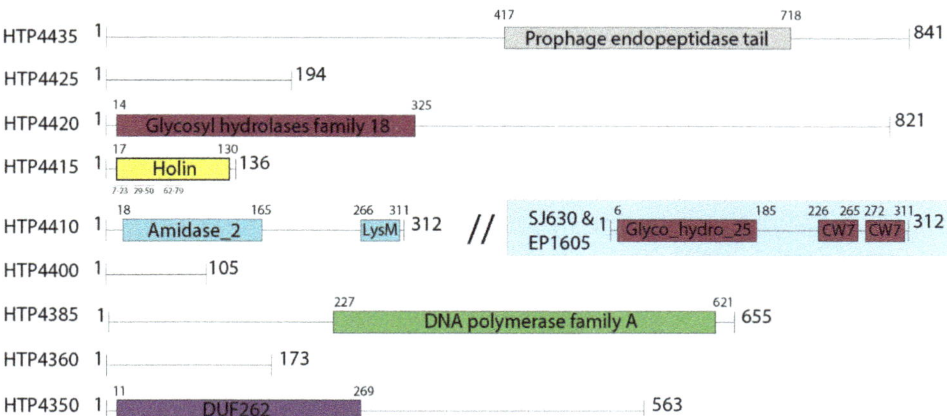

Figure 3. Illustration depicting sequence features of chosen candidate proteins predicted by HMMER [82]. Black lines show non-annotated amino acid sequences, grey boxes show predicted Pfam domains, purple lines mark transmembrane domains and numbers flanking each feature show their respective amino acid residue number ranges. The blue box shows the HTP4410 analogue found in *Streptococcus* phage Javan630 (SJ630) and *Erysipelothrix* phage phi1605 (EP1605).

3.2. Selection of Genes for Expression Trials

HTH1 genes with annotations related to roles in lysis and DNA replication were examined further, examining protein domain structures through comparisons to multiple sequence databases (Supplementary Materials Tables S1 and S2). A set of nine genes were chosen for protein expression trials, as shown in Figure 3, with their designations and associated domain structures. Gene targets associated with the prophage lytic cassette (defined in Section 3.1), including holin (HTP4415), glycosyl hydrolase (HTP4420) and the amidase with a LysM domain (HTP4410), were selected for their putative role in cell lysis, in addition to the phage tail protein (HTP4435) with associations to endopeptidase activity. The hypothetical gene HTP4425 neighbouring the glycosyl hydrolase (HTP4420) was also picked for its potential connection to the lysis-related cluster. Two genes annotated with nucleotide cleavage and production activities were also selected: the rRNA biogenesis protein RRP5 (HTP4400) with putative endonucleolytic activity towards rRNA, and the DNA polymerase I (HTP4385). Furthermore, two genes flanking the HNH endonuclease, HTP4360 and HTP4350, were picked for their potential associations with nucleolytic activity. The gene HTP4350 (GenBank: WP_134112775.1) was annotated as "DUF262 domain containing protein" by the NCBI pipeline; however, a putative DNase activity was also suggested when analysed with HHpred (Supplementary Materials Table S2), and it is upstream of the prophage gene region in the *H. thermotrophus* genome.

Searches made against PDB for structural insight pertaining to the nine HTH1 proteins revealed only low similarity hits for three proteins, HTP4420, HTP4410 and HTP4350, to PDB entries 4S3J, 3HMB and 1D9D, respectively (Supplementary Materials Table S5). However, all three structures reported associations with the expected functions in the HTH1 proteins, such as peptidoglycan lysis for HTP4410 and HTP4350, and DNA polymerase for HTP4350 (Supplementary Materials Table S4).

3.3. Expression of Target Codon-Adjusted Gene Variants

The codon frequencies of the HTH1 gene sequences were analysed, estimating CAI for the native host *H. thermotrophus*. All target protein genes demonstrated CAI values of approximately 0.4–0.5 (Table 1). Estimated CAI values indicated that HTH1 gene sequences were moderately adapted for expression in the native host, predicting comparatively moderate native expression level of the target proteins. Target genes were subsequently processed to generate codon-optimized and codon-harmonized gene sequence variants, adjusted from the *H. thermotrophus* codon usage bias towards compatibility with the expression host *E. coli* BL21(DE3). Quantitative analysis of codon-adjusted sequence variants confirmed the expected levels of codon adaptation (Table 1). The CAI of codon-optimized gene sequences varied between 0.84 and 0.89, indicating high adaptation towards heterologous expression in *E. coli*. Codon-harmonized sequences, as expected, were less adapted to be expressed in the selected strain, with CAI varying between 0.58 and 0.74. It was noted that CAI of codon-harmonized sequences showed higher variation compared to CAI of codon-optimized sequences. The CHI values of codon-optimized variants were 0.12–0.13 below (Table 1) the estimated CHI values from codon-harmonized sequences, confirming an expected trend for more substantial changes imposed on codon-optimized variants. Moreover, the CHI value of each codon-harmonized gene variant was similar and between 0.43 and 0.48. Even though CHI comparison indicated that codon-harmonized variants were closer to native codon sequences of target protein genes, the "harmonization" effect observed could be interpreted as moderate [51].

All nine codon-optimized gene variants were successfully expressed in *E. coli*, at different levels (data not shown). However, the hypothetical protein (HTP4425), glycosyl hydrolase (HTP4420), holin (HTP4415) and the DNA polymerase I (HTP4385) were not detected in the soluble protein fraction, as estimated by SDS-PAGE. Insolubility was particularly expected for the holin because of the multiple transmembrane helices present in the structure (Figure 3), and no significant difference was observed from the use of either codon adjustment approach. Among the codon-harmonized set of genes, expression in *E. coli* could not be observed for the genes encoding the holin (HTP4415) as well as the hypothetical protein (HTP4360). For the other seven genes, only four were found to yield soluble proteins. These proteins were the endopeptidase tail protein (HTP4435), amidase (HTP4410), rRNA biogenesis protein RRP5 (HTP4400) and DUF262 / DNase (HTP4350) (Table 1).

In total, implementation of codon adjustment approaches for selected HTH1 genes resulted in the soluble protein production from five codon-optimized and four codon-harmonized gene variants (Figure 4). The set of soluble proteins expressed from codon-optimized and codon-harmonized variants differed by the hypothetical protein (HTP4360) that was not found expressed as soluble from its codon-harmonized variant. As typically expected [50,93], expression levels estimated by densitometry analyses for the five common soluble protein targets revealed higher yields from codon-optimized variants (Figure 4). Exemplifying this trend, the relative soluble abundance of the rRNA biogenesis protein RRP5 (HTP4400) was found nearly three times higher when expressed from its codon-optimized variant compared to its harmonized equivalent (Figure 4); corresponding to a yield difference of ~110 mg/L (Table 1). The codon-optimized gene variant of hypothetical protein (HTP4360) was also expressed at a high level, with an estimated yield of ~150 mg/L soluble protein. Endopeptidase tail protein (HTP4435), amidase (HTP4410) and DUF262/DNase (HTP4350) expressed from codon-optimized gene sequences demonstrated only slightly higher relative abundance (by 2–5%, respectively,) compared to respective codon-harmonized variants (Figure 4). The soluble yields of HTP4435, HTP4410 and HTP4350 from codon-optimized variants were also found to be ~8–16 mg/L higher than the yields of the corresponding codon-harmonized variant (Table 1). Following this step, target proteins from both variants, which were noted as soluble, were up-scaled to be produced in 1 L expression cultures.

Table 1. Codon usage parameters and soluble production yield estimation of target HTH1 proteins. CAI—codon adaptation index, CHI—codon harmonization index, CO—codon-optimized, CH—codon-harmonized, ND—target protein not detected in total soluble protein fraction.

Identifier	Proposed Protein Function	CAI for Expression Host		CHI for Expression Host		Codon Native Gene Sequence CAI for Native Host	Soluble Produced Protein Yield * (mg/L)	
		CO Gene Variant	CH Gene Variant	CO Gene Variant	CH Gene Variant		Expressed from CO Gene Variant	Expressed from CH Gene Variant
HTP4435	Endopeptidase tail	0.89	0.64	0.61	0.48	0.50	18.7 ± 1.3	13.8 ± 1.5
HTP4425	Hypothetical protein	0.87	0.67	0.59	0.48	0.51	ND	ND
HTP4420	Glycosyl hydrolase 18	0.89	0.66	0.60	0.45	0.51	ND	ND
HTP4415	Holin, toxin secretion/ phage lysis	0.87	0.58	0.60	0.47	0.40	ND	ND
HTP4410	N-acetylmuramoyl-L-alanine amidase	0.88	0.64	0.61	0.47	0.46	40.7 ± 5	30.8 ± 3.4
HTP4400	rRNA biogenesis protein rrp5, putative	0.84	0.64	0.58	0.48	0.48	135.80 ± 2.49	27.9 ± 3.2
HTP4385	DNA Polymerase	0.86	0.62	0.60	0.46	0.47	ND	ND
HTP4360	hypothetical protein	0.84	0.74	0.55	0.43	0.56	151.7 ± 10.2	ND
HTP4350	DUF262 / DNase	0.86	0.62	0.59	0.45	0.44	32.3 ± 1.2	15.6 ± 2.7

* Values represent mean ± standard error of three independent expressions.

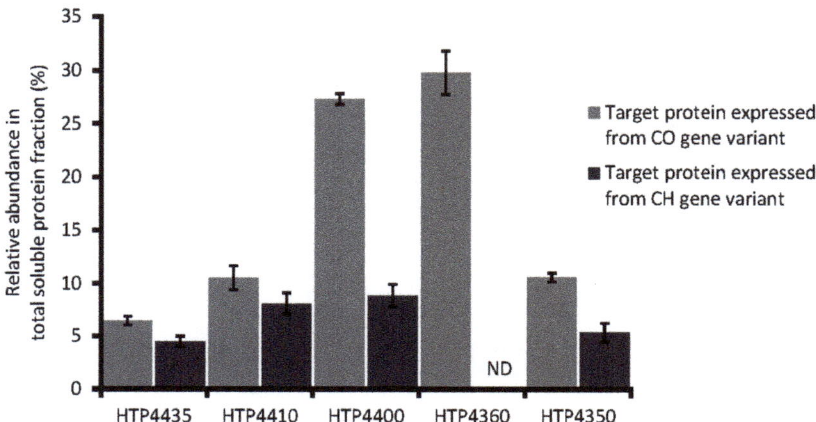

Figure 4. Relative abundance of target HTH1 proteins produced after expression from codon-optimized (CO) and codon-harmonized (CH) gene variants in total soluble protein fraction. ND—target protein not detected in total soluble protein fraction. Values represent relative abundance mean in percent of total proteins in total soluble protein fraction ± standard error of three independent expressions.

3.4. Protein Purification

Soluble proteins produced from 1 L cultures were purified to near homogeneity by nickel affinity chromatography. An optimized affinity chromatography purification protocol ensured high purity of the target proteins as was visualized by SDS-PAGE (Figure 5), where target proteins were observed at bands corresponding to their expected sizes. Purified endopeptidase tail protein (HTP4435), expressed from both types of codon-adjusted gene variants, were aggregation-prone, while the other target HTH1 proteins remained stably soluble after purification. The single step purification strategy led to generally high purification yields (Table 2). Comparison of the obtained yields of amidase (HTP4410) as well as DUF262/DNase (HTP4350) expressed from codon-optimized and codon-harmonized gene sequences did not differ, whereas the purification yield of codon-harmonized rRNA biogenesis protein RRP5 (HTP4400) was approximately 20% higher compared with the yield of its codon-optimized gene counterpart. In general, the purification yields confirmed a comparatively high affinity of heterologous proteins towards the chromatography resin and were in the expected range for the method [94,95].

Table 2. Purification yield of target HTH1 proteins. Protein concentrations were measured spectrophotometrically estimating total amount of target recombinant protein in clarified lysate by combining densitometry calculation results and total soluble protein quantification results. CO—codon-optimized, CH—codon-harmonized, ND—target protein not detected in total soluble protein fraction.

Identifier	Proposed Protein Function	Protein Purification Yield * (%)	
		Target Protein Expressed from CO Gene Variant	Target Protein Expressed from CH Gene Variant
HTP4410	N-acetylmuramoyl-L-alanine amidase	85.6 ± 1.4	85.9 ± 1.9
HTP4400	rRNA biogenesis protein rrp5, putative	38.6 ± 7.2	58 ± 3.7
HTP4360	hypothetical protein	75 ± 3.8	ND
HTP4350	DUF262/DNase	83.5 ± 5.2	92.1 ± 2.1

* Values represent mean ± standard error of three independent purifications.

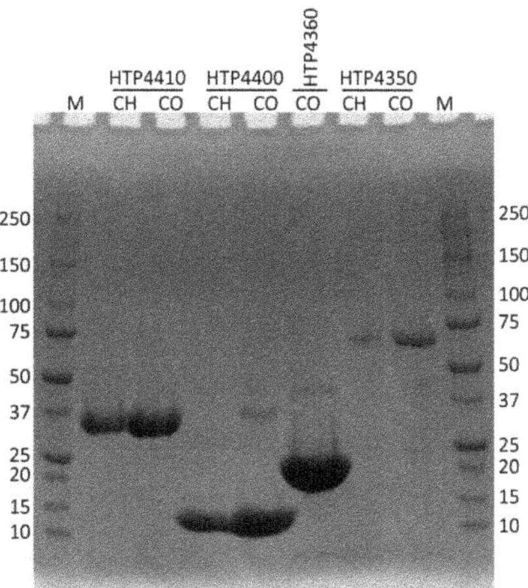

Figure 5. SDS-PAGE image of purified proteins produced from codon-harmonized (CH) and codon-optimized (CO) genes. The HTP prefix and the numbers above the lanes correspond to the identifiers of the genes tested. M indicates the protein marker (Bio-Rad Precision Plus Dual Color). Numbers next to each protein marker lane show the respective molecular weight labels in kDa.

3.5. Crystallization and Thermostability of Target Proteins

Purified, stably soluble target HTH1 proteins expressed from the optimized and harmonized types of codon-adjusted gene variants were subjected to both crystallization trials and analysis of thermostability. As a higher thermal unfolding temperature has been indirectly connected to an improved fold, that may affect the possibility to crystallize the target protein. In crystallization trials, amidase (HTP4410) as well as DUF262/DNase (HTP4350) expressed from codon-harmonized gene variants (Supplementary Materials Figure S5) and rRNA biogenesis protein RRP5 (HTP4400) from both codon sequence adjustment variants were observed to form protein crystals (M. Håkansson and S. Al-Karadaghi, SARomics Biostructures, personal communication).

In the thermostability assessment with differential scanning fluorometry, which was performed to compare melting temperatures (T_m) of target recombinant proteins expressed from both types of codon-adjusted gene sequence variants, an increase in unfolding temperature was observed from the codon-harmonized variants of the three target proteins where crystal formation was observed. The in vitro thermostability (T_m) of the target HTH1 proteins amidase (HTP4410), rRNA biogenesis protein RRP5 (HTP4400) and DUF262/DNase (HTP4350) varied between approximately 51 and 73 °C. Remarkably, recombinant proteins expressed from the codon-harmonized gene variants were all observed to unfold at higher T_m values (3–7 °C) than corresponding codon-optimized gene variants (Table 3). A T_m of approximately 61 °C was determined for DUF262/DNase (HTP4350) expressed from a codon-optimized gene variant, which was an almost 3 °C lower unfolding temperature compared with the T_m observed for this hypothetical protein expressed from the codon-harmonized version. Amidase (HTP4410) and rRNA biogenesis protein RRP5 (HTP4400) expressed from codon-harmonized gene sequence versions demonstrated a T_m at 73 °C and 56 °C, respectively—increases of almost 7 and 5 °C compared to the T_m of proteins expressed from codon-optimized genes.

Table 3. Thermal unfolding estimation with differential scanning fluorimetry of stably soluble target HTH1 proteins. CO—codon-optimized, CH—codon-harmonized.

Target Protein	Proposed Protein Function	Melting Temperature (T_m, °C)	
		Target Protein Expressed from CO Gene Variant	Target Protein Expressed from CH Gene Variant
HTP4410	N-acetylmuramoyl-L-alanine amidase	66.23 ± 0.07 *	73.03 ± 0.10
HTP4400	rRNA biogenesis protein rrp5, putative	51.57 ± 0.34	55.70 ± 0.22
HTP4350	DUF262 / DNase	61.57 ± 1.47	65.24 ± 0.43

* Values represent mean ± standard error of three independent differential scanning fluorimetry assays.

4. Discussion

Marine bacteriophages remain a largely unexplored resource for enzyme bioprospecting. As a part of the Virus-X consortium (http://virus-x.eu/, accessed on 1 May 2021), successful expression of genes from bacteriophage genomes was identified as a key step towards discovering enzymes from various marine niches. Crystallization of novel viral proteins to collect structural knowledge was another aim of the consortium, as recently exemplified for the proteins XepA and YomS from a *Bacillus subtilis* prophage [96]. Hence, significant research interest currently exists for the analysis of new phage genes that may hold interest both in basic and structural research and for applications in biotechnology.

In this context, a novel prophage, designated HTH1, was identified via the study of the Gram-negative hydrothermal vent bacterium *H. thermotrophus*, which is classified in the phylum *Fusobacteria*. The relationship between *H. thermotrophus* and HTH1 can be considered fitting, as lysogeny is suggested to be prevalent in physiochemically demanding environments. These include deep-sea biomes [97] and diffuse-flow hydrothermal vent communities [22], where temperate phages may provide benefits to host fitness via various mechanisms [98–100].

Taxonomic analyses placed HTH1 within the family *Siphoviridae*, which contains dsDNA viruses defined by their long, non-contractile tails, as opposed to the contractile tails of the *Myoviridae* and the short and non-contractile tails of the *Podoviridae* [101]. The genome size of HTH1 was 41571 bp, indicating it to be smaller compared to the average genome size of *Siphoviridae* at ~53 kb [102]. Interestingly, phylogeny (Figure 1), Virfam [66] and sequence homology analyses of HTH1 genes (Supplementary Materials Figure S1) all suggested closest similarity of HTH1 to siphoviruses that infect Gram-positive bacteria, mainly of the phylum *Firmicutes*.

HTH1 was annotated to contain a suite of expected viral backbone genes, such as structural elements for the viral head, neck, capsid and tail, core viral enzymes such as integrases, terminases, the viral lytic enzymes, and DNA modifying enzymes such as DNA polymerase, endonuclease and recombinases (Supplementary Materials Tables S1 and S2, Figure 3. However, further studies including the lytic induction and isolation of viral particles would be required to confidently determine whether the presented genome of HTH1 corresponds to the complete and functional phage genome infecting *H. thermotrophus*.

Closer inspection of the HTH1 lytic cassette revealed three main genes related to cell lysis: a glycosyl hydrolase putatively capable of chitin and peptidoglycan-degrading activities specific to endo-β-N-acetylglucosamine residues [103,104]; a holin crucial for the perforation of the cell membrane [105,106]; and an N-acetylmuramoyl-L-alanine amidase featuring a membrane binding lysin motif (LysM), with an expected activity of cleaving bonds between N-acetylmuramoyl residues and L-amino acids in the bacterial cell wall (Figures 2 and 3). However, no genes related to spanins, rod-like viral lysis proteins considered essential to disrupt the cell membranes of Gram-negative hosts, were detected [106,107].

The enzymes of the HTH1 lytic cassette, containing the genes annotated to encode glycosyl hydrolase, holin and amidase, were of obvious interest as their peptidoglycan-degrading capabilities could be utilized against pathogenic bacteria as bactericidal agents [108].

In addition, the hypothetical protein HTP4435 was selected for testing due to the presence of a tail-associated endopeptidase domain (Pfam PF06605, MEROPS M23) (Figure 3). Such peptidases may find a broad range of potential uses in industrial, medical or scientific applications [109–111]. The DNA polymerase I (HTP4385) was also of direct interest for its potential as an enzymatic tool in many modern molecular biological methods such as PCR, genome sequencing and more [112]. As *H. thermotrophus* was reported to grow optimally at 48 °C [55], the proteins encoded by HTH1 may possess elevated thermostability and thermal activity, which are desirable traits in many industrial or scientific applications [113,114]. Furthermore, only limited structural similarity was observed for the chosen HTH1 proteins to structures present in PDB (Supplementary Materials Table S4), suggesting novel features could potentially be revealed with their future structural analyses.

The heterologous expression of native phage proteins has been reported to be challenging [115]. To aid in this process, codon optimization [49] and codon harmonization [50] approaches were considered for the heterologous production of proteins encoded by HTH1. Here, these two approaches were tested, and compared over their effects towards obtaining and increasing soluble protein yields, and also for their effects on the thermostability of the proteins produced. While codon optimization is commercially offered as an option during gene-synthesis services [116], codon harmonization must be carried out manually, and so a deeper understanding of the native viral host is required. As bacteriophages can naturally use their host's machinery to express their genes, they are understood to adapt the same codon usage frequency (CUF) as the host [117]. Therefore, while preparing sequences for codon harmonization, the genome of *H. thermotrophus* was used to calculate and compare CUFs between itself and *E. coli* as the expression host.

Codon analysis of selected native HTH1 genes suggested the target proteins are naturally produced in moderate amounts in *H. thermotrophus*. As expected, heterologous target proteins were produced more readily from codon-optimized gene variants than comparable codon-harmonized genes (Table 2, Figure 4 and Supplementary Materials Figure S4), which were adjusted to mimic the gene native codon landscape, sacrificing overall codon adaptation to the expression host in the process [50]. The codon optimization approach for selected HTH1 proteins was successful, as quantitatively confirmed by estimated CAI values and also by observed soluble expression yields. The CAI for the codon-harmonized variants of selected genes were comparatively high and varied substantially, indicating that the codon harmonization algorithms used [83,118] were suitable and specific for each of the HTH1 genes.

Protein folding quality is typically reflected by a higher thermal unfolding temperature and a higher thermostability [119]. While the codon harmonization approach did not result in the soluble expression of a greater variety of HTH1 proteins than codon optimization, it yielded proteins with comparatively higher melting temperatures (T_m) determined by differential scanning fluorimetry, suggesting a higher folding quality. Assayed under identical conditions, higher unfolding temperatures were observed for all HTH1 target proteins expressed from codon-harmonized gene variants compared to corresponding proteins from codon-optimized variants. The melting temperatures determined were in an expected range for HTH1 proteins natively produced within the host cells, fitting with the optimal growth temperature of *H. thermotrophus* [55]. Furthermore, ongoing crystallization trials also confirmed better crystal-forming properties of target HTH1 proteins expressed from codon-harmonized genes as an indicator of improved folding quality (M. Håkansson and S. Al-Karadaghi, SARomics Biostructures, personal communication).

The CHI values estimated for codon-harmonized variants of the selected gene set were comparatively high and did not differ substantially between the different genes in the set, indicating moderate, if not limited harmonization of codons (Table 1). These results could partially explain why target proteins produced from codon-harmonized variants were not persistently more soluble than codon-optimized variants after production in *E. coli*. In theory, production of soluble proteins should be ensured by codon harmonization [84], even though further optimization of physiochemical heterologous expression parameters is recommended to enhance the expression level of soluble protein from codon-harmonized gene variants [120]. Preliminary experiments to express selected HTH1 genes in *E. coli* were carried out under the recommended conditions for the expression vector and strain used [85]. Further optimization of the process could be implemented to achieve soluble production of target proteins, which remained insoluble despite codon harmonization. As the current codon adjustment algorithm was mainly developed using non-viral genome sequences, its efficacy could be limited for the adjustment of viral genes. With the limited data available for the implementation of codon adjustment for viral genes [121,122], the results presented herein may aid the further development of codon adjustment algorithms.

5. Conclusions

In this work, complementary application of bioinformatics and molecular methods allowed the identification, description and protein-level study of a novel marine prophage. Here, we describe the first genome sequence of a prophage discovered in *H. thermotrophus*, a Gram-negative, moderately thermophilic bacterium isolated from the Seven Sisters hydrothermal vent field. The *H. thermotrophus* phage H1 (HTH1) showed similarity to phages infecting Gram-positive bacteria of the genus *Firmicutes*, but in our study, it was found within the genome of a Gram-negative host. A set of nine genes were identified with putative functions, including cell lysis, nucleotide lysis and replication—interesting for both ecological studies and potential biotechnology applications. To facilitate the soluble heterologous production of HTH1 proteins in *E. coli*, codon optimization, and harmonization approaches were tested in parallel. Valuable data regarding production yield, solubility and folding quality of heterologous HTH1 proteins were gathered following expression of codon-adjusted gene variants, which may be useful in improving the application of codon adjustment strategies for viral genes. In the context of the proteins tested, codon optimization was found to lead to higher protein yields, whereas codon harmonization was underlined as more beneficial for the production of proteins with higher stability and folding quality.

Supplementary Materials: The following are available online at https://www.mdpi.com/article/10.3390/v13071215/s1: Table S1: The list of prophage-associated CDs identified in the *Hypnocyclicus thermotrophus* genome and their putative functions, predicted by NCBI and PHASTER annotation pipelines; Table S2: HHpred-suggested annotations of prophage-associated genes; Table S3: Highest identities of HTH1 proteins with at least 3 protein hits in Aclame, via Virfam analysis; Table S4: Amino acid sequence comparisons of target HTH1 proteins to their homologues in the three chosen viral gene clusters; Table S5: Detailed information on the HTH1 proteins chosen for expression trials; Figure S1: Taxonomical distribution of phage hosts; Figure S2: Extended holin phylogeny analysis; Figure S3: DNA polymerase phylogeny analysis; Figure S4: SDS-PAGE gel images; Figure S5: Protein crystal of HTP4350 produced from a codon-harmonized gene variant; File S1: Nucleic acid sequences of all genes chosen for protein expression, in Fasta format.

Author Contributions: Conceptualization, H.A., A.J. and I.H.S.; formal analysis, H.A. and A.J.; funding acquisition, R.-A.S., E.N.K. and I.H.S.; methodology, A.J., R.S. and I.H.S.; project administration, I.H.S.; resources, E.N.K. and I.H.S.; software, R.S.; supervision, I.H.S.; validation, A.J.; writing—original draft, H.A.; writing—review and editing, A.J., H.D., R.-A.S., R.S., E.N.K. and I.H.S. All authors have read and agreed to the published version of the manuscript.

Funding: Generous funding was received from the European Research Council (ERC) under the European Union's Horizon 2020 research and innovation programme Virus-X project: Viral Metagenomics for Innovation Value (grant no. 685778), from the Research Council of Norway within the MARINFORSK programme, project: VirVar (project number 294363) and the Kristian Gerard Jebsen Foundation.

Institutional Review Board Statement: Not applicable.

Informed Consent Statement: Not applicable.

Data Availability Statement: All relevant data for the study is provided within the article, and its supplements.

Acknowledgments: The sequence data of Marine anoxygenic phototropic community R3 (MAPCR3) (IMG scaffold ID: Ga0071011_100294) were produced by the US Department of Energy Joint Genome Institute (https://www.jgi.doe.gov/, accessed on 1 May 2021) in collaboration with the user community and was used with permission of the P.I. (Jean J. Huang).

Conflicts of Interest: The authors declare no conflict of interest. The funders had no role in the design of the study; in the collection, analyses or interpretation of data; in the writing of the manuscript; or in the decision to publish the results.

References

1. Steen, I.H.; Dahle, H.; Stokke, R.; Roalkvam, I.; Daae, F.-L.; Rapp, H.T.; Pedersen, R.B.; Thorseth, I.H. Novel barite chimneys at the Loki's Castle vent field shed light on key factors shaping microbial communities and functions in hydrothermal systems. *Front. Microbiol.* **2016**, *6*, 1510. [CrossRef] [PubMed]
2. Nakamura, K.; Takai, K. Theoretical constraints of physical and chemical properties of hydrothermal fluids on variations in chemolithotrophic microbial communities in seafloor hydrothermal systems. *Prog. Earth Planet. Sci.* **2014**, *1*, 1–24. [CrossRef]
3. Corliss, J.B.; Dymond, J.; Gordon, L.I.; Edmond, J.M.; Von Herzen, R.P.; Ballard, R.D.; Green, K.; Williams, D.; Bainbridge, A.; Crane, K.; et al. Submarine thermal springs on the Galápagos Rift. *Science* **1979**, *203*, 1073. [CrossRef] [PubMed]
4. Spiess, F.N.; Macdonald, K.C.; Atwater, T.; Ballard, R.; Carranza, A.; Cordoba, D.; Cox, C.; Diaz Garcia, V.M.; Francheteau, J.; Guerrero, J.; et al. East Pacific Rise: Hot springs and geophysical experiments. *Science* **1980**, *207*, 1421–1433. [CrossRef]
5. Ferrer, M.; Beloqui, A.; Timmis, K.N.; Golyshin, P.N. Metagenomics for mining new genetic resources of microbial communities. *J. Mol. Microbiol. Biotechnol.* **2009**, *16*, 109–123. [CrossRef]
6. Ferrer, M.; Golyshina, O.; Beloqui, A.; Golyshin, P.N. Mining enzymes from extreme environments. *Curr. Opin. Microbiol.* **2007**, *10*, 207–214. [CrossRef]
7. Vester, J.K.; Glaring, M.A.; Stougaard, P. Improved cultivation and metagenomics as new tools for bioprospecting in cold environments. *Extremophiles* **2015**, *19*, 17–29. [CrossRef]
8. Stokke, R.; Reeves, E.P.; Dahle, H.; Fedøy, A.-E.; Viflot, T.; Lie Onstad, S.; Vulcano, F.; Pedersen, R.B.; Eijsink, V.G.H.; Steen, I.H. Tailoring hydrothermal vent biodiversity toward improved biodiscovery using a novel in situ enrichment strategy. *Front. Microbiol.* **2020**, *11*, 249. [CrossRef]
9. Delbarre-Ladrat, C.; Salas, M.L.; Sinquin, C.; Zykwinska, A.; Colliec-Jouault, S. Bioprospecting for exopolysaccharides from deep-sea hydrothermal vent bacteria: Relationship between bacterial diversity and chemical diversity. *Microorganisms* **2017**, *5*, 63. [CrossRef]
10. Zykwinska, A.; Marchand, L.; Bonnetot, S.; Sinquin, C.; Colliec-Jouault, S.; Delbarre-Ladrat, C. Deep-sea hydrothermal vent bacteria as a source of glycosaminoglycan-mimetic exopolysaccharides. *Molecules* **2019**, *24*, 1703. [CrossRef]
11. Luo, E.; Aylward, F.O.; Mende, D.R.; Delong, E.F. Bacteriophage distributions and temporal variability in the ocean's interior. *MBio* **2017**, *8*, e01903-17. [CrossRef]
12. Ray, J.; Dondrup, M.; Modha, S.; Steen, I.H.; Sandaa, R.A.; Clokie, M. Finding a needle in the virus metagenome haystack-micro-metagenome analysis captures a snapshot of the diversity of a bacteriophage armoire. *PLoS ONE* **2012**, *7*, e34238. [CrossRef]
13. Gregory, A.C.; Zayed, A.A.; Conceição-Neto, N.; Temperton, B.; Bolduc, B.; Alberti, A.; Ardyna, M.; Arkhipova, K.; Carmichael, M.; Cruaud, C.; et al. Marine DNA Viral Macro- and Microdiversity from Pole to Pole. *Cell* **2019**, *177*, 1109.e14–1123.e14. [CrossRef]
14. Castelán-Sánchez, H.G.; Lopéz-Rosas, I.; García-Suastegui, W.A.; Peralta, R.; Dobson, A.D.W.; Batista-García, R.A.; Dávila-Ramos, S. Extremophile deep-sea viral communities from hydrothermal vents: Structural and functional analysis. *Mar. Genom.* **2019**, *46*, 16–28. [CrossRef]
15. Lossouarn, J.; Dupont, S.; Gorlas, A.; Mercier, C.; Bienvenu, N.; Marguet, E.; Forterre, P.; Geslin, C. An abyssal mobilome: Viruses, plasmids and vesicles from deep-sea hydrothermal vents. *Res. Microbiol.* **2015**, *166*, 742–752. [CrossRef]
16. Hatfull, G.F. Dark matter of the biosphere: The amazing world of bacteriophage diversity. *J. Virol.* **2015**, *89*, 8107–8110. [CrossRef]
17. Youle, M.; Haynes, M.; Rohwer, F. Scratching the surface of biology's dark matter. In *Viruses: Essential Agents of Life*; Springer: Dordrecht, The Netherlands, 2012; pp. 61–81. ISBN 9789400748996.
18. Fernandes, S.; São-José, C. Enzymes and mechanisms employed by tailed bacteriophages to breach the bacterial cell barriers. *Viruses* **2018**, *10*, 396. [CrossRef]
19. Briers, Y.; Walmagh, M.; Van Puyenbreck, V.; Cornelissen, A.; Cenens, W.; Aertsen, A.; Oliveira, H.; Azeredo, J.; Verween, G.; Pirnay, J.P.; et al. Engineered endolysin-based "Artilysins" to combat multidrug-resistant gram-negative pathogens. *MBio* **2014**, *5*, 1379–1393. [CrossRef]
20. Lwoff, A. Lysogeny. *Bacteriol. Rev.* **1953**, *17*, 269. [CrossRef]
21. Hobbs, Z.; Abedon, S.T. Diversity of phage infection types and associated terminology: The problem with "Lytic or lysogenic". *FEMS Microbiol. Lett.* **2016**, *363*, 47. [CrossRef]

22. Williamson, S.J.; Cary, S.C.; Williamson, K.E.; Helton, R.R.; Bench, S.R.; Winget, D.; Wommack, K.E. Lysogenic virus-host interactions predominate at deep-sea diffuse-flow hydrothermal vents. *ISME J.* **2008**, *2*, 1112–1121. [CrossRef]

23. Levin, B.R.; Lenski, R.E. Coevolution in bacteria and their viruses and plasmids. In *Coevolution*; Futuyama, D.J., Statkin, M., Eds.; Sinauer Associates: Sunderland, MA, USA, 1983.

24. Sandaa, R.A. Burden or benefit? Virus-host interactions in the marine environment. *Res. Microbiol.* **2008**, *159*, 374–381. [CrossRef]

25. Bondy-Denomy, J.; Davidson, A.R. When a virus is not a parasite: The beneficial effects of prophages on bacterial fitness. *J. Microbiol.* **2014**, *52*, 235–242. [CrossRef]

26. Engler, M.J.; Richardson, C.C. DNA Ligases. *Enzymes* **1982**, *15*, 3–29. [CrossRef]

27. Dale, R.M.K.; McClure, B.A.; Houchins, J.P. A rapid single-stranded cloning strategy for producing a sequential series of overlapping clones for use in DNA sequencing: Application to sequencing the corn mitochondrial 18 S rDNA. *Plasmid* **1985**, *13*, 31–40. [CrossRef]

28. Doherty, A.J.; Ashford, S.R.; Subramanya, H.S.; Wigley, D.B. Bacteriophage T7 DNA ligase: Overexpression, purification, crystallization, and characterization. *J. Biol. Chem.* **1996**, *271*, 11083–11089. [CrossRef]

29. Hori, K.; Mark, D.F.; Richardson, C.C. Deoxyribonucleic acid polymerase of bacteriophage T7. Characterization of the exonuclease activities of the gene 5 protein and the reconstituted polymerase. *J. Biol. Chem.* **1979**, *254*, 11598–11604. [CrossRef]

30. Engler, M.J.; Lechner, R.L.; Richardson, C.C. Two forms of the DNA polymerase of bacteriophage T7. *J. Biol. Chem.* **1983**, *258*, 11165–11173. [CrossRef]

31. Sayers, J.R.; Eckstein, F. A single-strand specific endonuclease activity copurifies with overexpressed T5 D15 exonuclease. *Nucleic Acids Res.* **1991**, *19*, 4127–4132. [CrossRef]

32. Islam, M.R.; Son, N.; Lee, J.; Lee, D.W.; Sohn, E.J.; Hwang, I. Production of bacteriophage-encoded endolysin, LysP11, in *Nicotiana benthamiana* and its activity as a potent antimicrobial agent against *Erysipelothrix rhusiopathiae*. *Plant Cell Rep.* **2019**, *38*, 1485–1499. [CrossRef]

33. Fischetti, V.A. Bacteriophage endolysins: A novel anti-infective to control Gram-positive pathogens. *Int. J. Med. Microbiol.* **2010**, *300*, 357–362. [CrossRef] [PubMed]

34. Witzenrath, M.; Schmeck, B.; Doehn, J.M.; Tschernig, T.; Zahlten, J.; Loeffler, J.M.; Zemlin, M.; Müller, H.; Gutbier, B.; Schütte, H.; et al. Systemic use of the endolysin Cpl-1 rescues mice with fatal pneumococcal pneumonia. *Crit. Care Med.* **2009**, *37*, 642–649. [CrossRef]

35. Gupta, R.; Prasad, Y. P-27/HP Endolysin as antibacterial agent for antibiotic resistant *Staphylococcus aureus* of human infections. *Curr. Microbiol.* **2011**, *63*, 39–45. [CrossRef]

36. Hermoso, J.A.; García, J.L.; García, P. Taking aim on bacterial pathogens: From phage therapy to enzybiotics. *Curr. Opin. Microbiol.* **2007**, *10*, 461–472. [CrossRef] [PubMed]

37. Zelcbuch, L.; Yahav, S.; Buchshtab, N.; Kahan-Hanum, M.; Vainberg-Slutskin, I.; Weiner, I.; Golembo, M.; Kredo-Russo, S.; Zak, N.; Gahali-Sass, I.; et al. Abstract PR07: Novel phages targeting the intratumor-associated bacteria *Fusobacterium nucleatum*. *Cancer Res.* **2020**, *80*, PR07. [CrossRef]

38. Liekniņa, I.; Kalniņš, G.; Akopjana, I.; Bogans, J.; Šišovs, M.; Jansons, J.; Rūmnieks, J.; Tārs, K. Production and characterization of novel ssRNA bacteriophage virus-like particles from metagenomic sequencing data. *J. Nanobiotechnology* **2019**, *17*, 61. [CrossRef]

39. Zhu, B.; Wang, L.; Mitsunobu, H.; Lu, X.; Hernandez, A.J.; Yoshida-Takashima, Y.; Nunoura, T.; Tabor, S.; Richardson, C.C. Deep-sea vent phage DNA polymerase specifically initiates DNA synthesis in the absence of primers. *Proc. Natl. Acad. Sci. USA* **2017**, *114*, E2310–E2318. [CrossRef]

40. Fernández-Garciá, J.L.; De Ory, A.; Brussaard, C.P.D.; De Vega, M. *Phaeocystis globosa* Virus DNA Polymerase X: A "swiss Army knife", Multifunctional DNA polymerase-lyase-ligase for base excision repair. *Sci. Rep.* **2017**, *7*, 1–13. [CrossRef]

41. Rosano, G.L.; Ceccarelli, E.A. Recombinant protein expression in *Escherichia coli*: Advances and challenges. *Front. Microbiol.* **2014**, *5*, 172. [CrossRef]

42. Kurland, C.G. Codon bias and gene expression. *FEBS Lett.* **1991**, *285*, 165–169. [CrossRef]

43. Makrides, S.C. Strategies for achieving high-level expression of genes in *Escherichia coli*. *Microbiol. Rev.* **1996**, *60*, 512–538. [CrossRef]

44. Rosano, G.L.; Morales, E.S.; Ceccarelli, E.A. New tools for recombinant protein production in *Escherichia coli*: A 5-year update. *Protein Sci.* **2019**, *28*, 1412–1422. [CrossRef]

45. Costa, S.; Almeida, A.; Castro, A.; Domingues, L. Fusion tags for protein solubility, purification and immunogenicity in *Escherichia coli*: The novel Fh8 system. *Front. Microbiol.* **2014**, *5*, 63. [CrossRef]

46. Kapust, R.B.; Waugh, D.S. *Escherichia coli* maltose-binding protein is uncommonly effective at promoting the solubility of polypeptides to which it is fused. *Protein Sci.* **1999**, *8*, 1668–1674. [CrossRef]

47. Esposito, D.; Chatterjee, D.K. Enhancement of soluble protein expression through the use of fusion tags. *Curr. Opin. Biotechnol.* **2006**, *17*, 353–358. [CrossRef]

48. Bjerga, G.E.K.; Arsın, H.; Larsen, Ø.; Puntervoll, P.; Kleivdal, H.T. A rapid solubility-optimized screening procedure for recombinant subtilisins in *E. coli*. *J. Biotechnol.* **2016**, *222*, 38–46. [CrossRef]

49. Elena, C.; Ravasi, P.; Castelli, M.E.; Peirú, S.; Menzella, H.G. Expression of codon optimized genes in microbial systems: Current industrial applications and perspectives. *Front. Microbiol.* **2014**, *5*, 21. [CrossRef]

50. Mignon, C.; Mariano, N.; Stadthagen, G.; Lugari, A.; Lagoutte, P.; Donnat, S.; Chenavas, S.; Perot, C.; Sodoyer, R.; Werle, B. Codon harmonization–going beyond the speed limit for protein expression. *FEBS Lett.* **2018**, *592*, 1554–1564. [CrossRef]
51. Claassens, N.J.; Siliakus, M.F.; Spaans, S.K.; Creutzburg, S.C.A.A.; Nijsse, B.; Schaap, P.J.; Quax, T.E.F.F.; Van Der Oost, J. Improving heterologous membrane protein production in *Escherichia coli* by combining transcriptional tuning and codon usage algorithms. *PLoS ONE* **2017**, *12*, e0184355. [CrossRef]
52. Gao, C.Y.; Xu, T.T.; Zhao, Q.J.; Li, C.L. Codon optimization enhances the expression of porcine β-defensin-2 in *Escherichia coli*. *Genet. Mol. Res.* **2015**, *14*, 4978–4988. [CrossRef]
53. Burgess-Brown, N.A.; Sharma, S.; Sobott, F.; Loenarz, C.; Oppermann, U.; Gileadi, O. Codon optimization can improve expression of human genes in *Escherichia coli*: A multi-gene study. *Protein Expr. Purif.* **2008**, *59*, 94–102. [CrossRef]
54. Fei, D.; Zhang, H.; Diao, Q.; Jiang, L.; Wang, Q.; Zhong, Y.; Fan, Z.; Ma, M. Codon optimization, expression in *Escherichia coli*, and immunogenicity of recombinant Chinese Sacbrood Virus (CSBV) structural proteins VP1, VP2, and VP3. *PLoS ONE* **2015**, *10*, e0128486. [CrossRef]
55. Roalkvam, I.; Bredy, F.; Baumberger, T.; Pedersen, R.B.; Steen, I.H. *Hypnocyclicus thermotrophus* gen. Nov., sp. nov. isolated from a microbial mat in a hydrothermal vent field. *Int. J. Syst. Evol. Microbiol.* **2015**, *65*, 4521–4525. [CrossRef]
56. Marques, A.F.A.; Roerdink, D.L.; Baumberger, T.; de Ronde, C.E.J.; Ditchburn, R.G.; Denny, A.; Thorseth, I.H.; Okland, I.; Lilley, M.D.; Whitehouse, M.J.; et al. The seven sisters hydrothermal system: First record of shallow hybrid mineralization hosted in mafic volcaniclasts on the arctic mid-ocean ridge. *Minerals* **2020**, *10*, 439. [CrossRef]
57. Zhou, Y.; Liang, Y.; Lynch, K.H.; Dennis, J.J.; Wishart, D.S. PHAST: A fast phage search tool. *Nucleic Acids Res.* **2011**, *39*, W347–W352. [CrossRef] [PubMed]
58. Arndt, D.; Grant, J.R.; Marcu, A.; Sajed, T.; Pon, A.; Liang, Y.; Wishart, D.S. PHASTER: A better, faster version of the PHAST phage search tool. *Nucleic Acids Res.* **2016**, *44*, W16–W21. [CrossRef] [PubMed]
59. Söding, J. Protein homology detection by HMM-HMM comparison. *Bioinformatics* **2005**, *21*, 951–960. [CrossRef] [PubMed]
60. Zimmermann, L.; Stephens, A.; Nam, S.Z.; Rau, D.; Kübler, J.; Lozajic, M.; Gabler, F.; Söding, J.; Lupas, A.N.; Alva, V. A completely reimplemented MPI bioinformatics toolkit with a new HHpred server at its core. *J. Mol. Biol.* **2018**, *430*, 2237–2243. [CrossRef]
61. Steinegger, M.; Meier, M.; Mirdita, M.; Vöhringer, H.; Haunsberger, S.J.; Söding, J. HH-suite3 for fast remote homology detection and deep protein annotation. *BMC Bioinform.* **2019**, *20*, 473. [CrossRef]
62. Huerta-Cepas, J.; Forslund, K.; Coelho, L.P.; Szklarczyk, D.; Jensen, L.J.; von Mering, C.; Bork, P. Fast genome-wide functional annotation through orthology assignment by eggNOG-Mapper. *Mol. Biol. Evol.* **2017**, *34*, 2115–2122. [CrossRef]
63. Huerta-Cepas, J.; Szklarczyk, D.; Heller, D.; Hernández-Plaza, A.; Forslund, S.K.; Cook, H.; Mende, D.R.; Letunic, I.; Rattei, T.; Jensen, L.J.; et al. eggNOG 5.0: A hierarchical, functionally and phylogenetically annotated orthology resource based on 5090 organisms and 2502 viruses. *Nucleic Acids Res.* **2019**, *47*, D309–D314. [CrossRef]
64. Huson, D.H.; Auch, A.F.; Qi, J.; Schuster, S.C. MEGAN analysis of metagenomic data. *Genome Res.* **2007**, *17*, 377–386. [CrossRef]
65. Huson, D.H.; Beier, S.; Flade, I.; Górska, A.; El-Hadidi, M.; Mitra, S.; Ruscheweyh, H.-J.; Tappu, R. MEGAN Community Edition-Interactive Exploration and Analysis of Large-Scale Microbiome Sequencing Data. *PLOS Comput. Biol.* **2016**, *12*, e1004957. [CrossRef]
66. Lopes, A.; Tavares, P.; Petit, M.A.; Guérois, R.; Zinn-Justin, S. Automated classification of tailed bacteriophages according to their neck organization. *BMC Genom.* **2014**, *15*, 1–17. [CrossRef]
67. Letunic, I. phyloT: A Phylogenetic Tree Generator. Available online: https://phylot.biobyte.de/index.cgi (accessed on 5 June 2020).
68. Altschul, S.F.; Gish, W.; Miller, W.; Myers, E.W.; Lipman, D.J. Basic local alignment search tool. *J. Mol. Biol.* **1990**, *215*, 403–410. [CrossRef]
69. Katoh, K.; Standley, D.M. MAFFT Multiple Sequence Alignment Software Version 7: Improvements in Performance and Usability. *Mol. Biol. Evol.* **2013**, *30*, 772–780. [CrossRef]
70. Capella-Gutiérrez, S.; Silla-Martínez, J.M.; Gabaldón, T. trimAl: A tool for automated alignment trimming in large-scale phylogenetic analyses. *Bioinformatics* **2009**, *25*, 1972–1973. [CrossRef]
71. Trifinopoulos, J.; Nguyen, L.-T.; von Haeseler, A.; Minh, B.Q. W-IQ-TREE: A fast online phylogenetic tool for maximum likelihood analysis. *Nucleic Acids Res.* **2016**, *44*, W232–W235. [CrossRef]
72. Kalyaanamoorthy, S.; Minh, B.Q.; Wong, T.K.F.; Von Haeseler, A.; Jermiin, L.S. ModelFinder: Fast model selection for accurate phylogenetic estimates. *Nat. Methods* **2017**, *14*, 587–589. [CrossRef]
73. Minh, B.Q.; Nguyen, M.A.T.; von Haeseler, A. Ultrafast approximation for phylogenetic bootstrap. *Mol. Biol. Evol.* **2013**, *30*, 1188–1195. [CrossRef]
74. Le, S.Q.; Gascuel, O. An improved general amino acid replacement matrix. *Mol. Biol. Evol.* **2008**, *25*, 1307–1320. [CrossRef]
75. Gu, X.; Fu, Y.X.; Li, W.H. Maximum likelihood estimation of the heterogeneity of substitution rate among nucleotide sites. *Mol. Biol. Evol.* **1995**, *12*, 546–557. [CrossRef]
76. Letunic, I.; Bork, P. Interactive Tree Of Life (iTOL) v4: Recent updates and new developments. *Nucleic Acids Res.* **2019**, *47*, W256–W259. [CrossRef] [PubMed]
77. Paez-Espino, D.; Chen, I.-M.A.; Palaniappan, K.; Ratner, A.; Chu, K.; Szeto, E.; Pillay, M.; Huang, J.; Markowitz, V.M.; Nielsen, T.; et al. IMG/VR: A database of cultured and uncultured DNA Viruses and retroviruses. *Nucleic Acids Res.* **2016**, *45*. [CrossRef] [PubMed]

78. Villar, E.; Vannier, T.; Vernette, C.; Lescot, M.; Cuenca, M.; Alexandre, A.; Bachelerie, P.; Rosnet, T.; Pelletier, E.; Sunagawa, S.; et al. The Ocean Gene Atlas: Exploring the biogeography of plankton genes online. *Nucleic Acids Res.* **2018**, *46*, W289–W295. [CrossRef] [PubMed]

79. Harrison, K.J.; de Crécy-Lagard, V.; Zallot, R. Gene Graphics: A genomic neighborhood data visualization web application. *Bioinformatics* **2017**, *34*, 1406–1408. [CrossRef] [PubMed]

80. Finn, R.D.; Clements, J.; Eddy, S.R. HMMER web server: Interactive sequence similarity searching. *Nucleic Acids Res.* **2011**, *39*, W29–W37. [CrossRef]

81. Finn, R.D.; Clements, J.; Arndt, W.; Miller, B.L.; Wheeler, T.J.; Schreiber, F.; Bateman, A.; Eddy, S.R. HMMER web server: 2015 update. *Nucleic Acids Res.* **2015**, *43*, W30–W38. [CrossRef]

82. Potter, S.C.; Luciani, A.; Eddy, S.R.; Park, Y.; Lopez, R.; Finn, R.D. HMMER web server: 2018 update. *Nucleic Acids Res.* **2018**, *46*, W200–W204. [CrossRef]

83. Angov, E.; Legler, P.; Mease, R. Adjustment of codon usage frequencies by codon harmonization improves protein expression and folding. *Methods Mol. Biol.* **2011**, *705*, 1–13. [CrossRef]

84. Angov, E. Codon usage: Nature's roadmap to expression and folding of proteins. *Biotechnol. J.* **2011**, *6*, 650–659. [CrossRef]

85. Novagen Inc. pET System Manual. In *TB055*; Merck: Darmstadt, Germany, 2006; pp. 19–24.

86. Hochuli, E.; Bannwarth, W.; Döbeli, H.; Gentz, R.; Stüber, D. Genetic approach to facilitate purification of recombinant proteins with a novel metal chelate adsorbent. *Nat. Biotechnol.* **1988**, *6*, 1321–1325. [CrossRef]

87. Kristensen, D.M.; Waller, A.S.; Yamada, T.; Bork, P.; Mushegian, A.R.; Koonin, E.V. Orthologous gene clusters and taxon signature genes for viruses of prokaryotes. *J. Bacteriol.* **2013**, *195*, 941–950. [CrossRef]

88. Rezaei Javan, R.; Ramos-Sevillano, E.; Akter, A.; Brown, J.; Brueggemann, A.B. Prophages and satellite prophages are widespread in *Streptococcus* and may play a role in pneumococcal pathogenesis. *Nat. Commun.* **2019**, *10*, 4852. [CrossRef]

89. Cochrane, K.; Manson McGuire, A.; Priest, M.E.; Abouelleil, A.; Cerqueira, G.C.; Lo, R.; Earl, A.M.; Allen-Vercoe, E. Complete genome sequences and analysis of the *Fusobacterium nucleatum* subspecies animalis 7-1 bacteriophage ΦFunu1 and ΦFunu2. *Anaerobe* **2016**, *38*, 125–129. [CrossRef]

90. Machuca, P.; Daille, L.; Vinés, E.; Berrocal, L.; Bittner, M. Isolation of a novel bacteriophage specific for the periodontal pathogen *Fusobacterium nucleatum*. *Appl. Environ. Microbiol.* **2010**, *76*, 7243–7250. [CrossRef]

91. Kabwe, M.; Brown, T.L.; Dashper, S.; Speirs, L.; Ku, H.; Petrovski, S.; Chan, H.T.; Lock, P.; Tucci, J. Genomic, morphological and functional characterisation of novel bacteriophage FNU1 capable of disrupting *Fusobacterium nucleatum* biofilms. *Sci. Rep.* **2019**, *9*, 1–12. [CrossRef]

92. Brennan, C.A.; Garrett, W.S. *Fusobacterium nucleatum*—Symbiont, opportunist and oncobacterium. *Nat. Rev. Microbiol.* **2019**, *17*, 156–166. [CrossRef]

93. Quax, T.E.F.; Claassens, N.J.; Söll, D.; van der Oost, J. Codon bias as a means to fine-tune gene expression. *Mol. Cell* **2015**, *59*, 149–161. [CrossRef]

94. Porath, J. Immobilized metal ion affinity chromatography. *Protein Expr. Purif.* **1992**, *3*, 263–281. [CrossRef]

95. Razdan, A. *Affinity Chromatography Handbook, Vol. 2: Tagged Proteins*; GE Healthcare Bio-Sciences AB: Uppsala, Sweden, 2000; Available online: https://www.cytivalifesciences.com/en/us/support/handbooks (accessed on 9 January 2021).

96. Freitag-Pohl, S.; Jasilionis, A.; Håkansson, M.; Svensson, L.A.; Kovačič, R.; Welin, M.; Watzlawick, H.; Wang, L.; Altenbuchner, J.; Płotka, M.; et al. Crystal structures of the *Bacillus subtilis* prophage lytic cassette proteins XepA and YomS. *Acta Crystallogr. Sect. D Struct. Biol.* **2019**, *75*, 1028–1039. [CrossRef]

97. Weinbauer, M.G.; Brettar, I.; Höfle, M.G. Lysogeny and virus-induced mortality of bacterioplankton in surface, deep, and anoxic marine waters. *Limnol. Oceanogr.* **2003**, *48*, 1457–1465. [CrossRef]

98. Weinbauer, M.G. Ecology of prokaryotic viruses. *FEMS Microbiol. Rev.* **2004**, *28*, 127–181. [CrossRef] [PubMed]

99. Harrison, E.; Brockhurst, M.A. Ecological and evolutionary benefits of temperate phage: What does or doesn't kill you makes you stronger. *BioEssays* **2017**, *39*, 1700112. [CrossRef] [PubMed]

100. Paul, J.H. Prophages in marine bacteria: Dangerous molecular time bombs or the key to survival in the seas? *ISME J.* **2008**, *2*, 579–589. [CrossRef] [PubMed]

101. Grose, J.H.; Casjens, S.R. Bacteriophage Diversity. In *Encyclopedia of Virology*; Elsevier: Amsterdam, The Netherlands, 2019.

102. Amarillas, L.; Rubí-Rangel, L.; Chaidez, C.; González-Robles, A.; Lightbourn-Rojas, L.; León-Félix, J. Isolation and characterization of phiLLS, a novel phage with potential biocontrol agent against multidrug-resistant *Escherichia coli*. *Front. Microbiol.* **2017**, *8*, 1355. [CrossRef]

103. Bokma, E.; Van Koningsveld, G.A.; Jeronimus-Stratingh, M.; Beintema, J.J. Hevamine, a chitinase from the rubber tree *Hevea brasiliensis*, cleaves peptidoglycan between the C-1 of N-acetylglucosamine and C-4 of N-acetylmuramic acid and therefore is not a lysozyme. *FEBS Lett.* **1997**, *411*, 161–163. [CrossRef]

104. Horn, S.J.; Sørbotten, A.; Synstad, B.; Sikorski, P.; Sørlie, M.; Vårum, K.M.; Eijsink, V.G.H. Endo/exo mechanism and processivity of family 18 chitinases produced by *Serratia marcescens*. *FEBS J.* **2006**, *273*, 491–503. [CrossRef]

105. Saier, M.H.; Reddy, B.L. Holins in bacteria, eukaryotes, and archaea: Multifunctional xenologues with potential biotechnological and biomedical applications. *J. Bacteriol.* **2015**, *197*, 7–17. [CrossRef]

106. Cahill, J.; Young, R. Phage Lysis: Multiple genes for multiple barriers. In *Advances in Virus Research*; Academic Press Inc.: Cambridge, MA, USA, 2019; Volume 103, pp. 33–70. ISBN 9780128177228.

107. Kongari, R.; Rajaure, M.; Cahill, J.; Rasche, E.; Mijalis, E.; Berry, J.; Young, R. Phage spanins: Diversity, topological dynamics and gene convergence. *BMC Bioinform.* **2018**, *19*, 326. [CrossRef]
108. Schmelcher, M.; Donovan, D.M.; Loessner, M.J. Bacteriophage endolysins as novel antimicrobials. *Future Microbiol.* **2012**, *7*, 1147–1171. [CrossRef]
109. Li, Q.; Yi, L.; Marek, P.; Iverson, B.L. Commercial proteases: Present and future. *FEBS Lett.* **2013**, *587*, 1155–1163. [CrossRef]
110. Chapman, J.; Ismail, A.; Dinu, C.; Chapman, J.; Ismail, A.E.; Dinu, C.Z. Industrial applications of enzymes: Recent Advances, techniques, and outlooks. *Catalysts* **2018**, *8*, 238. [CrossRef]
111. Gomes, J.; Steiner, W. The biocatalytic potential of extremophiles and extremozymes. *Food Technol. Biotechnol.* **2004**, *42*, 223–225.
112. Aschenbrenner, J.; Marx, A. DNA polymerases and biotechnological applications. *Curr. Opin. Biotechnol.* **2017**, *48*, 187–195. [CrossRef]
113. Vieille, C.; Burdette, D.S.; Zeikus, J.G. Thermozymes. *Biotechnol. Annu. Rev.* **1996**, *2*, 1–83.
114. Ishino, S.; Ishino, Y. DNA polymerases as useful reagents for biotechnology - The history of developmental research in the field. *Front. Microbiol.* **2014**, *5*, 465. [CrossRef]
115. Aevarsson, A.; Kaczorowska, A.-K.; Adalsteinsson, B.T.; Ahlqvist, J.; Al-Karadaghi, S.; Altenbuchner, J.; Arsın, H.; Átlasson, Ú.Á.; Brandt, D.; Cichowicz-Cieślak, M.; et al. Going to extremes—A metagenomic journey into the dark matter of life. *FEMS Microbiol. Lett.* **2021**. [CrossRef]
116. GenScript GenSmartTM Codon Optimization Tool-GenScript. Available online: https://www.genscript.com/gensmart-free-gene-codon-optimization.html (accessed on 15 September 2020).
117. Chen, F.; Wu, P.; Deng, S.; Zhang, H.; Hou, Y.; Hu, Z.; Zhang, J.; Chen, X.; Yang, J.-R. Dissimilation of synonymous codon usage bias in virus–host coevolution due to translational selection. *Nat. Ecol. Evol.* **2020**, *4*, 589–600. [CrossRef]
118. Angov, E.; Hillier, C.J.; Kincaid, R.L.; Lyon, J.A. Heterologous protein expression is enhanced by harmonizing the codon usage frequencies of the target gene with those of the expression host. *PLoS ONE* **2008**, *3*, e2189. [CrossRef]
119. Wen, J.; Lord, H.; Knutson, N.; Wikström, M. Nano differential scanning fluorimetry for comparability studies of therapeutic proteins. *Anal. Biochem.* **2020**, *593*, 113581. [CrossRef] [PubMed]
120. Pellizza, L.; Smal, C.; Rodrigo, G.; Arán, M. Codon usage clusters correlation: Towards protein solubility prediction in heterologous expression systems in *E. coli*. *Sci. Rep.* **2018**, *8*, 1–12. [CrossRef]
121. Gould, N.; Hendy, O.; Papamichail, D. Computational tools and algorithms for designing customized synthetic genes. *Front. Bioeng. Biotechnol.* **2014**, *2*, 41. [CrossRef]
122. Sen, A.; Kargar, K.; Akgün, E.; Pınar, M.C. Codon optimization: A mathematical programing approach. *Bioinformatics* **2020**, *36*, 4012–4020. [CrossRef] [PubMed]

Review

Plant Viruses: From Targets to Tools for CRISPR

Carla M. R. Varanda [1,*], Maria do Rosário Félix [2], Maria Doroteia Campos [1], Mariana Patanita [1] and Patrick Materatski [1,*]

1   MED—Mediterranean Institute for Agriculture, Environment and Development, Instituto de Investigação e Formação Avançada, Universidade de Évora, Pólo da Mitra, Ap. 94, 7006-554 Évora, Portugal; mdcc@uevora.pt (M.D.C.); mpatanita@uevora.pt (M.P.)
2   MED—Mediterranean Institute for Agriculture, Environment and Development & Departamento de Fitotecnia, Escola de Ciências e Tecnologia, Universidade de Évora, Pólo da Mitra, Ap. 94, 7006-554 Évora, Portugal; mrff@uevora.pt
*   Correspondence: carlavaranda@uevora.pt (C.M.R.V.); pmateratski@uevora.pt (P.M.)

Abstract: Plant viruses cause devastating diseases in many agriculture systems, being a serious threat for the provision of adequate nourishment to a continuous growing population. At the present, there are no chemical products that directly target the viruses, and their control rely mainly on preventive sanitary measures to reduce viral infections that, although important, have proved to be far from enough. The current most effective and sustainable solution is the use of virus-resistant varieties, but which require too much work and time to obtain. In the recent years, the versatile gene editing technology known as CRISPR/Cas has simplified the engineering of crops and has successfully been used for the development of viral resistant plants. CRISPR stands for 'clustered regularly interspaced short palindromic repeats' and CRISPR-associated (Cas) proteins, and is based on a natural adaptive immune system that most archaeal and some bacterial species present to defend themselves against invading bacteriophages. Plant viral resistance using CRISPR/Cas technology can been achieved either through manipulation of plant genome (plant-mediated resistance), by mutating host factors required for viral infection; or through manipulation of virus genome (virus-mediated resistance), for which CRISPR/Cas systems must specifically target and cleave viral DNA or RNA. Viruses present an efficient machinery and comprehensive genome structure and, in a different, beneficial perspective, they have been used as biotechnological tools in several areas such as medicine, materials industry, and agriculture with several purposes. Due to all this potential, it is not surprising that viruses have also been used as vectors for CRISPR technology; namely, to deliver CRISPR components into plants, a crucial step for the success of CRISPR technology. Here we discuss the basic principles of CRISPR/Cas technology, with a special focus on the advances of CRISPR/Cas to engineer plant resistance against DNA and RNA viruses. We also describe several strategies for the delivery of these systems into plant cells, focusing on the advantages and disadvantages of the use of plant viruses as vectors. We conclude by discussing some of the constrains faced by the application of CRISPR/Cas technology in agriculture and future prospects.

Keywords: CRISPR/Cas systems; viral vectors; gene editing; plant genome engineering; viral resistance

Citation: Varanda, C.M.R.; Félix, M.d.R.; Campos, M.D.; Patanita, M.; Materatski, P. Plant Viruses: From Targets to Tools for CRISPR. *Viruses* **2021**, *13*, 141. https://doi.org/10.3390/v13010141

Academic Editor: Henryk Czosnek
Received: 21 December 2020
Accepted: 17 January 2021
Published: 19 January 2021

1. Introduction

Plant viruses are known to infect and cause devastating diseases in many agricultural systems, leading to significant losses in crop quality and yield, with extreme economic impacts worldwide, being a serious threat for the provision of adequate nourishment to a continuous growing population [1,2]. Climate change has been rapidly causing aggravation of viral disease impacts, with existing virus showing pandemic behavior, and with the appearance of new emergent viruses, making the development of efficient long term disease management approaches difficult [3].

Plant viruses are obligate intracellular pathogens and at present there are no chemical products that directly target the virus, that can be used in agronomic context, making preventive sanitary measures the only way to hamper infections. Preventive sanitary measures consist mostly of good sanitation techniques during cultural practices, that include the immediate removal and destruction of infected plants, the limitation of the virus vector organisms populations and the development of legislative measures concerning the commercialization and trade of virus free plant material [4]. Many of these conventional strategies are unsafe for the environment and have proved to be far from enough. The use of viral resistant plants is currently the most efficient and sustainable solution to reduce viral infections. Thus, it is essential to develop effective and durable virus resistant varieties to face the increasingly severe viral diseases and viral variants [5–8]. For many years, classical breeding for crop improvement involved the selection of plants with certain agronomic characteristics and absence of viral symptoms, a very laborious and time-consuming strategy [9].

Advances in biotechnology have provided new knowledge on molecular mechanisms of plant virus interactions, which accelerated the process of breeding through approaches based on molecular marker-assisted breeding, genomic selection, gene silencing, pathogen-derived resistance (PDR), etc., and has provided many resistant varieties to agriculture [10–12]. However, the rapid evolution and emergence of new viruses makes the durability of the resistance a major drawback and creates the need of rapid and efficient techniques for obtaining resistant plants.

In recent years, the versatile gene editing technology known as CRISPR/Cas has simplified the engineering of crops and has already been used for the development of resistance to viral pathogens, overcoming many difficulties of the techniques used to date [13–15].

Moreover, viruses can be manipulated to be beneficial and useful for several purposes as they present an efficient machinery and a comprehensive genome structure. They have been used in biotechnology as molecular tools in several areas such as medicine, materials industry, and agriculture with different purposes including the production of proteins and being targets and vectors of many materials [16,17]. Due to all this potential, it is not surprising that viruses have also been used in this revolutionary genome editing technique.

In this review, we start by describing the basic principles of CRISPR/Cas technology, with special focus on the advances of CRISPR/Cas to engineer plant resistance against RNA and DNA viruses. We demonstrate that, for the successful use of this technology, it is imperative that the CRISPR/Cas system is efficiently delivered and expressed in the targeted cells, and we describe several strategies for the delivery of these systems into plant cells. In a different perspective, we show how viruses can be manipulated to be used as tools for the delivery of CRISPR/Cas systems into plant cells, focusing on the advantages and disadvantages of the use of viruses as vectors of CRISPR systems into plant cells. We conclude by discussing the constrains faced by CRISPR/Cas technology and the future prospects.

2. CRISPR: From a Natural Bacterial Immune System to a Gene Editing Tool

Clustered regularly interspaced short palindromic repeats (CRISPR) and CRISPR-associated (Cas) proteins is a natural adaptive immune system that some bacterial and most archaeal species present to defend themselves against invading bacteriophages, which works on the basis of sequence complementarity via cleavage [18,19].

CRISPR systems may be divided into two main classes (I and II) and six different types (I to VI), defined by the nature of the nucleases complex and the mechanism of targeting, each presenting a unique nuclease Cas protein. Class I systems are multicomponent systems composed of multiple effectors; these systems are subdivided into types I, III, and IV. Class II systems include the types II, V, and VI and are single-component systems consisting of a single effector guided by the CRISPR RNA (crRNA) [20].

The CRISPR/Cas9, belonging to class II, is based on the immune system of *Streptococcus pyogenes*. It consists of the capacity of the bacteria to acquire pieces of DNA from an invading phage or plasmid and incorporating them in their own DNA, which will further serve to guide Cas9 to cleave homologous RNA, leading to immediate RNA disruption and further specific RNA disruption in subsequent invasions, thus providing immunity to the bacterial cell [21]. The mechanism involved in this natural immune system is very simple and has been the basis for the most developed CRISPR/Cas genome-editing platform.

The first steps of CRISPR/Cas9 as a successful editing tool, started with the possibility of engineering into a single RNA chimera (sgRNA), two noncoding RNAs essential for CRISPR, crRNA, and trans-activating crRNA (tracrRNA) [22]. crRNA is the genomic complementary region, i.e., the target for Cas (the programmable portion defined by the user) and tracrRNA is the RNA sequence that provides the stem loop structure to bound Cas. This has simplified gene editing using CRISPR/Cas9, which can now be accomplished by introducing two components in the same cell: the sgRNA and the Cas protein [22] and led to efficient genetic manipulation in a wide array of plants, becoming the most promising, versatile, and powerful tool for plant improvement [23].

In CRISPR/Cas9 system (Figure 1), first Cas9 binds to the sgRNA to create the Cas9-sgRNA duplex which becomes catalytically active and directs the RNA-guided DNA endonuclease Cas9 to target. For target recognition and cleavage, it is also required the presence of a Protospacer Adjacent Motif (PAM) positioned 3–4 nucleotides downstream of the 3′ end of the target sequence, which differs depending on the species of Cas9 (this sequence consists of NGG in *S. pyogenes*) [22,24]. Once the PAM sequence is recognized by the Cas9-sgRNA complex, and the crRNA portion within the sgRNA (the 5′ most 20 nts) anneals to the genomic DNA through Watson–Crick base pairing, it will cleave both DNA strands, three bases upstream of the PAM, creating sequence-specific blunt end double-stranded breaks (DSBs) at target site. When a DSB in the DNA is created, the host cell repairs it via evolutionary conserved DNA pathways such as error-prone non-homologous end-joining (NHEJ) and homology-directed repair (HDR).

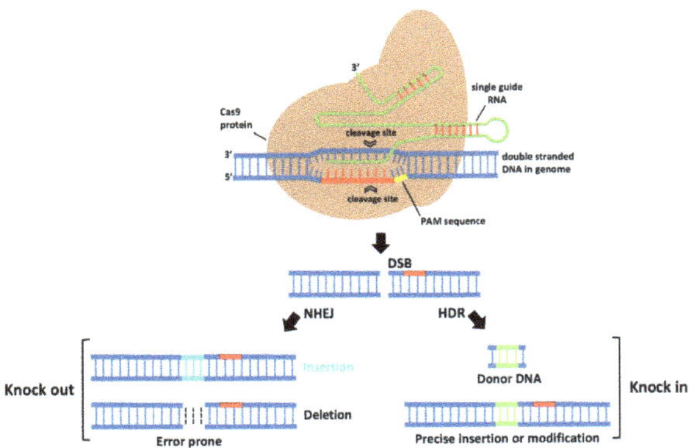

Figure 1. The mechanism of CRISPR-Cas9-mediated genome engineering in plants. A single guide RNA recognizes a region in the genome followed by a PAM sequence, and recruits a Cas9 protein that will cleave DNA, creating a double-stranded break that is repaired by error-prone non-homologous end-joining (NHEJ) and homology-directed repair (HDR).

NHEJ creates insertions or deletions (indels) at the target site that, if within the protein coding region, can cause a frameshift mutation that eliminates gene expression, leading to gene knock out [25]. HDR is a more precise method for DSB repair; it requires, besides sgRNA and Cas, a donor repair template with ends homologous to each border of the target site sequence. When a repair template is provided, HDR will result in the introduction of new sequences at breaking site and a knock in occurs [25]. For producing specific desired mutations and genomic replacement, DSBs should be repaired by HDR pathway. More recently, a new generation of CRISPR is being developed by fusing nuclease DNA targeting proteins with deactivated nuclease domains, with enzymes to enable direct conversion of a single DNA nucleotide into another [26] without the need of DSB formation.

Genetic engineering using CRISPR/Cas systems enables accurate and precise genomic modifications. Moreover, this strategy can be used to target different sequences simultaneously with high efficiency [27], achieving a broader result, as for example immunity against different pathogens.

The easiness and rapidity of execution, low cost, reproducibility and efficiency turns understandable why it is the system of choice for many genome engineering applications in several fields using different organisms. The possibility of using Cas proteins with deactivated nuclease domains can contribute to a broader application of CRISPR such as regulating gene transcription and inducing targeted epigenic modifications [28]. In addition, CRISPR has shown to have potential for other applications besides genome engineering, such as studies on gene functions and diagnostics. CRISPR/LwaCas13a system was able to highly select and detect up to a single copy of RNA [29], which may be a very interesting starting point to develop a far more sensitive method than currently available methods, for the detection of RNA viruses, including qPCR [30].

In plants, this technology has been used for plant breeding including nutrition enhancement and plant resistance against several agents such as fungi, bacteria, and viruses in many crop plants—including rice [31], tomato [32], citrus [33,34], wheat [35], and maize [36,37]—proving its potential to transform agriculture and enhancing world food safety.

3. CRISPR to Engineer Plant Virus Resistance

Due to the devastating losses that plant viruses cause, it is not surprising that CRISPR/Cas technologies have been applied to develop plant resistance against viral pathogens.

Plant viral resistance using CRISPR/Cas systems can been achieved either through manipulation of plant genome (plant-mediated resistance), or virus genome (virus-mediated resistance).

The CRISPR/Cas technology was initially thought to be exclusively applied to DNA, which, in terms of its use for plant viral resistance through manipulation of viral genome, would be restricted to DNA viruses. However, thanks to the discovery of RNA-targeting CRISPR/Cas effectors that efficiently target and cleave single-stranded RNAs, an exciting opportunity has been opened for achieving plant resistance also against RNA viruses, which are most of the plant viruses known [38,39].

Below we present several studies that report the use of the CRISPR/Cas system to engineer plant resistance against several viruses, either by acting on plant genome (plant mediated resistance) or on viral genomes (virus mediated resistance). These studies have shown the capacity of CRISPR to confer efficient and durable molecular immunity to plants against viruses that rely on the integrity of their genome at some point of their replication cycle [15,40–43].

3.1. CRISPR for Plant Mediated Resistance

Plant viruses are dependent on the host's machinery for their replication, since they interact with many host factors required for viral replication and movement inside plants, essential to complete their cycle of infection [44]. CRISPR/Cas allows the mutation/deletion of recessive genes that encode critical host factors for viral infection, conferring recessive resistance, which, as an inherited characteristic is very durable [45].

Considerable knowledge has been generated on the genetics of plant disease resistance and many plant genes have been discovered as essential for viral infections and have been the focus for the development of plant resistance using transgenic approaches [12,46,47]. These studies have provided many valuable potential targets for genome editing and genes—such as the translation initiation-like factors elF4E, elF4G, and their isoforms—that have shown to be directly involved in the infection process of viruses. Those genes are being subjected to targeted mutations introduced by CRISPR to engineer plant resistance [48]. In fact, any host gene encoding a factor required by the virus is a potential target for CRISPR.

This approach is interesting as it allows that Cas9, as well as other endonucleases which target DNA, to be used to provide plant resistance to RNA viruses by mutating host factors/genes associated to viral pathogenesis in the plant [49]. In addition, CRISPR for plant mediated resistance does not require the maintenance of a transgene for Cas9 and sgRNA in the plant genome, engineering transgenic-free virus-resistant plants [14,42,49].

Several studies have achieved plant mediated resistance against viruses using CRISPR/Cas9 (Table 1). For example, specific mutations were introduced in *Arabidopsis thaliana*, causing the knock out of elF(iso)4E gene, which resulted in a stable resistance against *Turnip mosaic virus* (TuMV) [42]. Macovei et al. [50] developed rice plants resistant to *Rice tungro spherical virus* (RTSV) through mutation of elF4G gene. Similarly, the disruption of the cucumber (*Cucumis sativus*) elF4E gene provided plant resistance to multiple members of the *Potyviridae*, namely the ipomovirus *Cucumber vein yellowing virus* (CVYV) and the potyviruses *Zucchini yellow mosaic virus* (ZYMV) and *Papaya ringspot mosaic virus* (PRSV) [49]. Resistance against *Clover yellow vein virus* (CYVV) was achieved in *A. thaliana* plants by targeting the elF4E1 gene using CRISPR/Cas9 [51]. Very recently, CRISPR/Cas9 has also allowed to perform double mutations on the novel cap-binding protein-1 and protein-2 (nCBP-1 and nCBP-2) belonging to the elF4E family, on cassava, which increased the resistance to *Cassava brown streak virus* (CBSV) [52].

It is a fact that modifications of plant genes may always face the risk to interfere with plant functions associated to those genes, with a fitness cost for the host, however these examples have demonstrated the success of CRISPR/Cas9 to produce genetic resistant plants through plant mediated resistance and without compromising plant functions.

3.2. CRISPR for Virus Mediated Resistance

Another approach to achieve plant viral resistance through CRISPR systems is by directly targeting viral genomes. In this approach, the problems that may arise by interfering with genes, that may also be associated to other plant functions—such as growth, reproduction, or others—are surpassed. However, for this type of mediated resistance, CRISPR/Cas systems must specifically directly target and cleave DNA of DNA viruses, or RNA of RNA viruses [43].

CRISPR for virus mediated resistance was first exploited to fight DNA viruses, as the discovery of CRISPR/Cas systems that can cleave RNA was more recent [27,39]. The discovery of such systems (class II, type VI Cas effectors, and Cas9 variants)—namely Cas13a (C2c2), Cas13b (C2c6), Cas13c (C2c7), Cas13d, FnCas9, and RCas9 (RNA targeting SpCas9) [20,27,53–56], was a great benefit—enabling direct targeting of RNA viruses which represent most plant pathogenic viruses.

Several studies have demonstrated the potential of CRISPR to impart plant resistance by targeting either DNA or RNA viral genomes, causing delayed or reduced accumulation of viruses and significantly attenuating symptoms of infection [57]. Some of those studies which directly mutate DNA and RNA viruses in plants expressing CRISPR/Cas machinery are described below (Table 1).

There are two major groups of plant DNA viruses, the double stranded caulimoviruses and the geminiviruses, the later which, although single stranded, replicate within the plant cell as double stranded DNA [58]. According to the latest report of the international Committee on Taxonomy of Viruses (ICTV), the *Geminiviridae* is the largest group of plant viruses, with 485 species [59]. Geminiviruses infect many economically important crops such as cassava, watermelon, squash, petunia, tobacco, pepper, potato, tomato, bean, soybean, cowpea, cotton, and others, leading to reduced crop yields worldwide [60,61]. Due to this reason, it is not surprising that most DNA virus mediated resistance studies have been applied to geminiviruses (Table 1). Ali et al. [62] used sgRNA molecules targeting coding (rep genes and coat proteins) and non-coding sequences (conserved intergenic region) of the *Tomato yellow leaf curl virus* (TYLCV) genome, that were delivered via *Tobacco rattle virus* (TRV) system into *Nicotiana benthamiana* plants expressing Cas9, causing a reduction of accumulation of viral DNA and reduction of symptoms in plants.

A subsequent study using CRISPR/Cas9 system with a sgRNA targeting a conserved region in multiple begomoviruses (CLCuKoV, TYLCV, TYLCSV, MeMV, BCTV-Worland and BCTV-Logan), simultaneously mediated interference and showed that the targeting of viral non-coding, intergenic sequences was more efficient, limiting the generation of recovered viral variants that evade CRISPR-mediated immunity by reverting the induced mutations through NHEJ [40]. Other studies have achieved plant viral resistance through the expression of sgRNAs complementary to sequences either within *Bean yellow dwarf virus* (BeYDV), *Wheat dwarf virus* (WDV) or *Beet severe curly top virus* (BSCTV) genomes, which reduced virus accumulation and symptoms in plants overexpressing Cas9 such as *N. benthamiana*, barley, and *A. thaliana* [41,63,64]. Similarly, CRISPR/Cas9 allowed to obtain resistance against banana streak disease by targeting endogenous *Banana streak virus* (eBSV) sequences [65].

Table 1. CRISPR/Cas for viral resistance in plants by targeting viral genome (virus mediated resistance) and host factors (plant mediated resistance).

Plant Species	Target Virus	Type of Resistance	Targeting Genomic Regions	Reference
N. benthamiana	BeYDV	DNA virus mediated	sgRNAs targeting LIR and rep/RepA	[63]
A. thaliana and *N. benthamiana*	BSCTV	DNA virus mediated	43 sgRNAs targeting BSCTV genome	[41]
N. benthamiana	TYLCV	DNA virus mediated	sgRNAs targeting Rep and CP	[62]
N. benthamiana	CLCuKoV TYLCV TYLCSV MeMV BCTV-Worland BCTV-Logan	DNA virus mediated	sgRNAs targeting non-coding IR, CP and Rep	[40]
A. thaliana	TuMV	Plant mediated	eIF(iso)4E knock out	[42]
Cucumis sativus	CVYV ZYMV PRSV	Plant mediated	eIF4E knock out	[49]
A. thaliana	CYVV	Plant mediated	eIF4E1	[51]
Oryza sativa	RTSV	Plant mediated	eIF4G knock out	[50]
A. thaliana	CaMV	DNA virus mediated	sgRNAs targeting CP	[38]
A. thaliana and *N. benthamiana*	CMV TMV	RNA virus mediated	22 sgRNAs targeting CMV genome and 3 sgRNAs targeting TMV genome	[13]
N. benthamiana	TuMV	RNA virus mediated	sgRNAs targeting HC-Pro and CP	[39]
Cassava	CBSV	Plant mediated	nCBP-1 and nCBP-2 (eIF4E family)	[52]
Barley	WDV	DNA virus mediated	sgRNAs targeting MP, CP, Rep(Rep A and LIR	[64]
N. benthamiana and *Oryza sativa*	TMV SRBDSV RSMV	RNA virus mediated	sgRNAs targeting 5 regions in TMV, 3 in SRBDSV and 3 in RSMV	[15]
Banana (*Gonja manjaya*)	eBSV	DNA virus mediated	sgRNAs targeting ORF1, ORF2 and ORF3	[65]
Potato (*Solanum tuberosum*)	PVY	RNA virus mediated	sgRNAs targeting P3, CI, NIb and CP	[66]

Plant resistance to a caulimovirus was achieved when Liu et al. [38] expressed multiple sgRNAs targeting the caulimovirus *Cauliflower mosaic virus* (CaMV) coat protein gene in Arabidopsis plants and 20 days after mechanical inoculation of the virus, 85–90% of the plants remained symptomless and showed no presence of CaMV.

Immunity against the RNA viruses *Cucumber mosaic virus* (CMV) and *Tobacco mosaic virus* (TMV) was achieved in *N. benthamiana* and *A. thaliana* transgenic plants expressing FnCas9 and a sgRNA complementary to viral genome delivered through a pCambia based vector [13]. Another study showed that *N. benthamiana* expressing Cas13a either transiently (using binary vector pK2WG$_7$) or constitutively, and expressing crRNAs complementary to different *Tulip mosaic virus* (TuMV) genomic regions, delivered through TRV system, interfered with viral replication and spread [39]. CRISPR/Cas13a (LshCas13a) system showed to target and degrade genomic RNA of TMV in *N. benthamiana* plants and to confer resistance to *Southern rice black-streaked dwarf virus* (SRBSDV) and *Rice stripe mosaic virus* (RSMV) in rice plants [15]. Zhan et al. [66] showed that transgenic potato lines expressing Cas13a/sgRNA constructs targeting conserved coding regions of different *Potato virus Y* (PVY) strains allowed to confer broad spectrum resistance against multiple PVY strains.

As stated above, many studies have shown the great versatility of the CRISPR technology towards plant virus resistance and have successfully shown the production of viral resistant plants. CRISPR has the potential to accelerate viral resistance breeding, since it is more effective and rapid than conventional breeding. In addition, CRISPR has the capacity to target virus directly and therefore to be applied to crops with limited genome sequence information.

There are also limitations of the use of CRISPR in virus plant resistance that must not be discarded. Knocking out essential host factors may always lead to the possibility of plant lethality or impaired growth [67,68]. Although many studies concerning mutations of host factors did not report any negative effects, the introduction of point mutations in host factor genes, instead of knocking out, should be considered, so that it does not interfere with plant growth but still prevents viral infection [69]. Another important limitation of CRISPR is the undesirable genomic modifications of plant genome, the off-targets. Although much less common to occur in plants than in other systems, off-target mutations may be avoided by the use of catalytically inactive Cas nucleases [70] or by using systems that only target RNA, which will be further destroyed by the plant silencing system.

CRISPR/Cas requires the optimal selection of sgRNA target sites to ensure that targeted viruses do not evolve mutations that escape from CRISPR/Cas cleavage, and that novel and more severe strains that cannot be cleaved again do not arise [40,71]. Additionally, multiplex targeting and targeting noncoding regions of viral genomes have shown to reduce viral mutation rates and minimize the formation of new viral strains capable of infection [40]. Also, CRISPR/Cas systems that target or bind RNA can be used together with Cas9 to reduce the RNA intermediates of DNA viruses, eliminating the viruses that may escape the CRISPR/Cas9 machinery [40]. FnCas9 has shown binding capacity to viral transcripts which probably provides even more durable resistance than nucleases that provide direct targeting [43].

There is still a long way to go concerning the full potential of CRISPR/Cas systems for engineering plant virus resistance, and more studies still need to be performed to improve their efficiency. However, it is clear that CRISPR is a milestone in plant virus resistance and the utilization of this technology in agriculture will certainly result in higher yields and quality of plants.

4. Delivery and Expression of CRISPR Systems in Plants

One crucial step in CRISPR for achieving a highly efficient genome engineering technology is the delivery and expression of CRISPR/Cas components within a plant cell [72], which greatly influences the editing efficiency.

If alien DNA is introduced in the host in a way that it gets incorporated into host genome (transgenic plants), a stable expression is provided and higher editing efficiencies may be obtained, but it is more likely that undesirable off-target mutations are originated [73]. On the other hand, if introduced DNA does not get incorporated into host genome and is expressed transiently, the host is considered free from the alien DNA or simply DNA-free.

Transient expression may be achieved by using ribonucleoproteins (RNP) or plasmids or other vectors delivered by agroinfiltration, carrying CRISPR/Cas components. Several studies have used CRISPR by expressing both Cas and sgRNA constitutively, both transiently or either Cas or sgRNA transiently and the other constitutively [72].

Transient expression of Cas endonuclease reduces off-target modifications, while maintaining a high expression of the sgRNAs that would be constitutively being expressed in the plant. However, this situation involves the use of two different plasmids (which would increase to three if a donor DNA was used for knock in). Transient expression of all CRISPR/Cas components (if no donor for DNA repair is used) can obtain DNA-free plants, avoiding the hurdles associated to transgenic plants.

Either way, it is desirable that CRISPR/Cas components are expressed in germline cells, which easily occurs in stable integration, as all cells in transgenic plants will express the CRISPR system, but which may not occur in transient expression. In this case, CRISPR/Cas components must be introduced directly into germline cells or be able to migrate to these cells, thus allowing mutations to be transmitted to the next generation of plants, without the need of tissue culture and all the labor and time consumption it implies.

Several methods have been used to introduce CRISPR/Cas components in plants, including Agrobacterium-mediated T-DNA transformation or physical means such as protoplast transfection and microprojectile bombardment. These methods rely on mediators such as plasmids, ribonucleoproteins or viruses to carry the sequences to be introduced.

Plant protoplasts can be obtained by digesting cell walls with enzymes and editing reagents, that can be delivered by electroporation or by polyethylene glycol (PEG) treatment. Transfection of CRISPR/Cas components into protoplasts with subsequent regeneration of plants allowed to successfully introduce mutations with editing efficiencies ranging from 3% to 46%, resulting in either stable or transient expression in several plants including rice, soybean, *A. thaliana*, potato, grapevine, wheat, and lettuce [74–81]. This method allowed the creation of DNA-free edited plants by delivering preassembled Cas9-sgRNA ribonucleoproteins (RNPs) [79,80,82], which cannot be delivered by Agrobacterium [83]. The delivery of Cas9-sgRNA RNPs instead of plasmids that encode Cas9-sgRNA avoids that plasmids are degraded in cells by nucleases, resulting in small DNA fragments that may undesirably be inserted in the host genome [84]. This method has the ability to deliver multiple components to a large number of transfectable cells and to obtain vector less or DNA-free plants, since regenerants are obtained from single genetically modified protoplasts. This is an important advantage as plants edited using transfection of protoplasts may not be subjected to the regulatory issues and ethical barriers associated to transgenic plants. However, if this technique is used for knock in, an exogenous DNA template is required and regulation may no longer be avoided. In addition, protoplast transfection is in many cases associated with problems with plant regeneration and presence of undesired somaclonal mutations.

Another method used to deliver CRISPR/Cas components in plants is biolistic bombardment. It consists of coating microprojectiles—generally gold, silver, or tungsten particles—with DNA constructions which are then fired into plant cells with high pressure to penetrate the cell wall. Biolistic bombardment has introduced targeted mutations into plants, by using gold particles to carry and deliver CRISPR/Cas9 reagents in plasmids, causing stable integration in rice, wheat and soybean genomes, with editing efficiencies ranging from 14.5% to 76% [31,85,86]. Other study achieved TECCDNA (transiently expressing CRISPR/Cas 9 DNA) in wheat with editing efficiency of 1–9.5% [35]. Edited plants, without alien DNA integration, were obtained by biolistic delivery of RNP in maize [87] and wheat [88] with editing efficiencies that range from 21.8% to 47%. A geminivirus *Wheat dwarf virus*-based vector, pWDV2, carrying both Cas9 and sgRNA was used for biolistic transformation in wheat, providing a 12-fold increase editing efficiency when compared to the delivery of this system by traditional vectors [81]. The use of viruses to deliver CRISPR/Cas components will be further discussed in this review. Biolistic bombardment is usually efficient, multiple constructs can be delivered simultaneously and it can be used for many plant species. The major disadvantage is that it leads to multiple copies of the introduced genes, with random integration within genomes, which can lead to phenomena such as gene suppression in the recovered transgenic plants. It is also more costly than other methods.

To date, the most common system used to obtain transgenic plants is based on *Agrobacterium tumefaciens*. This approach has been widely used to deliver CRISPR/Cas components into plant cells of a variety of plant species. Agrobacterium has the ability to transfer a piece of its genome (T-DNA) to the cell nucleus, where it randomly integrates the plant genome [89]. Cas9 and sgRNA expression cassettes can be easily cloned into Ti plasmid, transformed into Agrobacterium and then introduced into plants. Many studies have used *A. tumefaciens* to deliver CRISPR/Cas components into plant cells, providing the insertions of T-DNA and achieved stable integration of transgenes in the genomes of many plant species—such as sorghum, *A. thaliana*, rice, tomato, maize, grapevine, aspen, rapeseed, and watermelon—with editing efficiencies that ranged from 23% to 100% [36,75,90–94].

Agrobacterium may also be used for transient expression of Cas9/sgRNA (agroinfiltration) [95]. This has been achieved in citrus with editing efficiency of 20% [33]. In *N. benthamiana*, rice and *A. thaliana*, viral transient expression resulted in editing efficiencies reaching 85% [23,62,96]. The use of viruses to deliver CRISPR/Cas components will be further discussed in the following section.

Agrobacterium rhizogenes has also been used for genome editing, resulting in stable integration of foreign DNA in soybean and a few other plant species, with editing efficiencies that range from 14.7% to 95% [97–99]. *A. rhizogenes* indicates a successful editing event by the appearance of hairy roots, however it requires regeneration of whole plants from these roots, which can be problematic for some species.

Agrobacterium-mediated delivery presents several advantages, it requires technology available in most laboratories, it is cheap, it allows multiplex editing as multiple binary vectors can be delivered into Agrobacterium and co-transformed into plant cells. Additionally, it can be used in transient assays, which may result in a non-transgenic plant and in a lower number of edited off-target sites.

The Use of Viruses to Carry CRISPR Components

Many viruses, including retroviruses, adenoviruses and adeno-associated virus, have already shown to achieve effective delivery of genome-engineering reagents in mammalian systems [100,101].

In plants, *Tobacco mosaic virus* (TMV) was the first virus to be manipulated as vector, resulting in virus-induced gene silencing (VIGS) of an endogenous gene in *N. benthamiana* [102]. Since then, many other viruses have been widely used as vectors of gene silencing and for expression of foreign proteins in plants. However, their specific use to deliver genetic material such as CRISPR/Cas components in plants is much more recent. The first reports of the use of viruses to assist CRISPR/Cas gene editing, were in 2014 and were based on geminiviruses [103]. Since then, studies have been focused not only on the use of the DNA geminiviruses [23,81,96,104,105] but also on RNA viruses [40,62,106–111] as sgRNA delivery systems.

The numerous studies on the use of geminiviruses as vectors, result mostly from their easy manipulation. Geminiviruses (family *Geminiviridae*) are widespread, insect-transmitted and infect a wide range of plants [60,112]. Geminiviruses have a single stranded circular DNA with monopartite or bipartite genomes that range between 2.5 kb to 3 kb, with four to six open reading frames (ORFs). Once inside a plant cell, their single stranded genome forms a double stranded intermediate which is then used as template for transcription and for rolling-circle replication. They require only one replication initiator protein, Rep (C1), to initiate rolling-circle replication inside the host. Following replication, single stranded genomes are either converted to double stranded intermediates to initiate another replication cycle, or encapsidated by the coat protein to produce virions which then move to adjacent cells through plasmodesmata. Their small sizes mean they are easy to manipulate but on the other hand, it physically limits their cargo capacity; as so, they are unable to carry long DNA fragments, such as genes encoding Cas nucleases (~4.2 kb) [113].

To retain most of the features required for movement and replication, the CP of some bipartite begomoviruses may be replaced by the desired heterologous sequence of up to

800 bp or up to 1000 bp with further modifications [96,103,114]. However, with this change, geminiviruses are still unable to carry long DNA fragments such as genes encoding Cas nucleases, but it is enough to express and produce high amounts of sgRNA. In fact, the number of double stranded intermediates during viral replication is higher in the absence of the CP, possibly because the CP sequesters and packages ssDNA to form viral particles.

To increase cargo capacity, geminiviruses have been manipulated into non-infectious replicons (GVRs) by removing movement protein (MP) and coat protein (CP) coding sequences, and thereby eliminating cell to cell movement and insect transmission. In these cases, viral vectors are not infectious on their own and must be delivered into plant cells using Agrobacterium mediated transformation, in contrast to the possibility of agroinfiltration or mechanical inoculation for virus-induced gene editing (VIGE). These deconstructed DNA replicons have been used to introduce large amounts of repair templates in plants, which are required for HDR to outcompete NHEJ, showing high efficiency of HDR in plants.

Several studies have shown the use of geminiviruses to assist CRISPR/Cas (Table 2). Baltes et al. [103] used *Bean yellow dwarf virus* (BeYDV) replicons to efficiently deliver a sequence-specific nuclease (Cas9) and a repair template to tobacco plants for gene targeting, showing a considerable cargo capacity and with gene targeting frequencies with two orders of magnitude increase over conventional Agrobacterium T-DNA transformation. The use of BeYDV replicons also allowed genome editing in potato, by causing mutations capable of supporting a reduced herbicide susceptibility phenotype, while Agrobacterium T-DNA transformation held no detectable mutations for the same phenotype [104]. Cermark et al. [105] used BeYDV replicons to insert a strong promotor upstream of a tomato (*Solanum lycopersicum*) gene that regulates anthocyanin synthesis (ANT1) and obtained efficiencies 12-fold higher than traditional Agrobacterium T-DNA delivery. Similar efficiencies were obtained by Yin et al. [96] who used *Cabbage leaf curl virus* (CaLCuv) for VIGE by replacing viral CP by sgRNA, to edit different genes (NbPDS3 and NblspH) in *N. benthamiana* plants. VIGE makes use of Cas9 overexpression in plants and transient delivery of geminivirus vectors carrying sgRNAs and can be used as an alternative to VIGS.

In 2017, *Wheat dwarf virus* (WDV) replicons were used for gene targeting in wheat and rice [23,81]. WDV replicons showed high gene targeting efficiency and allowed to target multiple genes within the same cell [81]. Using this WDV-based system, Wang et al. [23] showed efficient HDR in rice.

In addition to geminiviruses, many RNA viruses have been used as vectors in plants (Table 2).

RNA virus-based vectors have the advantage of not integrating plant genome accidentally, so resulting in DNA-free plants, which avoids raising additional regulatory and ethical issues.

One of such virus-based vector, also widely used for VIGS, is *Tobacco rattle virus* (TRV) [115]. TRV belongs to genus *Tobravirus*, family *Virgaviridae*; it infects over 400 plant species and is transmitted by nematodes of the family *Trichodoridae*. It has a bipartite genome with two positive sense single stranded RNAs, RNA1 (TRV1), and RNA2 (TRV2). TRV1 is essential for virus replication and movement and TRV2 genome has genes encoding the CP and nonstructural proteins involved in nematode transmission. For its use as vector, these non-structural proteins in TRV2 can be replaced for the fragments of interest [116].

The first application of TRV as vector for genome engineering was in a non-transgenic approach for zinc-finger nucleases (ZFN) delivery in plants, by replacing RNA2 with RNA for the Zif268: FokI ZFN. In this system, targeted genome modifications were recovered at an integrated reporter gene in somatic tobacco and petunia cells, and transmission of mutations to next generation confirmed the stability of the ZFN induced changes [117].

The first use of TRV as a vector for CRISPR was in 2015, when TRV was developed as a vehicle for delivery of sgRNAs to modify genomes of *N. benthamiana* and *A. thaliana* [115]. A TRV vector containing sgRNA for phytoene desaturase gene (PDS) was introduced into leaves of *N. benthamiana* transgenic lines overexpressing Cas9, via agroinfection, which showed modification of the PDS gene [115]. In addition, TRV showed the ability to infect germline cells, as TRV-mediated delivery of sgRNA was not limited to infiltrated plants, allowing to successfully recover the desired modification in the next generation [115]. TRV can carry DNA fragments up to 3000 bp, however it is still not enough for the Cas gene, having been used only for sgRNA delivery into transgenic plants stably expressing Cas nuclease, thereby requiring that all genome edited plants are transgenic.

TMV, as mentioned previously, was the first virus to be manipulated as vector in plants, and has shown high level of accumulation and gene expression in several hosts, as well as prolonged integrity of its derived gene vectors [107,118]. Based on this potential, TMV was also developed as a vehicle for delivering sgRNA by partially substituting the CP with a sgRNA [107]. TMV showed to mediate target gene editing by showing the ability to deliver high concentrations of sgRNA and to efficient edit the target host gene in *N. benthamiana* plants, that was previously infiltrated with a plasmid expressing Cas9 [107].

Ali et al. [106] demonstrated that *Pea early browning virus* (PEBV) was able to deliver sgRNAs, resulting in mutagenesis of the targeted genomic loci in *N. benthamiana* plants, constitutively overexpressing the Cas9, in a more efficient way than TRV. In addition, like TRV, PEBV can infect meristematic tissues [119] which may allow the recovery of seeds with the desired mutations and obviate the need for tissue culture to generate heritable targeted mutations. *Barley stripe mosaic virus* (BSMV) has also been engineered as a sgRNA delivery system for CRISPR/Cas9 mediated targeted mutagenesis in wheat and maize, both transformed constitutively with Cas9 [108]. Recently, *Beet necrotic yellow vein virus* (BNYVV)-based vectors were designed to allow simultaneous expression of multiple foreign proteins and used for efficient sgRNA delivery for genome editing in transgenic *N. benthamiana* plants expressing Cas9 [109].

Foxtail mosaic virus (FoMV) has also showed to express sgRNAs in *N. benthamiana*, *Setaria viridis* and maize plants constitutively expressing Cas9, demonstrating that FoMV can enable gene editing [110].

All these previous attempts using plant RNA viruses for expression of sgRNA were able to express sgRNAs and introduce mutations into plant genomes that were overexpressing Cas9.

Until recently, there were no reports of delivery of the entire CRISPR/Cas system into plants through viral vectors due to their small capacity for carrying DNA/RNA fragments [120]. This was overcome when technical breakthroughs in delivering all CRISPR/Cas components into plant cells using negative-strand viruses were reported [121,122]. The negative-strand viruses, *Barley yellow striate mosaic virus* (BYSMV) and *Sonchus yellow net rhabdovirus* (SYNV), were used to successfully deliver CRISPR/Cas reagents and sgRNAs into plant cells. Ma et al. [122] showed that SYNV was able to knock out different genes in plants, achieving highly efficient DNA-free genome editing. This study also showed the multiplex editing ability of virus-delivered CRISPR/Cas9 system by designing sgRNAs for different genes without affecting the efficiency, and confirmed that genome-edited plants pass the genome alteration to subsequent generations. However, rhabdoviruses rarely infect germline cells, and SYNV mediated genome editing only works efficiently in somatic cells being plant tissue culture required to obtain an individual genome edited plant.

Table 2. Viruses used to carry CRISPR sequences into plants and type of delivery.

Virus Type	Virus Family/ Genus	Virus Vector	Type of Delivery	Plant Species	Reference
ssDNA	*Geminiviridae/ Mastrevirus*	BeYDV	Agrobacterium	*Nicotiana tabacum*	[103]
ssDNA	*Geminiviridae/ Mastrevirus*	BeYDV	Agrobacterium	Potato (*Solanum tuberosum*)	[104]
ssDNA	*Geminiviridae/ Begomovirus*	CaLCuV	Agrobacterium	*N. benthamiana*	[96]
ssDNA	*Geminiviridae/ Mastrevirus*	BeYDV	Agrobacterium	Tomato (*Solanum lycopersicum*)	[105]
ssDNA	*Geminiviridae/ Mastrevirus*	WDV	Protoplasts transfection	Wheat (*Triticum aestivum*)	[81]
ssDNA	*Geminiviridae/ Mastrevirus*	WDV	Agrobacterium	Rice (O. sativa)	[23]
+ ssRNA	*Virgaviridae/ Tobravirus*	TRV	Agrobacterium	*N. benthamiana; A. thaliana*	[40,106,115]
+ ssRNA	*Virgaviridae/ Tobamovirus*	TMV	Agrobacterium	*N. benthamiana*	[107]
+ ssRNA	*Virgaviridae/ Tobravirus*	PEBV	Agrobacterium	*N. benthamiana; A. thaliana*	[106]
+ ssRNA	*Virgaviridae/ Hordeivirus*	BSMV	Agrobacterium	*N. benthamiana;* Wheat (*Triticum aestivum*)	[108]
+ ssRNA	*Benyviridae/ Benyvirus*	BNYVV	Agrobacterium	*N. benthamiana*	[109]
+ ssRNA	*Alphaflexiviridae/ Potexvirus*	FoMV	Agrobacterium	Maize (*Zea mays*), Foxtail (*Setaria viridis*), and N. benthamiana	[110]
- ssRNA	*Rhabdoviridae/ Cytorhabdovirus*	BYSMV	Agrobacterium	*N. benthamiana*	[121]
+ ssRNA	*Alphaflexiviridae/ Potexvirus*	PVX	Agrobacterium Mechanical inoculation	*N. benthamiana*	[111]
- ssRNA	*Rhabdoviridae/ Betanucleorhabdovirus*	SYNV	Agrobacterium	*N. benthamiana*	[122]

More recently, *Potato virus X* (PVX) has also been used to efficiently deliver both Cas9 and sgRNA into *N. benthamiana* plants [111]. PVX has a filamentous flexible structure with a 6345 nt (+) ssRNA, and each particle contains ~1350 coat protein subunits [123]. In opposition to what happens to small viruses, it is not likely that gene insert size is physically limited in PVX. Cas9 and sgRNA were placed between Triple Gene Block (movement proteins MP1, MP2, and MP3) and the CP of PVX and virus vector was both agroinfiltrated and mechanically inoculated in *N. benthamiana* plants. PVX-Cas9 RNA showed to infect most cells and express a large amount of Cas9 protein, while T-DNA integration into *N. benthamiana* genome occurred at low frequency. In addition, the mutation introduced was inherited by the next generation, but no PVX RNA was detected in these plants, showing that PVX was not transmitted through seed, leading to the suggestion that transgenerational transmission of PVX is unlikely to occur, resulting in DNA-free genome edited plants [111]. The possibility of such as simple and efficient virus-vector mediated delivery as the mechanical inoculation of a virus carrying the entire CRISPR/Cas system greatly facilitates transgene free gene editing in plants.

5. Challenges in the Use of Viruses for CRISPR

Virus mediated delivery of CRISPR/Cas is an easy way to deliver Cas nuclease and sgRNAs into plants, that overcome many challenges of transgene delivery, with no additional requirements, allowing to edit a desired feature into a plant, in laboratory or in the field, to obtain an improved DNA-free plant. They present several advantages such as they are easy to manipulate; viral genome can be used as repair template; they replicate to high copy number and accumulate at high levels (including sgRNAs and repair template) and systemically spread in a large number of plants leading high level expression and genome editing efficiency; multiple sgRNAs can be expressed from a single viral genome, allowing multi targeted genome editing; VIGE phenotypic alterations appear in plants in a relatively short time. In fact, VIGE is a promising tool for transgene integration-free genome editing, as it may not require the production of transgenic lines or simplify this operation, which is often laborious and time consuming, expensive, and raises public concerns and extra regulations [124,125].

In addition, some viruses have shown the capacity of invading meristems when used as CRISPR/Cas vectors, by systemically deliver sgRNAs and therefore enabling the recovery of progeny carrying the targeted genomic modification, overcoming the need of tissue culture—i.e., start from leaf tissue and regenerate the whole plant and then genotype for the presence of the modification [62,106], and opens new possibilities for producing plants with desired characteristics without the need of laborious and time consuming steps. Therefore, as a vector for genome engineering, it is highly desirable that viral vector infects germline cells, so that it will be possible to harvest mutant seeds from infected plants.

VIGE, especially RNA-based, may also contribute to decrease off-target activities, a major issue in CRISPR that occurs due to sgRNA mismatches and continuous expression of Cas nucleases, that result of editing unintended sites in the genome [126]. When viruses are used to express CRISPR/Cas systems, these will only be expressed when viruses invade plant cells, limiting the concentration of Cas and thus more likely that no off-target effect is detected [127].

Despite all these advantages, the limited cargo capacity that many viruses present (typically <1 kb) is a major drawback for their use for delivery of all gene editing reagents such as Cas9 (approx. 4.2 kb), as excess cargo results in the loss of systemic movement or loss of the cargo DNA [128].

For this reason, viruses have been developed to deliver sgRNAs to transgenic plants expressing Cas9 or have been deconstructed into non-infectious replicons or, more recently, a negative sense RNA virus and PVX showed to be able to carry the entire CRISPR/Cas system. All these studies show the huge possibilities and great potential of the use of plant viruses as vectors to efficiently target and deliver CRISPR/Cas reagents.

Further research may result in new discoveries that may allow positive-strand RNA or DNA viruses to be engineered to carry large DNA/RNA sequences without affecting their infectivity and with even greater editing efficiencies.

6. Concluding Remarks and Future Prospects for CRISPR in Agriculture

CRISPR/Cas technology has definitely simplified gene engineering showing great potential on improving several traits in plants, not only on the development of resistance to viral pathogens, but also to fungi, bacteria and insects, as well as tolerance to abiotic stresses and increase in yield [129–131] overcoming many difficulties of the techniques used until now [13–15]. This innovative technology at the disposal of plant breeding holds promise for protecting crops against abiotic and biotic stresses, so that farmers can meet consumers expectations for healthful and affordable products obtained by using few natural resources.

There are still technological improvements needed, such as precise editing and strategies to bypass the need for tissue culture. When using genome editing strategies, the possibility of editing unintended sites in the genome, off-targets, can never be ignored. As mentioned before, viruses as vectors of CRISPR systems may be used to decrease these collateral effects. In addition, a CRISPR/Cas technology in which a single nucleotide is chemically modified instead of producing DSB may also be widely used to prevent off-target effects [26].

Another constraint of the implementation of CRISPR as a plant breeding technique, is the difficulty to obtain new edited plants without tissue culture. Regeneration of plants through tissue culture is a time-consuming process, and there is the possibility of producing random somatic mutations. In addition, some crops are recalcitrant to regeneration through tissue culture. Delivery of CRISPR components in plant apical meristems so that seeds harvested will carry the mutations is desirable and already showed to be possible. However, many crop plants will lose valuable traits when propagated by seed.

Besides the technical and scientific aspects that must be overcome, CRISPR will also have to deal with social and political aspects such as the public concerns and government regulations mostly associated with transgenic plants. It is essential to provide clear information on CRISPR to the public and government to gain their acceptance and to influence regulatory policies on the use of CRISPR technologies in agriculture. The first clarification that must be done is that CRISPR may be applied to rapidly produce plants with traits that might easily also result from conventional plant breeding, as deletions and small insertions may also occur naturally or be induced during conventional plant breeding; or, in alternative, it can be used to introduce exogenous genes in plants and, only so, it would be equated with genetically modified organisms (GMO).

Plants subjected to CRISPR/Cas have gained extreme attention in terms of regulation. The United States Department of Agriculture (USDA) has recently regulated genome edited plants as safe for human consumption and the environment, as long as the resulting mutations are indistinguishable from mutations that occur naturally or by traditional breeding techniques. USDA has considered genome editing as an expansion of traditional plant breeding that can introduce new traits in plants more quickly and precisely, saving years or decades to bring needed new varieties to farmers, which is a great advance in the application of CRISPR in agriculture (Code of Federal Regulations, Vol. 7, part 340). This view has been adopted by most of the world, with the exception of the European Union, where, in 2018, the European Court of Justice (ECJ) ruled that genome edited organisms are GMOs until clarification of their legal status and, as so, are at present, subjected to the same obligations as transgenic organisms (Judgement in case C-528/16) and therefore fall under the European GMO Directive (2001/18/EC). The European Commission is currently carrying out a study on the potential of new genomic techniques that may play a role in sustainability, provided that resulting products they are safe for consumers and environment, as stated on the communication of 'A Farm and Fork Strategy for a fair, healthy and environmentally-friendly food system' (COM/2020/381), which is expected to be concluded in April 2021, and a different perception may be achieved. However, the current regulation is a clear obstacle to European agricultural innovation as greatly makes it difficult for genome-edited products to reach the market and has a huge impact in terms of competitivity with other countries with less restrictions.

CRISPR is a powerful plant breeding tool, which can contribute to provide food security to the ever-growing world population and to a sustainable agriculture, and discussions concerning the risks associated with genome editing should be driven more by scientific principles than by socio-political factors.

Author Contributions: Conceptualization, C.M.R.V. and P.M.; Resources, C.M.R.V., M.d.R.F., and P.M.; Writing—original draft preparation, C.M.R.V. and P.M.; Writing—review and editing, C.M.R.V., P.M., M.d.R.F., M.D.C., and M.P.; Funding acquisition, C.M.R.V. and P.M. All authors have read and agreed to the published version of the manuscript.

Funding: This work was funded by the projects "Control of olive anthracnose through gene silencing and gene expression using a plant virus vector" with the references ALT20-03-0145-FEDER-028263 and PTDC/ASP-PLA/28263/2017 and "Development of a new virus-based vector to control TSWV in tomato plants" with the references ALT20-03-0145-FEDER-028266 and PTDC/ASP-PLA/28266/2017, both co-financed by the European Union through the European Regional Development Fund, under the ALENTEJO 2020 (Regional Operational Program of the Alentejo), ALGARVE 2020 (Regional Operational Program of the Algarve) and through the Foundation for Science and Technology, in its national component. M.P. was supported by the FCT research grant SFRH/BD/145321/2019. This work was also funded by the National Funds through FCT—Foundation for Science and Technology under project no. UIDB/05183/2020.

Institutional Review Board Statement: Not applicable.

Informed Consent Statement: Not applicable.

Conflicts of Interest: The authors declare no conflict of interest.

References

1. Anderson, P.K.; Cunningham, A.A.; Patel, N.G.; Morales, F.J.; Epstein, P.R.; Daszak, P. Emerging infectious diseases of plants: Pathogen pollution, climate change and agrotechnology drivers. *Trends Ecol. Evol.* **2004**, *19*, 535–544. [CrossRef] [PubMed]
2. Tilman, D.; Balzer, C.; Hill, J.; Befort, B.L. Global food demand and the sustainable intensification of agriculture. *Proc. Natl. Acad. Sci. USA* **2011**, *108*, 20260–20264. [CrossRef] [PubMed]
3. Jones, R.A.C. Control of Plant Virus Diseases. *Adv. Virus Res.* **2006**, *67*, 205–244. [CrossRef] [PubMed]
4. Boualem, A.; Dogimont, C.; Bendahmane, A. The battle for survival between viruses and their host plants. *Curr. Opin. Virol.* **2016**, *17*, 32–38. [CrossRef] [PubMed]
5. Meziadi, C.; Blanchet, S.; Geffroy, V.; Pflieger, S. Genetic resistance against viruses in *Phaseolus vulgaris* L.: State of the art and future prospects. *Plant Sci.* **2017**, *265*, 39–50. [CrossRef]
6. Varanda, C.M.R.; Nolasco, G.; Clara, M.I.; Félix, M.R. Genetic diversity of the coat protein of olive latent virus 1 isolates. *Arch. Virol.* **2014**, *159*, 1351–1357. [CrossRef]
7. Varanda, C.M.R.; Machado, M.; Martel, P.; Nolasco, G.; Clara, M.I.E.; Félix, M.R. Genetic diversity of the coat protein of olive mild mosaic virus (OMMV) and tobacco necrosis virus D (TNV-D) isolates and its structural implications. *PLoS ONE* **2014**, *9*. [CrossRef]
8. Félix, M.R.; Cardoso, J.M.S.; Varanda, C.M.R.; Oliveira, S.; Clara, M.I.E. Complete nucleotide sequence of an Olive latent virus 1 isolate from olive trees. *Arch. Virol.* **2005**, *150*, 2403–2406. [CrossRef]
9. Gómez, P.; Rodríguez-Hernández, A.M.; Moury, B.; Aranda, M. Genetic resistance for the sustainable control of plant virus diseases: Breeding, mechanisms and durability. *Eur. J. Plant Pathol.* **2009**, *125*, 1–22. [CrossRef]
10. Galvez, L.C.; Banerjee, J.; Pinar, H.; Mitra, A. Engineered plant virus resistance. *Plant Sci.* **2014**, *228*, 11–25. [CrossRef]
11. Duan, C.G.; Wang, C.H.; Guo, H.S. Application of RNA silencing to plant disease resistance. *Silence* **2012**, *3*, 1–8. [CrossRef] [PubMed]
12. de Ronde, D.; Butterbach, P.; Kormelink, R. Dominant resistance against plant viruses. *Front. Plant Sci.* **2014**, *5*, 1–17. [CrossRef] [PubMed]
13. Zhang, T.; Zheng, Q.; Yi, X.; An, H.; Zhao, Y.; Ma, S.; Zhou, G. Establishing RNA virus resistance in plants by harnessing CRISPR immune system. *Plant Biotechnol. J.* **2018**, *16*, 1415–1423. [CrossRef] [PubMed]
14. Borrelli, V.M.G.; Brambilla, V.; Rogowsky, P.; Marocco, A.; Lanubile, A. The enhancement of plant disease resistance using crispr/cas9 technology. *Front. Plant Sci.* **2018**, *9*. [CrossRef] [PubMed]
15. Zhang, T.; Zhao, Y.; Ye, J.; Cao, X.; Xu, C.; Chen, B.; An, H.; Jiao, Y.; Zhang, F.; Yang, X.; et al. Establishing CRISPR/Cas13a immune system conferring RNA virus resistance in both dicot and monocot plants. *Plant Biotechnol. J.* **2019**, *17*, 1185–1187. [CrossRef]
16. Rybicki, E.P. Plant-produced vaccines: Promise and reality. *Drug Discov. Today* **2009**, *14*, 16–24. [CrossRef]
17. Bruckman, M.A.; Czapar, A.E.; VanMeter, A.; Randolph, L.N.; Steinmetz, N.F. Tobacco mosaic virus-based protein nanoparticles and nanorods for chemotherapy delivery targeting breast cancer. *J. Control. Release* **2016**, *231*, 103–113. [CrossRef]
18. Jansen, R.; Van Embden, J.D.A.; Gaastra, W.; Schouls, L.M. Identification of genes that are associated with DNA repeats in prokaryotes. *Mol. Microbiol.* **2002**, *43*, 1565–1575. [CrossRef]
19. Wright, A.V.; Nuñez, J.K.; Doudna, J.A. Biology and Applications of CRISPR Systems: Harnessing Nature's Toolbox for Genome Engineering. *Cell* **2016**, *164*, 29–44. [CrossRef]
20. Shmakov, S.; Smargon, A.; Scott, D.; Cox, D.; Pyzocha, N.; Yan, W.; Abudayyeh, O.O.; Gootenberg, J.S.; Makarova, K.S.; Wolf, Y.I.; et al. Diversity and evolution of class 2 CRISPR-Cas systems. *Nat. Rev. Microbiol.* **2017**, *15*, 169–182. [CrossRef]
21. Rath, D.; Amlinger, L.; Rath, A.; Lundgren, M. The CRISPR-Cas immune system: Biology, mechanisms and applications. *Biochimie* **2015**, *117*, 119–128. [CrossRef] [PubMed]
22. Jinek, M.; Chylinski, K.; Fonfara, I.; Hauer, M.; Doudna, J.A.; Charpentier, E. A Programmable Dual-RNA-dual-RNA-guided DNA endonuclease in adaptive bacterial immunity. *Science* **2012**, *337*, 816–821. [CrossRef] [PubMed]

23. Wang, M.; Lu, Y.; Botella, J.R.; Mao, Y.; Hua, K.; Zhu, J. Gene Targeting by Homology-Directed Repair in Rice Using a Geminivirus-Based CRISPR/Cas9 System. *Mol. Plant* **2017**, *10*, 1007–1010. [CrossRef] [PubMed]
24. Mali, P.; Yang, L.; Esvelt, K.M.; Aach, J.; Guell, M.; DiCarlo, J.E.; Norville, J.E.; Church, G.M. RNA-guided human genome engineering via Cas9. *Science* **2013**, *339*, 823–826. [CrossRef]
25. Yang, H.; Ren, S.; Yu, S.; Pan, H.; Li, T.; Ge, S.; Zhang, J.; Xia, N. Methods favoring homology-directed repair choice in response to crispr/cas9 induced-double strand breaks. *Int. J. Mol. Sci.* **2020**, *21*, 6461. [CrossRef]
26. Gaudelli, N.M.; Komor, A.C.; Rees, H.A.; Packer, M.S.; Badran, A.H.; Bryson, D.I.; Liu, D.R. Programmable base editing of A•T to G•C in genomic DNA without DNA cleavage. *Nature* **2018**, *551*, 464–471. [CrossRef]
27. Abudayyeh, O.O.; Gootenberg, J.S.; Konermann, S.; Joung, J.; Slaymaker, I.M.; Cox, D.B.T.; Shmakov, S.; Makarova, K.S.; Semenova, E.; Minakhin, L.; et al. C2c2 is a single-component programmable RNA-guided RNA-targeting CRISPR effector. *Science* **2016**, *353*, aaf5573. [CrossRef]
28. Gao, C. The future of CRISPR technologies in agriculture. *Nat. Rev. Mol. Cell Biol.* **2018**, *19*, 275–276. [CrossRef]
29. Gootenberg, J.S.; Abudayyeh, O.O.; Lee, J.W.; Essletzbichler, P.; Dy, A.J.; Joung, J.; Verdine, V.; Donghia, N.; Daringer, N.M.; Freije, C.A.; et al. Nucleic acid detection with CRISPR-Cas13a/C2c2. *Science* **2017**, *356*, 438–442. [CrossRef]
30. Khan, M.Z.; Amin, I.; Hameed, A.; Mansoor, S. CRISPR–Cas13a: Prospects for Plant Virus Resistance. *Trends Biotechnol.* **2018**, *36*, 1207–1210. [CrossRef]
31. Shan, Q.; Wang, Y.; Li, J.; Gao, C. Genome editing in rice and wheat using the CRISPR/Cas system. *Nat. Protoc.* **2014**, *9*, 2395–2410. [CrossRef] [PubMed]
32. Brooks, C.; Nekrasov, V.; Lipppman, Z.B.; Van Eck, J. Efficient gene editing in tomato in the first generation using the clustered regularly interspaced short palindromic repeats/CRISPR-associated9 system. *Plant Physiol.* **2014**, *166*, 1292–1297. [CrossRef] [PubMed]
33. Jia, H.; Nian, W. Targeted genome editing of sweet orange using Cas9/sgRNA. *PLoS ONE* **2014**, *9*. [CrossRef] [PubMed]
34. Peng, A.; Chen, S.; Lei, T.; Xu, L.; He, Y.; Wu, L.; Yao, L.; Zou, X. Engineering canker-resistant plants through CRISPR/Cas9-targeted editing of the susceptibility gene CsLOB1 promoter in citrus. *Plant Biotechnol. J.* **2017**, *15*, 1509–1519. [CrossRef] [PubMed]
35. Zhang, Y.; Liang, Z.; Zong, Y.; Wang, Y.; Liu, J.; Chen, K.; Qiu, J.L.; Gao, C. Efficient and transgene-free genome editing in wheat through transient expression of CRISPR/Cas9 DNA or RNA. *Nat. Commun.* **2016**, *7*, 12617. [CrossRef]
36. Char, S.N.; Neelakandan, A.K.; Nahampun, H.; Frame, B.; Main, M.; Spalding, M.H.; Becraft, P.W.; Meyers, B.C.; Walbot, V.; Wang, K.; et al. An Agrobacterium-delivered CRISPR/Cas9 system for high-frequency targeted mutagenesis in maize. *Plant Biotechnol. J.* **2017**, *15*, 257–268. [CrossRef] [PubMed]
37. Shi, J.; Gao, H.; Wang, H.; Lafitte, H.R.; Archibald, R.L.; Yang, M.; Hakimi, S.M.; Mo, H.; Habben, J.E. ARGOS8 variants generated by CRISPR-Cas9 improve maize grain yield under field drought stress conditions. *Plant Biotechnol. J.* **2017**, *15*, 207–216. [CrossRef]
38. Liu, H.; Soyars, C.L.; Li, J.; Fei, Q.; He, G.; Peterson, B.A.; Meyers, B.C.; Nimchuk, Z.L.; Wang, X. CRISPR/Cas9-mediated resistance to cauliflower mosaic virus. *Plant Direct* **2018**, *2*, e00047. [CrossRef]
39. Aman, R.; Ali, Z.; Butt, H.; Mahas, A.; Aljedaani, F.; Khan, M.Z.; Ding, S.; Mahfouz, M. RNA virus interference via CRISPR/Cas13a system in plants. *Genome Biol.* **2018**, *19*, 1. [CrossRef]
40. Ali, Z.; Ali, S.; Tashkandi, M.; Zaidi, S.S.E.A.; Mahfouz, M.M. CRISPR/Cas9-Mediated Immunity to Geminiviruses: Differential Interference and Evasion. *Sci. Rep.* **2016**, *6*, 26912. [CrossRef]
41. Ji, X.; Zhang, H.; Zhang, Y.; Wang, Y.; Gao, C. Establishing a CRISPR–Cas-like immune system conferring DNA virus resistance in plants. *Nat. Plants* **2015**, *1*, 15144. [CrossRef] [PubMed]
42. Pyott, D.E.; Sheehan, E.; Molnar, A. Engineering of CRISPR/Cas9-mediated potyvirus resistance in transgene-free Arabidopsis plants. *Mol. Plant Pathol.* **2016**, *17*, 1276–1288. [CrossRef] [PubMed]
43. Zaidi, S.S.E.A.; Tashkandi, M.; Mansoor, S.; Mahfouz, M.M. Engineering plant immunity: Using CRISPR/Cas9 to generate virus resistance. *Front. Plant Sci.* **2016**, *7*, 1673. [CrossRef] [PubMed]
44. Hull, R. *Matthews' Plant Virology*, 4th ed.; Academic Press: New York, NY, USA, 2001.
45. Hashimoto, M.; Neriya, Y.; Yamaji, Y.; Namba, S. Recessive resistance to plant viruses: Potential resistance genes beyond translation initiation factors. *Front. Microbiol.* **2016**, *7*, 1695. [CrossRef] [PubMed]
46. Wittmann, S.; Chatel, H.; Fortin, M.G.; Laliberté, J.F. Interaction of the viral protein genome linked of turnip mosaic potyvirus with the translational eukaryotic initiation factor (iso) 4E of Arabidopsis thaliana using the yeast two-hybrid system. *Virology* **1997**, *234*, 84–92. [CrossRef] [PubMed]
47. Cillo, F.; Palukaitis, P. *Transgenic Resistance*; Academic Press: New York, NY, USA, 2014; Volume 90, ISBN 9780128012468.
48. Sanfaçon, H. Plant translation factors and virus resistance. *Viruses* **2015**, *7*, 3392–3419. [CrossRef]
49. Chandrasekaran, J.; Brumin, M.; Wolf, D.; Leibman, D.; Klap, C.; Pearlsman, M.; Sherman, A.; Arazi, T.; Gal-On, A. Development of broad virus resistance in non-transgenic cucumber using CRISPR/Cas9 technology. *Mol. Plant Pathol.* **2016**, *17*, 1140–1153. [CrossRef]
50. Macovei, A.; Sevilla, N.R.; Cantos, C.; Jonson, G.B.; Slamet-Loedin, I.; Čermák, T.; Voytas, D.F.; Choi, I.R.; Chadha-Mohanty, P. Novel alleles of rice eIF4G generated by CRISPR/Cas9-targeted mutagenesis confer resistance to Rice tungro spherical virus. *Plant Biotechnol. J.* **2018**, *16*, 1918–1927. [CrossRef]

51. Bastet, A.; Lederer, B.; Giovinazzo, N.; Arnoux, X.; German-Retana, S.; Reinbold, C.; Brault, V.; Garcia, D.; Djennane, S.; Gersch, S.; et al. Trans-species synthetic gene design allows resistance pyramiding and broad-spectrum engineering of virus resistance in plants. *Plant Biotechnol. J.* **2018**, *16*, 1569–1581. [CrossRef]

52. Gomez, M.A.; Lin, Z.D.; Moll, T.; Chauhan, R.D.; Hayden, L.; Renninger, K.; Beyene, G.; Taylor, N.J.; Carrington, J.C.; Staskawicz, B.J.; et al. Simultaneous CRISPR/Cas9-mediated editing of cassava eIF4E isoforms nCBP-1 and nCBP-2 reduces cassava brown streak disease symptom severity and incidence. *Plant Biotechnol. J.* **2019**, *17*, 421–434. [CrossRef]

53. Cox, D.B.T.; Gootenberg, J.S.; Abudayyeh, O.O.; Franklin, B.; Kellner, M.J.; Joung, J.; Zhang, F. RNA editing with CRISPR-Cas13. *Science* **2017**, *358*, 1019–1027. [CrossRef] [PubMed]

54. Yan, W.X.; Chong, S.; Zhang, H.; Makarova, K.S.; Koonin, E.V.; Cheng, D.R.; Scott, D.A. Cas13d Is a Compact RNA-Targeting Type VI CRISPR Effector Positively Modulated by a WYL-Domain-Containing Accessory Protein. *Mol. Cell* **2018**, *70*, 327–339.e5. [CrossRef]

55. O'Connell, M.R.; Oakes, B.L.; Sternberg, S.H.; East-Seletsky, A.; Kaplan, M.; Doudna, J.A. Programmable RNA recognition and cleavage by CRISPR/Cas9. *Nature* **2014**, *516*, 263–266. [CrossRef] [PubMed]

56. Price, A.A.; Sampson, T.R.; Ratner, H.K.; Grakoui, A.; Weiss, D.S. Cas9-mediated targeting of viral RNA in eukaryotic cells. *Proc. Natl. Acad. Sci. USA* **2015**, *112*, 6164–6169. [CrossRef] [PubMed]

57. Mahas, A.; Mahfouz, M. Engineering virus resistance via CRISPR–Cas systems. *Curr. Opin. Virol.* **2018**, *32*, 1–8. [CrossRef]

58. van Regenmortel, M.H.V.; Fauquet, C.M.; Bishop, D.H.L.; Carstens, E.B.; Estes, M.K.; Lemon, S.M.; Maniloff, J.; Mayo, M.A.; McGeoch, D.J.; Pringle, C.R.; et al. *Virus Taxonomy Seventh Report of the International Committee on Taxonomy of Viruses*; Academic Press: San Diego, CA, USA, 2000; ISBN 0-12-370200-3.

59. ICTV. Available online: https://talk.ictvonline.org/taxonomy/ (accessed on 14 September 2020).

60. Moffat, A.S. Geminiviruses Emerge as Serious Crop Threat. *Science* **1999**, *286*, 1835. [CrossRef]

61. Zaidi, S.S.A.; Tashkandi, M.; Mahfouz, M.M.; Zaidi, S.S.A. Engineering Molecular Immunity Against Plant Viruses. *Prog. Mol. Biol. Transl. Sci.* **2017**, *149*, 167–186. [CrossRef]

62. Ali, Z.; Abulfaraj, A.; Idris, A.; Ali, S.; Tashkandi, M.; Mahfouz, M.M. CRISPR/Cas9-mediated viral interference in plants. *Genome Biol.* **2015**, *16*. [CrossRef]

63. Baltes, N.J.; Hummel, A.W.; Konecna, E.; Cegan, R.; Bruns, A.N.; Bisaro, D.M.; Voytas, D.F. Conferring resistance to geminiviruses with the CRISPR-Cas prokaryotic immune system. *Nat. Plants* **2015**, *1*. [CrossRef]

64. Kis, A.; Hamar, É.; Tholt, G.; Bán, R.; Havelda, Z. Creating highly efficient resistance against wheat dwarf virus in barley by employing CRISPR/Cas9 system. *Plant Biotechnol. J.* **2019**, *17*, 1004–1006. [CrossRef]

65. Tripathi, J.N.; Ntui, V.O.; Ron, M.; Muiruri, S.K.; Britt, A.; Tripathi, L. CRISPR/Cas9 editing of endogenous banana streak virus in the B genome of Musa spp. overcomes a major challenge in banana breeding. *Commun. Biol.* **2019**, *2*, 46. [CrossRef] [PubMed]

66. Zhan, X.; Zhang, F.; Zhong, Z.; Chen, R.; Wang, Y.; Chang, L.; Bock, R.; Nie, B.; Zhang, J. Generation of virus-resistant potato plants by RNA genome targeting. *Plant Biotechnol. J.* **2019**, *17*, 1814–1822. [CrossRef] [PubMed]

67. Callot, C.; Gallois, J.L. Pyramiding resistances based on translation initiation factors in Arabidopsis is impaired by male gametophyte lethality. *Plant Signal. Behav.* **2014**, *9*. [CrossRef] [PubMed]

68. Gauffier, C.; Lebaron, C.; Moretti, A.; Constant, C.; Moquet, F.; Bonnet, G.; Caranta, C.; Gallois, J.L. A TILLING approach to generate broad-spectrum resistance to potyviruses in tomato is hampered by eIF4E gene redundancy. *Plant J.* **2016**, *85*, 717–729. [CrossRef]

69. Rees, H.A.; Liu, D.R. Base editing: Precision chemistry on the genome and transcriptome of living cells. *Nat. Rev. Genet.* **2018**, *19*, 770–788. [CrossRef]

70. Naeem, M.; Majeed, S.; Hoque, M.Z.; Ahmad, I. Latest Developed Strategies to Minimize the Off-Target Effects in CRISPR-Cas-Mediated Genome Editing. *Cells* **2020**, *9*, 1608. [CrossRef]

71. Mehta, D.; Stürchler, A.; Anjanappa, R.B.; Zaidi, S.S.E.A.; Hirsch-Hoffmann, M.; Gruissem, W.; Vanderschuren, H. Linking CRISPR-Cas9 interference in cassava to the evolution of editing-resistant geminiviruses. *Genome Biol.* **2019**, *20*, 80. [CrossRef]

72. Kuluev, B.R.; Gumerova, G.R.; Mikhaylova, E.V.; Gerashchenkov, G.A.; Rozhnova, N.A.; Vershinina, Z.R.; Khyazev, A.V.; Matniyazov, R.T.; Baymiev, A.K.; Baymiev, A.K.; et al. Delivery of CRISPR/Cas Components into Higher Plant Cells for Genome Editing. *Russ. J. Plant Physiol.* **2019**, *66*, 694–706. [CrossRef]

73. Gao, J.; Wang, G.; Ma, S.; Xie, X.; Wu, X.; Zhang, X.; Wu, Y.; Zhao, P.; Xia, Q. CRISPR/Cas9-mediated targeted mutagenesis in Nicotiana tabacum. *Plant Mol. Biol.* **2015**, *87*, 99–110. [CrossRef]

74. Xie, K.; Yang, Y. RNA-Guided genome editing in plants using a CRISPR-Cas system. *Mol. Plant* **2013**, *6*, 1975–1983. [CrossRef]

75. Feng, Z.; Mao, Y.; Xu, N.; Zhang, B.; Wei, P.; Yang, D.L.; Wang, Z.; Zhang, Z.; Zheng, R.; Yang, L.; et al. Multigeneration analysis reveals the inheritance, specificity, and patterns of CRISPR/Cas-induced gene modifications in Arabidopsis. *Proc. Natl. Acad. Sci. USA* **2014**, *111*, 4632–4637. [CrossRef] [PubMed]

76. Butt, H.; Eid, A.; Ali, Z.; Atia, M.A.M.; Mokhtar, M.M.; Hassan, N.; Lee, C.M.; Bao, G.; Mahfouz, M.M. Efficient CRISPR/Cas9-mediated genome editing using a chimeric single-guide RNA molecule. *Front. Plant Sci.* **2017**, *8*, 1441. [CrossRef] [PubMed]

77. Malnoy, M.; Viola, R.; Jung, M.H.; Koo, O.J.; Kim, S.; Kim, J.S.; Velasco, R.; Kanchiswamy, C.N. DNA-free genetically edited grapevine and apple protoplast using CRISPR/Cas9 ribonucleoproteins. *Front. Plant Sci.* **2016**, *7*, 1904. [CrossRef]

78. Andersson, M.; Turesson, H.; Nicolia, A.; Fält, A.S.; Samuelsson, M.; Hofvander, P. Efficient targeted multiallelic mutagenesis in tetraploid potato (Solanum tuberosum) by transient CRISPR-Cas9 expression in protoplasts. *Plant Cell Rep.* **2017**, *36*, 117–128. [CrossRef] [PubMed]

79. Woo, J.W.; Kim, J.; Kwon, S.I.; Corvalán, C.; Cho, S.W.; Kim, H.; Kim, S.G.; Kim, S.T.; Choe, S.; Kim, J.S. DNA-free genome editing in plants with preassembled CRISPR-Cas9 ribonucleoproteins. *Nat. Biotechnol.* **2015**, *33*, 1162–1164. [CrossRef]

80. Kim, H.; Kim, S.T.; Ryu, J.; Kang, B.C.; Kim, J.S.; Kim, S.G. CRISPR/Cpf1-mediated DNA-free plant genome editing. *Nat. Commun.* **2017**, *8*, 14406. [CrossRef]

81. Gil-Humanes, J.; Wang, Y.; Liang, Z.; Shan, Q.; Ozuna, C.V.; Sánchez-León, S.; Baltes, N.J.; Starker, C.; Barro, F.; Gao, C.; et al. High-efficiency gene targeting in hexaploid wheat using DNA replicons and CRISPR/Cas9. *Plant J.* **2017**, *89*, 1251–1262. [CrossRef]

82. Andersson, M.; Turesson, H.; Olsson, N.; Fält, A.S.; Ohlsson, P.; Gonzalez, M.N.; Samuelsson, M.; Hofvander, P. Genome editing in potato via CRISPR-Cas9 ribonucleoprotein delivery. *Physiol. Plant.* **2018**, *164*, 378–384. [CrossRef]

83. Sandhya, D.; Jogam, P.; Allini, V.R.; Abbagani, S.; Alok, A. The present and potential future methods for delivering CRISPR/Cas9 components in plants. *J. Genet. Eng. Biotechnol.* **2020**, *18*. [CrossRef]

84. Cho, S.W.; Kim, S.; Kim, J.M.; Kim, J.S. Targeted genome engineering in human cells with the Cas9 RNA-guided endonuclease. *Nat. Biotechnol.* **2013**, *31*, 230–232. [CrossRef]

85. Li, Z.; Liu, Z.-B.; Xing, A.; Moon, B.P.; Koellhoffer, J.P.; Huang, L.; Ward, R.T.; Clifton, E.; Falco, S.C.; Cigan, A.M. Cas9-guide RNA directed genome editing in soybean. *Plant Physiol.* **2015**, *169*, 960–970. [CrossRef] [PubMed]

86. Wang, Y.; Cheng, X.; Shan, Q.; Zhang, Y.; Liu, J.; Gao, C.; Qiu, J.L. Simultaneous editing of three homoeoalleles in hexaploid bread wheat confers heritable resistance to powdery mildew. *Nat. Biotechnol.* **2014**, *32*, 947–951. [CrossRef] [PubMed]

87. Svitashev, S.; Schwartz, C.; Lenderts, B.; Young, J.K.; Mark Cigan, A. Genome editing in maize directed by CRISPR-Cas9 ribonucleoprotein complexes. *Nat. Commun.* **2016**, *7*, 13274. [CrossRef] [PubMed]

88. Liang, Z.; Chen, K.; Li, T.; Zhang, Y.; Wang, Y.; Zhao, Q.; Liu, J.; Zhang, H.; Liu, C.; Ran, Y.; et al. Efficient DNA-free genome editing of bread wheat using CRISPR/Cas9 ribonucleoprotein complexes. *Nat. Commun.* **2017**, *8*, 14261. [CrossRef] [PubMed]

89. Nester, E.W. Agrobacterium: Nature's Genetic Engineer. *Front. Plant Sci.* **2015**, *5*, 730. [CrossRef] [PubMed]

90. Tian, S.; Jiang, L.; Cui, X.; Zhang, J.; Guo, S.; Li, M.; Zhang, H.; Ren, Y.; Gong, G.; Zong, M.; et al. Engineering herbicide-resistant watermelon variety through CRISPR/Cas9-mediated base-editing. *Plant Cell Rep.* **2018**, *37*, 1353–1356. [CrossRef] [PubMed]

91. Miao, J.; Guo, D.; Zhang, J.; Huang, Q.; Qin, G.; Zhang, X.; Wan, J.; Gu, H.; Qu, L.J. Targeted mutagenesis in rice using CRISPR-Cas system. *Cell Res.* **2013**, *23*, 1233–1236. [CrossRef]

92. Pan, C.; Ye, L.; Qin, L.; Liu, X.; He, Y.; Wang, J.; Chen, L.; Lu, G. CRISPR/Cas9-mediated efficient and heritable targeted mutagenesis in tomato plants in the first and later generations. *Sci. Rep.* **2016**, *6*, 24765. [CrossRef]

93. Nakajima, I.; Ban, Y.; Azuma, A.; Onoue, N.; Moriguchi, T.; Yamamoto, T.; Toki, S.; Endo, M. CRISPR/Cas9-mediated targeted mutagenesis in grape. *PLoS ONE* **2017**, *12*, e0177966. [CrossRef]

94. Zhou, X.; Jacobs, T.B.; Xue, L.J.; Harding, S.A.; Tsai, C.J. Exploiting SNPs for biallelic CRISPR mutations in the outcrossing woody perennial Populus reveals 4-coumarate: CoA ligase specificity and redundancy. *New Phytol.* **2015**, *208*, 298–301. [CrossRef]

95. Nekrasov, V.; Staskawicz, B.; Weigel, D.; Jones, J.D.G.; Kamoun, S. Targeted mutagenesis in the model plant Nicotiana benthamiana using Cas9 RNA-guided endonuclease. *Nat. Biotechnol.* **2013**, *31*, 691–693. [CrossRef] [PubMed]

96. Yin, K.; Han, T.; Liu, G.; Chen, T.; Wang, Y.; Yu, A.Y.L.; Liu, Y. A geminivirus-based guide RNA delivery system for CRISPR/Cas9 mediated plant genome editing. *Sci. Rep.* **2015**, *5*, 14926. [CrossRef] [PubMed]

97. Cai, Y.; Chen, L.; Liu, X.; Sun, S.; Wu, C.; Jiang, B.; Han, T.; Hou, W. CRISPR/Cas9-mediated genome editing in soybean hairy roots. *PLoS ONE* **2015**, *10*, e0136064. [CrossRef] [PubMed]

98. Jacobs, T.B.; LaFayette, P.R.; Schmitz, R.J.; Parrott, W.A. Targeted genome modifications in soybean with CRISPR/Cas9. *BMC Biotechnol.* **2015**, *15*, 16. [CrossRef] [PubMed]

99. Sun, X.; Hu, Z.; Chen, R.; Jiang, Q.; Song, G.; Zhang, H.; Xi, Y. Targeted mutagenesis in soybean using the CRISPR-Cas9 system. *Sci. Rep.* **2015**, *5*, 10342. [CrossRef]

100. Ellis, J.; Bernstein, A. Gene targeting with retroviral vectors: Recombination by gene conversion into regions of nonhomology. *Mol. Cell. Biol.* **1989**, *9*, 1621–1627. [CrossRef]

101. Chen, X.; Gonçalves, M.A.F.V. Engineered viruses as genome editing devices. *Mol. Ther.* **2016**, *24*, 447–457. [CrossRef]

102. Kumagai, M.H.; Donson, J.; Della-Cioppa, G.; Harvey, D.; Hanley, K.; Grill, L.K. Cytoplasmic inhibitoin of carotenoid biosynthesis with virus-derived RNA. *Proc. Natl. Acad. Sci. USA* **1995**, *92*, 1679–1683. [CrossRef]

103. Baltes, N.J.; Gil-Humanes, J.; Cermak, T.; Atkins, P.A.; Voytas, D.F. DNA replicons for plant genome engineering. *Plant Cell* **2014**, *26*, 151–163. [CrossRef]

104. Butler, N.M.; Baltes, N.J.; Voytas, D.F.; Douches, D.S. Geminivirus-mediated genome editing in potato (*Solanum tuberosum* L.) using sequence-specific nucleases. *Front. Plant Sci.* **2016**, *7*, 1045. [CrossRef]

105. Čermák, T.; Baltes, N.J.; Čegan, R.; Zhang, Y.; Voytas, D.F. High-frequency, precise modification of the tomato genome. *Genome Biol.* **2015**, *16*. [CrossRef] [PubMed]

106. Ali, Z.; Eid, A.; Ali, S.; Mahfouz, M.M. Pea early-browning virus-mediated genome editing via the CRISPR/Cas9 system in Nicotiana benthamiana and Arabidopsis. *Virus Res.* **2018**, *244*, 333–337. [CrossRef] [PubMed]

107. Cody, W.B.; Scholthof, H.B.; Mirkov, T.E. Multiplexed gene editing and protein overexpression using a tobacco mosaic virus viral vector. *Plant Physiol.* **2017**, *175*, 23–35. [CrossRef] [PubMed]
108. Hu, J.; Li, S.; Li, Z.; Li, H.; Song, W.; Zhao, H.; Lai, J.; Xia, L.; Li, D.; Zhang, Y. A barley stripe mosaic virus-based guide RNA delivery system for targeted mutagenesis in wheat and maize. *Mol. Plant Pathol.* **2019**, *20*, 1463–1474. [CrossRef] [PubMed]
109. Jiang, N.; Zhang, C.; Liu, J.Y.; Guo, Z.H.; Zhang, Z.Y.; Han, C.G.; Wang, Y. Development of Beet necrotic yellow vein virus-based vectors for multiple-gene expression and guide RNA delivery in plant genome editing. *Plant Biotechnol. J.* **2019**, *17*, 1302–1315. [CrossRef] [PubMed]
110. Mei, Y.; Beernink, B.M.; Ellison, E.E.; Konečná, E.; Neelakandan, A.K.; Voytas, D.F.; Whitham, S.A. Protein expression and gene editing in monocots using foxtail mosaic virus vectors. *Plant Direct* **2019**, *3*, e00181. [CrossRef] [PubMed]
111. Ariga, H.; Toki, S.; Ishibashi, K. Potato virus X Vector-Mediated DNA-Free genome editing in plants. *Plant Cell Physiol.* **2020**, *61*, 1946–1953. [CrossRef]
112. Shan-E-Ali Zaidi, S.; Mansoor, S. Viral vectors for plant genome engineering. *Front. Plant Sci.* **2017**, *8*, 539. [CrossRef]
113. Gilbertson, R.L.; Sudarshana, M.; Jiang, H.; Rojas, M.R.; Lucas, W.J. Limitations on Geminivirus Genome Size Imposed by Plasmodesmata and Virus-Encoded Movement Protein: Insights into DNA Trafficking. *Plant Cell* **2003**, *15*, 2578–2591. [CrossRef]
114. Gardiner, W.E.; Sunter, G.; Brand, L.; Elmer, J.S.; Rogers, S.G.; Bisaro, D.M. Genetic analysis of tomato golden mosaic virus: The coat protein is not required for systemic spread or symptom development. *EMBO J.* **1988**, *7*, 899–904. [CrossRef]
115. Ali, Z.; Abul-Faraj, A.; Piatek, M.; Mahfouz, M.M. Activity and specificity of TRV-mediated gene editing in plants. *Plant Signal. Behav.* **2015**, *10*, e1044191. [CrossRef] [PubMed]
116. Senthil-Kumar, M.; Mysore, K.S. Tobacco rattle virus-based virus-induced gene silencing in Nicotiana benthamiana. *Nat. Protoc.* **2014**, *9*, 1549–1562. [CrossRef] [PubMed]
117. Marton, I.; Zuker, A.; Shklarman, E.; Zeevi, V.; Tovkach, A.; Roffe, S.; Ovadis, M.; Tzfira, T.; Vainstein, A. Nontransgenic genome modification in plant cells. *Plant Physiol.* **2010**, *154*, 1079–1087. [CrossRef] [PubMed]
118. Gleba, Y.; Klimyuk, V.; Marillonnet, S. Viral vectors for the expression of proteins in plants. *Curr. Opin. Biotechnol.* **2007**, *18*, 134–141. [CrossRef] [PubMed]
119. Constantin, G.D.; Krath, B.N.; MacFarlane, S.A.; Nicolaisen, M.; Johansen, E.; Lund, O.S. Virus-induced gene silencing as a tool for functional genomics in a legume species. *Plant J.* **2004**, *40*, 622–631. [CrossRef] [PubMed]
120. Liu, H.; Zhang, B. Virus-Based CRISPR/Cas9 Genome Editing in Plants. *Trends Genet.* **2020**, *36*, 810–813. [CrossRef]
121. Gao, Q.; Xu, W.Y.; Yan, T.; Fang, X.D.; Cao, Q.; Zhang, Z.J.; Ding, Z.H.; Wang, Y.; Wang, X.B. Rescue of a plant cytorhabdovirus as versatile expression platforms for planthopper and cereal genomic studies. *New Phytol.* **2019**, *223*, 2120–2133. [CrossRef]
122. Ma, X.; Zhang, X.; Liu, H.; Li, Z. Highly efficient DNA-free plant genome editing using virally delivered CRISPR–Cas9. *Nat. Plants* **2020**, *6*, 773–779. [CrossRef]
123. Kendall, A.; McDonald, M.; Bian, W.; Bowles, T.; Baumgarten, S.C.; Shi, J.; Stewart, P.L.; Bullitt, E.; Gore, D.; Irving, T.C.; et al. Structure of Flexible Filamentous Plant Viruses. *J. Virol.* **2008**, *82*, 9546–9554. [CrossRef]
124. Hanley-Bowdoin, L.; Bejarano, E.R.; Robertson, D.; Mansoor, S. Geminiviruses: Masters at redirecting and reprogramming plant processes. *Nat. Rev. Microbiol.* **2013**, *11*, 777–788. [CrossRef]
125. Cody, W.B.; Scholthof, H.B. Plant Virus Vectors 3.0: Transitioning into Synthetic Genomics. *Annu. Rev. Phytopathol.* **2019**, *57*, 211–230. [CrossRef] [PubMed]
126. Zhang, X.H.; Tee, L.Y.; Wang, X.G.; Huang, Q.S.; Yang, S.H. Off-target effects in CRISPR/Cas9-mediated genome engineering. *Mol. Ther. Nucleic Acids* **2015**, *4*, e264. [CrossRef] [PubMed]
127. Ji, X.; Si, X.; Zhang, Y.; Zhang, H.; Zhang, F.; Gao, C. Conferring DNA virus resistance with high specificity in plants using virus-inducible genome-editing system. *Genome Biol.* **2018**, *19*, 197. [CrossRef] [PubMed]
128. Shivprasad, S.; Pogue, G.P.; Lewandowski, D.J.; Hidalgo, J.; Donson, J.; Grill, L.K.; Dawson, W.O. Heterologous sequences greatly affect foreign gene expression in tobacco mosaic virus-based vectors. *Virology* **1999**, *255*, 312–323. [CrossRef]
129. Wang, H.; La Russa, M.; Qi, L.S. CRISPR/Cas9 in Genome Editing and beyond. *Annu. Rev. Biochem.* **2016**, *85*, 227–264. [CrossRef]
130. Osakabe, Y.; Watanabe, T.; Sugano, S.S.; Ueta, R.; Ishihara, R.; Shinozaki, K.; Osakabe, K. Optimization of CRISPR/Cas9 genome editing to modify abiotic stress responses in plants. *Sci. Rep.* **2016**, *6*, 26685. [CrossRef]
131. Jiang, Y.; Qian, F.; Yang, J.; Liu, Y.; Dong, F.; Xu, C.; Sun, B.; Chen, B.; Xu, X.; Li, Y.; et al. CRISPR-Cpf1 assisted genome editing of Corynebacterium glutamicum. *Nat. Commun.* **2017**, *8*, 15179. [CrossRef]

 *viruses*

Article

Relocation of the attTn7 Transgene Insertion Site in Bacmid DNA Enhances Baculovirus Genome Stability and Recombinant Protein Expression in Insect Cells

Gorben P. Pijlman [1,*], Carissa Grose [2], Tessy A. H. Hick [1], Herman E. Breukink [1], Robin van den Braak [1], Sandra R. Abbo [1], Corinne Geertsema [1], Monique M. van Oers [1], Dirk E. Martens [3] and Dominic Esposito [2]

1   Laboratory of Virology, Wageningen University, Droevendaalsesteeg 1, 6708PB Wageningen, The Netherlands; tessy.hick@wur.nl (T.A.H.H.); herman.breukink@agilent.com (H.E.B.); robinvandenbraak@hotmail.com (R.v.d.B.); sandra.abbo@wur.nl (S.R.A.); corinne.geertsema@wur.nl (C.G.); monique.vanoers@wur.nl (M.M.v.O.)
2   Protein Expression Laboratory, Cancer Research Technology Program, Frederick National Laboratory for Cancer Research, Leidos Biomedical Research, Inc. PO Box B, Frederick, MD 21702, USA; carissa.grose@nih.gov (C.G.); dom.esposito@nih.gov (D.E.)
3   Bioprocess Engineering, Wageningen University, Droevendaalsesteeg 1, 6708PB Wageningen, The Netherlands; dirk.martens@wur.nl
*   Correspondence: gorben.pijlman@wur.nl; Tel.: +31-317-484498

Academic Editors: Carla Varanda and Patrick Materatski
Received: 23 November 2020; Accepted: 14 December 2020; Published: 16 December 2020

Abstract: Baculovirus expression vectors are successfully used for the commercial production of complex (glyco)proteins in eukaryotic cells. The genome engineering of single-copy baculovirus infectious clones (bacmids) in *E. coli* has been valuable in the study of baculovirus biology, but bacmids are not yet widely applied as expression vectors. An important limitation of first-generation bacmids for large-scale protein production is the rapid loss of gene of interest (GOI) expression. The instability is caused by the mini-F replicon in the bacmid backbone, which is non-essential for baculovirus replication in insect cells, and carries the adjacent GOI in between attTn7 transposition sites. In this paper, we test the hypothesis that relocation of the attTn7 transgene insertion site away from the mini-F replicon prevents deletion of the GOI, thereby resulting in higher and prolonged recombinant protein expression levels. We applied lambda red genome engineering combined with SacB counterselection to generate a series of bacmids with relocated attTn7 sites and tested their performance by comparing the relative expression levels of different GOIs. We conclude that GOI expression from the *odv-e56* (*pif-5*) locus results in higher overall expression levels and is more stable over serial passages compared to the original bacmid. Finally, we evaluated this improved next-generation bacmid during a bioreactor scale-up of Sf9 insect cells in suspension to produce enveloped chikungunya virus-like particles as a model vaccine.

Keywords: baculovirus; insect cells; bacmid; Tn7; genome stability; protein expression; chikungunya virus; VLPs; bioreactor

1. Introduction

The baculovirus expression vector system (BEVS) is one of the most versatile platform technologies developed in the past 30 years to produce complex (glyco)proteins in eukaryotic cells, because it frequently results in high yields of purified proteins and allows for similar post-translational modifications as mammalian expression systems [1]. An increasing number of commercial BEVS products are becoming available, and these include human vaccines against influenza A virus and human papilloma virus, gene therapy products, and veterinary vaccines [2,3]. More BEVS bio-pharmaceuticals are currently in (pre)clinical trials and these include prototype COVID-19 vaccines that are aimed at protection against SARS-CoV-2 infection [4]. Traditional baculovirus expression vectors are generated by homologous recombination in insect cells between wildtype baculovirus DNA and a so-called transfer vector (plasmid DNA), which carries the gene of interest (GOI) under a suitable baculovirus promoter, typically the very strong polyhedrin or p10 promoter [5,6]. There have been sophisticated improvements throughout the years to streamline the generation and isolation of recombinant baculoviruses, and numerous easy-to-use commercial baculovirus kits are available [7].

One of the major breakthroughs in baculovirus technology was the generation of an *Autographa californica* multiple nucleopolyhedrovirus (AcMNPV) infectious clone that could be maintained as a ~140 kb large, single-copy bacterial artificial chromosome (BAC) in *E. coli* [8]. This so-called "bacmid" carries a mini-F replicon for single-copy replication in *E. coli*, a kanamycin resistance gene (kan^R) for selection, and a LacZ-attTn7 transposon integration site (inserted between kan^R and mini-F) for insertion of the GOI by Tn7 transposition (Figure 1). A commercial kit based on these bacmids was successfully launched by Invitrogen Life Technologies (now ThermoFisher) as the "Bac-to-Bac baculovirus expression system", which uses so-called "pFastBac" transfer vectors in which the GOI is cloned in between attL and attR sites, and downstream of the strong polyhedrin (polh) or p10 promoter. After Tn7 transposition, the recombinant bacmids are selected for antibiotic resistance (kan^R and gen^R) and, upon transfection of insect cells, this yields, at least in theory, a genetically homogeneous recombinant baculovirus stock that does not require further plaque purification.

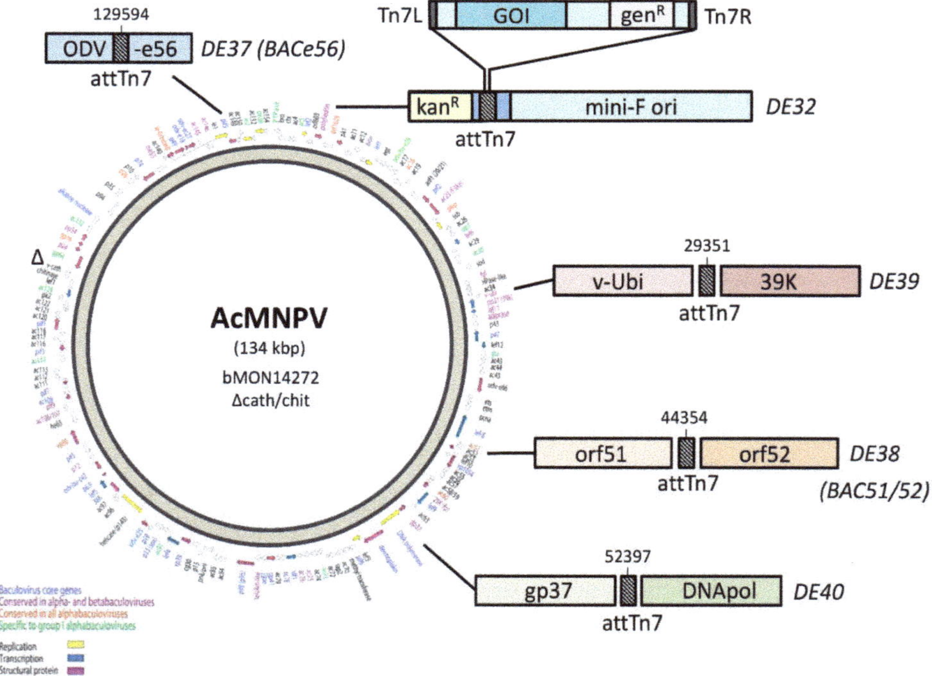

Figure 1. Bacmid strains with relocated attTn7 sites. Schematic diagram of new attTn7 insertion sites in the modified bacmid strains. The crosshatched boxes represent the attTn7 sites introduced into different loci in the parental DE32 bacmid. Insertion site location numbering is based on the wildtype *Autographa californica* multiple nucleopolyhedrovirus (AcMNPV) E2 genome sequence.

While bacmids are efficient for rapidly creating recombinant baculoviruses, the introduction of genes is limited to a single insertion site. MultiBac (Geneva Biotech) has improved the system to allow for the introduction of several genes, with the multigene construct being shuttled into the same attTn7 site [9]. The original AcMNPV bacmid (bMON14272, [8]) can be easily manipulated in *E. coli* and this has been the driving force for initiating numerous knock out studies to study baculovirus gene functions. Bacmids have been constructed for baculoviruses other than AcMNPV [10–12] and this has tremendously accelerated functional studies on previously uncharacterized baculovirus genes, including baculovirus core genes or those involved in oral infectivity [13–16].

Despite the major contribution of bacmids in the understanding of baculovirus biology, most commercial baculovirus vectors are still based on the traditional method of homologous recombination in insect cells between a transfer plasmid carrying the GOI and a (linearized) AcMNPV genome, followed by several rounds of plaque purification and master seed production.

For large-scale protein expression, bacmid-based expression vectors have not been applied, despite improvements that have increased the efficiency of GOI transposition and recombinant bacmid selection [17]. This may change soon, however, since Novavax appears to produce its COVID-19 vaccine NVX-CoV2373 using a pFastBac vector containing a SARS-CoV-2 spike gene [18]. An important reason for not using bacmids for large-scale production is the relative rapid loss of GOI expression upon serial passage [19] resulting from the genomic instability of the mini-F replicon in the bacmid backbone [20]. The exact reason for the rapid deletion of the mini-F and the adjacent GOI is unknown, but its bacterial origin and large size (>8 kbp) likely account for a negative selection pressure on recombinant baculovirus genomes during replication in insect cells.

We hypothesize that relocation of the attTn7 away from the mini-F replicon would prevent or at least delay GOI deletion, thereby resulting in higher recombinant protein expression levels upon serial passage. To test this hypothesis, we generate a series of bacmids with attTn7 insertions at different locations on the AcMNPV genome that had previously been shown to tolerate insertions [21]. We analyze the performance of the different bacmids by comparing the relative expression levels of different GOIs. The best performing bacmid is subsequently tested for the stability of recombinant protein expression upon serial undiluted passages in insect cells. Finally, we demonstrate the viability of this novel bacmid for industrial vaccine scale-up by producing chikungunya virus-like particles (CHIKV VLPs) in Sf9 suspension cells in shaker flasks and a bioreactor [22–25].

2. Materials and Methods

2.1. Construction of ΔattTn7 Bacmid

Before moving the attTn7 binding region to alternate locations, the original attTn7 site had to be removed. To knock out the attTn7 binding region, a linear cassette was generated by a Gibson Isothermal Assembly, which included ~300 bp homology arms in both the pLac and LacZ regions along with the counter-selectable cassette Cat-SacB (chloramphenicol acetyltransferase-sucrose sensitivity), which was amplified from pELO4 (gift from Don Court, NCI); both homology arms included 40 bp of sequence homology at the appropriate ends of the Cat-SacB cassette. Twenty-five femtomoles of each amplicon were included in a 20 µL reaction with 2× Gibson Assembly Master Mix (New England Biolabs, Ipswich, MA, USA) and incubated at 50 °C for 30 min. To amplify the assembled cassette, 1 µL of the assembly reaction was used as a template in a 100 µL PCR reaction, including 0.4 µM of each primer and 2× Phusion HF Mastermix (60 °C annealing temperature, 4 min extension, 25 cycles). The linear cassette was purified using a QiaQuick PCR Purification Kit (Qiagen, Hilden, Germany).

The original strain for this work was λRed *E. coli* SW106 (gift from Don Court, NCI) harboring an AcMNPV bacmid, from which the chitinase and cathepsin genes had previously been deleted (bMON14272Δvcath-chiA or DE32). The strain was grown in LB with 50 µg/mL kanamycin at 30 °C with shaking to OD_{600} of 0.4. The culture was agitated in a 42 °C water bath for 15 min to induce the temperature-sensitive lambda prophage. The cells were then chilled and made electrocompetent.

One hundred nanograms of the knock-out cassette were added to 50 µL of the cells, which were subsequently electroporated. The transformation was incubated with shaking for two hours at 30 °C in 1 mL LB before plating 100 µL on LB agar with 50 µg/mL kanamycin and 20 µg/mL chloramphenicol at 30 °C. Four colonies were selected on the second morning and diluted in 50 µL water, with 1 µL being used as a template for colony PCR with primers that bound just outside the insertion.

Once this intermediate bacmid was confirmed, the Cat-SacB selection cassette was seamlessly removed. Similar ~300 bp arms, homologous to the end of pLac and the beginning of LacZ, were generated by PCR, where the LacZ arm product included 40 bp homology to the end of pLac. The two amplicons were assembled by overlap PCR (0.2 µL of each amplicon, 2× HF mastermix, 0.4 µM primers, 100 µL total volume) (60 °C annealing, 30 s extension, 25 cycles) to produce the marker knock-out cassette and were purified using a QiaQuick PCR Purification Kit. Lambda Red was induced in the SW106 bMON14272Δvcath-chiAΔ*attTn7*:Cat-SacB strain and electrocompetent cells were made as described above. One hundred nanograms of the marker removal cassette was added to 50 µL of electrocompetent cells and transformed by electroporation. The cells were allowed to recover by shaking at 30 °C for 4 h in 10 mL LB. Cells from 1 mL of the transformation were pelleted by centrifugation for 30 s at 13,000× *g* and resuspended in 1 volume 1× M9 salts twice to wash the cells. After the second wash, the cells were resuspended in 100 µl M9 salts and plated on LB agar with no salt, 6% sucrose, and 50 µg/mL kanamycin at 37 °C, to select cells that no longer contained the Cat-SacB selection marker. Bacmid DNA was isolated from four colonies by alkaline lysis and the recombined region was amplified by PCR to confirm the deletion of the Cat-SacB marker. The resulting bacmid with the correctly sized amplicon was verified by sequencing the entire cloned region.

The confirmed bMON14272Δvcath-chiAΔ*attTn7* was used to transform *E. coli* DE25 cells, a DH10B-derived cell line compatible with the Bac-to-Bac system for recombinant baculovirus production. The strain contains an optimized helper plasmid that encodes Tn7 transposition functions to generate recombinant baculovirus and confers resistance to tetracycline [17]. The transformation was plated on LB with 50 µg/mL kanamycin and 12.5 µg/mL tetracycline at 37 °C. A single colony was selected from the plate to generate competent cells for the new deletion strain.

2.2. Construction of Bacmids with Relocated attTn7

Bacmids with relocated attTn7 sites Strains DE37, DE38, DE39, and DE40 were generated in a similar fashion. The knock-in cassettes for odv-e56 (pif-5) and the intergenic regions of orf51/52, v-ubiquitin (v-ubi)/39 k, and gp37/DNApol included 4 amplicons, which contained a ~300 bp region of homology to the immediate left of the desired insertion point with a 40 bp homology to attTn7 (Forward: 5′- TGTGGAATTGTGAGCGGATA; Reverse: 5′-TCCTGTGACGGAAGATCACTTCGCAGAATAAAT AAATCCTGGTGCTGCAAGGCGATTAAGT), Cat-SacB cassette, and ~300 bp region of homology to the right of the insertion including 40 bp homology to the end of SacB (Table 1).

Table 1. attTn7 location in different strains.

Strain	Location	Left Flanking Sequence	Right Flanking Sequence
DE32	polh locus	TGCTTCCGGCTCGTATGTTG	TCCCCCTTTCGCCAGCTGGC
DE37	mid odv-e56	TTGACGACAAGTGCGCTGCA	ATAACAAGCAGGCCTCGGCG
DE38	between orf51 and orf52	GTTTTTTTCTAGTGTCGTACTT	TTTTACAATGCGTCTGTTGTCC
DE39	between v-ubiquitin and 39k	AATAATAAAAACCATTAAAT	ACAAAAGTTTTTTATT
DE40	between gp37 and DNApol	TTGGTCAAAAACGTTATGTT	GAAACATAATAACACCTTAC

The amplicons were assembled by Isothermal Assembly with 25 femtomoles of each piece in a 20 µL reaction with 2× Gibson Assembly Master Mix (New England Biolabs) and incubated for 30 min at 50 °C. The cassette was then amplified with 1 µL of the assembly reaction as a template in a 100 uL PCR reaction, including 0.4 µM of each primer and 2× Phusion HF Mastermix (60 °C annealing temperature, 4 min extension, 25 cycles). The linear cassette was purified using a QiaQuick PCR Purification Kit (Qiagen).

The previously generated strain, SW106 bMON14272Δvcath-chiA; ΔattTn7, was induced as described previously and electrocompetent cells were made. One hundred nanograms of the attTn7 knock-in cassette was added to 50 μL of electrocompetent cells and electroporated at 1.8 kV. The cells were recovered in 1 mL LB at 30 °C for 2 h and 100 μL was then plated on LB with 50 ug/mL kanamycin and 20 ug/mL chloramphenicol. Four colonies were selected on the second morning and diluted in 50 uL water, with 1ul being used as template for colony PCR, with primers landing just outside the insertion. The remaining diluted cells were then used to seed cultures to generate glycerol stocks for the selected positive construct. The Cat-SacB marker was removed from these intermediate strains in a manner similar to that described above. The loss of the marker was confirmed by PCR amplification of the insertion region and sequencing of the entire region for each new bacmid.

The confirmed bacmids were transformed into *E. coli* DE25 cells as described above. A single colony was selected from each plate to generate competent cells for DE37, DE38, DE39, and DE40. The entire bacmid for each strain was analyzed by PacBio sequencing to be certain that there was no unintended rearrangement during recombination.

2.3. Generation of Recombinant Baculoviruses

To test the expression levels at each location, four GOIs were tested: mouse A15 extracellular domain (NP_062608.2), human NDUFA13 (NP_057049.5), human FLCN (NP_659434.2), and enhanced green fluorescent protein (eGFP). Expression clones were generated for each by gateway recombination of previously validated entry clones into baculovirus expression backbones (pDest-636), including the polh promoter and N-terminal His6-MBP (maltose-binding protein) tag. To facilitate shuttling into the bacmid, the expression backbone contained Tn7 left and right arms as well as the gentamycin resistance marker. One microliter of the verified expression clone was used to transform 50 μL of chemically competent cells of each of the new strains (DE37-DE40) as well as the DE32 (Δvcath-chiA) control. One ml of LB was added to the transformants and cultures were allowed to grow at 37 °C for 4 h with shaking. Ten microliters of the outgrowth were diluted into 190 μL LB and the resulting 200 μL was plated onto LB agar plates with gentamycin (7 ug/mL), kanamycin (50 ug/mL), tetracycline (10 ug/mL), IPTG (40 ug/mL), and Bluo-gal (100 ug/mL), and grown overnight at 37 °C. White colonies were selected from each plate and grown overnight in 3 mL LB with gentamycin and kanamycin at 37 °C with shaking. Two milliliter aliquots were pelleted and bacmid DNA was prepared by alkaline lysis. Junction PCR was performed to verify the correct inserts using the lysis as template. Fifty milliliters of Sf9s at a density of 1.5×10^6 cells/mL was transfected with 25 μL of bacmid complexed with 125 μL of Insect GeneJuice (Sigma-Aldrich, Saint Louis, MO, USA). The complexes were allowed to form for 20 min and were then added to the culture. The cultures were incubated for 5 days and then harvested by centrifugation at 2000 RPM for 10 min and the supernatants were then collected in 50 mL conical tubes.

2.4. Recombinant Protein Production and Quantitation

For small-scale protein expression, 50 mL of Tni-FNL cells were seeded at 1.5×10^6 cells/mL and were infected with the generated viruses at a multiplicity of infection (MOI) of 3. The infected cultures were allowed to grow at 27 °C for 72 h. The cells were then harvested by centrifugation at 2500 rpm for 20 min in 50 mL conical tubes. The pellets were each resuspended in 5 mL buffer (20 mM Hepes pH 7.3, 300 mM NaCl, 1 mM TCEP) by gently vortexing, and then were lysed using an LV-1 microfluidizer (Microfluidics, Inc., Westwood, MA, USA) set at a pressure of 7000 psi for 2 passes per lysate. Lysates were then clarified using ultracentrifugation at 33,100 rpm for 30 min, and the supernatant was collected in a 15 mL conical tube. Clarified cell extracts of A15, NDUFA13, and FLCN were purified using immobilized metal affinity chromatography on Phynexus tips with elution in 500 mM imidazole. Final proteins were analyzed by SDS–PAGE chromatography to ensure purity, and quantitated in a Nanodrop spectrophotometer at 280 nm. Cell extracts expressing eGFP were quantified by measuring GFP fluorescence (excitation = 485 nm, emission = 510 nm) on a BMG Omega plate reader. Data presented are averages of three replications of the purification process with the standard deviation shown via error bars.

2.5. Serial Undiluted Baculovirus Passage on Sf21 Cells

Bacmid-derived viruses expressing CHIKV VLPs were passaged ten times in duplicate in Sf21 cells, as described before [19,25]. Sf21 cells were cultured in supplemented Grace's insect medium (Gibco, Life Technologies, Carlsbad, CA, USA) with 10% fetal bovine serum (FBS, Invitrogen, Carlsbad, CA, USA) and 50 μg/mL gentamycin (Gibco, Life Technologies) in a T25 cell culture flask (Greiner, Alphen aan den Rijn, the Netherlands). For serial passage, 1 mL of each virus suspension was added to a T25 flask that contained healthy Sf21 cells with a confluency of 50–60%. After three hours, the virus suspension was replaced with 4 mL fresh medium. The cells were incubated for 3 days at 27 °C. Cells were detached and the cell suspension was centrifuged for 5 min at 4000 rpm. The cell pellet was washed once with 500 μL PBS and resuspended in PBS with protease inhibitor (Roche, Basel, Switserland). For the next virus passage, 1 mL of the supernatant was added to new healthy Sf-21 cells. The remaining supernatant was stored at 4 °C. This procedure was repeated ten times in duplicate for each virus.

2.6. SDS–PAGE and Western Blot Immunodetection

Protein samples containing equal numbers of cells were analyzed in 12.5% SDS–PAGE gels (Mini-PROTEAN® Tetra System, Bio-Rad Laboratories, Hercules, CA, USA). Proteins were transferred to an Immobilon-P membrane (Millipore) using a Tris–Glycine buffer (25 mM Tris, 192 mM glycine, 10% (*v/v*) methanol, pH 8.3). Membranes were blocked in 3% low-fat milk powder (Campina, The Netherlands) in PBS 0.1% (*v/v*) Tween-20 (PBS-T) (Merck). Blots were washed with 2.5 mL PBS-T for 5 min at RT and the primary polyclonal antibody rabbit-anti-CHIKV-E2 or -E1 immunoglobulin was added (dilution 1:10,000). For detection of the AcMNPV VP39 major capsid protein, a polyclonal mouse anti-VP39 immunoglobulin was used (dilution 1:1000). Secondary antibodies were polyclonal goat anti-rabbit immunoglobulin (AP-conjugated) or polyclonal goat anti-mouse immunoglobulin (AP-conjugated) (both Sigma-Aldrich). After incubation and washing, the blot was stained with NBT/BCIP (Roche Diagnostics GmbH, Basel, Switserland).

2.7. Shake Flask and Bioreactor Experiments

In the shake flask experiments, Sf9 suspension cells were cultured in Sf-900 II SFM medium (Gibco, Life Technologies, Carlsbad, CA, USA) in 125 mL flasks (Nalgene, Rochester, NY, USA) with a working volume of 25 mL. The culture conditions of the cells were identical to the culture conditions of the maintenance culture (27 °C, 100 rpm, inoculation at 0.5×10^6 cells/mL from an exponential phase cell culture at $\sim 4 \times 10^6$ cells/mL). When the desired cell concentration was reached in the shake flasks, infection with the recombinant baculovirus was performed at a defined MOI. Samples were obtained at specific time points to analyze the infection process. A 1-litre DASGIP bioreactor (Eppendorf, Hamburg, Germany) was inoculated at 0.5×10^6 cells/mL by the Sf9 culture stock in the exponential growth phase. The bioreactors had a working volume of 500 mL and cells were maintained in Sf-900 II SFM medium (Gibco, Life Technologies, Carlsbad, CA, USA) at 27 °C and an agitation speed of 150 rpm. The DO (DO sensor, Broadley James, Irvine, CA, USA) was controlled at 30% air saturation by headspace aeration. The pH (pH sensor, Mettler Toledo, Tiel, the Netherlands) was retained at a value of 6.3 by automatic base and CO_2 addition. Once the cells reached the desired CCI, infection with recombinant baculoviruses was performed at a defined MOI. Samples were obtained at several time points during the infection process.

3. Results

3.1. Construction of AcMNPV Bacmids with a Relocated attTn7 Integration Site

Four different bacmids were constructed, each with the attTn7 integration site at different locations in the AcMNPV genome (Figure 1). A modified bacmid lacking the chitinase/cathepsin loci was used as a parental bacmid (bMON14272Δvcath-chiA or DE32). First, the attTn7 site was removed from its original location (polh locus) by a two-step homologous recombination in *E. coli* (Figure 2).

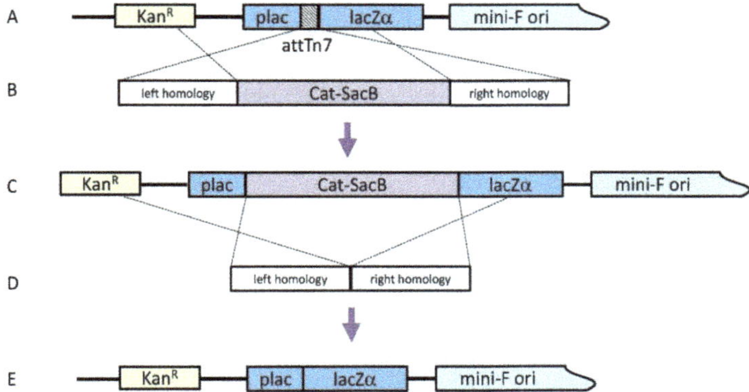

Figure 2. Construction of ΔattTn7 bacmid. (**A**) Parent strain; (**B**) linear attTn7 removal cassette; (**C**) intermediate Cat-SacB strain; (**D**) linear removal marker; (**E**) final strain.

Briefly, a Cat-SacB positive/negative selection cassette was generated by PCR and electroporated into *E. coli* harboring DE32 (Figure 2A,B). Chloramphenicol-resistant colonies were selected (Figure 2C) and another PCR product, consisting of attTn7 flanking regions, was then introduced by electroporation (Figure 2D). Bacmids were counterselected against SacB and the resulting bacmids containing a scarless deletion of attTn7 were selected (Figure 2E).

Next, the attTn7 site was inserted by homologous recombination at four different loci in the AcMNPV genome: in the odv-e56 ORF (DE37 or BACe56), in between orf51 and orf52 (DE38 or BAC51/52), in between v-Ubi and 39k (DE39) and in between DNApol and gp37 (DE40) (Figure 3).

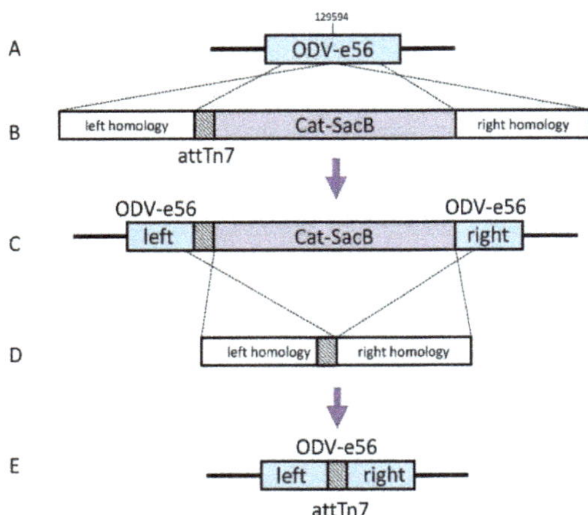

Figure 3. Construction of bacmids with relocated attTn7. (**A**) Parent strain; (**B**) linear attTn7 removal cassette; (**C**) intermediate Cat-SacB strain; (**D**) linear removal marker; (**E**) final strain.

3.2. Heterologous Protein Expression Levels with Modified Bacmids

After the attTn7 site had been relocated, four different heterologous genes known to be challenging to produce in *E. coli* were inserted into the four new bacmids. Recombinant baculoviruses were generated and cells were infected with an MOI of 3 tissue culture infective doses of 50% ($TCID_{50}$)/cell. After 72 h, cells were harvested and protein was purified from a cell extract. Protein levels relative to those of the parental bacmid were assessed (Figure 4). Two of the new bacmids (DE39 and DE40) with the attTn7 inserted in between gp37 and DNApol, and in between v-Ubi and 39 k, displayed an overall lower expression level for all GOIs than the parental bacmid (DE32). The new bacmid with attTn7 inserted in between ORF51 and ORF52 (DE38 or BAC51/52) had a similar expression (90–110%) as the parental bacmid, whereas the new bacmid with attTn7 inserted in the odv-e56 ORF (DE37 or BACe56) displayed enhanced expression levels (110–140%) for all tested GOIs.

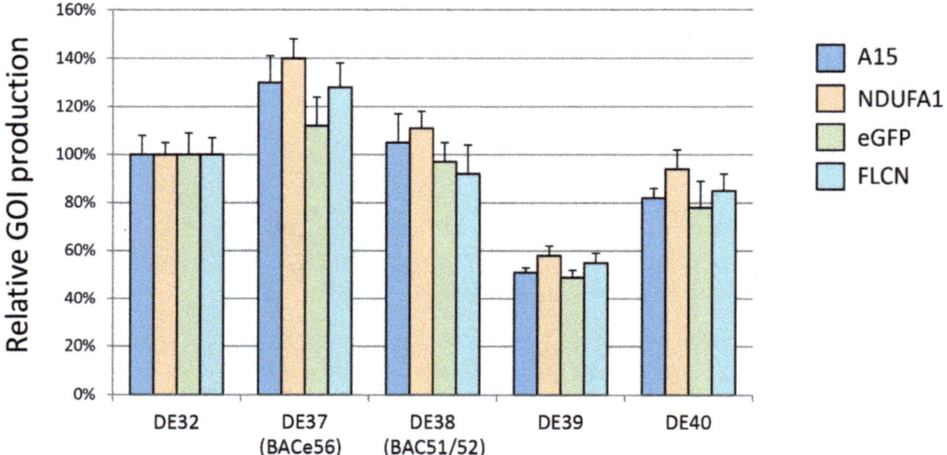

Figure 4. Protein expression in Tni-FNL cells using modified bacmid strains. Protein expression in the modified bacmid strains was measured by green fluorescent protein (GFP) fluorescence (eGFP) or protein quantitation of small-scale IMAC-purified proteins (A15, NDUFA1, FLCN). In all cases, the protein levels were normalized to the level of protein produced in the DE32 control strain. Data represent triplicate measurements and standard deviation is noted with error bars.

3.3. Performance of Novel Bacmids Expressing Chikungunya VLPs upon Serial Undiluted Passage

The two bacmids (BAC51/52 and BACe56) with expression equal to or higher than their parental bacmid were investigated for the stability of heterologous protein expression during serial undiluted passages. As GOI, we chose to express an enveloped virus-like particle (VLP) of chikungunya virus (CHIKV), since this is a glycoprotein-containing complex, capable of self-assembly and inducing a protective immune response in animal trials [22,23,25]. The CHIKV structural genes (capsid, envelope proteins E3, E2, 6K and E1) were expressed from the polh promoter in order to produce CHIKV-enveloped VLPs (Figure 5A) [25]. The recombinant baculoviruses (BAC-CHIKV, BAC51/52-CHIKV and BACe56-CHIKV) were generated by bacmid transfection of Sf21 cells. Serial undiluted passages in Sf21 cells were conducted ten times and in duplicate for each construct (Figure 5B). CHIKV VLP expression was measured by Western bot analysis using an antibody specific to the structural glycoprotein E2, which also recognizes the uncleaved precursor E3E2 [22,26]. The results show that viral titers fluctuated between 10^6 and 10^8 $TCID_{50/mL}$ during serial undiluted passages (Figure 5C). This is similar to what was seen in other studies [19,27]. Cell lysates of selected passages (P1, P4 and P10) were checked for GOI expression levels (Figure 5D).

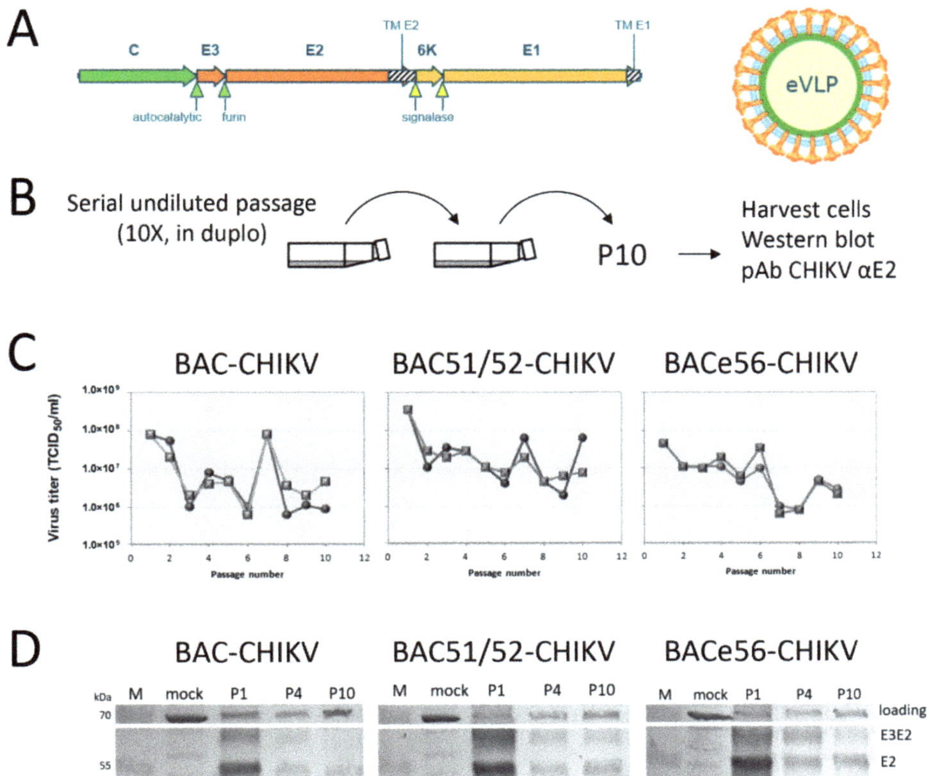

Figure 5. Effect of serial undiluted baculovirus passage on chikungunya virus-like particle (VLP) production. (**A**) Schematic representation of the chikungunya virus (CHIKV) structural genes for insect cell expression of enveloped (e)VLPs using recombinant baculoviruses. Shaded areas represent transmembrane domains. Arrows indicate protease cleavage sites. (**B**) Experimental set-up for serial undiluted baculovirus passage in Sf21 insect cells. (**C**) Viral titers of serial undiluted passaging experiment. Each recombinant baculovirus was passaged in duplicate for 10 passages. Viral titers were determined by end point dilution assay and are expressed as tissue culture infective dose 50% per ml (TCID$_{50}$/$_{\text{mL}}$). (**D**) Chikungunya VLP expression upon serial passage. Western blots detected with anti-E2 polyclonal antiserum show CHIKV glycoprotein E2 and the precursor E3E2. M: protein marker; loading: host cell protein; mock: healthy cells.

Equal amounts of protein were loaded for all samples, and the CHIKV structural proteins E3E2 and E2 were detected by Western blot analysis. For BAC-CHIKV, the expression level of CHIKV (E3)E2 was high at P1, but dramatically decreased at P4 and P10. BAC51/52-CHIKV also had high expression levels at P1, but this was reduced at P4 and P10, although not as low as for BAC-CHIKV. BACe56-CHIKV showed the highest expression at P1 and the least reduction in expression levels at P4 and P10.

3.4. BACe56 Displays Increased Genome Stability and Retains GOI Expression

To better compare the expression levels between the parental bacmid BAC-CHIKV and the novel bacmid BACe56-CHIKV, an infection experiment was conducted in duplicate and at an MOI of 0.5 TCID$_{50}$ per cell, using P9 as inoculum (Figure 6A). In this way, a fair comparison of expression levels can be made. High expression levels of CHIKV envelope proteins (E3E2, E2 and E1) were detected for BACe56-CHIKV, but with the same substrate development time it was not possible to detect CHIKV envelope proteins with the parental bacmid BAC-CHIKV (Figure 6B).

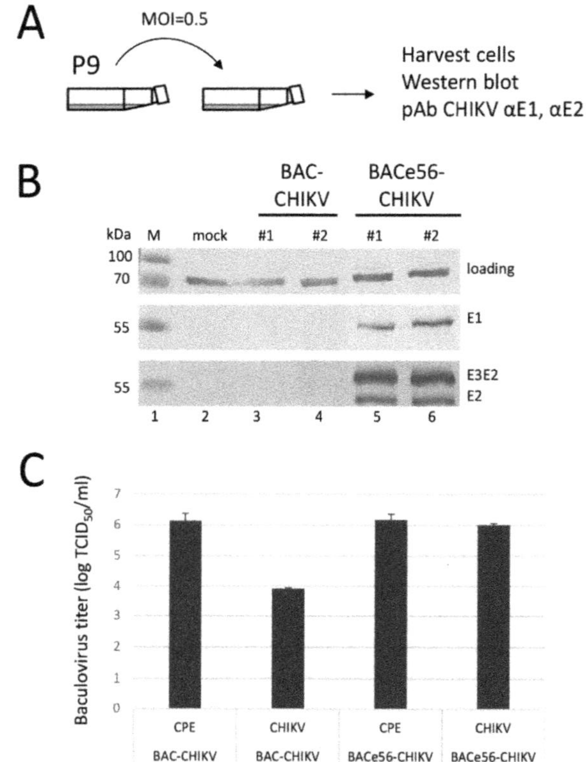

Figure 6. Stability of chikungunya VLP production with novel recombinant baculovirus BACe56. (**A**) Experimental set-up to determine VLP production after 10 duplicate passages. (**B**) Western blots detected with anti-E1 and anti E2 polyclonal antiserum show CHIKV glycoproteins E1, E2 and the precursor E3E2. #1 and #2 represent duplicate serial passage experiments. M: protein marker; loading: host cell protein; mock: healthy cells. (**C**) Titers of the recombinant baculovirus at passage 10. Viral titers were determined by end point dilution assay and are expressed as tissue culture infective doses 50% per ml ($TCID_{50/mL}$). Titers were scored on cytopathic effect (CPE) or on reactivity with anti-E2 polyclonal antiserum (CHIKV).

We hypothesized that the loss of CHIKV expression was the result of GOI deletion from BAC-CHIKV. In order to check whether the GOI had indeed been lost, the viral titers at P10 were determined in two ways: 1) by scoring the microtiter plate by cytopathic effects (CPE) and 2) by scoring after staining with anti-CHIKV E2 antibodies. If there is no loss of the GOI, the CPE and antibody-based titers are the same. If the GOI is lost, the antibody-based titers are lower than the CPE-based titers. The average titer of BAC-CHIKV was determined at 1.5×10^6 (CPE) and 8.2×10^3 (antibody) $TCID_{50/mL}$, whereas BACe56-CHIKV titers were determined at 1.6×10^6 (CPE) and 1.0×10^6 (antibody) $TCID_{50/mL}$ (Figure 6C). These results clearly demonstrate that, for BAC-CHIKV, the antibody-based titer is much lower than the CPE-based titer, meaning the relative proportion of baculoviruses with the GOI retained is less than 1%, whereas for BACe56-CHIKV the CPE- and antibody-based titers are similar. We conclude that BACe56-CHIKV is the most optimal bacmid generated, when both expression levels (Figures 4 and 5) and stability of expression during serial passages (Figures 5 and 6) are taken into account.

3.5. BACe56 Expression Dynamics of a Chikungunya VLP Prototype Vaccine in Suspension Sf9 Cells

To further evaluate the potential of the improved BACe56 bacmid, its performance was investigated at different MOIs. Infections with BACe56-CHIKV were performed at MOIs of 0.01, 0.1, 1 and 5 $TCID_{50}$/cell in shake flasks at a concentration 3×10^6 Sf9 cells/mL (Figure 7). It was observed that the cultures infected with the lowest MOI reached higher maximum cell concentrations (of up to 5×10^6 cells/mL), because uninfected cells were still able to divide (Figure 7A). Furthermore, infected cells had increased cell diameters and showed formation of CHIKV capsid bodies that appear in the nuclei as a result of CHIKV structural gene overexpression [28]. The baculovirus titers were also determined at several time points, which demonstrated that titer development was influenced by the initial MOI (Figure 7B). The infections at the higher MOIs of 1 and 5 $TCID_{50}$/cell reached maximum baculovirus titers from 10 to 28 h post infection. The infections at MOI 0.01 and 0.1 $TCID_{50}$/cell displayed a slight lag in the development of maximum baculovirus titers, which can be explained by the fact that the baculovirus needs additional round(s) of infection to infect all cells. The maximum titers are less dependent on the MOI and range from 10^7–10^8 $TCID_{50}$/mL.

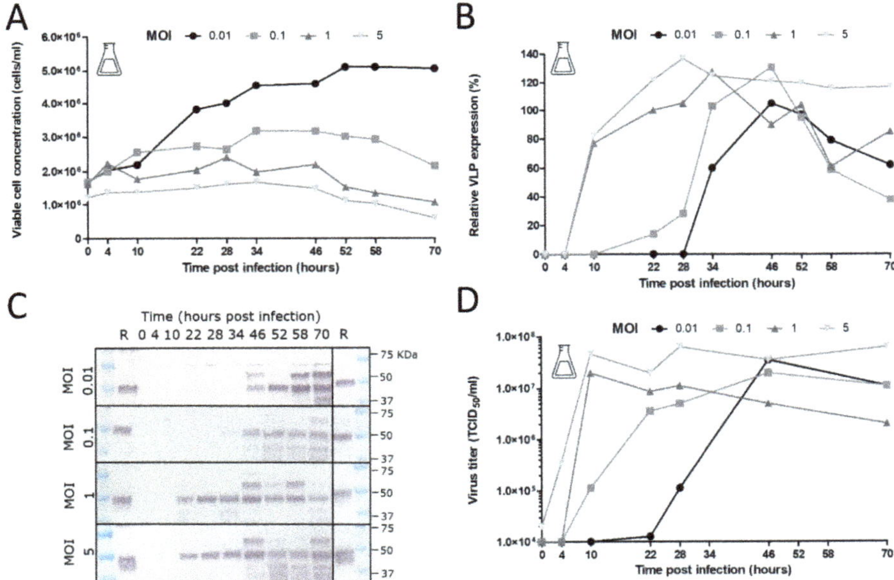

Figure 7. Effect of multiplicity of infection (MOI) on cell growth, baculovirus production and chikungunya VLP expression. (**A**) Cell growth after BACe56-CHIKV infection of Sf9 cells in shake flasks. Baculovirus infection was performed at different MOIs ranging from 0.01 to 5 $TCID_{50}$/cell. (**B**) Baculovirus titers upon BACe56-CHIKV infection as function of MOI. Viral titers were determined by end point dilution assay and are expressed as tissue culture infective dose 50% per ml (TCID50/mL). (**C**) CHIKV VLP expression determined by Western blot detection with anti-E2 polyclonal antiserum. R; reference VLPs (**D**) Relative CHIKV VLP expression levels as function of MOI.

In order to determine heterologous protein expression levels, the CHIKV VLPs in the culture fluid were quantified by Western blot using anti-CHIKV-E2 polyclonal antiserum (Figure 7C). The infections at the higher MOIs (1 and 5 $TCID_{50}$/cell) showed comparable expression of CHIKV VLPs over time with earlier and slightly higher maximum relative expression levels than with the lower MOIs (0.01 and 0.1 $TCID_{50}$/cell) (Figure 7D). It was observed that the peak of CHIKV VLP accumulation was followed by a decrease, likely as a result of cell lysis and released proteases. This influenced the optimum time of harvesting, which was approximately 34 h post infection for the higher MOIs (of 1 and 5 $TCID_{50}$/cell), and 46 h post infection for the lower MOIs (of 0.01 and 0.1 $TCID_{50}$/cell). The CHIKV VLP concentrations at the optimum point of harvesting ranged between 2.6 and 3.8 mg/L.

Next, the effect of cell concentration at time of infection (CCI) on CHIKV VLP expression was investigated (Figure 8). These experiments were performed with a low MOI since comparable product yields were reached for higher MOIs. The use of a low MOI is especially preferred for the development of a large-scale industrial production process as smaller quantities of virus are needed for the final production reactor. However, the use of low MOI requires multiple viral infection cycles to infect the entire cell population as not all cells become initially infected. Shake flask infection experiments with BACe56-CHIKV at aN MOI of 0.01 $TCID_{50}$/cell were performed at a range of CCIs (2×10^6, 3×10^6 and 6×10^6 cells/mL) to gain more insight into the interaction of the infection parameters for this particular production system.

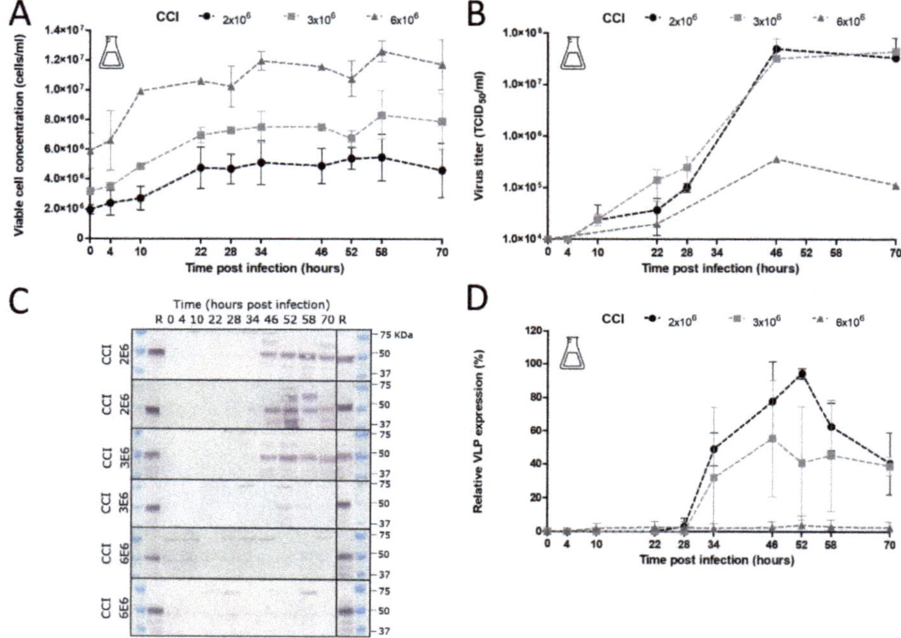

Figure 8. Effect of cell concentration at point of infection (CCI) on cell growth, baculovirus production and chikungunya VLP expression. (**A**) Cell growth after BACe56-CHIKV infection of Sf9 cells in shake flasks. Baculovirus infection was performed at MOI 0.01 $TCID_{50}$/cell and different CCIs ranging from 2×10^6 till 6×10^6 cells/mL in shake flasks. Viral titers were determined by end point dilution assay and are expressed as tissue culture infective dose 50% per ml (TCID50/mL). Error bars are from duplicate experiments. (**B**) Baculovirus titers upon BACe56-CHIKV infection as function of CCI (**C**) CHIKV VLP expression determined by Western blot detection with anti-E2 polyclonal antiserum. R; reference VLPs. (**D**) Relative CHIKV VLP expression levels as function of CCI.

Similar patterns in cell growth were observed with specific growth rates and comparable cell-doubling factors for CCIs of 2×10^6 and 3×10^6 cells/mL (Figure 8A). The effect of CCI on baculovirus production and VLP expression was significant. Infections at a CCI of 2×10^6 and 3×10^6 cells/mL were quite similar, and both conditions demonstrated high baculovirus titers at 46 hpi (Figure 8B) and optimum VLP concentrations between 46 and 52 hpi (Figure 8C,D). In sharp contrast, the infection at a CCI of 6×10^6 cells/mL displayed strongly reduced baculovirus titers (Figure 8B). No VLP expression was observed, neither in the medium fraction nor in the cell fraction (Figure 8C,D).

We conclude that no successful baculovirus infection took place at a CCI of 6×10^6 cells/mL and an MOI of 0.01 TCID$_{50}$/cell.

3.6. BACe56 for Scale Up of CHIKV VLP Production in an Insect Cell Bioreactor

After the optimal MOI, CCI and VLP harvest time had been determined in shake flasks, the translation of these parameters to an Sf9 suspension cell bioreactor was investigated. First, uninfected Sf9 cells were cultured in the bioreactor. The controlled cultivation on larger scale did not have any influence on the cell growth as the uninfected cell culture showed the same growth characteristics and maximum cell density of roughly 1.1×10^7 cells/mL as in shake flask cultivation. Next, an infection experiment was performed with BACe56-CHIKV at an MOI 0.01 TCID$_{50}$/cell and a CCI of 2×10^6 cells/mL. An initial lower growth rate was observed in the bioreactor as compared to the shake flasks, but the cell concentrations were similar from 46 hpi onwards (Figure 9A). The virus titer in the bioreactor developed at the same rate as in shake flasks until 28 hpi. After that, it somewhat slowed down and reached a lower maximum titer as compared to the shake flask of just under 10^7 TCID$_{50}$/mL at 70 hpi (Figure 9B), which is probably a result of the reduced cell growth. Despite these minor differences in cell growth and baculovirus titers, the CHIKV VLP expression in the culture fluid was highly similar between bioreactor and shake flask experiments (Figure 9C,D). The optimum time of harvest in the bioreactor was estimated around 52 hpi. The CHIKV VLP concentration measured in the bioreactor was 2.1 mg/L. This experiment demonstrated that BACe56 can be successfully used for upscaling of the BEVS in an insect cell bioreactor.

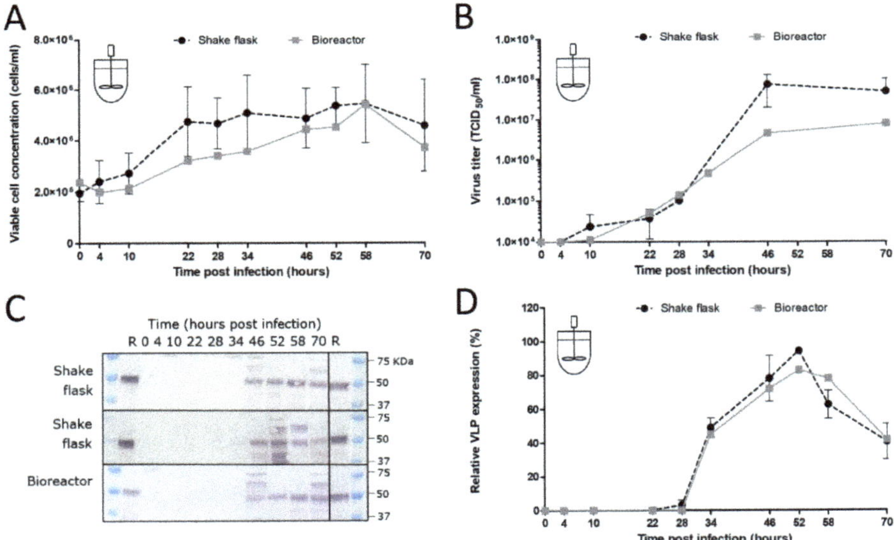

Figure 9. CHIKV VLP production in an insect cell bioreactor (**A**) Cell growth after BACe56-CHIKV infection of Sf9 cells in bioreactor compared to shake flask. Infections were performed at MOI 0.01 TCID$_{50}$/cell and CCI 2×10^6 cells/mL. (**B**) Baculovirus titers upon BACe56-CHIKV infection cells in bioreactor compared to shake flask. Viral titers were determined by end point dilution assay and are expressed as tissue culture infective dose 50% per ml (TCID50/mL). (**C**) CHIKV VLP expression determined by Western blot detection with anti-E2 polyclonal antiserum. R; reference VLPs (**D**) Relative CHIKV VLP expression levels in bioreactor compared to shake flask, determined by Western blot detection with anti-E2 polyclonal antiserum.

4. Discussion

In this study, we scouted the AcMNPV genome for additional landing pads for the attTn7 transposition insertion site to present new options for the insertion of heterologous genes. These regions were initially selected based on data from a previous manuscript, which looked at direct insertion of recombinant protein coding regions into the baculovirus genome [21]. We chose several of the most interesting regions from that work to examine in the context of the Tn7 transposition system. Each region had a unique expression capacity and is, therefore, useful for future experimentation. Most notably, BACe56 (DE37) consistently produced higher levels of protein expression than the control bacmid strain, with some proteins generating 1.4 times the original levels of recombinant protein. This suggests that some aspects of this location, whether due to the stability of the insertion or to some unknown transcriptional enhancer, lead to higher levels of protein production. While a 1.4 time increase in yield might appear to be minimal at a first glance, such an improvement in the production of proteins for therapeutic usage or as a vaccine product could carry significant reductions in cost of operations or the scale of production required.

Two of the other strains, DE38 (BAC51/52) and DE40 (v-Ubi/39k), with similar expression levels to the original polh locus, could potentially be used as alternative sites for the introduction of heterologous genes, or might prove useful as secondary insertion locations if the expression of multiple genes is desired. While the 2x reduced expression from the DE39 (gp37/DNApol) strain may seem to be a negative attribute of this location, in the case of slightly toxic proteins, or proteins which suffer from aggregation or solubility issues when overexpressed, the reduced expression level of DE39 might prove a good solution.

The serial undiluted passaging experiments very clearly showed that it was beneficial to separate the GOI from the unstable mini-F replicon in the bacmid. Based on the results, we could conclude that GOI expression from the odv-e56 (pif-5) locus resulted in the highest overall expression levels and increased stability upon serial passage compared to the original bacmid. The loss of GOI expression upon serial passage of original bacmid constructs was more rapid than what was seen in an earlier study [19]. However, the size of the GOI in this study (CHIKV structural genes, 3.8 kbp) is much larger than the GFP gene (0.7 kbp), which increases the chances of deletion. Given the non-essential nature of the miniF replicon for baculovirus infection, the enhanced GOI stability in BACe56 vs. the original bacmid most likely relates to the relocation of the attTn7 site to a more stable location in the genome. In fact, odv-e56 is located in between baculovirus essential genes (ie-1 and ie-2) that cannot be deleted. Furthermore, instability of certain regions may correlate with a low density of baculovirus homologous repeat regions (hrs), which are dispersed throughout the baculovirus genome and are believed to act as origins of viral DNA replication (oris) [29]. The odv-e56 gene is located quite close to hr1, which may enhance its stability in the genome and perhaps influences the transcriptional activity of nearby genes by functioning as an enhancer. This may explain why BACe56 gives slightly higher recombinant protein yields.

As the demand for the production of recombinant protein complexes increases, the four new bacmids generated in this study will be useful tools for optimizing expression and quality, while remaining the advantages of expression during the very late stage of the lytic cycle. Although these next-generation bacmids have been developed to be compatible with the Bac-to-Bac system, they can be used in combination with any Tn7-based baculovirus cloning method.

The added stability of GOI expression upon serial undiluted passage of BACe56 could have major impacts on its utility for large-scale protein production, which we examined in shake flask and bioreactor experiments. The interplay between MOI, CCI and time of harvest (TOH) of the new bacmid BACe56-expressing CHIKV VLPs in Sf9 suspension cells was investigated. Overall, the reported results showed a clear difference between infections at high MOIs (1 and 5) and low MOIs (0.01 and 0.1), which is in line with other studies [30–33]. Infection of the insect cells at lower MOIs lead to 1–2 infection cycles in the culture and to the proliferation of the uninfected cells and thus higher maximum cell concentrations. Infection at an MOI of five immediately inhibited cell growth because all cells were infected simultaneously. The differences in cell growth, viral infection process and VLP expression observed between infections at diverse MOIs highlights the importance of this infection parameter in the establishment of a production process for recombinant baculovirus-expressed VLPs. Infection at an MOI of 0.01 is preferred for further process developments, since comparable product yields were reached because higher MOIs and smaller quantities of virus are required, which may obliviate the use of additional bioreactors for virus production.

The choice of CCI is also very important since it can cause significant alterations in product yield, especially when low-MOI infection is applied. Baculovirus infection above a certain optimum CCI leads to an ineffective infection with poor baculovirus titers and low product concentrations [32,34,35]. Indeed, we observed unsuccessful infection at a CCI above 4×10^6 cells/mL with no detectable CHIKV VLP yields. The optimum VLP production was achieved at a CCI of 2×10^6 cells/mL, whereas infections at CCIs higher than 3×10^6 cells/mL resulted in a reduction in baculovirus infection and VLP production. With regard to specific protein yields, we generated overall CHIKV VLP yields between 2–4 mg/L. In similar systems, the yields varied from 0.2 mg/L for SARS coronavirus VLPs [36] to 662 mg/L for rotavirus VLPs [37]. Most likely, further optimization of CHIKV VLP yields can be accomplished since CHIKV VLPs up to 40 mg/L were produced in adherent Sf21 cells [22], while 28 mg/L CHIKV VLPs were produced in in Sf21-derived suspension Sf-basic cells [38]. We conclude that BACe56 is a stable vector with improved expression characteristics that is suitable for the production of complex VLP vaccines in insect cell bioreactors.

Author Contributions: Conceptualization, G.P.P., D.E.M. and D.E.; methodology, C.G. (Carissa Grose), T.A.H.H., H.E.B., R.v.d.B., S.R.A., C.G. (Corinne Geertsema), D.E.; formal analysis, C.G. (Carissa Grose), T.A.H.H., H.E.B., R.v.d.B., S.R.A., C.G. (Corinne Geertsema), D.E.; investigation, C.G. (Corinne Geertsema), T.A.H.H., H.E.B., R.v.d.B., S.R.A., C.G. (Corinne Geertsema); writing—original draft preparation, G.P.P., C.G. (Carissa Grose), D.E.; writing—review and editing, G.P.P., M.M.v.O., T.A.H.H., S.R.A., D.E.M., C.G. (Carissa Grose) and D.E.; funding acquisition, G.P.P. and D.E. All authors have read and agreed to the published version of the manuscript.

Funding: This project was funded completely or in part by federal funds from the National Cancer Institute, National Institutes of Health, under contract number HHSN261200800001E. The content of this publication does not necessarily reflect the views or policies of the Department of Health and Human Services, nor does any mention of trade names, commercial products, or organizations imply endorsement by the U.S. Government.

Acknowledgments: We acknowledge the technical assistance of Els Roode (Laboratory of Virology at WU), Fred van den End and Wendy Evers (Bioprocess Engineering at WU), Jennifer Mehalko and Matthew Drew (Protein Expression Laboratory at FNLCR).

Conflicts of Interest: The authors declare no conflict of interest.

References

1.   van Oers, M.M.; Pijlman, G.P.; Vlak, J.M. Thirty years of baculovirus-insect cell protein expression: From dark horse to mainstream technology. *J. Gen. Virol.* **2015**, *96*, 6–23. [CrossRef] [PubMed]

2.   Kost, T.A.; Kemp, C.W. Fundamentals of Baculovirus Expression and Applications. *Adv. Exp. Med. Biol.* **2016**, *896*, 187–197. [CrossRef] [PubMed]

3.   Felberbaum, R.S. The baculovirus expression vector system: A commercial manufacturing platform for viral vaccines and gene therapy vectors. *Biotechnol. J.* **2015**, *10*, 702–714. [CrossRef] [PubMed]

4.   Amanat, F.; Krammer, F. SARS-CoV-2 Vaccines: Status Report. *Immunity* **2020**, *52*, 583–589. [CrossRef]

5.   Summers, M.D. Milestones leading to the genetic engineering of baculoviruses as expression vector systems and viral pesticides. *Adv. Virus Res.* **2006**, *68*, 3–73. [CrossRef]

6.   Possee, R.D.; King, L.A. Baculovirus Transfer Vectors. *Methods Mol. Biol.* **2016**, *1350*, 51–71. [CrossRef]

7.   van Oers, M.M. Opportunities and challenges for the baculovirus expression system. *J. Invertebr. Pathol.* **2011**, *107*, S3–S15. [CrossRef]

8.   Luckow, V.A.; Lee, S.C.; Barry, G.F.; Olins, P.O. Efficient generation of infectious recombinant baculoviruses by site-specific transposon-mediated insertion of foreign genes into a baculovirus genome propagated in Escherichia coli. *J. Virol.* **1993**, *67*, 4566–4579. [CrossRef]

9.   Berger, I.; Fitzgerald, D.J.; Richmond, T.J. Baculovirus expression system for heterologous multiprotein complexes. *Nat. Biotechnol.* **2004**, *22*, 1583–1587. [CrossRef]

10. Pijlman, G.P.; Dortmans, J.C.; Vermeesch, A.M.; Yang, K.; Martens, D.E.; Goldbach, R.W.; Vlak, J.M. Pivotal role of the non-hr origin of DNA replication in the genesis of defective interfering baculoviruses. *J. Virol.* **2002**, *76*, 5605–5611. [CrossRef]

11. Wang, H.; Deng, F.; Pijlman, G.P.; Chen, X.; Sun, X.; Vlak, J.M.; Hu, Z. Cloning of biologically active genomes from a Helicoverpa armigera single-nucleocapsid nucleopolyhedrovirus isolate by using a bacterial artificial chromosome. *Virus Res.* **2003**, *97*, 57–63. [CrossRef] [PubMed]

12. Serrano, A.; Pijlman, G.P.; Vlak, J.M.; Munoz, D.; Williams, T.; Caballero, P. Identification of Spodoptera exigua nucleopolyhedrovirus genes involved in pathogenicity and virulence. *J. Invertebr. Pathol.* **2015**, *126*, 43–50. [CrossRef] [PubMed]

13. Pijlman, G.P.; Pruijssers, A.J.; Vlak, J.M. Identification of pif-2, a third conserved baculovirus gene required for per os infection of insects. *J. Gen. Virol.* **2003**, *84*, 2041–2049. [CrossRef] [PubMed]

14. Boogaard, B.; Evers, F.; van Lent, J.W.M.; van Oers, M.M. The baculovirus Ac108 protein is a per os infectivity factor and a component of the ODV entry complex. *J. Gen. Virol.* **2019**, *100*, 669–678. [CrossRef]

15. Peng, K.; van Lent, J.W.; Vlak, J.M.; Hu, Z.; van Oers, M.M. In situ cleavage of baculovirus occlusion-derived virus receptor binding protein P74 in the peroral infectivity complex. *J. Virol.* **2011**, *85*, 10710–10718. [CrossRef]

16. Javed, M.A.; Biswas, S.; Willis, L.G.; Harris, S.; Pritchard, C.; van Oers, M.M.; Donly, B.C.; Erlandson, M.A.; Hegedus, D.D.; Theilmann, D.A. Autographa californica Multiple Nucleopolyhedrovirus AC83 is a Per Os Infectivity Factor (PIF) Protein Required for Occlusion-Derived Virus (ODV) and Budded Virus Nucleocapsid Assembly as well as Assembly of the PIF Complex in ODV Envelopes. *J. Virol.* **2017**, *91*. [CrossRef]

17. Mehalko, J.L.; Esposito, D. Engineering the transposition-based baculovirus expression vector system for higher efficiency protein production from insect cells. *J. Biotechnol.* **2016**, *238*, 1–8. [CrossRef]

18. Guebre-Xabier, M.; Patel, N.; Tian, J.H.; Zhou, B.; Maciejewski, S.; Lam, K.; Portnoff, A.D.; Massare, M.J.; Frieman, M.B.; Piedra, P.A.; et al. NVX-CoV2373 vaccine protects cynomolgus macaque upper and lower airways against SARS-CoV-2 challenge. *Vaccine* **2020**, *38*, 7892–7896. [CrossRef]

19. Pijlman, G.P.; van den Born, E.; Martens, D.E.; Vlak, J.M. Autographa californica baculoviruses with large genomic deletions are rapidly generated in infected insect cells. *Virology* **2001**, *283*, 132–138. [CrossRef]

20. Pijlman, G.P.; van Schijndel, J.E.; Vlak, J.M. Spontaneous excision of BAC vector sequences from bacmid-derived baculovirus expression vectors upon passage in insect cells. *J. Gen. Virol.* **2003**, *84*, 2669–2678. [CrossRef]

21. Noad, R.J.; Stewart, M.; Boyce, M.; Celma, C.C.; Willison, K.R.; Roy, P. Multigene expression of protein complexes by iterative modification of genomic Bacmid DNA. *BMC Mol. Biol.* **2009**, *10*, 87. [CrossRef] [PubMed]

22. Metz, S.W.; Gardner, J.; Geertsema, C.; Le, T.T.; Goh, L.; Vlak, J.M.; Suhrbier, A.; Pijlman, G.P. Effective chikungunya virus-like particle vaccine produced in insect cells. *PLoS Negl. Trop. Dis.* **2013**, *7*, e2124. [CrossRef] [PubMed]

23. Metz, S.W.; Martina, B.E.; van den Doel, P.; Geertsema, C.; Osterhaus, A.D.; Vlak, J.M.; Pijlman, G.P. Chikungunya virus-like particles are more immunogenic in a lethal AG129 mouse model compared to glycoprotein E1 or E2 subunits. *Vaccine* **2013**, *31*, 6092–6096. [CrossRef] [PubMed]

24. Metz, S.W.; Pijlman, G.P. Arbovirus vaccines; opportunities for the baculovirus-insect cell expression system. *J. Invertebr. Pathol.* **2011**, *107*, S16–S30. [CrossRef] [PubMed]

25. Metz, S.W.; Pijlman, G.P. Production of Chikungunya Virus-Like Particles and Subunit Vaccines in Insect Cells. *Methods Mol. Biol.* **2016**, *1426*, 297–309. [CrossRef]

26. Metz, S.W.; Geertsema, C.; Martina, B.E.; Andrade, P.; Heldens, J.G.; van Oers, M.M.; Goldbach, R.W.; Vlak, J.M.; Pijlman, G.P. Functional processing and secretion of Chikungunya virus E1 and E2 glycoproteins in insect cells. *Virol. J.* **2011**, *8*, 353. [CrossRef]

27. Zwart, M.P.; Pijlman, G.P.; Sardanyes, J.; Duarte, J.; Januario, C.; Elena, S.F. Complex dynamics of defective interfering baculoviruses during serial passage in insect cells. *J. Biol. Phys.* **2013**, *39*, 327–342. [CrossRef]

28. Hikke, M.C.; Geertsema, C.; Wu, V.; Metz, S.W.; van Lent, J.W.; Vlak, J.M.; Pijlman, G.P. Alphavirus capsid proteins self-assemble into core-like particles in insect cells: A promising platform for nanoparticle vaccine development. *Biotechnol. J.* **2016**, *11*, 266–273. [CrossRef]

29. Pijlman, G.P.; de Vrij, J.; van den End, F.J.; Vlak, J.M.; Martens, D.E. Evaluation of baculovirus expression vectors with enhanced stability in continuous cascaded insect-cell bioreactors. *Biotechnol. Bioeng.* **2004**, *87*, 743–753. [CrossRef]

30. Maranga, L.; Brazao, T.F.; Carrondo, M.J. Virus-like particle production at low multiplicities of infection with the baculovirus insect cell system. *Biotechnol. Bioeng.* **2003**, *84*, 245–253. [CrossRef]

31. Radford, K.M.; Cavegn, C.; Bertrand, M.; Bernard, A.R.; Reid, S.; Greenfield, P.F. The indirect effects of multiplicity of infection on baculovirus expressed proteins in insect cells: Secreted and non-secreted products. *Cytotechnology* **1997**, *24*, 73–81. [CrossRef] [PubMed]

32. Wong, K.T.; Peter, C.H.; Greenfield, P.F.; Reid, S.; Nielsen, L.K. Low multiplicity infection of insect cells with a recombinant baculovirus: The cell yield concept. *Biotechnol. Bioeng.* **1996**, *49*, 659–666. [CrossRef]

33. Zheng, Y.Z.; Greenfield, P.F.; Reid, S. Optimized production of recombinant bluetongue core-like particles produced by the baculovirus expression system. *Biotechnol. Bioeng.* **1999**, *65*, 600–604. [CrossRef]

34. Doverskog, M.; Bertram, E.; Ljunggren, J.; Ohman, L.; Sennerstam, R.; Haggstrom, L. Cell cycle progression in serum-free cultures of Sf9 insect cells: Modulation by conditioned medium factors and implications for proliferation and productivity. *Biotechnol. Prog.* **2000**, *16*, 837–846. [CrossRef]

35. Radford, K.M.; Reid, S.; Greenfield, P.F. Substrate limitation in the baculovirus expression vector system. *Biotechnol. Bioeng.* **1997**, *56*, 32–44. [CrossRef]

36. Mortola, E.; Roy, P. Efficient assembly and release of SARS coronavirus-like particles by a heterologous expression system. *FEBS Lett.* **2004**, *576*, 174–178. [CrossRef]

37. Vieira, H.L.; Estevao, C.; Roldao, A.; Peixoto, C.C.; Sousa, M.F.; Cruz, P.E.; Carrondo, M.J.; Alves, P.M. Triple layered rotavirus VLP production: Kinetics of vector replication, mRNA stability and recombinant protein production. *J. Biotechnol.* **2005**, *120*, 72–82. [CrossRef]

38. Wagner, J.M.; Pajerowski, J.D.; Daniels, C.L.; McHugh, P.M.; Flynn, J.A.; Balliet, J.W.; Casimiro, D.R.; Subramanian, S. Enhanced production of Chikungunya virus-like particles using a high-pH adapted spodoptera frugiperda insect cell line. *PLoS ONE* **2014**, *9*, e94401. [CrossRef]

Publisher's Note: MDPI stays neutral with regard to jurisdictional claims in published maps and institutional affiliations.

 *viruses*

Article

Recombinant Lactococcus Expressing a Novel Variant of Infectious Bursal Disease Virus VP2 Protein Can Induce Unique Specific Neutralizing Antibodies in Chickens and Provide Complete Protection

Zhihao Wang [†], Jielan Mi [†], Yulong Wang, Tingting Wang, Xiaole Qi, Kai Li, Qing Pan, Yulong Gao, Li Gao, Changjun Liu, Yanping Zhang, Xiaomei Wang * and Hongyu Cui *

State Key Laboratory of Veterinary Biotechnology, Harbin Veterinary Research Institute, Chinese Academy of Agricultural Sciences, Harbin 150069, China; a17854265231@163.com (Z.W.); mijielan1007@163.com (J.M.); xyylong@126.com (Y.W.); wangtingting0852@163.com (T.W.); qxl@hvri.ac.cn (X.Q.); likaihvri@163.com (K.L.); panqing@caas.cn (Q.P.); ylgao8@163.com (Y.G.); gaoli0820@163.com (L.G.); liucj93711@hvri.ac.cn (C.L.); zhangyanping03@caas.cn (Y.Z.)

* Correspondence: xmw@hvri.ac.cn (X.W.); cuihongyu@caas.cn (H.C.); Tel.: +86-0451-5105-1693 (H.C.)

† These authors contributed equally to this work.

Academic Editor: Carla Varanda

Received: 30 October 2020; Accepted: 23 November 2020; Published: 25 November 2020

Abstract: Recent reports of infectious bursal disease virus (IBDV) infections in China, Japan, and North America have indicated the presence of variant, and the current conventional IBDV vaccine cannot completely protect against variant IBDV. In this study, we constructed recombinant *Lactococcus lactis* (r-*L. lactis*) expressing a novel variant of IBDV VP2 (avVP2) protein along with the Salmonella resistance to complement killing (RCK) protein, and Western blotting analysis confirmed that r-*L. lactis* successfully expressed avVP2-RCK fusion protein. We immunized chickens with this vaccine and subsequently challenged them with the very virulent IBDV (vvIBDV) and a novel variant wild IBDV (avIBDV) to evaluate the immune effect of the vaccine. The results show that the r-*L. lactis*-avVP2-RCK-immunized group exhibited a 100% protection rate when challenged with avIBDV and 100% survival rate to vvIBDV. Furthermore, this immunization resulted in the production of unique neutralizing antibodies that cannot be detected by conventional ELISA. These results indicate that r-*L. lactis*-avVP2-RCK is a promising candidate vaccine against IBDV infections, which can produce unique neutralizing antibodies that cannot be produced by other vaccines and protect against IBDV infection, especially against the variant strain.

Keywords: infectious bursal disease virus; immunization; recombinant Lactococcus lactis; variant strain; vaccine

1. Introduction

Infectious bursal disease virus (IBDV), also known as Gumboro, is the causative agent of a highly infectious disease in chickens. The main features of infectious bursal disease (IBD) is an infection of the central organ of the immune system and the damage of B lymphocytes in the bursa of Fabricius [1,2]. IBD can cause strong immunosuppression in chickens [3], affect the immunological effects of multiple vaccines, such as Newcastle disease, Avian infectious bronchitis, and Chicken Infectious Anemia, and increase the infection rates of bacteria such as *Escherichia coli*, *Salmonella*, and *Staphylococcus aureus* [4–6]. Hence, IBD has received significant attention from the poultry industry, where chickens are protected by vaccination [1]. However, recent reports have suggested that most of the IBDVs popular in Japan, North America, and China were variant IBDVs, and caused huge economic losses [7–11]. Since conventional commercial vaccines cannot provide complete protection against such variant strains [9,12], there is an urgent need to develop a variant IBDV vaccine.

Although attenuated live vaccines offer good prospects for the prevention and treatment of very virulent IBDV (vvIBDV), live IBD vaccines present risks of virulence enhancement and immunosuppression if they are widely used [3,13]. Therefore, these issues have generated increasing interest in the development of subunit vaccines. The neutralizing escape epitope located in the highly variable region of the VP2 protein in IBDV can induce the body to produce neutralizing antibodies to protect the host from IBDV infection [14,15]. VP2 was found to be expressed as a target antigen protein in baculoviruses [16,17], *E. coli* [18,19], yeasts [20,21], and plant and insect cell lines [22]. Thus, VP2 is commonly purified or processed to form a virus-like particle (VP2-VLP) [23], which provides complete immune protection to chickens against IBDV upon immunization [16,24–26]. Our previous studies showed that a recombinant Lactococcus co-expressing the outer membrane protein (Omp) H of the microfold (M) cell-targeting ligand and the major vvIBDV antigens VP2 and a recombinant Lactococcus co-expressing the major vvIBDV antigens VP2 and resistance to complement killing (RCK) protein of Salmonella enterica were promising candidate vaccines to prevent vvIBDV infection [24,25]. However, live vaccines and inactivated vaccines against vvIBDV, vvIBDV VP2 subunit vaccine cannot completely protect against the variant IBDV strains [12]. Therefore, in this study, we designed a subunit vaccine against the variant IBDV strain.

We used *Lactococcus lactis* (*L. lactis*) as the host strain for producing recombinant proteins. Lactic acid bacteria are Gram-positive bacteria that can ferment carbohydrates to lactic acid [27]. *L. lactis* is a food-grade probiotic with non-pathogenic, non-invasive, and non-colonizing properties. Therefore, *L. lactis* is an ideal host for the production of recombinant proteins [28,29]. As a promising candidate for use as antigen carriers [30,31], a major advantage of this system is the ability to safely deliver antigens to the immune system [32]. Therefore, we used *L. lactis* as a vector to express heterologous proteins, which is a common practice [24,25,33–38]. Moreover, to enhance the antigen presentation of avVP2 and improve the immune protection efficiency, we utilized fusion expression avVP2-RCK following our previous study. In our previous study, we co-expressed vvIBDV-VP2-RCK fusion protein and detected high levels of specific neutralizing antibodies against vvIBDV after immunization [24].

In this study, we report the expression of a novel variant IBDV (avIBDV) antigen avVP2-RCK fusion protein in *L. lactis*. Using the nisin-controlled gene expression system, recombinant *L. lactis* was used for injection immunization of chickens. We found that r-*L. lactis*-avVP2-RCK could induce the body to produce a high level of unique specific neutralizing antibodies, which provided complete immune protection against avIBDV, and the survival rate could reach 100% after the challenge with vvIBDV.

2. Materials and Methods

2.1. Experimental Materials

L. lactis NZ3900 and the expression vector pNZ8149 were procured from MoBiTec (MoBiTec, Goettingen, Germany). Chinese vvIBDV reference strain HLJ0504 (GenBank accession: GQ451330 (Segment A); GQ451331 (Segment B), vvIBDV-HLJ0504) and a novel variant IBDV wild strain SHG19 (GenBank accession: MN393076 (Segment A); MN393077 (Segment B), avIBDV-SHG19) were stored at the Harbin Veterinary Research Institute (HVRI) of the Chinese Academy of Agricultural Sciences (CAAS) at −70 °C [39]. VP2 specific mouse monoclonal antibody (MAb) was prepared in our laboratory according to standard procedures [40]. Commercial live vaccine Gt (the licensed attenuated live vaccine) was purchased from the Weike Biotechnology Development Company of China, and commercial live vaccine B87 (the licensed medium virulent live vaccine) was purchased from the Howe Biotechnology Company of China. Lipopolysaccharide (LPS) was purchased from Sigma (Sigma-Aldrich, St. Louis, MO, USA). Phorbol 12-myristate 13-acetate (PMA) and concanavalin A (ConA) were purchased from Invivogen (Invivogen, Toulouse, France). Cell Counting Kit-8 was purchased from Dojindo (Dojindo Laboratories, Kumamoto, Japan).

2.2. Construction of Recombinant Plasmid and Cell Transformation

The avVP2-RCK gene was amplified using the forward primer 5′-AGGCACTCACCATGACAAA TTTAC-3′ and the reverse primer 5′-GTTCAAAGAAAGCTTAAACAACATT-3′ from the plasmid pUC57-avIBDV-VP2-RCK (kindly codon-optimized and synthesized by the Nanjing GenScript Biotechnology Corporation, China). The linear fragment of pNZ8149 was amplified by PCR using the forward primer 5′-GCTTTCTTTGAACCAAAATTAG-3′ and the reverse primer 5′-GGTGAGTGCCTCCTTATAATTTATT-3′. Homologous recombination of the avVP2 fragment and pNZ8149 linear vector was performed using a one-step cloning kit (Vazyme Biotech Co., Nanjing, China). The recombinant vector pNZ8149-avVP2-RCK was transformed into *L. lactis* NZ3900 by electroporation and r-*L. lactis*-avVP2-RCK was selected on Elliker culture agar plates supplemented with 0.5% lactose, and grown at 30 °C according to standard protocols [41]. Subsequently, avVP2-RCK fragments were detected by PCR to identify positive strains.

2.3. Nisin-Induced Expression and Western Blotting Analysis

Nisin-induced expression system was used according to the protocol described previously [24,25]. Briefly, the recombinant lactic acid bacteria cultured overnight were inoculated in Elliker culture solution at a 1:100 ratio and cultured at 30 °C until the optical density at 600 nm (OD600) reached values of 0.4 to 0.5. Nisin (10 ng/mL) was then added to induce expression for 4 to 5 h. After centrifugation at 8000× *g* for 5 min, the cells were resuspended at a 1:10 ratio in sterile water, and the cells were destroyed with the EpiShear probe sonicator for subsequent analysis. The same method was followed for the r-*L. lactis*-pNZ8149 control.

Western blotting was performed to verify the expression of the target protein as described previously [24,25]. In short, the ultrasonically broken protein samples were electrophoretically separated by 8–12% SDS-PAGE. Subsequently, mouse monoclonal anti-VP2 antibody and IRDye 800CW goat anti-mouse secondary antibody (LI-COR Biosciences, Lincoln, NE, USA) were used for detection. Finally, we used an Odyssey imaging system (LI-COR Biosciences) to analyze the protein bands.

2.4. Recombinant VP2 Protein Quantitative Analysis

To quantify the content of avVP2-RCK in the supernatant, a double dilution standard of 2 mg/mL bovine serum albumin was used. The r-*L. lactis*-pNZ8149 lysate (negative control) and r-*L. lactis*-avVP2-RCK lysate (sample) were electrophoretically separated by 8–12% SDS-PAGE, and the gel was stained with Coomassie Brilliant Blue. By drawing the relationship between the gray value of the gradient bovine serum albumin on the *x*-axis and the concentration value on the *y*-axis, an optimal standard curve was generated. The recombinant protein concentration was calculated after subtracting the background effect of the negative control [25].

2.5. Experimental Chickens

Specific-pathogen-free (SPF) chickens were purchased from the Experimental Animal Center of the Harbin Veterinary Research Institute of the Chinese Academy of Agricultural Sciences (HVRI of CAAS, Harbin, China) and raised in a negative-pressure isolator fitted with an air filter. All animal experiments were approved by the HVRI of CAAS and carried out by animal ethics guidelines and approved protocols (SYXK (Hei) 2017-009). In this study, the animal experimental permission date was 04-23-2020, and the number of experimental animals was 93.

2.6. Immunoprotection Experiment

r-*L. lactis*-avVP2-RCK was inoculated and induced with nisin as described above. The collected cells were washed twice with sterile phosphate buffer saline (PBS) and then suspended in sterile PBS to immunize at an appropriate concentration (1×10^9 CFU/mL r-*L. lactis*-avVP2-RCK and controls in a 500 µL volume) and then inactivated at 70 °C for 10 min.

The 15-day-old SPF chickens were randomly divided into four groups. In group one, 23 chickens were immunized with 500 µL inactivated r-*L. lactis*-avVP2-RCK by intramuscular injection. In group two, 20 chickens were immunized with the Gt live vaccine by intraocular-nasal route. In group three, 20 chickens were immunized with the B87 live vaccine by intraocular-nasal route. In group four, 30 chickens were used as a non-immunized healthy control group. After 15 days of immunization, SPF chickens in each group were regrouped to challenge with Chinese vvIBDV reference strain HLJ0504 (vvIBDV-HLJ0504) and a novel variant IBDV wild strain SHG19 (avIBDV-SHG19). In group one, 10 chickens were challenged with vvIBDV-HLJ0504, 10 chickens challenged with avIBDV-SHG19, and 3 were not challenged with any virus. In groups two and three, 10 chickens from each group were challenged with vvIBDV-HLJ0504, and 10 chickens from each group were challenged with avIBDV-SHG19. In group four, 10 chickens were challenged with vvIBDV-HLJ0504, 10 chickens were challenged with avIBDV-SHG19, and 10 chickens were not challenged with the virus. The challenge dose of vvIBDV-HLJ0504 was 10^3 ELD$_{50}$ (median embryo lethal dose, ELD$_{50}$) and the challenge dose of the avIBDV-SHG19 was 10 BAD$_{50}$ (50% buysae atrophy dose, BAD$_{50}$). After 7 days of challenge, the surviving animals were put to death, and autopsies were performed. We then isolated peripheral blood mononuclear cells (PBMCs) from fresh anticoagulant blood to detect their proliferative activity. The bursa/body weight index (BBIX) and bursa/body weight were calculated as described previously [25]. Serums were collected two weeks and three weeks after immunization, and one week after challenge, and then stored at −80 °C.

2.7. Serological ELISA Antibody Detection and Neutralization Test

Neutralization tests to detect serum neutralizing antibodies have been described previously [9,42]. Briefly, the serum collected after two weeks of immunization was diluted twice (briefly, 100 µL of serum was mixed with 100 µL of RPMI 1640 culture medium in the first hole, then 100 µL of the mixture was mixed with 100 µL of RPMI 1640 culture medium in the second hole, starting from 1:2^1 to 1:2^{12}), and 100 µL of the diluted serum was mixed with 100 µL of 200 TCID$_{50}$ (50% tissue culture infective dose, TCID$_{50}$) vvIBDV-HLJ0504 or avIBDV-SHG19 and incubated for 1 h at 37 °C in a cell culture incubator. Thereafter, this serum and virus mixture (100 µL) was added to DT40 cells and incubated at 37 °C for 24 h. The infection-positive wells were detected by immunofluorescence assays. Meanwhile, the serum samples after two weeks of immunization were tested with IBDV VP2-coated ELISA kit (IDEXX IBD-XR Ab Tests kit, Westbrook, ME, USA) and IBD whole virus-coated ELISA (IDEXX IBD Ab Tests kit, Westbrook, ME, USA).

2.8. Measurement of PBMC Cell Proliferation Activity

The proliferative activity of PBMCs stimulated by concanavalin A (ConA) and Phorbol 12-myristate 13-acetate (PMA) was measured as described previously [42]. Isolated chicken PBMCs were obtained using a peripheral chicken blood mononuclear cell isolate kit (TBD, Tianjin, China). PBMCs were diluted to 1×10^6 cells/mL with RPMI 1640 culture medium containing 10% fetal bovine serum (Gibco, Gaithersburg, MD, USA) and penicillin-streptomycin (Wisent, Saint-Jean-Baptiste, QC, Canada). Then, 100 µL PBMCs were added to each well of the 96-well plates. Next, 5 µg/mL ConA and 100 ng/mL PMA were added to each well and maintained at 37 °C in a cell incubator for 48 h. Subsequently, 10 µL CCK-8 was added for 4 h, and absorbance was measured at 450 nm.

The proliferative activity of PBMCs stimulated with Lipopolysaccharide (LPS) was measured as described previously [43]. PBMC was diluted to 1×10^6 cells/mL with RPMI 1640 culture medium containing 10% fetal bovine serum (Gibco, Gaithersburg, MD, USA) and penicillin-streptomycin (Wisent, Saint-Jean-Baptiste, QC, Canada). Subsequently, 100 µL of this solution were added to a 96-well plate, along with LPS at a final concentration of 1 µg/mL per well, and maintained at 37 °C for

72 h in a cell culture incubator. Next, 10 µL CCK-8 were added for 4 h, and absorbance was measured at 450 nm.

2.9. Cytokine ELISA

The serum collected two weeks after immunization and one week after challenge with the virus was diluted to a 1:2 ratio, and interleukins (IL) IL-4, IL-10, and IL-12, and interferon-γ (IFN-γ) were detected using Indirect fluorescence ELISA kit (Solarbio, Beijing, China) according to the manufacturer's instructions. Cytokine concentration was calculated according to the standard curve.

2.10. Statistical Analysis

Data were analyzed using Prism 6 software (GraphPad Inc., La Jolla, CA, USA). One-way ANOVA was used to evaluate the statistical significance of the differences among different groups, and results with $p < 0.05$ were considered statistically significant.

3. Result

3.1. Construction of Recombinant Lactic Acid Bacteria and Expression and Quantitative Analysis of Target Protein

First, we amplified the 2505 bp pNZ8149 linear vector and the 1491 bp avVP2-RCK target fragment with homology arms (Figure 1A Lanes 2 and 3), and then transferred the recombinant plasmid into *L. lactis*. The r-*L. lactis* expressing the target fragment avVP2-RCK were successfully screened by PCR (Figure 1B Lanes 2–6). After sonicating the r-*L. lactis*, the expression of target protein avVP2-RCK (about 52 kDa) was verified in r-*L. lactis*-avVP2-RCK NZ3900 (Figure 1C, Lane 2; Figure 1D), whereas the target protein avVP2-RCK was not expressed in the empty vector strain (Figure 1C Lane 3). Quantitative analysis showed that the expression level of the target protein was 55.4 µg/mL in r-*L. lactis*-avVP2-RCK.

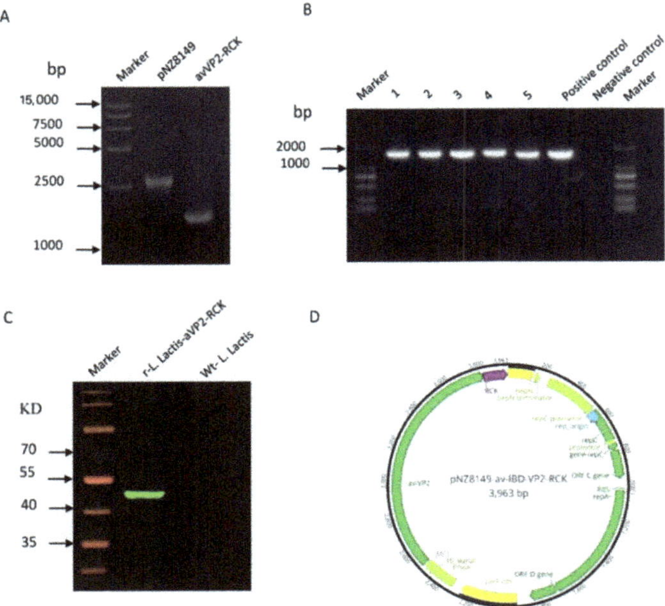

Figure 1. Construction of the plasmid pNZ8149-avVP2-RCK expressing the avVP2-RCK fusion protein

and identification of recombinant proteins via Western blotting analysis. (**A**) PCR amplification of pNZ8149 (2505 bp, lane 2) and avVP2-RCK (1491 bp, lane 3) optimization gene. (**B**) Amplified avVP2-RCK fragments (1491 bp, lanes 2–6) by bacterial solution PCR to screen positive strains. (**C**) Immunoblot analysis of total whole-cell protein extracts from r-*L. lactis*-avVP2-RCK. (**D**) Schematic diagrams of the recombinant plasmid pNZ8149-avVP2-RCK with the avVP2-RCK gene fusion.

3.2. Recombinant Lactococcus Induces Unique and Specific Neutralizing Antibodies

Two weeks after immunization, the level of the neutralizing antibody of the r-*L. lactis*-avVP2-RCK group serum against avIBDV-SHG19 reached $1:2^{4-6}$ (Figure 2A), extremely significantly higher than the Gt group ($1:2^0$), B87 Group ($1:2^0$), and the blank control group ($1:2^0$) ($p < 0.0001$) (Figure 2A). The level of neutralizing antibodies of the r-*L. lactis*-avVP2-RCK group serum against vvIBDV-HLJ0504 reached $1:2^{1-2}$ (Figure 2B), which was extremely significantly lower than the Gt group ($1:2^{4-6}$) ($p < 0.0001$) and B87 group ($1:2^{3-4}$) ($p < 0.0001$), but extremely significantly higher than the blank control group ($1:2^0$) ($p < 0.0001$) (Figure 2B).

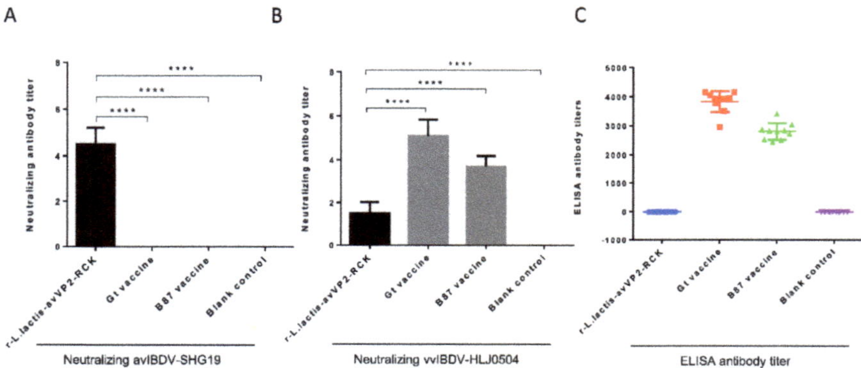

Figure 2. Serum antibody response after two weeks of immunization. Detection and comparison of virus neutralization antibodies against avIBDV-SHG19 (**A**) and vvIBDV-HLJ0504 (**B**) in chickens. (**C**) Detection of Elisa antibodies against IBDV virus in chickens two weeks after vaccination. Statistical significance was set at **** $p < 0.0001$. IBDV: infectious bursal disease virus.

Of particular importance is that the serum antibody titers in the r-*L. lactis*-avVP2-RCK group is negative when using the whole virus coated ELISA (Figure 2C), while the serum antibody titers in the Gt group reached close to 4000 (Figure 2C). The ELISA serum antibody titers in the B87 group reached approximately 3000 (Figure 2C); similar results were obtained using the IBDV VP2-coated ELISA (data not shown), and we repeated the ELISA test multiple times and got the same results. The above results show that injection of r-*L. lactis*-avVP2-RCK can produce a specific immune response against the avIBDV-VP2, but it induces a unique and specific neutralizing antibody.

3.3. Immunoprotective Effects against vvIBDV and avIBDV

One week after being challenged with avIBDV-SHG19, the survival rates of chickens from the r-*L. lactis*-avVP2-RCK, Gt live vaccine, B87 live vaccine, blank control, and challenge control groups were all 100% (Figure 3A). Necropsy revealed that the bursa/body ratio of chickens from the r-*L. lactis*-avVP2-RCK group was extremely significantly higher than that of chickens from the B87 live vaccine group ($p < 0.001$) and the challenge control group ($p < 0.0001$). However, no significant difference was observed in the bursa/body ratio of chickens from the r-*L. lactis*-avVP2-RCK group, and those from the Gt live vaccine group and the blank control group ($p > 0.05$) (Figure 4A). The BBIX of all chickens in the r-*L. lactis*-avVP2-RCK group was greater than 0.7 (Figure 4B), which was extremely significantly higher than that of chickens from the B87 live vaccine group ($p < 0.001$) and the challenge control group ($p < 0.0001$). However, the index of the r-*L. lactis*-avVP2-RCK group chickens was not significantly different from that of the Gt live vaccine group and the blank control group chickens ($p > 0.05$) (Figure 4B). These results show that r-*L. lactis*-avVP2-RCK immunization can completely protect against avIBDV-SHG19.

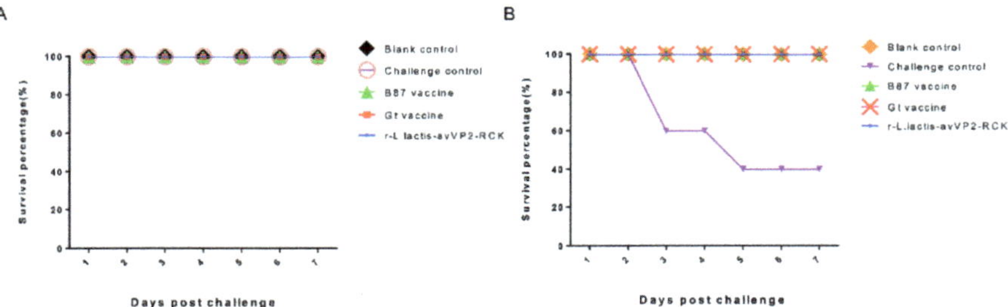

Figure 3. Protective efficacy of every group against avIBDV-SHG19 (**A**) and vvIBDV-HLJ0504 (**B**) in immunized chicken and survival rate over 7 days.

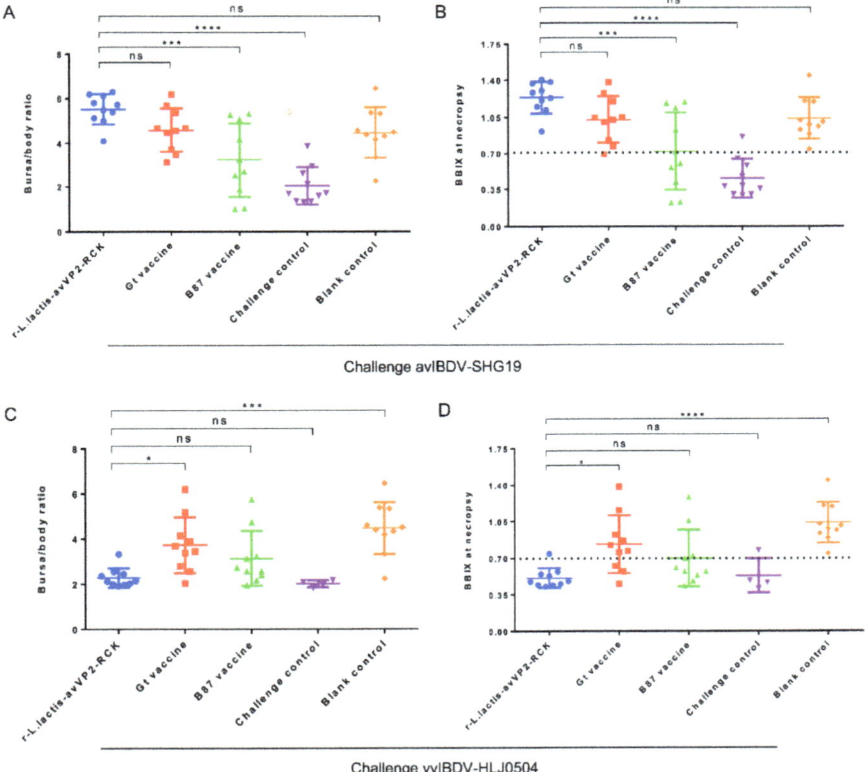

Figure 4. Analysis of protection against IBDV challenge determined by calculating BBIX values and bursa/body ratio. Bursa/body ratio (**A**) and BBIX (**B**) after challenge avIBDV-SHG19. (**C**,**D**) Bursa/body ratio (**C**) and BBIX (**D**) after challenge vvIBDV-HLJ0504. Surviving birds were euthanized and analyzed after the observation period on day 7 post challenge. Statistical significance was set at * $p < 0.05$, *** $p < 0.001$, **** $p < 0.0001$. NS, no significance; BBIX, bursa/body weight index.

One week after the challenge of vvIBDV-HLJ0504, the survival rates of chickens from the r-*L. lactis*-avVP2-RCK, Gt live vaccine, B87 live vaccine, and blank control groups were 100%, while the survival rate of the chickens from the challenge control group was 40% (Figure 3B). Necropsy revealed that the bursa/body ratio of chickens from the r-*L. lactis*-avVP2-RCK group was extremely significantly lower than that of chickens from the blank control group ($p < 0.001$) and significantly lower than that of the Gt live vaccine group ($p < 0.05$). However, the bursa/body ratio of chickens of the r-*L. lactis*-avVP2-RCK group was not significantly different from that of chickens from the B87 blank control group or the challenge control group ($p > 0.05$) (Figure 4C). The BBIX of chickens from the r-*L. lactis*-avVP2-RCK group was less than 0.7 (Figure 4D), which was extremely significantly lower than that of chickens from the blank control group ($p < 0.0001$); significantly lower than the Gt live vaccine group ($p < 0.05$). However, it was not significantly different from chickens from the B87 blank control group and the challenge control group ($p > 0.05$) (Figure 4D). These results suggest that, although the survival rate of the immunized group was 100%, immunization did not protect the bursa of these chickens.

3.4. Peripheral PBMCs Proliferation Activity

PBMCs collected two weeks after immunization and one week after challenge were stimulated by ConA and PMA. The results showed that the cell proliferation activity of PBMCs was not significantly different among the different groups after two weeks of immunization ($p > 0.05$). After challenge with avIBDV-SHG19 for one week, cell proliferation activity of peripheral PBMCs of the r-*L. lactis*-avVP2-RCK group was significantly higher than that of the challenge control group ($p < 0.01$) and the blank control group ($p < 0.0001$) (Figure 5A). After challenge with vvIBDV-HLJ0504, cell proliferation activity of peripheral PBMCs in the r-*L. lactis*-avVP2-RCK group was significantly lower than that of the challenge control group ($p < 0.01$) and significantly higher than that of the blank control group ($p < 0.01$) (Figure 5B). These results show that r-*L. lactis*-avVP2-RCK can significantly increase the proliferation activity of specific T cells in PBMCs, which is beneficial for the immune response against infection.

PBMCs collected two weeks after immunization and one week after challenge were also stimulated by LPS. The results showed that the cell proliferation activity of PBMCs was not significantly different among the different groups after two weeks of immunization ($p > 0.05$). After challenge with avIBDV-SHG19 for one week, the cell proliferation activity of peripheral PBMCs in the r-*L. lactis*-avVP2-RCK group was extremely significantly higher than that of the challenge control group ($p < 0.001$) and the blank control group ($p < 0.0001$) (Figure 5C). After challenge with vvIBDV-HLJ0504, the proliferation activity of peripheral PBMCs in the r-*L. lactis*-avVP2-RCK group was extremely significantly higher than that of the challenge control group ($p < 0.0001$) and the blank control group ($p < 0.0001$) (Figure 5D). Thus, these results show that immunization with r-*L. lactis*-avVP2-RCK can significantly increase the proliferation activity of antigen-presenting cells in PBMCs, which is beneficial for the immune response against infection.

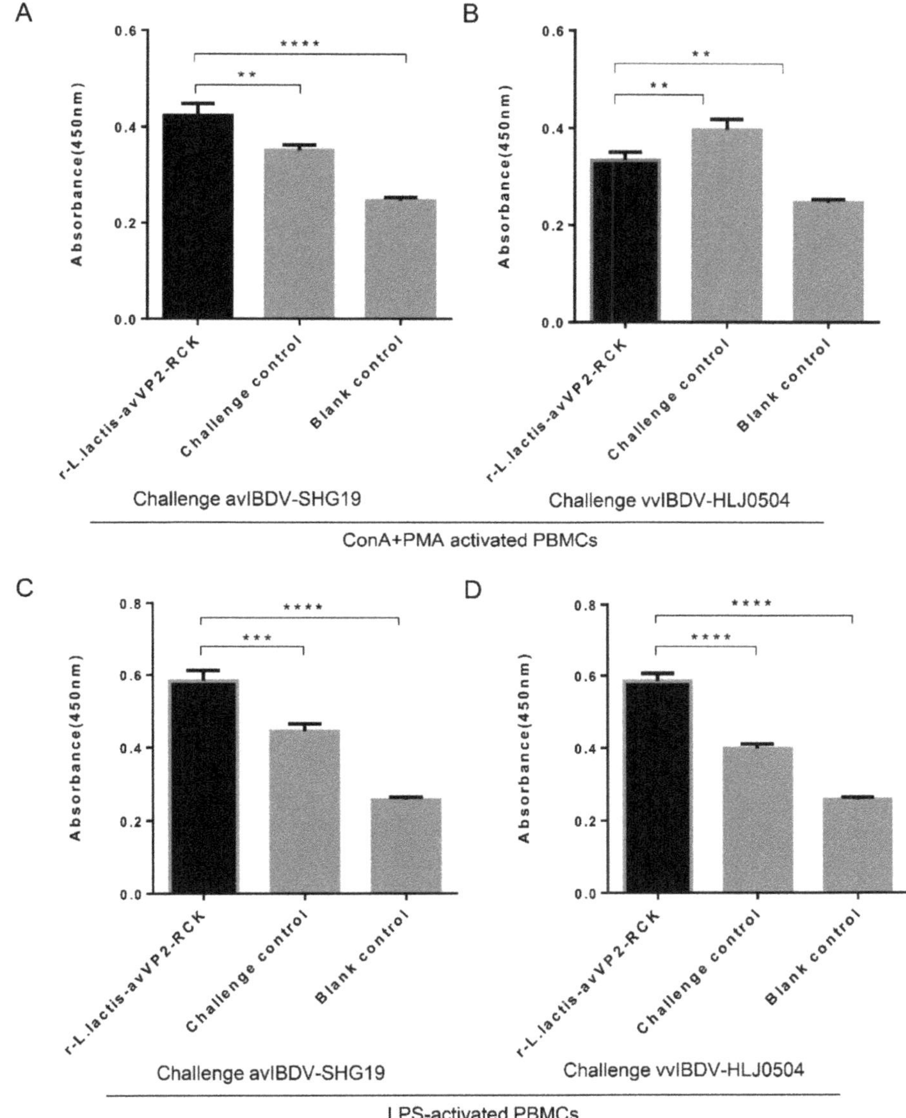

Figure 5. The peripheral PBMCs in ConA+PMA; LPS-stimulated cell proliferation activity after two weeks after immunization and one week of challenge IBDV. The peripheral PBMCs in ConA+PMA-stimulated cell proliferation activity after challenge avIBDV-SHG19 (**A**) and vvIBDV-HLJ0504 (**B**). The peripheral PBMCs in LPS-stimulated cell proliferation activity after challenge avIBDV-SHG19 (**C**) and vvIBDV-HLJ0504 (**D**). Statistical significance was set at ** $p < 0.01$, *** $p < 0.001$, **** $p < 0.0001$.

3.5. ELISA Detects Serum Cytokines

Serum cytokines were tested one week after the challenge with avIBDV-SHG19, and the results showed that the level of IL-4 in the serum of the r-*L. lactis*-avVP2-RCK group was extremely significantly lower than that of the challenge control group ($p < 0.0001$), the immune non-challenge control group

($p < 0.01$), but not significantly different from that of the blank control group ($p > 0.05$) (Figure 6A). However, the level of IL-10 in the serum of the r-*L. lactis*-avVP2-RCK group was extremely significantly lower than that of the challenge control group ($p < 0.0001$) and significantly lower than the blank control group ($p < 0.01$), but significantly higher than that of the immune non-challenge control group ($p < 0.01$) (Figure 6B). The level of IL-12 in the serum of the r-*L. lactis*-avVP2-RCK group was significantly lower than that of the challenge control group ($p < 0.01$) and the blank control group ($p < 0.01$), but extremely significantly higher than that of the immune non-challenge control group ($p < 0.001$) (Figure 6C). Similarly, the level of IFN-γ in the serum of r-*L. lactis*-avVP2-RCK group was extremely significantly higher than that of the challenge control group ($p < 0.0001$), the blank control group ($p < 0.0001$), and the immune non-challenge control group ($p < 0.0001$) (Figure 6D). These results show that vaccination with r-*L. lactis*-avVP2-RCK significantly improved the immune response against infection without causing adverse inflammatory reactions.

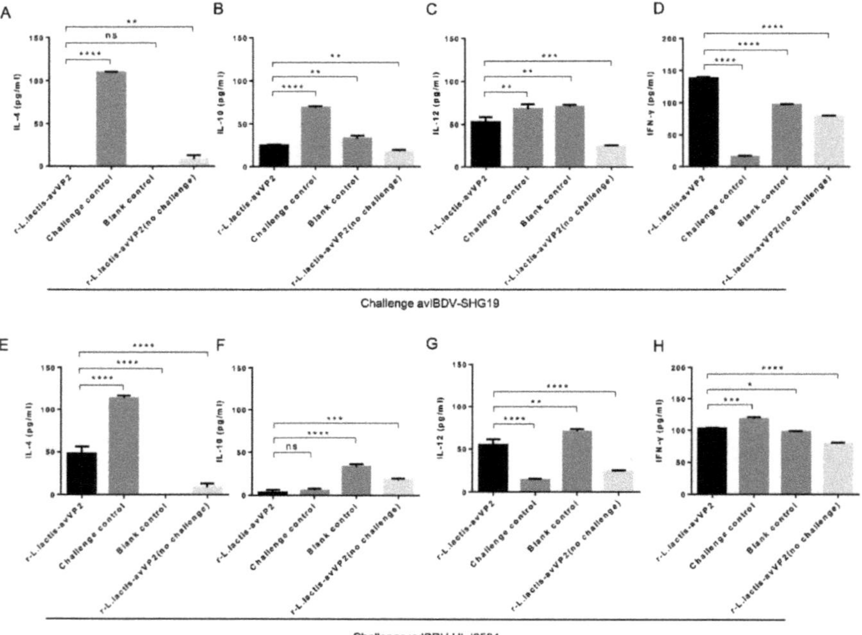

Figure 6. Elisa was used to detect cytokines in serum collected one week after challenge. Detection of IL-4 (**A**), IL-10 (**B**), IL-12 (**C**), and IFN-γ (**D**) antibody in serum after challenge variant IBDV-SHG19. Detection of IL-4 (**E**), IL-10 (**F**), IL-12 (**G**), and IFN-γ (**H**) after challenge vvIBDV-HLJ0504. Statistical significance was set at * $p < 0.05$, ** $p < 0.01$, *** $p < 0.001$, **** $p < 0.0001$. NS, no significance.

The serum cytokines were also tested one week after challenge with vvIBDV-HLJ0504. The results showed that the level of IL-4 in the serum of the r-*L. lactis*-avVP2-RCK group was extremely significantly lower than that of the challenge control group ($p < 0.0001$), but extremely significantly higher than that of the blank control group ($p < 0.0001$) and the immune non-challenge control group ($p < 0.0001$) (Figure 6E). When we examined IL-10, we found that the level of IL-10 in the serum of the r-*L. lactis*-avVP2-RCK group was extremely significantly lower than that of the blank control group ($p < 0.0001$) and the immune non-challenge control group ($p < 0.001$), but not significantly different from that of the challenge control group ($p > 0.05$) (Figure 6F). The level of IL-12 in the serum of the r-*L. lactis*-avVP2-RCK group was extremely significantly higher than that of the challenge control group ($p < 0.0001$) and the immune non-challenge control group ($p < 0.0001$), but significantly lower than that of the blank control group ($p < 0.01$) (Figure 6G). In contrast, the level of IFN-γ in the serum of the r-*L. lactis*-avVP2-RCK group was extremely significantly lower than that of the challenge control group ($p < 0.001$), but significantly higher than that of the blank control group ($p < 0.05$), and extremely significantly higher than the immune non-challenge control group ($p < 0.0001$) (Figure 6H). These results show that the immunization with r-*L. Lactis*-avVP2-RCK can also improve the anti-infective immune response to typical virulent toxins without causing adverse inflammation.

4. Discussion

In this research, a very important discovery was that the injection of inactivated r-*L. lactis*-avVP2-RCK into SPF chickens induces to produce unique specific neutralizing antibodies, which provide complete protection of avIBDV-SHG19. However, more importantly, this unique neutralizing antibody cannot be detected by conventional IBD antibody ELISA kits, and ELISA cannot detect unique neutralizing antibodies. In contrast, many live IBDV vaccines, inactivated vaccines, and VP2-VLP subunit vaccines can produce detectable IBDV ELISA antibodies in the serum of immunized chickens [24,25,44–46]. The high neutralization titer of avIBDV-SHG19 was detected by neutralization test in chickens immunized with r-*L. lactis*-avVP2-RCK. In order to verify that the neutralization titer was indeed produced by neutralizing antibodies, we prepared r-*L. lactis*-avVP2-RCK for SDS-PAGE detection; the specific protein bands of avVP2-RCK were observed in chicken serum 2 weeks after immunization as the primary antibody for Western blots, because such antibodies were not detected in ELISA kits coated with both full virus and VP2. These results suggest that r-*L. lactis*-avVP2-RCK induces a unique VP2 neutralizing antibody that is different from those produced by other vaccines. The novel neutralizing antibody provides an additional protective barrier for the prevention and control of IBDV. We speculate that this phenomenon may have multiple reasons. First, the VP2-RCK fusion protein may form a new spatial and topological structure in lactic acid bacteria, further affecting its processing and presentation pathways, and forming unique antibodies following different B cell epitopes. Alternatively, the processing and presentation of VP2 by lactic acid bacteria as a particle antigen may contribute to this phenomenon. However, further research is needed to distinguish these possibilities.

L. lactis is an important microorganism used in industry as a homozygous bacterium in the fermentation of many dairy products, and the host strain NZ3900 used in this study is a derivative strain of *L. lactis* MG1363 [28,47–49]. We expressed the VP2 protein of avIBDV in *L. lactis* strain NZ3900 and used r-*L. lactis*-avVP2-RCK for injection immunization of chickens. Since the NZ3900 strain is absent in the intestinal tract of chickens, the balance of intestinal flora would not be disturbed when the chicken immune system reacts to this strain. Additionally, the nucleic acids and cell wall components of lactic acid bacteria have good immune adjuvant activity, which promotes the immune effect of r-*L. lactis*-avVP2-RCK in chickens [28,47]. However, further studies are needed to determine whether some of the proteins in r-*L. lactis*-avVP2-RCK had any effect on chickens after injection. While screening host strains, we did not choose the lactic acid bacteria in the chicken intestine as the host strain, because we feel that immune responses of the chickens against this injected bacteria may also target intestinal probiotics, which may lead to an imbalance of intestinal flora. However, further studies are needed to prove this effect.

5. Conclusions

In this study, we constructed r-*L. lactis* expressing avIBDV VP2 as a subunit vaccine. We combined the RCK protein of *Salmonella enterica* and the VP2 gene of the variant IBDV within the pNZ8149 vector, transferred it into the *L. lactis* NZ3900 strain, and successfully expressed the avVP2-RCK protein. We showed that r-*L. lactis*-avVP2-RCK can induce unique and specific neutralizing antibodies against avVP2, and can completely protect chickens from avIBDV-SHG19, such that chickens challenged with vvIBDV-HLJ0504 exhibit a 100% survival rate. Additionally, after challenging with avIBDV, the cell proliferation activity of peripheral PBMCs was enhanced and the body was induced to produce high levels of IFN-γ, which reduced the body's inflammatory response. This indicates that *L. lactis* is a potentially useful tool for vaccine development. However, the ELISA titers of serum antibodies were all negative, indicating that the immune delivery system of recombinant lactic acid bacteria is different from other conventional IBDV vaccines. It is suggested that immunization of our recombinant lactic acid bacteria can produce unique antibodies that other vaccines cannot produce, can compensate for the immune protection that other vaccines cannot produce, thus establishing a new immune protection barrier, so it is of great practical significance for IBD prevention and control. Hence, further studies are required to elaborate on the major mechanisms of immune protection of the recombinant Lactococcus.

Author Contributions: H.C. and X.W. conceived of and designed the study; Z.W., J.M., Y.W., T.W., K.L., Q.P., X.Q., L.G., Y.G., C.L., and Y.Z. performed the experiments and analyzed the data; H.C. contributed technical knowledge; Z.W. and H.C. wrote the paper; Y.G. and X.Q. were involved in the interpretation of the results and critical reading of the manuscript. All authors have read and agreed to the published version of the manuscript.

Funding: This research was supported by grants from "13th Five-year" National Key Research and Development Plan (No. 2017YFD0500802; 2016YFE0203200); the National Natural Science Foundation of China (No. 31570930); the China Agriculture Research System (CARS-41-G15); and the Natural Science Foundation of Heilongjiang Province (TD2019C003).

Conflicts of Interest: The authors declare no conflict of interest.

References

1. Mahgoub, H.A.; Bailey, M.; Kaiser, P. An overview of infectious bursal disease. *Arch. Virol.* **2012**, *157*, 2047–2057. [CrossRef]
2. Berg, T.P.; Gonze, M.; Meulemans, G. Acute infectious bursal disease in poultry: Isolation and characterisation of a highly virulent strain. *Avian Pathol.* **1991**, *20*, 133–143. [CrossRef]
3. Saif, Y.M. Immunosuppression induced by infectious bursal disease virus. *Vet. Immunol. Immunopathol.* **1991**, *30*, 45–50. [CrossRef]
4. Allan, W.H.; Faragher, J.T.; Cullen, G.A. Immunosuppression by the infectious bursal agent in chickens immunised against Newcastle disease. *Vet. Rec.* **1972**, *90*, 511–512. [CrossRef]
5. Santivatr, D.; Maheswaran, S.K.; Newman, J.A.; Pomeroy, B.S. Effect of infectious bursal disease virus infection on the phagocytosis of Staphylococcus aureus by mononuclear phagocytic cells of susceptible and resistant strains of chickens. *Avian Dis.* **1981**, *25*, 303–311. [CrossRef]
6. Rosenberger, J.K.; Gelb, J. Response to several avian respiratory viruses as affected by infectious bursal disease virus. *Avian Dis.* **1978**, *22*, 95–105. [CrossRef] [PubMed]
7. Zachar, T.; Popowich, S.; Goodhope, B.; Knezacek, T.; Ojkic, D.; Willson, P.; Ahmed, K.A.; Gomis, S. A 5-year study of the incidence and economic impact of variant infectious bursal disease viruses on broiler production in Saskatchewan, Canada. *Can. J. Vet. Res.* **2016**, *80*, 255–261. [PubMed]
8. Ojkic, D.; Martin, E.; Swinton, J.; Binnington, B.; Brash, M. Genotyping of Canadian field strains of infectious bursal disease virus. *Avian Pathol.* **2007**, *36*, 427–433. [CrossRef] [PubMed]
9. Fan, L.; Wu, T.; Hussain, A.; Gao, Y.; Zeng, X.; Wang, Y.; Gao, L.; Li, K.; Wang, Y.; Liu, C.; et al. Novel variant strains of infectious bursal disease virus isolated in China. *Vet. Microbiol.* **2019**, *230*, 212–220. [CrossRef] [PubMed]
10. Xu, A.; Pei, Y.; Zhang, K.; Xue, J.; Ruan, S.; Zhang, G. Phylogenetic analyses and pathogenicity of a variant infectious bursal disease virus strain isolated in China. *Virus Res.* **2020**, *276*, 197833. [CrossRef] [PubMed]
11. Myint, O.; Suwanruengsri, M.; Araki, K.; Izzati, U.Z.; Pornthummawat, A.; Nueangphuet, P.; Fuke, N.; Hirai, T.; Jackwood, D.J.; Yamaguchi, R. The bursa atrophy at 28 days old by the variant infectious bursal disease virus makes a negative economic impact on broiler farms in Japan. *Avian Pathol.* **2020**, 1–41. [CrossRef] [PubMed]
12. Kurukulasuriya, S.; Ahmed, K.A.; Ojkic, D.; Gunawardana, T.; Goonewardene, K.; Gupta, A.; Chow-Lockerbie, B.; Popowich, S.; Willson, P.; Tikoo, S.K.; et al. Modified live infectious bursal disease virus (IBDV) vaccine delays infection of neonatal broiler chickens with variant IBDV compared to turkey herpesvirus (HVT)-IBDV vectored vaccine. *Vaccine* **2017**, *35*, 882–888. [CrossRef] [PubMed]
13. Qin, Y.; Zheng, S.J. Infectious Bursal Disease Virus-Host Interactions: Multifunctional Viral Proteins that Perform Multiple and Differing Jobs. *Int. J. Mol. Sci.* **2017**, *18*, 161. [CrossRef] [PubMed]

14. Grgacic, E.V.; Anderson, D.A. Virus-like particles: Passport to immune recognition. *Methods* **2006**, *40*, 60–65. [CrossRef] [PubMed]
15. Roldão, A.; Mellado, M.C.; Castilho, L.R.; Carrondo, M.J.; Alves, P.M. Virus-like particles in vaccine development. *Expert Rev. Vaccines* **2010**, *9*, 1149–1176. [CrossRef] [PubMed]
16. Jackwood, D.J. Multivalent virus-like-particle vaccine protects against classic and variant infectious bursal disease viruses. *Avian Dis.* **2013**, *57*, 41–50. [CrossRef]
17. Xu, X.G.; Tong, D.W.; Wang, Z.S.; Zhang, Q.; Li, Z.C.; Zhang, K.; Li, W.; Liu, H.J. Baculovirus virions displaying infectious bursal disease virus VP2 protein protect chickens against infectious bursal disease virus infection. *Avian Dis.* **2011**, *55*, 223–229. [CrossRef]
18. Pradhan, S.N.; Prince, P.R.; Madhumathi, J.; Roy, P.; Narayanan, R.B.; Antony, U. Protective immune responses of recombinant VP2 subunit antigen of infectious bursal disease virus in chickens. *Vet. Immunol. Immunopathol.* **2012**, *148*, 293–301. [CrossRef]
19. Wang, Y.S.; Ouyang, W.; Liu, X.J.; He, K.W.; Yu, S.Q.; Zhang, H.B.; Fan, H.J.; Lu, C.P. Virus-like particles of hepatitis B virus core protein containing five mimotopes of infectious bursal disease virus (IBDV) protect chickens against IBDV. *Vaccine* **2012**, *30*, 2125–2130. [CrossRef]
20. Taghavian, O.; Spiegel, H.; Hauck, R.; Hafez, H.M.; Fischer, R.; Schillberg, S. Protective oral vaccination against infectious bursal disease virus using the major viral antigenic protein VP2 produced in Pichia pastoris. *PLoS ONE* **2013**, *8*, e83210. [CrossRef]
21. Macreadie, I.G.; Vaughan, P.R.; Chapman, A.J.; McKern, N.M.; Jagadish, M.N.; Heine, H.G.; Ward, C.W.; Fahey, K.J.; Azad, A.A. Passive protection against infectious bursal disease virus by viral VP2 expressed in yeast. *Vaccine* **1990**, *8*, 549–552. [CrossRef]
22. Gómez, E.; Lucero, M.S.; Chimeno Zoth, S.; Carballeda, J.M.; Gravisaco, M.J.; Berinstein, A. Transient expression of VP2 in Nicotiana benthamiana and its use as a plant-based vaccine against infectious bursal disease virus. *Vaccine* **2013**, *31*, 2623–2627. [CrossRef] [PubMed]
23. Wang, M.; Pan, Q.; Lu, Z.; Li, K.; Gao, H.; Qi, X.; Gao, Y.; Wang, X. An optimized, highly efficient, self-assembled, subvirus-like particle of infectious bursal disease virus (IBDV). *Vaccine* **2016**, *34*, 3508–3514. [CrossRef] [PubMed]
24. Wang, W.; Song, Y.; Liu, L.; Zhang, Y.; Wang, T.; Zhang, W.; Li, K.; Qi, X.; Gao, Y.; Gao, L.; et al. Neutralizing-antibody-mediated protection of chickens against infectious bursal disease via one-time vaccination with inactivated recombinant Lactococcus lactis expressing a fusion protein constructed from the RCK protein of Salmonella enterica and VP2 of infectious bursal disease virus. *Microb. Cell Fact.* **2019**, *18*, 21.
25. Liu, L.; Zhang, W.; Song, Y.; Wang, W.; Zhang, Y.; Wang, T.; Li, K.; Pan, Q.; Qi, X.; Gao, Y.; et al. Recombinant Lactococcus lactis co-expressing OmpH of an M cell-targeting ligand and IBDV-VP2 protein provide immunological protection in chickens. *Vaccine* **2018**, *36*, 729–735. [CrossRef]
26. Yamazaki, K.; Ohta, H.; Kawai, T.; Yamaguchi, T.; Obi, T.; Takase, K. Characterization of variant infectious bursal disease virus from a broiler farm in Japan using immunized sentinel chickens. *J. Vet. Med. Sci.* **2017**, *79*, 175–183. [CrossRef]
27. Caggianiello, G.; Kleerebezem, M.; Spano, G. Exopolysaccharides produced by lactic acid bacteria: From health-promoting benefits to stress tolerance mechanisms. *Appl. Microbiol. Biotechnol.* **2016**, *100*, 3877–3886. [CrossRef]
28. Wells, J.M.; Mercenier, A. Mucosal delivery of therapeutic and prophylactic molecules using lactic acid bacteria. *Nat. Rev. Microbiol.* **2008**, *6*, 349–362. [CrossRef]
29. Bahey-El-Din, M.; Gahan, C.G.; Griffin, B.T. Lactococcus lactis as a cell factory for delivery of therapeutic proteins. *Curr. Gene Ther.* **2010**, *10*, 34–45. [CrossRef]
30. Trombert, A. Recombinant lactic acid bacteria as delivery vectors of heterologous antigens: The future of vaccination. *Benef. Microbes* **2015**, *6*, 313–324. [CrossRef]
31. Bahey-El-Din, M. Lactococcus lactis-based vaccines from laboratory bench to human use: An overview. *Vaccine* **2012**, *30*, 685–690. [CrossRef] [PubMed]
32. Ythier, M.; Resch, G.; Waridel, P.; Panchaud, A.; Gfeller, A.; Majcherczyk, P.; Quadroni, M.; Moreillon, P. Proteomic and transcriptomic profiling of Staphylococcus aureus surface LPXTG-proteins: Correlation with agr genotypes and adherence phenotypes. *Mol. Cell Proteom.* **2012**, *11*, 1123–1139. [CrossRef] [PubMed]

33. Liu, X.; Qi, L.; Lv, J.; Zhang, Z.; Zhou, P.; Ma, Z.; Wang, Y.; Zhang, Y.; Pan, L. The immune response to a recombinant Lactococcus lactis oral vaccine against foot-and-mouth disease virus in mice. *Biotechnol. Lett.* **2020**, 1–11. [CrossRef]

34. Cho, S.W.; Yim, J.; Seo, S.W. Engineering Tools for the Development of Recombinant Lactic Acid Bacteria. *Biotechnol. J.* **2020**, *15*, e1900344. [CrossRef] [PubMed]

35. Xu, P.; Wang, Y.; Tao, L.; Wu, X.; Wu, W. Recombinant lactococcus lactis secreting viral protein 1 of enterovirus 71 and its immunogenicity in mice. *Biotechnol. Lett.* **2019**, *41*, 867–872. [CrossRef] [PubMed]

36. Taghinezhad-S, S.; Mohseni, A.H.; Keyvani, H.; Razavi, M.R. Phase 1 Safety and Immunogenicity Trial of Recombinant Lactococcus lactis Expressing Human Papillomavirus Type 16 E6 Oncoprotein Vaccine. *Mol. Ther. Methods Clin. Dev.* **2019**, *15*, 40–51. [CrossRef]

37. Song, S.; Li, P.; Zhang, R.; Chen, J.; Lan, J.; Lin, S.; Guo, G.; Xie, Z.; Jiang, S. Oral vaccine of recombinant Lactococcus lactis expressing the VP1 protein of duck hepatitis A virus type 3 induces mucosal and systemic immune responses. *Vaccine* **2019**, *37*, 4364–4369. [CrossRef]

38. Lahiri, A.; Sharif, S.; Mallick, A.I. Intragastric delivery of recombinant Lactococcus lactis displaying ectodomain of influenza matrix protein 2 (M2e) and neuraminidase (NA) induced focused mucosal and systemic immune responses in chickens. *Mol. Immunol.* **2019**, *114*, 497–512. [CrossRef]

39. Fan, L.; Wu, T.; Wang, Y.; Hussain, A.; Jiang, N.; Gao, L.; Li, K.; Gao, Y.; Liu, C.; Cui, H.; et al. Novel variants of infectious bursal disease virus can severely damage the bursa of fabricius of immunized chickens. *Vet. Microbiol.* **2020**, *240*, 108507. [CrossRef]

40. Li, K.; Gao, H.; Gao, L.; Qi, X.; Gao, Y.; Qin, L.; Wang, Y.; Wang, X. Recombinant gp90 protein expressed in Pichia pastoris induces a protective immune response against reticuloendotheliosis virus in chickens. *Vaccine* **2012**, *30*, 2273–2281. [CrossRef]

41. Platteeuw, C.; van Alen-Boerrigter, I.; van Schalkwijk, S.; de Vos, W.M. Food-grade cloning and expression system for Lactococcus lactis. *Appl. Environ. Microbiol.* **1996**, *62*, 1008–1013. [CrossRef] [PubMed]

42. Miyamoto, T.; Min, W.; Lillehoj, H.S. Lymphocyte proliferation response during Eimeria tenella infection assessed by a new, reliable, nonradioactive colorimetric assay. *Avian Dis.* **2002**, *46*, 10–16. [CrossRef]

43. Liu, Q.; Yang, W.; Luo, N.; Liu, J.; Wu, Y.; Ding, J.; Li, C.; Cheng, Z. LPS and IL-8 activated umbilical cord blood-derived neutrophils inhibit the progression of ovarian cancer. *J. Cancer* **2020**, *11*, 4413–4420. [CrossRef] [PubMed]

44. Mierau, I.; Kleerebezem, M. 10 years of the nisin-controlled gene expression system (NICE) in Lactococcus lactis. *Appl. Microbiol. Biotechnol.* **2005**, *68*, 705–717. [CrossRef]

45. Guo, M.; Yi, S.; Guo, Y.; Zhang, S.; Niu, J.; Wang, K.; Hu, G. Construction of a Recombinant Lactococcus lactis Strain Expressing a Variant Porcine Epidemic Diarrhea Virus S1 Gene and Its Immunogenicity Analysis in Mice. *Viral Immunol.* **2019**, *32*, 144–150. [CrossRef]

46. Liu, S.; Li, Y.; Deng, B.; Xu, Z. Recombinant Lactococcus lactis expressing porcine insulin-like growth factor I ameliorates DSS-induced colitis in mice. *BMC Biotechnol.* **2016**, *16*, 25. [CrossRef]

47. Rueda, F.; Cano-Garrido, O.; Mamat, U.; Wilke, K.; Seras-Franzoso, J.; García-Fruitós, E.; Villaverde, A. Production of functional inclusion bodies in endotoxin-free Escherichia coli. *Appl. Microbiol. Biotechnol.* **2014**, *98*, 9229–9238. [CrossRef]

48. Gasson, M.J. Plasmid complements of Streptococcus lactis NCDO 712 and other lactic streptococci after protoplast-induced curing. *J. Bacteriol.* **1983**, *154*, 1–9. [CrossRef]

49. De Ruyter, P.G.; Kuipers, O.P.; de Vos, W.M. Controlled gene expression systems for Lactococcus lactis with the food-grade inducer nisin. *Appl. Environ. Microbiol.* **1996**, *62*, 3662–3667. [CrossRef]

 *viruses*

Review

Plant Virology Delivers Diverse Toolsets for Biotechnology

Mo Wang [1,2,*], **Shilei Gao** [1], **Wenzhi Zeng** [2], **Yongqing Yang** [3], **Junfei Ma** [4] **and Ying Wang** [4,*] 

[1] Fujian University Key Laboratory for Plant-Microbe Interaction, Fujian Agriculture and Forestry University, Fuzhou 350002, China; gaoshilei1996@163.com

[2] Key Laboratory of Ministry of Education for Genetics, Breeding and Multiple Utilization of Crops, College of Agriculture, Fujian Agriculture and Forestry University, Fuzhou 350002, China; fafuzwz@163.com

[3] Root Biology Center, Fujian Agriculture and Forestry University, Fuzhou 350002, China; yyq287346@163.com

[4] Department of Biological Sciences, Mississippi State University, Starkville, MS 39759, USA; jm5026@msstate.edu

* Correspondence: wangmo108@163.com (M.W.); wang@biology.msstate.edu (Y.W.)

Received: 7 November 2020; Accepted: 19 November 2020; Published: 23 November 2020

Abstract: Over a hundred years of research on plant viruses has led to a detailed understanding of viral replication, movement, and host–virus interactions. The functions of vast viral genes have also been annotated. With an increased understanding of plant viruses and plant–virus interactions, various viruses have been developed as vectors to modulate gene expressions for functional studies as well as for fulfilling the needs in biotechnology. These approaches are invaluable not only for molecular breeding and functional genomics studies related to pivotal agronomic traits, but also for the production of vaccines and health-promoting carotenoids. This review summarizes the latest progress in these forefronts as well as the available viral vectors for economically important crops and beyond.

Keywords: plant virus; viroid; viral vector; virus-induced gene silencing (VIGS); CRISPR/Cas9; genome editing; carotenoid biosynthesis; vaccine; circular RNA

1. Introduction

In 1898, the discovery of tobacco mosaic virus (TMV) as the causative agent for the tobacco mosaic disease marked the birth of virology and expanded the knowledge of life domains [1]. In 1939, TMV was observed under an electron microscope, providing the first image of a virion in history [2–4]. In 1957, Fraenkel-Conrat and coworkers elegantly demonstrated that RNA, akin to DNA, can serve as genetic material, using TMV infecting tobacco plants as a model system [5]. In 1971, the discovery of potato spindle tuber viroid as the causative agent for potato spindle tuber disease further expanded the knowledge of pathogens and established the minimal inheritable genome in biology [6]. Beyond those milestone discoveries, studies on plant viruses and viroids have significantly contributed to the development of numerous recent research forefronts, including, but not limited to, epigenetics [7] and RNA silencing [8,9].

With mounting knowledge about plant viral gene functions and plant–virus interactions, many plant viruses have been successfully developed as biotechnology tools (Figure 1A). For instance, plant viruses have been harnessed as RNA silencing vectors for functional studies on genes underlying desired crop traits [10–14]. Recently, numerous plant viral vectors have been developed for CRISPR/Cas9-based genome editing of model and crop plants [15]. Furthermore, plant viral vectors have been developed to express endogenous and foreign polypeptides in controlling agronomic traits or producing vaccines and valuable carotenoids benefiting human beings [16–18]. Compared with traditional transgenic approaches, plant viral vectors can markedly reduce the time and cost in modulating gene expression, thereby having great potential in agricultural and biomedical applications [16].

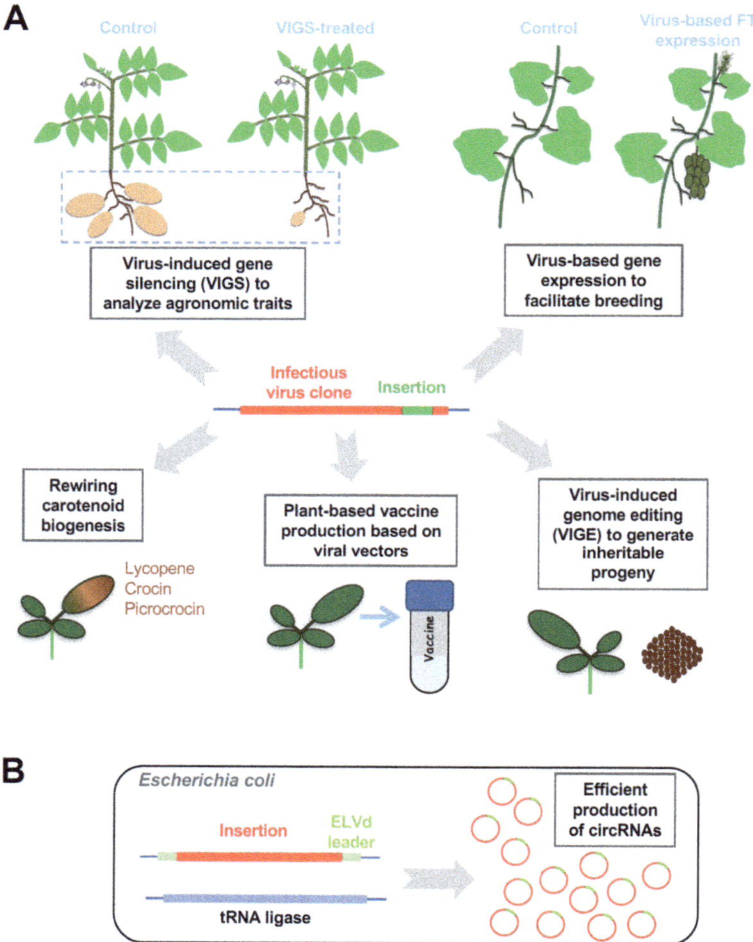

Figure 1. Plant virus/viroid-based technology. (**A**) Application of plant virus vectors in agriculture and production of carotenoids and vaccines. Virus-induced gene silencing (VIGS) has been used to characterize genes controlling important crop traits, exemplified by tuber formation (highlighted by blue dashed lines) in potato plants. Virus-based gene expression of FT (Flowering Locus T) can induce early flowering in grapevines, shortening the time for molecular breeding. VIGE in plants can shorten the time in generating stable transgenic progeny. Viral-based expression platform can be used for the production of vaccines and health-promoting carotenoids. (**B**) Viroid-based platform for circular RNA production. ELVd, eggplant latent viroid. circRNA, circular RNA.

In this review, we summarize the plant viral vectors designed for virus-induced gene silencing (VIGS), genome editing, and exogenous protein expression in major crops. In addition, we introduce the current status and progress of the plant virus-based production of vaccines and health-promoting carotenoids. Furthermore, we outline the viroid-based production of circular RNAs for research applications.

2. A brief Overview of Plant-Virus Interactions

Plant diseases caused by viruses are economically important, as they seriously affect the quality and yield of cereals, vegetables, and fruits. All the food, feed, fiber, ornamental, and industrial crops are threatened by at least one virus, and the great losses caused by plant virus diseases are second only to that by fungal diseases [19]. Plant viruses are obligate parasites, which lack protein-synthesizing and energy-producing apparatuses, and extensively depend on the host machinery for their replication [20]. Virus particles, also called virions, consist of two basic components: the nucleic acid genome and a protective protein coat. In general, viral genomes encode the minimal set of genes critical for infection, such as polymerases, coat proteins, movement proteins, etc. Interestingly, a peculiar group of noncoding RNAs, termed viroids, can cause plant disease without encoding any protein or being encapsidated.

The infection cycle of plant viruses starts from their penetration into host cells. Because plant viruses and viroids by themselves cannot breach the plant physical barriers (i.e., cuticle and cell wall), they are only able to enter the host cells passively through opportunistic mechanical wounds or with the aid of insect vectors (e.g., aphids or whiteflies) [20–22]. After they successfully enter cells, the following infection process can be artificially divided into four major steps [23]. The first step is the disassembly of viral particles, which is the partial or complete removal of coat proteins to release viral genomes into host cells [24]. The second step is the host cell-dependent replication of viral genomes and the translation of viral proteins [25–27]. In this process, plant viruses must recruit and utilize the host's translation apparatus [28] as well as the host's energy resources [29]. Some plant viruses, particularly single-stranded DNA (ssDNA) geminiviruses, rely on host polymerases for genome replication [30,31]. In the third step, viral genome encapsidation occurs to form new virions [32,33]. The last step is the cell-to-cell movement and long-distance trafficking to successfully colonize an entire plant [34–38].

In plants, RNA silencing plays a major role in defending viral infections [39], in addition to innate immunity [40]. It is generally accepted that viral replication intermediates form double-stranded RNAs (dsRNA) to activate plant RNA silencing. Viral dsRNAs are processed to viral short interfering RNAs (vsiRNAs) by plant dsRNA-specific RNases, Dicer-like enzymes (DCLs). VsiRNAs are then efficiently loaded into Argonaute proteins (AGOs) to form the antiviral RNA-induced silencing complexes (RISCs), which subsequently target viral RNAs via slicing or translational arrest [39,41]. Successful viral infection relies on the activity of virus-encoded viral suppressors of RNA silencing (VSRs) [42]. Despite the fact that viroids do not encode any proteins nor possess VSR activity, they can establish successful infections, which is likely attributable to their highly structured RNA genomes and differential subcellular localization of sense and antisense viroid RNAs [43].

3. Engineering VIGS Vectors

Taking advantage of the robust production of vsiRNAs, multiple infectious clones of plant viruses have been engineered to include fragments of endogenous genes for RNA silencing, termed VIGS [10–14]. As listed in Table 1, there are multiple strategies for generating viral vectors. The engineered viruses should retain infectivity and incite mild symptoms. In line with this consideration, non-structural genes or pathogenicity determinant factors are often replaced with cloning sites. For instance, the tobacco rattle virus (TRV)-based VIGS vector was engineered by removing two non-structural genes in RNA2 [44], whereas the tomato yellow leaf curl China virus (ToLCCNV)-based VIGS vector was engineered by removing the pathogenicity determinant factor βC1 in DNA β [45]. Coat/capsid protein genes are popular choices for modifications as well, by either completely being replaced by a multiple cloning site [46] or being partially truncated for insertion of cloning sites [47]. These modifications generally have minimal impacts on viral infectivity. For viruses expressing subgenomic RNAs, it is common to duplicate the subgenomic RNA promoters to flank a multiple cloning site [48–50]. This strategy can lead to the production of new subgenomic viral RNAs that have less impact on viral infectivity. Notably, the insertions can be designed to form a hairpin structure composed of inverted duplication of sense and antisense sequences of target genes to enhance silencing effects, if a duplicated subgenomic RNA promoter is harnessed [49]. Some viruses belonging to the same family may be engineered via the same or similar strategy. For instance, it is common to duplicate the protease cleavage site in the polyprotein to flank an inserted multiple cloning site for viruses of the family *Secoviroidae* [51–53].

Table 1. Strategies to engineer VIGS vectors for major crops.

Family	Virus	Strategies to Design Vectors
Alphaflexiviridae	potato virus X [48]	Duplication of the subgenomic (sg) RNA promoter of the coat protein (CP) to flank a multiple cloning site between two CP sgRNA promoters
	foxtail mosaic virus [54,55]	Insertion of the *Xba*I and *Xho*I sites immediately after the stop codon of the capsid protein gene [54] Or Duplication of CP subgenomic promoter to flank a multiple cloning site [55]
Betaflexiviridae	citrus leaf blotch virus [56]	Inserting a subgenomic RNA promoter followed by a *Pml*I site for inserting foreign sequences in the linear form or in the hairpin fashion
	grapevine virus A [50]	Duplication of Movement Protein (MP) subgenomic RNA promoter to flank a multiple cloning site
Bromoviridae	cucumber mosaic virus [57]	Replacing a portion at the 3′-end of ORF2b with a multiple cloning site in RNA-2
	prunus necrotic ringspot virus [58]	Inserting foreign sequences at the 3′end of the CP gene in RNA3; Combining RNA1 and RNA2 in the same binary vector to increase the efficiency
	brome mosaic virus [59]	Using the *Hind*III site in the RNA3 3′ untranslated region (UTR) for insertion; Replacing the *Bcl*I/*Bss*HII flanked RNA3 intergenic region of the *Festuca arundinacea* strain with that from the Russian strain
Caulimoviridae	rice tungro bacilliform virus [60]	Selectively keeping ORFIII and a 50 bp 3′-truncated ORF IV flanked by two constitutive promoters; adding a tRNA binding site essential for replication immediately after the first promoter near the 5′-end; adding a multiple cloning site immediately before the second promoter near the 3′end
Geminiviridae	tomato yellow leaf curl China virus [45]	Replacing the βC1 (pathogenic factor/VSR) ORF with a multiple cloning site in DNA β
	african cassava mosaic virus [47]	Replacing a portion of the capsid protein (AV1) ORF with a multiple cloning site in DNA-A
	cotton leaf crumple virus [46]	Replacing the CP gene with a multiple cloning site in DNA-A
Secoviridae	broad bean wilt virus [61]	Inserting a cloning site immediately after the stop codon of the RNA2 ORF in the 3′UTR of RNA2
	bean pod mottle virus [52]	Duplication of the protease site between MP and L-CP in RNA2 to flank a multiple cloning site
	tobacco ringspot virus [53]	Duplication of the C/A protease site between MP and CP in RNA2 to flank a multiple cloning site
	apple latent spherical virus [51]	Duplication of the Q/G protease site between 42KP and Vp25 in RNA2 to flank a multiple cloning site
Tymoviridae	turnip yellow mosaic virus [62]	Inserting a cloning site immediately downstream the CP protein for inserting foreign sequence in the hairpin fashion; Duplicating the CP stop codon to keep the tRNA-like structure for infectivity
Virgaviridae	tobacco rattle virus [44]	Replacing non-structural genes in RNA2 with a multiple cloning site
	pea early browning virus [63]	Replacing non-structural genes in RNA2 with a multiple cloning site
	barley stripe mosaic virus [64]	Inserting a multiple cloning site downstream of the γb (pathogenic factor/VSR) corresponding to the γ subgenomic RNA
	cucumber green mottle mosaic virus [49]	Duplication of the CP subgenomic RNA promoter to flank a *Bam*HI site

The engineered VIGS constructs are commonly delivered to plants via agro-infiltration or mechanically inoculation of in vitro transcribed RNA. Infection of engineered viruses results in abundant small RNAs from the inserted fragments that suppress the expression of host genes being targeted, based on the sequence homology. Because of its high efficiency and ease of handling, VIGS has been extensively used in plant functional studies, particularly in the species where the stable transformants are not easy to obtain [10–12].

4. VIGS for Rapid and Transient Gene Silencing in Plants/Crops

The host range of the parental wild-type viruses determines the plant species where these VIGS vectors can be used. VIGS vectors developed in the early days are mainly derived from TMV, potato virus X (PVX), and TRV, which are initially utilized in silencing genes in *Nicotiana benthamiana* and tomato (*Solanum lycopersicum*). Over the last two decades, more than 40 viruses have been developed as VIGS vectors for dicot and monocot plant species. Among those, more than 20 have been used for economically important crops, as summarized in Table 2. These available tools markedly shorten the time for functional assays in identifying genes related to desired traits in economically important crops, greatly facilitating the breeding efforts. Readers are referred to the collection of extensive reviews for more details [11,14,65].

Although VIGS provides a convenient approach to manipulate gene expressions in plants, it is important to note that these vectors represent infectious viruses. Despite the fact that most of them do not cause drastic phenotypic alterations, they may still affect gene expression in hosts. A recent study showed that TRV-based viral VIGS vectors alone can trigger changes in the alternative splicing of host genes, slicing activity of a plant microRNA (miR167), as well as the expression of plant mRNAs, phased secondary siRNAs, and long noncoding RNAs [66]. Thus, proper controls and considerable cautions need to be taken into account when analyzing experimental data based on viral silencing vectors. Furthermore, off-target effects can occasionally render the data annotation complicated [67].

Table 2. Viral VIGS vectors for major crops.

Major Crops	Viral VIGS Vectors
Tomato	apple latent spherical virus [51], tomato yellow leaf curl China virus [45], tobacco rattle virus [68]
Pepper	apple latent spherical virus [69], tobacco rattle virus [70], broad bean wilt virus2 [61]
Potato	tobacco rattle virus [71], potato virus X [48]
Cassava	african cassava mosaic virus [47,72]
Legume	apple latent spherical virus [51], bean pod mottle virus [52,73,74], cucumber mosaic virus [57], pea early browning virus [63], tobacco ringspot virus [53]
Cucurbits	apple latent spherical virus [51], tobacco ringspot virus [53], cucumber green mottle mosaic virus [49]
Spinach	cucumber mosaic virus [75]
Cabbage	turnip yellow mosaic virus [62]
Cotton	tobacco rattle virus [76], cotton leaf crumple virus [46]
Citrus	citrus leaf blotch virus [56]
Banana	cucumber mosaic virus [77]
Strawberry	tobacco rattle virus [78], apple latent spherical virus [79]
Apple	apple latent spherical virus [80]
Pear	apple latent spherical virus [80]
Peach	prunus necrotic ringspot virus [58]
Grape	grapevine virus A [50]
Wheat	foxtail mosaic virus [55], barley stripe mosaic virus [81]

Table 2. *Cont.*

Major Crops	Viral VIGS Vectors
Barley	foxtail mosaic virus [55], barley stripe mosaic virus [64], Brome mosaic virus [59]
Rice	brome mosaic virus [59], rice tungro bacilliform virus [60]
Maize	brome mosaic virus [59], foxtail mosaic virus [54]
Sorghum	brome mosaic virus [82]
Foxtail millet	foxtail mosaic virus [55]
Ginger	barley stripe mosaic virus [83]

5. Plant Virus-Based Tools for Plant Genome Editing

CRISPR/Cas nucleases-based genome editing technologies have provided unprecedented power in plant breeding to accelerate the manipulation of desired crop traits. The single-guide RNA (sgRNA) directs the Cas nucleases to target the genome regions introducing designed mutations. The traditional experimental process requires extensive tissue culture handlings and prolonged selection to remove the transgenic copy of the CRISPR/Cas cassette. The tissue culture handlings for many major crops are technically challenging and demanding. More importantly, the limited expression of sgRNA expression in tissue cultures leads to low efficiency in genome editing [15].

Geminiviruses, a group of ssDNA viruses that replicate in the nucleus, were soon exploited to express sgRNAs to boost the efficiency after the CRISPR/Cas nucleases-based technology became available [84]. This type of virus-based strategy in gene editing is termed virus-induced genome editing (VIGE). Interestingly, TRV, an RNA virus that replicates in the cytoplasm, was also capable of delivering sgRNA for genome editing in the nucleus [85]. Since 2014, various viral vectors have been developed for genome editing of important crops, such as potato, tomato, wheat, rice, maize, etc. (Table 3). Because the Cas nuclease genes are too large (~4.2 Kb) to insert into many viral vectors, most of these efforts rely on introducing the Cas nucleases into plants via traditional transgenic approaches or expressing Cas nucleases in a different vector. There are several approaches to incorporate sgRNAs into viral vectors. A popular choice is to place the sgRNA scaffold under the control of plant U6 gene promoter [84,86–90]. However, the U6 promoter occasionally results in weak expression of sgRNAs [91]. Recently, several reports used tRNAs to flank the sgRNA scaffold [92,93], and the tRNAs were subsequently removed by the activity of endogenous tRNA processing enzymes (RNase P and RNase Z) [94]. Notably, tRNA-flanking may not be needed based on the experimental practice when using some viral vectors [93,95,96].

Most viral vectors, by and large, only perform gene editing in local infection sites or protoplasts. Therefore, they do not lead to inheritable traits in the progeny. To circumvent this shortcoming, a recent study used Agrobacteria harboring the foxtail mosaic virus constructs to inoculate *N. benthamiana* seeds with the seed coat manually cracked, which resulted in the progeny inheriting the edits [86]. One critical factor to effectively generate inheritable genome-edited plants relies on the delivery of sgRNAs to germlines. A recent attempt employed a portion of the Flowering Locus T (FT) mRNA to promote the entry of sgRNAs to reproductive organs, thereby increasing the efficiency of the inheritable genome edits [97]. Although this approach indeed increased the frequency of the inheritable genome edits, the mechanism remains to be further elucidated, as the protein product of the FT gene, but not its transcripts, are the mobile signal [98–103]. Very recently, sonchus yellow net rhabdovirus (SYNV) was employed as a VIGE vector for tobacco plants [92]. This system, by far, provides the easiest and most robust DNA-free approach in generating plants bearing inheritable genome edits through simple leaf inoculations. The analysis also showed that the off-target effects are minimal through this approach. Moreover, the viral vector is stable through mechanical transmission/passages and can be easily eliminated after seed set, therefore preventing potential deleterious effects caused by the vectors [92]. Despite the fact that the host range restriction of SYNV limits its application in various crop species, this progress already demonstrates the great promise of VIGE in application.

Table 3. Plant virus-induced genome editing system.

	Viral Vectors	Guide RNA Design	Edited Plants	Inheritable
Dicot	cabbage leaf curl virus	U6p::gRNAScaffold::U6t inserted to the cloning site downstream of AL3	Transgenic *Nicotiana benthamiana* over-expressing Cas9 [87]	No
	tobacco rattle virus	PEBV::gRNAScaffold-Rz inserted to pTRV2 vector [85]; A FT fragment inserted at the 3′-end of gRNA resulting in a mobile sgRNA [97]	Transgenic *Nicotiana benthamiana* over-expressing Cas9 [85,97]	Yes
	bean yellow dwarf virus	Replacing MP and CP with U6::gRNAscaffold::U6t and 35S::Cas9; Agrobacterium-based transformation required for delivery	Wildtype *Nicotiana tabacum* [84]	NA
			Wildtype potato (Tetraploid and diploid) [88,89]	Via tissue culture [88]
			Wildtype tomato [90]	NA
	tobacco mosaic virus	A fragment containing the gRNAScaffold with or without a Rz inserted to the TRBO vector; 35S::Cas9 expressed from a different binary vector	*Nicotiana benthamiana* 16C [91]	NA
	potato virus X	gRNAScaffold driven by PVX CP promoter; tRNA flanking not needed	Transgenic *Nicotiana benthamiana* over-expressing Cas9 [93]	Via tissue culture
	sonchus yellow net rhabdovirus	gRNAScaffold (flanked by tRNAs) and Cas9 inserted between N and P genes under the control of duplicated N/P junction sequences	Wildtype *Nicotiana benthamiana* [92]	Yes
	beet necrotic yellow vein virus	gRNAScaffold fused to the 3′-end of the p31 ORF	Transgenic *Nicotiana benthamiana* over-expressing Cas9 [95]	NA
	foxtail mosaic virus vectors	U6p::gRNAScaffold or Cas9 inserted between duplicated CP subgenomic promoters; Mixing of gRNA and Cas9 clones for infection [86]	Transgenic *Nicotiana benthamiana* over-expressing Cas9 [104] or tomato bushy stunt virus P19 [86]	Yes if directly inoculating seeds [86]
	barley stripe mosaic virus	See below	Transgenic *Nicotiana benthamiana* over-expressing Cas9 [96]	Via tissue culture
Monocot	foxtail mosaic virus vectors	Inserting gRNAScaffold after a duplicated ORF5 promoter	Transgenic maize over-expressing Cas9 [104]	NA
			Transgenic *Setaria viridis* over-expressing Cas9 [104]	No
	wheat dwarf virus (WDV)	Replacing the MP and CP genes with Ubi::Cas9 and U6p::gRNAscaffold; T-DNA insertion procedures required	Wildtype wheat [105]	NA
		Replacing MP and CP with U6p::gRNAscaffold; Adding Ubi::Cas9::NOS in the binary vector but outside of the WDV replicon	Wildtype rice and transgenic rice over-expressing Cas9 [106]	NA
	barley stripe mosaic virus	Replacing CP with sgγ::gRNAScaffold in RNAβ or inserting gRNAScaffold immediately downstream of γb in RNAγ	Transgenic wheat over-expressing Cas9 [96]	NA
			Transgenic maize over-expressing Cas9 [96]	NA

NOTE: U6p: U6 promoter; U6t: U6 terminator; PEBV: pea early-browning virus; Rz: ribozyme; FT: flowering locus T; MP: movement protein; CP: coat protein; Ubi: ubiquitin. NA: Not Assessed.

Recent studies demonstrated that the ectopic expression of CRISPR/Cas nucleases in plants is subject to negative regulation by the RNA silencing machinery, hindering genome editing efficacy [107,108]. Notably, genome editing efficiency can be improved by including an artificial microRNA cassette in vectors to down-regulate the expression of key players in post-transcriptional gene silencing (e.g., RDR6 and AGO1) [107,108]. Similarly, RNA silencing suppressors cloned from plant viruses can also increase genome editing efficacy of either VIGE [86,104] or the transgene-based approach [108].

6. Plant Virus-Based Gene Expression Vectors

In addition to serving as the VIGS and VIGE vectors, most viral vectors listed in Tables 1 and 3 can be exploited for expressing heterogeneous proteins in plants. In the early days, plant viral vectors were based on the "full-virus" vector strategy to express genes-of-interest fused with a viral gene (e.g., coat protein gene in TMV) [109]. These viral vectors retain the full capacity of replication, assembly of virions, cell-to-cell movements, and resistance to host gene silencing [109]. The non-cell-autonomous nature of viruses can turn almost the entire plant into a factory for foreign protein synthesis. The expression level of foreign peptides can reach up to 10% of total soluble protein [109]. However, the insertion size limitation restrains the application of many viral vectors. Proteins larger than 30 kDa are difficult to express using the "full-virus" vector strategy [109]. To circumvent this shortcoming, it is common to replace certain viral non-structure genes or pathogenicity determinant factors with a multiple cloning site for large insertions as aforementioned. Another strategy employs a recombination system to deconstruct viral genes for generating a set of expression vectors [110]. In this system, the viral sequence is engineered to replace the coat protein gene with a LoxP recombination site. The gene to be expressed is placed in a separate vector downstream of another LoxP site. Both viral vectors and the plasmid harboring the expressing gene are mixed and co-infiltrated with a third vector to express the Cre recombinase [111]. This system further increases protein yield up to nearly 50% of total soluble proteins and facilitates the expression of larger foreign genes. However, since the viral elements are kept to a minimum, the deconstructed viral vectors can only be expressed in local leaves [111].

Plant virus-based protein expression vectors have been widely used in basic sciences to understand plant gene functions [11,112]. Moreover, these vectors have great application in altering agronomic traits as well. For instance, apple latent spherical virus was engineered to express the FT gene, which successfully promotes the early flowering of grapevine [113] and strawberry [79]. Similarly, citrus leaf blotch virus was used to express FT and prompt the early flowering of citrus plants [114]. This approach significantly accelerates the breeding process. More importantly, the capacity of plant viral vectors to promptly alter agronomic traits opens up many possibilities for precision agriculture.

7. Rewiring Plant Metabolic Pathways for the Production of Health-Promoting Carotenoids

Many plant secondary metabolites are useful nutrients or health-promoting molecules. However, those beneficial metabolites are often accumulated at low levels. For instance, crocins, which are carotenoid derivatives serving as a valuable spice and potent pain reliever, are mainly accumulated in the stigma of *Crocus sativus* L. flowers or, to a lesser extent, in gardenia fruits [115]. Due to the labor-intense procedures in collecting flowers, it is costly to produce crocins and the related molecules, such as picrocrocin. Using a tobacco etch virus (TEV)-based vector, specific cleavage dioxygenase enzymes (CCDs) in crocins biogenesis cloned from *C. sativus* or *Buddleja davidii* were expressed in *N. benthamiana* plants, resulting in a significant accumulation of crocins and picrocrocin [115]. The yield was further improved when the CCD from *C. sativus* was co-expressed with other carotenogenic enzymes (e.g., phytoene synthase from *Pantoea ananatis* and β-carotene hydroxylase 2 from saffron). This unique TEV vector removes the essential viral gene NIb (nuclear inclusion b) to gain the capacity for large insertions [116]. Consequently, this viral vector can only infect transgenic plants expressing NIb, which prevents the viral vector from entering the environment. Using the same TEV system, a soil bacterial gene cassette consisting of GGPP synthase, phytoene synthase, and phytoene desaturase was expressed in *N. benthamiana*, resulting in significantly increased production of lycopene in the cytoplasm [117]. Lycopene is a major carotenoid in human blood protecting against oxidative damage. Given the difficulties in rewiring carotenoid metabolism using traditional transgene approaches [118–120], viral vector-based production provides a plausible solution for engineering plant metabolic pathways with low cost and excellent performance.

8. Plant Virus-Based Production of Vaccines

Plant viral vectors have also been successfully harnessed in producing vaccines against devastating pathogens infecting human beings and livestock [18,121]. The surface antigen of human hepatitis B virus expressed in transgenic tobacco can form virus-like particles (VLPs) in plants [122], and those VLPs are capable of inducing potent B-cell and T-cell immune responses in mice [123,124]. Encouraged by this finding, viral vectors have been developed and employed for the rapid and robust production of various vaccines [18,121]. The target antigens can be expressed using plant viral vectors as either epitope presentation by displaying the recombinant epitope-coat protein on the surface of the chimeric virions or the polypeptides alone [18]. In 1995, the antigenic peptide of canine parvovirus VP2 protein was successfully expressed in plants, which can elicit high levels of neutralizing antibodies in mice and rabbits [125]. Since then, over a dozen antigenic peptides have been successfully expressed in plants against various pathogens, such as influenza virus, West Nile virus, hepatitis A and B viruses, human immunodeficiency virus, etc. [18,126]. A few vaccines or pharmaceutical proteins synthesized in plants using plant viral vector systems, such as the Newcastle virus subunit vaccine, have been approved for markets [126]. During the current COVID-19 pandemic, plant viral vector-based vaccine production may provide a convenient platform for production when some COVID-19 vaccine candidates prove to be safe and effective [127].

Notably, vectors based on plant RNA viruses are popular choices for expressing antigens. TMV [128], cowpea mosaic virus [129], potato virus X [130], TRV [131], and several more [18] have been successfully exploited in recent years. These viruses mostly possess positive-sense RNA genomes. In addition to RNA viruses, geminiviruses have also been used for vaccine production [132].

9. Viroids for Generating Circular RNA

Viroids are the first group of circular RNAs identified in nature [133]. Increasing evidence revealed that circular noncoding RNAs widely exist in many organisms across the Tree of Life [134–137]. Importantly, many endogenous circular RNAs have been shown to play regulatory roles in gene expression, development, disease, etc. [134,135,138,139]. It is noteworthy that the delivery of synthetic circular RNAs has led to the suppression of gastric carcinoma cell proliferation, as a novel means of therapy [140]. It is desirable to develop a robust expression system for generating circular RNA in large quantities for functional studies and potentially for clinical therapy. Although several methods have been developed using either a cascade of enzymatic reactions [141] or the intron backsplicing mechanism [142], these systems can only reach a moderate production rate.

As single-stranded circular noncoding RNAs, viroids can co-opt host machinery to achieve replication and systemic trafficking [43,143]. Interestingly, members in the family *Avsunviroidae* possess ribozyme activity, which is among the first groups of ribozymes identified in nature [144]. The hammerhead ribozyme in those viroids is critical to complete the infection cycle in chloroplasts [43,143,145]. Studies showed that the hammerhead ribozyme cleaves viroid RNA to generate 5′-hydroxyl and 2′,3′-phosphodiester termini, which are subsequently ligated by the chloroplast-localized tRNA ligase [146].

Based on this pathway, co-expressing the eggplant latent viroid (ELVd)-based construct and the recombinant tRNA ligase in bacteria resulted in high yield of circular ELVd RNA [147]. Interestingly, inserting exogenous sequences at a particular position of the ELVd molecule allowed the production of chimeric circular RNAs to desirable concentrations (Figure 1B) [148,149]. This circular RNA expression system provides higher production of desired RNAs in the circular form that exceeds the expression level in vivo, which will facilitate studies on circular RNA biology and its application [148,149].

10. Summary and Perspectives

In the mid-1980s, the need to purify large quantities of viroids for structural studies led to the development of a silica gel-based method [150,151]. This method was also demonstrated to be suitable for separating supercoiled plasmids from crude bacterial extracts, leading to the most popular commercial miniprep kit of Qiagen [145]. Along with the progress in nucleic acid purification techniques, structural analysis on viroid RNAs during the same period led to the recognition of suboptimal structures when certain base pairs did not belong to the deduced structure with the minimum free energy [152]. This concept constitutes a critical component in computational programs for the in silico prediction of RNA secondary structures [153], which greatly enhances the capacity and accuracy [145]. This is simply one of the many stories in history demonstrating that plant virology research has markedly contributed not only to basic sciences but also biotechnology. Recent progress in high throughput sequencing and bioinformatic tools has provided unprecedented power using small RNA sequencing to identify novel viruses and viroids from biological samples without pre-existing knowledge of viral sequences [154,155], which will uncover novel viruses to engineer suitable viral vectors for economically important crops. As plant virology research centers around the major questions in agriculture and basic sciences, it is certain that new discoveries will continue to deliver promising tools for biotechnology in the future.

Author Contributions: Conceptualization, M.W. and Y.W.; writing—original draft preparation, M.W., and Y.W.; writing—review and editing, M.W., S.G., W.Z., Y.Y., J.M., and Y.W. All authors have read and agreed to the published version of the manuscript.

Funding: This work was supported by the Startup fund from Mississippi State University to Y.W.

Acknowledgments: We are grateful for the constructive comments from Zhenghe Li at Zhejiang University, China. We thank Shachinthaka D. Dissanayaka Mudiyanselage and Bansi Patel at Mississippi State University for critical reading of the manuscript. We apologize to colleagues whose work was not cited in this review due to the page limit.

Conflicts of Interest: The authors declare no conflict of interest. The funders had no role in the design of the study; in the collection, analyses, or interpretation of data; in the writing of the manuscript, or in the decision to publish the results.

References

1. Lecoq, H. Discovery of the first virus, Tobacco mosaic virus: 1892 or 1898? *C. R. l'Acad. Sci. Ser. III* **2001**, *324*, 929–933.
2. Kausche, G.A.; Ruska, H. Die Sichtbarmachung von Adsorption von Metallkolloiden an Eiweißkörper—I. Die Reaktion kolloides Gold—Tabakmosaikvirus. *Kolloid-Zeitschrift* **1939**, *89*, 21–26.
3. Stanley, W.M.; Anderson, T.F. A study of purified viruses with the electron microscope. *J. Biol. Chem.* **1941**, *139*, 325–338.
4. Kausche, G.A.; Ruska, H. Die Struktur der "kristallinen Aggregate" des Tabakmosaikvirusproteins. *Biochem. Z.* **1939**, *303*, 211.
5. Fraenkel-Conrat, H.; Singer, B.; Williams, R.C. Infectivity of viral nucleic acid. *Biochim. Biophys. Acta* **1957**, *25*, 87–96.
6. Diener, T.O. Potato spindle tuber "virus": IV. A replicating, low molecular weight RNA. *Virology* **1971**, *45*, 411–428.
7. Wassenegger, M.; Heimes, S.; Riedel, L.; Sänger, H.L. RNA-directed de novo methylation of genomic sequences in plants. *Cell* **1994**, *76*, 567–576.
8. Baulcombe, D.C. RNA as a target and an initiator of post-transcriptional gene silencing in trangenic plants. In *Post-Transcriptional Control of Gene Expression in Plants*; Springer: Berlin/Heidelberg, Germany, 1996; pp. 79–88.
9. Voinnet, O.; Baulcombe, D.C. Systemic signalling in gene silencing. *Nature* **1997**, *389*, 553.
10. Burch-Smith, T.M.; Anderson, J.C.; Martin, G.B.; Dinesh-Kumar, S.P. Applications and advantages of virus-induced gene silencing for gene function studies in plants. *Plant J.* **2004**, *39*, 734–746.

11. Dommes, A.B.; Gross, T.; Herbert, D.B.; Kivivirta, K.I.; Becker, A. Virus-induced gene silencing: Empowering genetics in non-model organisms. *J. Exp. Bot.* **2019**, *70*, 817–833.

12. Ramanna, H.; Ding, X.S.; Nelson, R.S. Rationale for developing new virus vectors to analyze gene function in grasses through virus-induced gene silencing. *Methods Mol. Biol.* **2013**, *975*, 15–32. [PubMed]

13. Huang, C.J.; Qian, Y.J.; Li, Z.H.; Zhou, X.P. Virus-induced gene silencing and its application in plant functional genomics. *Sci. China Life Sci.* **2012**, *55*, 99–108. [PubMed]

14. Lange, M.; Yellina, A.L.; Orashakova, S.; Becker, A. Virus-induced gene silencing (VIGS) in plants: An overview of target species and the virus-derived vector systems. *Methods Mol. Biol.* **2013**, *975*, 1–14. [PubMed]

15. Syed Shan-E-Ali Zaidi, S.M. Viral vectors for plant genome engineering. *Front. Plant Sci.* **2017**, *8*, 539.

16. Cody, W.B.; Scholthof, H.B. Plant virus vectors 3.0: Transitioning into synthetic genomics. *Annu. Rev. Phytopathol.* **2019**, *57*, 211–230.

17. Hefferon, K.L. Plant virus expression vector development: New perspectives. *Biomed. Res. Int.* **2014**, *2014*, 785382.

18. Salazar-González, J.A.; Bañuelos-Hernández, B.; Rosales-Mendoza, S. Current status of viral expression systems in plants and perspectives for oral vaccines development. *Plant Mol. Biol.* **2015**, *87*, 203–217.

19. Zhao, L.; Feng, C.; Wu, K.; Chen, W.; Chen, Y.; Hao, X.; Wu, Y. Advances and prospects in biogenic substances against plant virus: A review. *Pestic. Biochem. Physiol.* **2017**, *135*, 15–26.

20. Gergerich, R.C.; Dolja, V.V. Introduction to plant viruses, the invisible foe. *Plant Health Instr.* **2006**. [CrossRef]

21. Whitfield, A.E.; Falk, B.W.; Rotenberg, D. Insect vector-mediated transmission of plant viruses. *Virology* **2015**, *479–480*, 278–289.

22. Caciagli, P.; Lemaire, O.; Lopez-Moya, J.; MacFarlane, S.; Peters, D.; Susi, P.; Torrance, L. Status and prospects of plant virus control through interference with vector transmission. *Annu. Rev. Phytopathol.* **2013**, *51*, 177–201.

23. Saxena, P.; Lomonossoff, G.P. Virus infection cycle events coupled to RNA replication. *Annu. Rev. Phytopathol.* **2014**, *52*, 197–212. [PubMed]

24. Bakker, S.E.; Ford, R.J.; Barker, A.M.; Robottom, J.; Saunders, K.; Pearson, A.R.; Ranson, N.A.; Stockley, P.G. Isolation of an asymmetric RNA uncoating intermediate for a single-stranded RNA plant virus. *J. Mol. Biol.* **2012**, *417*, 65–78. [PubMed]

25. Hull, R. Replication of plant viruses. In *Plant Virology*; Academic Press: Cambridge, MA, USA, 2014; pp. 341–421.

26. Medina-Puche, L.; Lozano-Duran, R. Tailoring the cell: A glimpse of how plant viruses manipulate their hosts. *Curr. Opin. Plant Biol.* **2019**, *52*, 164–173. [PubMed]

27. Heinlein, M. Plant virus replication and movement. *Virology* **2015**, *479–480*, 657–671.

28. Li, S. Regulation of ribosomal proteins on viral infection. *Cells* **2019**, *8*, 508.

29. Nagy, P.D.; Lin, W. Taking over cellular energy-metabolism for TBSV replication: The high ATP requirement of an RNA virus within the viral replication organelle. *Viruses* **2020**, *12*, 56.

30. Guerrero, J.; Regedanz, E.; Lu, L.; Ruan, J.; Bisaro, D.M.; Sunter, G. Manipulation of the plant host by the Geminivirus AC2/C2 protein, a central player in the infection cycle. *Front. Plant Sci.* **2020**, *11*, 591.

31. Wu, M.; Wei, H.; Tan, H.; Pan, S.; Liu, Q.; Bejarano, E.R.; Lozano-Durán, R. Plant DNA polymerases alpha and delta mediate replication of geminiviruses. *bioRxiv* **2020**. [CrossRef]

32. Chaturvedi, S.; Rao, A.L.N. Molecular and biological factors regulating the genome packaging in single-strand positive-sense tripartite RNA plant viruses. *Curr. Opin. Virol.* **2018**, *33*, 113–119.

33. Rao, A.; Chaturvedi, S.; Garmann, R.F. Integration of replication and assembly of infectious virions in plant RNA viruses. *Curr. Opin. Virol.* **2014**, *9*, 61–66. [PubMed]

34. Reagan, B.C.; Burch-Smith, T.M. Viruses reveal the secrets of plasmodesmal cell biology. *Mol. Plant Microbe Interact.* **2020**, *33*, 26–39. [PubMed]

35. Reagan, B.C.; Ganusova, E.E.; Fernandez, J.C.; McCray, T.N.; Burch-Smith, T.M. RNA on the move: The plasmodesmata perspective. *Plant Sci.* **2018**, *275*, 1–10. [PubMed]

36. Folimonova, S.Y.; Tilsner, J. Hitchhikers, highway tolls and roadworks: The interactions of plant viruses with the phloem. *Curr. Opin. Plant Biol.* **2018**, *43*, 82–88. [PubMed]

37. Tilsner, J.; Oparka, K.J. Missing links?—The connection between replication and movement of plant RNA viruses. *Curr. Opin. Virol.* **2012**, *2*, 705–711. [PubMed]

38. Harries, P.; Ding, B. Cellular factors in plant virus movement: At the leading edge of macromolecular trafficking in plants. *Virology* **2011**, *411*, 237–243. [PubMed]

39. Ding, S.W.; Voinnet, O. Antiviral immunity directed by small RNAs. *Cell* **2007**, *130*, 413–426.

40. Mandadi, K.K.; Scholthof, K.B.G. Plant immune responses against viruses: How does a virus cause disease? *Plant Cell* **2013**, *25*, 1489–1505.

41. Ding, S.W. RNA-based antiviral immunity. *Nat. Rev. Immunol.* **2010**, *10*, 632–644.

42. Wu, Q.; Wang, X.; Ding, S.W. Viral suppressors of RNA-Based viral immunity: Host targets. *Cell Host Microbe* **2010**, *8*, 12–15.

43. Ding, B. The biology of viroid-host interactions. *Annu. Rev. Phytopathol.* **2009**, *47*, 105–131. [PubMed]

44. Liu, Y.; Schiff, M.; Marathe, R.; Dinesh-Kumar, S.P. Tobacco Rar1, EDS1 and NPR1/NIM1 like genes are required for N-mediated resistance to tobacco mosaic virus. *Plant J.* **2002**, *30*, 415–429. [PubMed]

45. He, X.; Jin, C.; Li, G.; You, G.; Zhou, X.; Zheng, S.J. Use of the modified viral satellite DNA vector to silence mineral nutrition-related genes in plants: Silencing of the tomato ferric chelate reductase gene, FRO1, as an example. *Sci. China Ser. C Life Sci.* **2008**, *51*, 402–409.

46. Tuttle, J.R.; Haigler, C.H.; Robertson, D. Method: Low-cost delivery of the cotton leaf crumple virus-induced gene silencing system. *Plant Methods* **2012**, *8*, 27.

47. Fofana, I.B.F.; Sangaré, A.; Collier, R.; Taylor, C.; Fauquet, C.M. A geminivirus-induced gene silencing system for gene function validation in cassava. *Plant Mol. Biol.* **2004**, *56*, 613–624.

48. Faivre-Rampant, O.; Gilroy, E.M.; Hrubikova, K.; Hein, I.; Millam, S.; Loake, G.J.; Birch, P.; Taylor, M.; Lacomme, C. Potato virus X-induced gene silencing in leaves and tubers of potato. *Plant Physiol.* **2004**, *134*, 1308–1316.

49. Liu, M.; Liang, Z.; Aranda, M.A.; Hong, N.; Liu, L.; Kang, B.; Gu, Q. A cucumber green mottle mosaic virus vector for virus-induced gene silencing in cucurbit plants. *Plant Methods* **2020**, *16*, 9.

50. Muruganantham, M.; Moskovitz, Y.; Haviv, S.; Horesh, T.; Fenigstein, A.; du Preez, J.; Stephan, D.; Burger, J.T.; Mawassi, M. Grapevine virus A-mediated gene silencing in Nicotiana benthamiana and Vitis vinifera. *J. Virol. Methods* **2009**, *155*, 167–174.

51. Igarashi, A.; Yamagata, K.; Sugai, T.; Takahashi, Y.; Sugawara, E.; Tamura, A.; Yaegashi, H.; Yamagishi, N.; Takahashi, T.; Isogai, M.; et al. Apple latent spherical virus vectors for reliable and effective virus-induced gene silencing among a broad range of plants including tobacco, tomato, Arabidopsis thaliana, cucurbits, and legumes. *Virology* **2009**, *386*, 407–416.

52. Zhang, C.; Ghabrial, S.A. Development of Bean pod mottle virus-based vectors for stable protein expression and sequence-specific virus-induced gene silencing in soybean. *Virology* **2006**, *344*, 401–411.

53. Zhao, F.; Lim, S.; Igori, D.; Yoo, R.H.; Kwon, S.Y.; Moon, J.S. Development of tobacco ringspot virus-based vectors for foreign gene expression and virus-induced gene silencing in a variety of plants. *Virology* **2016**, *492*, 166–178. [PubMed]

54. Mei, Y.; Zhang, C.; Kernodle, B.M.; Hill, J.H.; Whitham, S.A. A foxtail mosaic virus vector for virus-induced gene silencing in maize. *Plant Physiol.* **2016**, *171*, 760–772. [PubMed]

55. Liu, N.; Xie, K.; Jia, Q.; Zhao, J.; Chen, T.; Li, H.; Wei, X.; Diao, X.; Hong, Y.; Liu, Y. Foxtail mosaic virus-induced gene silencing in monocot plants. *Plant Physiol.* **2016**, *171*, 1801–1807. [PubMed]

56. Agüero, J.; Vives, M.d.C.; Velázquez, K.; Pina, J.A.; Navarro, L.; Moreno, P.; Guerri, J. Effectiveness of gene silencing induced by viral vectors based on Citrus leaf blotch virus is different in Nicotiana benthamiana and citrus plants. *Virology* **2014**, *460–461*, 154–164.

57. Nagamatsu, A.; Masuta, C.; Senda, M.; Matsuura, H.; Kasai, A.; Hong, J.S.; Kitamura, K.; Abe, J.; Kanazawa, A. Functional analysis of soybean genes involved in flavonoid biosynthesis by virus-induced gene silencing. *Plant Biotechnol. J.* **2007**, *5*, 778–790.

58. Cui, H.; Wang, A. An efficient viral vector for functional genomic studies of Prunus fruit trees and its induced resistance to Plum pox virus via silencing of a host factor gene. *Plant Biotechnol. J.* **2017**, *15*, 344–356.

59. Ding, X.S.; Schneider, W.L.; Chaluvadi, S.R.; Mian, M.A.R.; Nelson, R.S. Characterization of a Brome mosaic virus strain and its use as a vector for gene silencing in monocotyledonous hosts. *Mol. Plant. Microbe Interact.* **2006**, *19*, 1229–1239.

60. Purkayastha, A.; Mathur, S.; Verma, V.; Sharma, S.; Dasgupta, I. Virus-induced gene silencing in rice using a vector derived from a DNA virus. *Planta* **2010**, *232*, 1531–1540.

61. Choi, B.; Kwon, S.J.; Kim, M.H.; Choe, S.; Kwak, H.R.; Kim, M.K.; Jung, C.; Seo, J.K. A Plant Virus-Based Vector System for Gene Function Studies in Pepper. *Plant Physiol.* **2019**, *181*, 867–880.

62. Yu, J.; Yang, X.D.; Wang, Q.; Gao, L.W.; Yang, Y.; Xiao, D.; Liu, T.K.; Li, Y.; Hou, X.L.; Zhang, C.W. Efficient virus-induced gene silencing in Brassica rapa using a turnip yellow mosaic virus vector. *Biol. Plant.* **2018**, *62*, 826–834.

63. Constantin, G.D.; Krath, B.N.; MacFarlane, S.A.; Nicolaisen, M.; Johansen, I.E.; Lund, O.S. Virus-induced gene silencing as a tool for functional genomics in a legume species. *Plant J.* **2004**, *40*, 622–631. [PubMed]

64. Holzberg, S.; Brosio, P.; Gross, C.; Pogue, G.P. Barley stripe mosaic virus-induced gene silencing in a monocot plant. *Plant J.* **2002**, *30*, 315–327. [PubMed]

65. Dhir, S.; Srivastava, A.; Yoshikawa, N.; Khurana, S.M.P. Plant viruses as virus induced gene silencing (VIGS) vectors. In *Plant Biotechnology: Progress in Genomic Era*; Springer: Singapore, 2019; pp. 517–526.

66. Zheng, Y.; Ding, B.; Fei, Z.; Wang, Y. Comprehensive transcriptome analyses reveal tomato plant responses to tobacco rattle virus-based gene silencing vectors. *Sci. Rep.* **2017**, *7*, 9771. [PubMed]

67. Oláh, E.; Pesti, R.; Taller, D.; Havelda, Z.; Várallyay, É. Non-targeted effects of virus-induced gene silencing vectors on host endogenous gene expression. *Arch. Virol.* **2016**, *161*, 2387–2393. [PubMed]

68. Liu, Y.; Schiff, M.; Dinesh-Kumar, S.P. Virus-induced gene silencing in tomato. *Plant J.* **2002**, *31*, 777–786.

69. Li, C.; Hirano, H.; Kasajima, I.; Yamagishi, N.; Yoshikawa, N. Virus-induced gene silencing in chili pepper by apple latent spherical virus vector. *J. Virol. Methods* **2019**, *273*, 113711.

70. Chung, E.; Seong, E.; Kim, Y.C.; Chung, E.J.; Oh, S.K.; Lee, S.; Park, J.M.; Joung, Y.H.; Choi, D. A method of high frequency virus-induced gene silencing in chili pepper (*Capsicum annuum* L. cv. Bukang). *Mol. Cells* **2004**, *17*, 377–380.

71. Brigneti, G.; Martín-Hernández, A.M.; Jin, H.; Chen, J.; Baulcombe, D.C.; Baker, B.; Jones, J.D.G. Virus-induced gene silencing in Solanum species. *Plant J.* **2004**, *39*, 264–272.

72. Lentz, E.M.; Kuon, J.E.; Alder, A.; Mangel, N.; Zainuddin, I.M.; McCallum, E.J.; Anjanappa, R.B.; Gruissem, W.; Vanderschuren, H. Cassava geminivirus agroclones for virus-induced gene silencing in cassava leaves and roots. *Plant Methods* **2018**, *14*, 73.

73. Zhang, C.; Bradshaw, J.D.; Whitham, S.A.; Hill, J.H. The development of an efficient multipurpose bean pod mottle virus viral vector set for foreign gene expression and RNA silencing. *Plant Physiol.* **2010**, *153*, 52–65.

74. Meziadi, C.; Blanchet, S.; Geffroy, V.; Pflieger, S. Virus-induced gene silencing (VIGS) and foreign gene expression in *Pisum sativum* L. using the "One-Step" Bean pod mottle virus (BPMV) viral vector. *Methods Mol. Biol.* **2017**, *1654*, 311–319. [PubMed]

75. Kim, H.; Onodera, Y.; Masuta, C. Application of cucumber mosaic virus to efficient induction and long-term maintenance of virus-induced gene silencing in spinach. *Plant Biotechnol.* **2020**, *37*, 83–88.

76. Gao, X.; Shan, L. Functional genomic analysis of cotton genes with agrobacterium-mediated virus-induced gene silencing. *Methods Mol. Biol.* **2013**, *975*, 157–165. [PubMed]

77. Tzean, Y.; Lee, M.C.; Jan, H.H.; Chiu, Y.S.; Tu, T.C.; Hou, B.H.; Chen, H.M.; Chou, C.N.; Yeh, H.H. Cucumber mosaic virus-induced gene silencing in banana. *Sci. Rep.* **2019**, *9*, 11553. [PubMed]

78. Chai, Y.M.; Jia, H.F.; Li, C.L.; Dong, Q.H.; Shen, Y.Y. FaPYR1 is involved in strawberry fruit ripening. *J. Exp. Bot.* **2011**, *62*, 5079–5089. [PubMed]

79. Li, C.; Yamagishi, N.; Kasajima, I.; Yoshikawa, N. Virus-induced gene silencing and virus-induced flowering in strawberry (Fragaria × ananassa) using apple latent spherical virus vectors. *Hortic. Res.* **2019**, *6*, 18.

80. Sasaki, S.; Yamagishi, N.; Yoshikawa, N. Efficient virus-induced gene silencing in apple, pear and Japanese pear using Apple latent spherical virus vectors. *Plant Methods* **2011**, *7*, 15.

81. Scofield, S.R.; Huang, L.; Brandt, A.S.; Gill, B.S. Development of a virus-induced gene-silencing system for hexaploid wheat and its use in functional analysis of the Lr21-mediated leaf rust resistance pathway. *Plant Physiol.* **2005**, *138*, 2165–2173.

82. Singh, D.K.; Lee, H.K.; Dweikat, I.; Mysore, K.S. An efficient and improved method for virus-induced gene silencing in sorghum. *BMC Plant Biol.* **2018**, *18*, 123.

83. Renner, T.; Bragg, J.; Driscoll, H.E.; Cho, J.; Jackson, A.O.; Specht, C.D. Virus-induced gene silencing in the culinary ginger (*Zingiber officinale*): An effective mechanism for down-regulating gene expression in tropical monocots. *Mol. Plant* **2009**, *2*, 1084–1094.

84. Baltes, N.J.; Gil-Humanes, J.; Cermak, T.; Atkins, P.A.; Voytas, D.F. DNA replicons for plant genome engineering. *Plant Cell* **2014**, *26*, 151–163. [PubMed]

85. Ali, Z.; Abul-faraj, A.; Li, L.; Ghosh, N.; Piatek, M.; Mahjoub, A.; Aouida, M.; Piatek, A.; Baltes, N.J.; Voytas, D.F.; et al. Efficient virus-mediated genome editing in plants using the CRISPR/Cas9 system. *Mol. Plant* **2015**, *8*, 1288–1291. [PubMed]

86. Zhang, X.; Kang, L.; Zhang, Q.; Meng, Q.; Pan, Y.; Yu, Z.; Shi, N.; Jackson, S.; Zhang, X.; Wang, H.; et al. An RNAi suppressor activates in planta virus-mediated gene editing. *Funct. Integr. Genom.* **2019**, *20*, 471–477.

87. Yin, K.; Han, T.; Liu, G.; Chen, T.; Wang, Y.; Yu, A.Y.L.; Liu, Y. A geminivirus-based guide RNA delivery system for CRISPR/Cas9 mediated plant genome editing. *Sci. Rep.* **2015**, *5*, 14926.

88. Butler, N.M.; Atkins, P.A.; Voytas, D.F.; Douches, D.S. Generation and Inheritance of Targeted Mutations in Potato (*Solanum tuberosum* L.) Using the CRISPR/Cas System. *PLoS ONE* **2015**, *10*, e0144591.

89. Butler, N.M.; Baltes, N.J.; Voytas, D.F.; Douches, D.S. Geminivirus-mediated genome editing in potato (*Solanum tuberosum* L.) using sequence-specific nucleases. *Front. Plant Sci.* **2016**, *7*, 1045.

90. Čermák, T.; Baltes, N.J.; Čegan, R.; Zhang, Y.; Voytas, D.F. High-frequency, precise modification of the tomato genome. *Genome Biol.* **2015**, *16*, 232.

91. Cody, W.B.; Scholthof, H.B.; Mirkov, T.E. Multiplexed gene editing and protein overexpression using a Tobacco mosaic virus viral vector. *Plant Physiol.* **2017**, *175*, 23–35.

92. Ma, X.; Zhang, X.; Liu, H.; Li, Z. Highly efficient DNA-free plant genome editing using virally delivered CRISPR–Cas9. *Nat. Plants* **2020**, *6*, 773–779.

93. Uranga, M.; Aragonés, V.; Selma, S.; Vázquez-Vilar, M.; Orzáez, D.; Darós, J.-A. Efficient Cas9 multiplex editing using unspaced gRNA arrays engineering in a Potato virus X vector. *bioRxiv* **2020**. [CrossRef]

94. Xie, K.; Minkenberg, B.; Yang, Y. Boosting CRISPR/Cas9 multiplex editing capability with the endogenous tRNA-processing system. *Proc. Natl. Acad. Sci. USA* **2015**, *112*, 3570–3575. [PubMed]

95. Jiang, N.; Zhang, C.; Liu, J.Y.; Guo, Z.H.; Zhang, Z.Y.; Han, C.G.; Wang, Y. Development of Beet necrotic yellow vein virus-based vectors for multiple-gene expression and guide RNA delivery in plant genome editing. *Plant Biotechnol. J.* **2019**, *17*, 1302–1315. [PubMed]

96. Hu, J.; Li, S.; Li, Z.; Li, H.; Song, W.; Zhao, H.; Lai, J.; Xia, L.; Li, D.; Zhang, Y. A barley stripe mosaic virus-based guide RNA delivery system for targeted mutagenesis in wheat and maize. *Mol. Plant Pathol.* **2019**, *20*, 1463–1474. [PubMed]

97. Ellison, E.E.; Nagalakshmi, U.; Gamo, M.E.; Huang, P.; Dinesh-Kumar, S.; Voytas, D.F. Multiplexed heritable gene editing using RNA viruses and mobile single guide RNAs. *Nat. Plants* **2020**, *6*, 620–624. [PubMed]

98. Corbesier, L.; Vincent, C.; Jang, S.; Fornara, F.; Fan, Q.; Searle, I.; Giakountis, A.; Farrona, S.; Gissot, L.; Turnbull, C.; et al. FT protein movement contributes to long-distance signaling in floral induction of Arabidopsis. *Science* **2007**, *316*, 1030–1033.

99. Jaeger, K.E.; Wigge, P.A. FT protein acts as a long-range signal in Arabidopsis. *Curr. Biol.* **2007**, *17*, 1050–1054. [PubMed]

100. Lin, M.K.; Belanger, H.; Lee, Y.J.; Varkonyi-Gasic, E.; Taoka, K.-I.; Miura, E.; Xoconostle-Cázares, B.; Gendler, K.; Jorgensen, R.A.; Phinney, B.; et al. FLOWERING LOCUS T protein may act as the long-distance florigenic signal in the Cucurbits. *Plant Cell* **2007**, *19*, 1488–1506.

101. Mathieu, J.; Warthmann, N.; Küttner, F.; Schmid, M. Export of FT protein from phloem companion cells is sufficient for floral induction in Arabidopsis. *Curr. Biol.* **2007**, *17*, 1055–1060.

102. Notaguchi, M.; Abe, M.; Kimura, T.; Daimon, Y.; Kobayashi, T.; Yamaguchi, A.; Tomita, Y.; Dohi, K.; Mori, M.; Araki, T. Long-distance, graft-transmissible action of Arabidopsis FLOWERING LOCUS T protein to promote flowering. *Plant Cell Physiol.* **2008**, *49*, 1645–1658.

103. Tamaki, S.; Matsuo, S.; Wong, H.L.; Yokoi, S.; Shimamoto, K. Hd3a Protein Is a Mobile Flowering Signal in Rice. *Science* **2007**, *316*, 1033–1036.

104. Mei, Y.; Beernink, B.M.; Ellison, E.E.; Konečná, E.; Neelakandan, A.K.; Voytas, D.F.; Whitham, S.A. Protein expression and gene editing in monocots using foxtail mosaic virus vectors. *Plant Direct* **2019**, *3*, e00181. [PubMed]

105. Gil-Humanes, J.; Wang, Y.; Liang, Z.; Shan, Q.; Ozuna, C.V.; Sánchez-León, S.; Baltes, N.J.; Starker, C.; Barro, F.; Gao, C.; et al. High-efficiency gene targeting in hexaploid wheat using DNA replicons and CRISPR/Cas9. *Plant J.* **2017**, *89*, 1251–1262. [PubMed]

106. Wang, M.; Lu, Y.; Botella, J.R.; Mao, Y.; Hua, K.; Zhu, J.K. Gene targeting by homology-directed repair in rice using a Geminivirus-based CRISPR/Cas9 system. *Mol. Plant* **2017**, *10*, 1007–1010. [PubMed]

107. Wang, X.; Lu, J.; Lao, K.; Wang, S.; Mo, X.; Xu, X.; Chen, X.; Mo, B. Increasing the efficiency of CRISPR/Cas9-based gene editing by suppressing RNAi in plants. *Sci. China Life Sci.* **2019**, *62*, 982–984. [PubMed]

108. Mao, Y.; Yang, X.; Zhou, Y.; Zhang, Z.; Botella, J.R.; Zhu, J.-K. Manipulating plant RNA-silencing pathways to improve the gene editing efficiency of CRISPR/Cas9 systems. *Genome Biol.* **2018**, *19*, 149. [PubMed]

109. Gleba, Y.; Marillonnet, S.; Klimyuk, V. Engineering viral expression vectors for plants: The 'full virus' and the 'deconstructed virus' strategies. *Curr. Opin. Plant Biol.* **2004**, *7*, 182–188. [PubMed]

110. Marillonnet, S.; Giritch, A.; Gils, M.; Kandzia, R.; Klimyuk, V.; Gleba, Y. In planta engineering of viral RNA replicons: Efficient assembly by recombination of DNA modules delivered by Agrobacterium. *Proc. Natl. Acad. Sci. USA* **2004**, *101*, 6852–6857.

111. Gleba, Y.; Klimyuk, V.; Marillonnet, S. Magnifection—A new platform for expressing recombinant vaccines in plants. *Vaccine* **2005**, *23*, 2042–2048.

112. Brewer, H.C.; Hird, D.L.; Bailey, A.M.; Seal, S.E.; Foster, G.D. A guide to the contained use of plant virus infectious clones. *Plant Biotechnol. J.* **2018**, *16*, 832–843.

113. Maeda, K.; Kikuchi, T.; Kasajima, I.; Li, C.; Yamagishi, N.; Yamashita, H.; Yoshikawa, N. Virus-induced flowering by Apple latent spherical virus vector: Effective use to accelerate breeding of grapevine. *Viruses* **2020**, *12*, 70.

114. Velázquez, K.; Agüero, J.; Vives, M.C.; Aleza, P.; Pina, J.A.; Moreno, P.; Navarro, L.; Guerri, J. Precocious flowering of juvenile citrus induced by a viral vector based on Citrus leaf blotch virus: A new tool for genetics and breeding. *Plant Biotechnol. J.* **2016**, *14*, 1976–1985. [PubMed]

115. Martí, M.; Diretto, G.; Aragonés, V.; Frusciante, S.; Ahrazem, O.; Gómez-Gómez, L.; Daròs, J.-A. Efficient production of saffron crocins and picrocrocin in Nicotiana benthamiana using a virus-driven system. *Metab. Eng.* **2020**, *61*, 238–250. [PubMed]

116. Daròs, J.A. Use of Potyvirus vectors to produce carotenoids in plants. In *Plant and Food Carotenoids*; Humana: New York, NY, USA, 2020; pp. 303–312.

117. Majer, E.; Llorente, B.; Rodríguez-Concepción, M.; Daròs, J.A. Rewiring carotenoid biosynthesis in plants using a viral vector. *Sci. Rep.* **2017**, *7*, 41645. [PubMed]

118. Giuliano, G. Plant carotenoids: Genomics meets multi-gene engineering. *Curr. Opin. Plant Biol.* **2014**, *19*, 111–117.

119. Cazzonelli, C.I.; Pogson, B.J. Source to sink: Regulation of carotenoid biosynthesis in plants. *Trends Plant Sci.* **2010**, *15*, 266–274.

120. Nisar, N.; Li, L.; Lu, S.; Khin, N.C.; Pogson, B.J. Carotenoid metabolism in plants. *Mol. Plant* **2015**, *8*, 68–82.

121. Hefferon, K.L. Plant virus expression vectors: A powerhouse for global health. *Biomedicines* **2017**, *5*, 44.

122. Mason, H.S.; Lam, D.M.; Arntzen, C.J. Expression of hepatitis B surface antigen in transgenic plants. *Proc. Natl. Acad. Sci. USA* **1992**, *89*, 11745–11749.

123. Thanavala, Y.; Yang, Y.F.; Lyons, P.; Mason, H.S.; Arntzen, C. Immunogenicity of transgenic plant-derived hepatitis B surface antigen. *Proc. Natl. Acad. Sci. USA* **1995**, *92*, 3358–3361.

124. Huang, Z.; Elkin, G.; Maloney, B.J.; Beuhner, N.; Arntzen, C.J.; Thanavala, Y.; Mason, H.S. Virus-like particle expression and assembly in plants: Hepatitis B and Norwalk viruses. *Vaccine* **2005**, *23*, 1851–1858.

125. Casal, J.I.; Langeveld, J.P.; Cortés, E.; Schaaper, W.W.; van Dijk, E.; Vela, C.; Kamstrup, S.; Meloen, R.H. Peptide vaccine against canine parvovirus: Identification of two neutralization subsites in the N terminus of VP2 and optimization of the amino acid sequence. *J. Virol.* **1995**, *69*, 7274–7277. [PubMed]

126. Gleba, Y.Y.; Tusé, D.; Giritch, A. Plant viral vectors for delivery by Agrobacterium. *Curr. Top. Microbiol. Immunol.* **2014**, *375*, 155–192. [PubMed]

127. Capell, T.; Twyman, R.M.; Armario-Najera, V.; Ma, J.K.-C.; Schillberg, S.; Christou, P. Potential Applications of Plant Biotechnology against SARS-CoV-2. *Trends Plant Sci.* **2020**, *25*, 635–643. [PubMed]

128. Lindbo, J.A. TRBO: A high-efficiency Tobacco mosaic virus RNA-based overexpression vector. *Plant Physiol.* **2007**, *145*, 1232–1240. [PubMed]

129. Peyret, H.; Lomonossoff, G.P. The pEAQ vector series: The easy and quick way to produce recombinant proteins in plants. *Plant Mol. Biol.* **2013**, *83*, 51–58.

130. Chapman, S.; Kavanagh, T.; Baulcombe, D. Potato virus X as a vector for gene expression in plants. *Plant J.* **1992**, *2*, 549–557.

131. Larsen, J.S.; Curtis, W.R. RNA viral vectors for improved Agrobacterium-mediated transient expression of heterologous proteins in Nicotiana benthamiana cell suspensions and hairy roots. *BMC Biotechnol.* **2012**, *12*, 21.

132. Hefferon, K.L. DNA virus vectors for vaccine production in plants: Spotlight on Geminiviruses. *Vaccines* **2014**, *2*, 642–653.

133. Sanger, H.L.; Klotz, G.; Riesner, D.; Gross, H.J.; Kleinschmidt, A.K. Viroids are single-stranded covalently closed circular RNA molecules existing as highly base-paired rod-like structures. *Proc. Natl. Acad. Sci. USA* **1976**, *73*, 3852–3856.

134. Kristensen, L.S.; Andersen, M.S.; Stagsted, L.V.W.; Ebbesen, K.K.; Hansen, T.B.; Kjems, J. The biogenesis, biology and characterization of circular RNAs. *Nat. Rev. Genet.* **2019**, *20*, 675–691.

135. Li, X.; Yang, L.; Chen, L.L. The biogenesis, functions, and challenges of circular RNAs. *Mol. Cell* **2018**, *71*, 428–442. [PubMed]

136. Jeck, W.R.; Sorrentino, J.A.; Wang, K.; Slevin, M.K.; Burd, C.E.; Liu, J.; Marzluff, W.F.; Sharpless, N.E. Circular RNAs are abundant, conserved, and associated with ALU repeats. *RNA* **2013**, *19*, 141–157.

137. Salzman, J.; Gawad, C.; Wang, P.L.; Lacayo, N.; Brown, P.O. Circular RNAs are the predominant transcript isoform from hundreds of human genes in diverse cell types. *PLoS ONE* **2012**, *7*, e30733.

138. Aufiero, S.; Reckman, Y.J.; Pinto, Y.M.; Creemers, E.E. Circular RNAs open a new chapter in cardiovascular biology. *Nat. Rev. Cardiol.* **2019**, *16*, 503–514.

139. Wang, Y.; Liu, B. Circular RNA in diseased heart. *Cells* **2020**, *9*, 1240.

140. Liu, X.; Abraham, J.M.; Cheng, Y.; Wang, Z.; Wang, Z.; Zhang, G.; Ashktorab, H.; Smoot, D.T.; Cole, R.N.; Boronina, T.N.; et al. Synthetic Circular RNA Functions as a miR-21 Sponge to Suppress Gastric Carcinoma Cell Proliferation. *Mol. Ther. Nucleic Acids* **2018**, *13*, 312–321. [PubMed]

141. Beaudry, D.; Perreault, J.P. An efficient strategy for the synthesis of circular RNA molecules. *Nucleic Acids Res.* **1995**, *23*, 3064–3066. [PubMed]

142. Wesselhoeft, R.A.; Kowalski, P.S.; Anderson, D.G. Engineering circular RNA for potent and stable translation in eukaryotic cells. *Nat. Commun.* **2018**, *9*, 2629.

143. Flores, R.; Minoia, S.; Carbonell, A.; Gisel, A.; Delgado, S.; López-Carrasco, A.; Navarro, B.; Di Serio, F. Viroids, the simplest RNA replicons: How they manipulate their hosts for being propagated and how their hosts react for containing the infection. *Virus Res.* **2015**, *209*, 136–145.

144. Hutchins, C.J.; Rathjen, P.D.; Forster, A.C.; Symons, R.H. Self-cleavage of plus and minus RNA transcripts of avocado sunblotch viroid. *Nucleic Acids Res.* **1986**, *14*, 3627–3640.

145. Steger, G.; Riesner, D. Viroid research and its significance for RNA technology and basic biochemistry. *Nucleic Acids Res.* **2018**, *46*, 10563–10576. [PubMed]

146. Nohales, M.A.; Molina-Serrano, D.; Flores, R.; Daros, J.A. Involvement of the chloroplastic isoform of tRNA ligase in the replication of viroids belonging to the family Avsunviroidae. *J. Virol.* **2012**, *86*, 8269–8276. [PubMed]

147. Cordero, T.; Ortolá, B.; Daròs, J.A. Mutational analysis of eggplant latent viroid RNA circularization by the eggplant tRNA ligase in Escherichia coli. *Front. Microbiol.* **2018**, *9*, 635. [PubMed]

148. Cordero, T.; Aragonés, V.; Daròs, J.A. Large-scale production of recombinant RNAs on a circular scaffold using a viroid-derived system in Escherichia coli. *J. Vis. Exp.* **2018**, *141*, 58472.

149. Daròs, J.A.; Aragonés, V.; Cordero, T. A viroid-derived system to produce large amounts of recombinant RNA in Escherichia coli. *Sci. Rep.* **2018**, *8*, 1904.

150. Colpan, M.; Schumacher, J.; Brüggemann, W.; Sänger, H.L.; Riesner, D. Large-scale purification of viroid RNA using Cs2SO4 gradient centrifugation and high-performance liquid chromatography. *Anal. Biochem.* **1983**, *131*, 257–265.

151. Colpan, M.; Riesner, D. High-performance liquid chromatography of high-molecular-weight nucleic acids on the macroporous ion exchanger, nucleogen. *J. Chromatogr. A* **1984**, *296*, 339–353.

152. Sieger, G.; Hofmann, H.; Förtsch, J.; Gross, H.J.; Randies, J.W.; Sänger, H.L.; Riesner, D. Conformational transitions in viroids and virusoids: Comparison of results from energy minimization algorithm and from experimental data. *J. Biomol. Struct. Dyn.* **1984**, *2*, 543–571.

153. Zuker, M. On finding all suboptimal foldings of an RNA molecule. *Science* **1989**, *244*, 48–52.

154. Wu, Q.; Wang, Y.; Cao, M.; Pantaleo, V.; Burgyan, J.; Li, W.-X.; Ding, S.-W. Homology-independent discovery of replicating pathogenic circular RNAs by deep sequencing and a new computational algorithm. *Proc. Natl. Acad. Sci. USA* **2012**, *109*, 3938–3943. [PubMed]

155. Zheng, Y.; Gao, S.; Padmanabhan, C.; Li, R.; Galvez, M.; Gutierrez, D.; Fuentes, S.; Ling, K.S.; Kreuze, J.; Fei, Z. VirusDetect: An automated pipeline for efficient virus discovery using deep sequencing of small RNAs. *Virology* **2017**, *500*, 130–138. [PubMed]

Publisher's Note: MDPI stays neutral with regard to jurisdictional claims in published maps and institutional affiliations.

Review

Application of Viral Vectors for Vaccine Development with a Special Emphasis on COVID-19

Kenneth Lundstrom

PanTherapeutics, CH1095 Lutry, Switzerland; lundstromkenneth@gmail.com; Tel.: +41-79-776-6351

Received: 12 October 2020; Accepted: 17 November 2020; Published: 18 November 2020

Abstract: Viral vectors can generate high levels of recombinant protein expression providing the basis for modern vaccine development. A large number of different viral vector expression systems have been utilized for targeting viral surface proteins and tumor-associated antigens. Immunization studies in preclinical animal models have evaluated the elicited humoral and cellular responses and the possible protection against challenges with lethal doses of infectious pathogens or tumor cells. Several vaccine candidates for both infectious diseases and various cancers have been subjected to a number of clinical trials. Human immunization trials have confirmed safe application of viral vectors, generation of neutralizing antibodies and protection against challenges with lethal doses. A special emphasis is placed on COVID-19 vaccines based on viral vectors. Likewise, the flexibility and advantages of applying viral particles, RNA replicons and DNA replicon vectors of self-replicating RNA viruses for vaccine development are presented.

Keywords: viral vaccines; infectious diseases; cancers; COVID-19 vaccines; self-replicating RNA vectors; DNA-based vaccines; RNA-based vaccines

1. Introduction

The recent coronavirus pandemic (COVID-19) has underlined the importance of vaccine development. It has also become clear to the general public that a number of competing approaches for vaccine candidates need to be developed in parallel to achieve success in the shortest possible time. The same strategy should be applied to any vaccine target albeit the global concern related to COVID-19 has drained resources from other important vaccine development initiatives. It should also be pointed out that vaccine development is not restricted to infectious diseases as quite a few approaches have focused on cancer vaccines as discussed below.

The traditional approach, which is still valid and plays an important role in COVID-19 vaccine development against viral infections, relates to the application of killed and live-attenuated vaccines [1]. Moreover, protein subunit and peptide vaccines have become popular, not least due to the development of efficient recombinant protein expression systems in the 1980s and 1990s [2]. The topic of this review is the utilization of viral vectors for vaccine development. In this context, a variety of viral expression systems have been engineered. Typically, expression vectors have been constructed for adenoviruses (Ads), alphaviruses, flaviviruses, measles viruses (MVs), rhabdoviruses, retroviruses (RVs), lentiviruses (LVs), and poxviruses [3,4]. Briefly, Ad vectors are non-enveloped double-stranded DNA (dsDNA) viruses with a packaging capacity of 7.5 kb foreign DNA providing transient episomal expression in a broad range of host cells [5]. Alphavirus- and flavivirus-based vectors are enveloped single-stranded RNA (ssRNA) viruses with a positive polarity, characterized for their self-replicating RNA property, which provides substantial amplification of foreign mRNA directly in infected host cells [6,7]. In contrast, MVs [8] and rhabdoviruses [9] possess an ssRNA genome of negative polarity, which requires reverse genetics to establish appropriate expression vectors. Among these self-amplifying RNA viral vectors, alphaviruses hold a packaging capacity of 8 kb of foreign genes, whereas for the others it is about 6 kb. RVs are ssRNA viruses, characterized by reverse transcription of their genome into DNA, which can be integrated into the host cell genome providing long-term transgene expression [10]. The chromosomal integration of RVs has posed some safety issues especially for gene therapy applications, where insertions in active oncogene loci has triggered the development of leukemia in patients with X-linked severe acute immunodeficiency (SCID-X1) [11]. However, this issue has been addressed by the engineering of self-inactivating RV vectors with targeted integration. Another issue with classic RVs is their inability to transduce non-diving cells. For this reason, many gene therapy and vaccine development activities have switched to LVs, also belonging to the genus of RVs, which otherwise provide the same properties as classic RVs including packaging of up to 8 kb of foreign sequences, but are able to infect both dividing and non-dividing cells [12]. Moreover, integration-defective LV vectors have been engineered based on targeted recombinase-mediated cassette exchange to provide safe episomal status [13]. Poxviruses are large dsDNA viruses with a packaging capacity of over 30 kb of foreign DNA, which have been frequently used for vaccine development [14]. Moreover, the small ssRNA Picornaviruses—especially coxsackieviruses—with the potential to insert 6 kb of foreign nucleic acids, have been engineered as expression vectors [15].

The application of different viral vector systems for vaccine development is reviewed below. The approaches of vaccine development for infectious diseases and cancer are presented in separate sections. Moreover, the accelerated efforts of virus-based vaccine development against COVID-19 are addressed in another section. Although viral vector-based vaccine development has in general relied on the expression of viral surface antigens and tumor-associated antigens for immunization, oncolytic viruses and viral vectors carrying reporter genes have been included in this review due to their capacity of tumor-specific replication, which can provide therapeutic activity similar to what has been discovered for viral vector-based vaccines.

2. Viral Vaccines for Infectious Diseases

A common strategy for vaccine development against infectious agents, mainly viruses, has been to introduce immunogenic full-length or truncated viral surface proteins into viral expression vectors for verification of antigen expression in vitro, followed by immunization studies in animal models to evaluate immune responses and potential protection against challenges with lethal doses of pathogenic infectious agents [16]. Due to the large number of preclinical and clinical vaccine studies using viral vectors, it is only possible to present some examples below, with a summary provided in Table 1. Moreover, the main focus is on viral diseases and although vaccines against other types of pathogens have been developed, these are only briefly described at the end of the section.

Although alphaviruses have been frequently used as vaccine vectors, some members of the family, such as Chikungunya virus (CHIKV), have been responsible for severe epidemics in the Republic of Congo [17] and in Reunion [18]. In this context, a chimeric vesicular stomatitis virus (VSV) vector was engineered to express the CHIKV envelope polyprotein (E3-E2-6K-E1) and the Zika virus (ZIKV) membrane-envelope protein (ME) [19]. A single immunization of mice with 1×10^7 pfu induced neutralizing antibodies and resulted in protection against challenges with both CHIKV and ZIKV. In another approach, an Ad-based vaccine strategy was applied for the expression of the Venezuelan equine encephalitis virus (VEE) structural proteins (E3-E2-6K) [20]. Improved codon usage showed a 10-fold increase in antibody responses in BALB/c mice, which also increased protection against challenges with VEE. Moreover, VEE, western equine encephalitis virus (WEE) and eastern equine encephalitis virus (EEE) have been targeted for vaccine development [21]. In this context, vectors for VEE, WEE and EEE have been engineered by removing the furin cleavage site between the E2 and E3 envelope proteins to prevent cleavage of the p62 precursor, which in turn will restrict formation of infectious particles and instead generate virus-like particles (VLPs) [22]. Immunization of mice with 1×10^7 IU of the VEE/WEE/EEE combination or individual VLPs elicited strong neutralizing antibody responses and provided protection against subcutaneous or aerosol challenges with VEE, WEE and EEE [22]. The VEE/WEE/EEE combination of 2×10^8 IU elicited robust neutralizing antibody responses in cynomolgus macaques and showed protection against challenges with VEE and EEE. However, the antibody response against WEE was poor, which also reflected the weak protection seen against WEE challenges. In another approach, the attenuated VEE V4020 strain was administered as a layered DNA/RNA vector into BALB/c mice resulting in a high titer of neutralizing antibodies and protection against challenges with wild-type VEE [23]. Moreover, intramuscular immunization of cynomolgus macaques with the VEE vaccine provided protection against aerosol challenges with wild-type VEE [24].

Related to arenavirus vaccines, Lassa virus (LASV) has been targeted by VSV-based expression of LASV glycoprotein (GPC) [25]. Protection against challenges with LASV strains from Liberia, Mali and Nigeria was obtained in guinea pigs and macaques vaccinated with 1×10^6 and 6×10^7 pfu, respectively. The engineering of an LASV-based replicon system, where the LASV GPC was supplied by Vero cell expression, provided protection in guinea pigs immunized with 5×10^5 focus forming units (ffu) [26]. Similarly, immunization of guinea pigs with 1×10^{10} pfu of Ad5-LASV-GPC and Ad5-LASV-NP vaccine candidates demonstrated protection against challenges with lethal doses of LASV [27]. Additionally, an MV-GPC vaccine also provided protection against LASV challenges after a single immunization with 6×10^6 pfu in macaques [28], which supported the initiation of a randomized, placebo-controlled, dose-finding phase I clinical trial in healthy volunteers [29]. Related to other filoviruses, VEE-based expression of Junin virus (JUNV) GPC and Machupo virus (MACV) GPC, respectively, induced humoral immune responses and provided protection in guinea pigs immunized with 1×10^7 pfu [30].

Ebola virus (EBOV), a member of filoviruses, has been an important target for vaccine development due to several Ebola virus disease (EVD) outbreaks, the most recent in 2014–2016 [31]. For instance, the flavivirus Kunjin virus (KUN) was utilized for the expression of the mutant EBOV glycoprotein GP/D637L, which displayed superior cleavability and shedding of GP compared to wild-type GP [32]. Subcutaneous administration of two doses of 1×10^9 KUN-GP/D637L VLPs provided protection in three out of four vaccinated primates. Moreover, immunization with 5×10^7 pfu of VSV-EBOV GP esulted in protection in macaques against challenges with the EBOV-Makona strain [33] and the Zaire strain (ZEBOV) [34]. Similarly, immunization of non-human primates with 1×10^{12} pfu of Ad5-EBOV-GP vaccine provided protection against lethal challenges with EBOV [35]. In another approach, a chimeric parainfluenza virus type 3 (HPIV3) with an EBOV-GP envelope showed strong immune responses in guinea pigs immunized with a single intranasal dose and protected them against challenges with guinea pig-adapted EBOV [36]. Due to the success from preclinical studies and the urgent needs for a functional vaccine in humans, VSV particles expressing the EBOV-GP from the Zaire strain (VSV-ZEBOV) were subjected to an open-label, cluster ring vaccination phase III trial [37]. In the trial, 4123 individuals with suspected EVD were immediately vaccinated, while 3528 participants received a delayed vaccination. There were no EVD cases discovered in the immediate vaccination group and only 16 EVD confirmed in the delayed vaccination group indicating that the immunization was efficient. Similar results were obtained from another phase III trial, where 2119 and 2041 participants received immediate and 21 days delayed vaccination, respectively [38]. The vaccination was efficient as no new EVD cases were recorded 10 days after the start of the trial. Related to other filoviruses such as Marburg virus (MARV), immunization of nonhuman primates with 1×10^7 pfu of VSV-MARV-GP particles resulted in protection against challenges with MARV [34]. Similarly, a single intramuscular injection of 1×10^{10} ffu of the VEE-based Sudan virus (SUDV) vaccine (VEE-SUDV-GP) provided complete protection in cynomolgus macaques [39]. Interestingly, VEE-SUDV-GP immunization also provided partial protection against challenges with EBOV. However, co-immunization with VEE-SUDV-GP and VEE-EBOV-GP resulted in complete protection against both SUDV and EBOV.

In addition to providing expression systems such as the one based on KUN [7], flaviviruses are known pathogens such as Dengue virus (DENV) and ZIKV causing diseases such a Dengue fever and Zika virus disease, respectively. In attempts to develop a vaccine against Dengue fever, VEE particles expressing the ectodomain of the DENV envelope protein E85 were subjected to a single injection of mice, which resulted in protective immunity against DENV challenges in BALB/c mice [40]. In another approach, administration of 2×10^6 pfu of MV-based vector expressing the DENV domain III of the envelope protein (ED3) to mice, induced DENV-specific immune responses and partial protection against DENV challenges [41]. In the context of clinical trials, DENV vaccine candidates have consisted of live-attenuated vaccines, the chimeric live-attenuated yellow fever-dengue virus tetravalent (CYF-TDV) vaccine or the DENV subunit (DEN-80E) vaccine produced in *Drosophila melanogaster* cells [42]. In the case of ZIKV, a VEE-based replicon RNA expressing the codon-optimized ZIKV *prM* and *E* genes was administered in nanostructured lipid carriers (NLCs) to C57BL/6 mice [43]. It was demonstrated that a single dose as low as 10 ng of the RNA replicon completely protected mice against challenges with ZIKV. As described above, co-expression of a CHIKV polyprotein and the ZIKV ME provided protection in immunized mice [19]. Most of the ZIKV clinical trials conducted relate to live-attenuated or DNA-based vaccines [44]. Although based on an attenuated DENV strain, the 2AA30 vaccine containing the ZIKV ME proteins showed good safety and ZIKV-specific neutralizing antibody responses in a phase I trial in 20 healthy volunteers [45].

Related to hepatotropic viruses, Ad7-based expression of hepatitis B virus (HBV) core antigen (HBcAg) and surface antigen (HBsAg) showed HBV-specific antibody responses in immunized dogs [46]. MV vectors have also been applied for the expression of HBsAg, which showed protection in 50% of rhesus monkeys immunized with 1×10^3 $TCID_{50}$ [47]. Moreover, Semliki Forest virus (SFV), an alphavirus, packaged into a VSV G envelope was used for the expression of the HBV middle surface envelope glycoprotein (MHB) and HBcAg [48]. Immunization of mice with 1×10^7 pfu demonstrated protection against HBV challenges for the SFV-G-MHB vaccine candidate, but not for the SFV-G-HBcAg. In the case of clinical trials, DNA-, live vector-, peptide-based vaccines and cell-based therapies have been preferred to viral-based vaccines [49], although a phase I trial has been initiated for an Ad5 vector expressing a fusion protein composed of a truncated HBV core, a modified HBV polymerase and two HBV envelope domains [50].

The annual influenza virus outbreaks have stressed the importance of the development of effective vaccines. In this context, Ad-based influenza A virus vaccines have been engineered by expression of different portions of the hemagglutinin (HA) protein [51]. BALB/c mice immunized with 5×10^{10} Ad particles expressing the full-length HA were protected from challenges with the lethal VN/1203/04 H5N1 influenza A virus strain. Similarly, a single subcutaneous immunization of 5×10^{10} Ad particles provided complete protection in chickens. In another approach, VEE particles expressing the HA gene from the Hong Kong influenza A virus isolate (A/HK/156/97) was evaluated in chicken [52]. A single dose of 1×10^7 pfu of VEE-HA provided complete protection in chickens. RNA-based immunization with 10 µg of SFV-HA RNA replicons elicited significant immune responses in BALB/c mice and provided protection in 90% of vaccinated animals [53]. In comparison to conventional mRNA immunization, only 1.25 µg of self-amplifying VEE-HA RNA was required to acquire protection against challenges with influenza A virus H1N1, H3N2 and B strains compared to 80 µg of synthetic mRNA [54]. Moreover, a replication-deficient modified vaccinia virus Ankara (MVA) expressing the *HA* gene from influenza virus A/HK/156/97 protected C57BL/6J mice from challenges with the three antigenically distinct strains A/HK/156/97, A/Vietnam/1194/04 and A/Indonesia/5/05 [55]. Most of the clinical development and approvals of influenza vaccines have relied on live-attenuated vaccines. However, limited clinical trials have been conducted with viral vector-based vaccines such as in a phase I/IIa study in 79 healthy volunteers receiving MVA-HA [56]. The vaccination was safe and induced significantly higher antibody titers in individuals receiving a higher dose of 1×10^8 pfu compared to 1×10^7 pfu.

HIV/AIDS has had a substantial impact globally, which has contributed to accelerated efforts to develop vaccines against HIV. Cytomegalovirus has the potential as an attractive candidate for vaccine development due to its feature of systematic induction and maintenance of high levels of effector memory T cells through the "memory inflation" mechanism [57]. This has also included applications of CMV for vaccine development against HIV, as T cell vaccines inducing noncanonical $CD8^+$ T cell responses could induce population-wide immunity against HIV [58]. Related to Ad-based

HIV vaccine development, it was demonstrated that replication-deficient Ad5 expressing HIV Gag elicited consistently strong, long-lived CD8$^+$ biased T cell responses in immunized baboons [59]. In the case of MV, live-attenuated MV expressing HIV-1 Gag like particles with a gp160DeltaV1V2 Env protein envelope elicited high levels of cellular and humoral activity against both MV and HIV with neutralizing activity in immunized mice [60]. Alphavirus vectors have also been subjected to HIV vaccine development, and for instance SFV-HIV-Env particles were compared to vaccines based on a DNA plasmid and a recombinant Env protein [61]. Immunized mice showed the highest antibody titers for the SFV particle-based vaccine. In another study, mice intramuscularly immunized with SFV replicon RNA expressing the HIV-1 Env gene elicited Env-specific antibody responses in four out of five mice [62]. In another approach, recombinant SFV particles and replicon RNA were compared for the expression of the Indian HIV-1C *Env/Gag/PolRT* genes in mice [63]. Significant T cell responses were detected for both particle- and RNA-based immunizations, although the titers were superior for SFV particles compared to RNA. Layered SFV DNA/RNA plasmid vectors expressing HIV Env and a Gag/Pol/Nef fusion protein have also been subjected to immunization studies in BALB/c mice, resulting in strong immune responses [64]. Moreover, alphavirus RNA replicons have been subjected to formulations with a cationic nanoemulsion (CNE) and compared to replicon particles and HIV Env formulated with MF59 adjuvant [65]. The replicon-vector, based on VEE included the HIV-1 glycoprotein 140 (gp140) and the packaging signal of Sindbis virus (SIN) and 3′ end untranslated region, was encapsulated in a CNE consisting of squalene, 1,2-dioleoyl-3-trimethylammonium-propane (DOTAP) and sorbitan trioleate. Intramuscular injection of 50 µg of encapsulated replicon RNA generated potent cellular immune responses in rhesus macaques, which were stronger than immunization with VEE particles or HIV gp140. Moreover, immunization of RNA replicons expressing the HIV glycoprotein 120 (gp120) and encapsulated in DOTAP-based lipid nanoparticles showed higher levels of HIV gp120 expression compared to modified conventional mRNA for 30 days in mice after intramuscular administration [66]. In the context of clinical trials of viral vector-based HIV vaccines, the Ad vaccine failed to show protection against infection in the STEP trial [67]. The vaccine, consisting of three Ad5 vectors expressing the HIV *Gag*, *Pol* and *Nef* genes, respectively, was administered to almost 3000 uninfected volunteers. Of even greater concern was the finding that the vaccine appeared to increase HIV infection rates in individuals with pre-existing immunity against Ad5, which resulted in the premature termination of the trial [68]. For this reason, other vaccine approaches such as DNA prime immunization followed by poxvirus boosting have been subjected to clinical trials [69]. Furthermore, a phase III clinical trial was conducted in Thailand with the Canarypox virus HIV vaccine (ALVAC), based on a canarypox virus, and the AIDSVAX B/E gp120 protein vaccine [70]. Vaccination of 16,402 subjects suggested a trend towards the prevention of HIV infection, but the efficacy was modest, only 32%. Another phase III HIV clinical trial, HVTN 702, in South Africa based on the ALVAC/gp120 vaccine was recently terminated by the National Institute of Allergy and Infectious Diseases following recommendations from an independent data and safety monitoring board indicating that the prime-boost vaccine was not efficacious at preventing HIV [71]. In addition, LV vectors have also been applied for prevention and treatment of HIV showing a high degree of immunogenicity in preclinical studies [72]. Moreover, a LV-based dendritic cell (DC) vaccine expressing the CD40 ligand (CD40L) and the HIV-1 SL9 epitope induced antigen-specific T cell proliferation and memory differentiation in humanized mice [73]. The viral load was reduced significantly (by 2 logs) in immunized mice challenged with HIV-1 and the antiviral response was superior when full-length HIV-1 proteins were expressed from the LV vector (Table 1).

Table 1. Examples of preclinical and clinical vaccine studies for infectious diseases.

Target	Antigen	Vector	Response	Reference
Alphaviruses				
CHIKV	E3-E2-6K-E1	VSV	Protection against CHIKV in mice	[19]
VEE	E3-E2-6K	Ad	Protection against VEE in mice	[20]
VEE	E3-E2-6K	VEE	Protection against VEE in mice, macaques	[21]

Table 1. *Cont.*

Target	Antigen	Vector	Response	Reference
EEE	E3-E2-6K	EEE	Protection against VEE in mice, macaques	[21]
WEE	E3-E2-6K	WEE	Only weak protection in macaques	[21]
VEE	V4020 strain	VEE DNA	Protection against VEE in mice	[23]
VEE	V4020 strain	VEE DNA	Protection against VEE in macaques	[24]
Arenaviruses				
LASV	LASV-GPC	VSV	LASV protection in guinea pigs, macaques	[25]
	LASV-GPC	LASV	Protection against LASV in guinea pigs	[26]
	LASV-GPC/NP	Ad5	Protection against LASV in guinea pigs	[27]
	LASV-GPC	MV	Protection against LASV in macaques	[28]
	LASV-GPC	MV	Phase I trial in progress (healthy volunteers)	[29]
JUNV	JUNV-GPC	VEE	Protection against JUNV in guinea pigs	[30]
MACV	MACV-GPC	VEE	Protection against MACV in guinea pigs	[30]
Filoviruses				
EBOV	GP/D637L	KUN	Protection against EBOV in 75% of primates	[32]
	EBOV-GP	VSV	Protection against EBOV in macaques	[33,34]
	EBOV-GP	Ad5	Protection against EBOV in primates	[35]
	EBOV-GP	HPIV3	Protection against EBOV in guinea pigs	[36]
	EBOV-GP	VSV	Good protection against EDV in phase III	[37,38]
	MARV-GP	VSV	Protection against MARV in macaques	[34]
	SUDV-GP	VEE	Protection against SUDV in macaques	[39]
Flaviviruses				
DENV	E85	VEE	Protection against DENV in mice	[40]
	ED3	MV	Partial protection against DENV in mice	[41]
ZIKV	prME	VEE-NLC-RNA	Protection against ZIKV with 10 ng NLC-RNA in mice	[43]
	ME	VSV	Protection against ZIKV in mice	[19]
	ME	DENV	Good safety, neutralizing Abs in volunteers	[45]
Hepatotropic				
HBV	HBsAg/HBcAg	Ad7	HBV-specific antibody responses in dogs	[46]
	HBsAg	MV	Partial protection against HBV in primates	[47]
	MHB	SFV-G	Protection against HBV challenges in mice	[48]
Influenza				
Influenza A	HA	Ad	Complete protection in mice and chickens	[51]
	HA	VEE	Protection in chicken	[52]
	HA	SFV RNA	Protection in chicken	[53]
	HA	VEE RNA	Protection in mice	[54]
	HA	MVA	Protection against 3 IVA strains in mice	[55]
	HA	MVA	High titer antibodies in phase I/II volunteers	[56]
Lentivirus				
HIV	HIV Gag	Ad5	Strong T cell responses in baboons	[59]
	HIV gp160 Env	MV	Neutralizing activity in mice	[60]
	HIV Env	SFV	Superior titers to DNA or protein vaccines	[61]
	HIV Env/Gag/Po	SFV	Particle-based response superior to RNA	[63]
	HIV Gag/Pol/Nef	SFV DNA	Strong immune responses in mice	[64]
	HIV TV1 gp140	VEE*RNA-NP	Stronger responses than for VEE, gp140	[65]
	HIV Env gp120	VEE RNA-NP	Superior response to conventional mRNA	[66]
	HIV Gag/Pol/Nef	3 Ad5	Failure to provide HIV protection in phase III, enhanced HIV rate for pre-existing Ad5	[68]
	HIV gp120	ALVAC/gp120Strong T cell responses in baboons	Modest HIV protection of 32% in phase III	[70]
	HIV-1, CD40L	LV-DCs	Reduced viral load in humanized mice	[73]

Ad5, adenovirus type 5; ALVAC, Canarypox virus HIV vaccine; CD40, CD40 ligand; CHIKV, Chikungunya virus; DENV, Dengue virus; E85, ectodomain of DENV envelope protein; EEE, eastern equine encephalitis virus; HA, hemagglutinin; HIV, human immunodeficiency virus; HPIV3, human parainfluenza virus type 3; IVA, Influenza virus A; JUNV, Junin virus; LASV, Lassa virus; LASV-GPC, Lassa virus glycoprotein; LV-CDs, lentivirus-transduced dendritic cells; MACV, Machupo virus; ME, membrane-envelope; MV, measles virus; MVA, modified vaccinia virus Ankara; NLC, nanostructured lipid carrier; SFV, Semliki Forest virus; SIN, Sindbis virus; VEE, Venezuelan equine encephalitis; VEE*, VEE vector with 3′ end untranslated region and packaging signal form SIN; VSV, vesicular stomatitis virus, WEE, western equine encephalitis virus; ZIKV, Zika virus.

Related to non-viral pathogens, Ad and alphavirus vectors have been applied for vaccine development. For instance, immunization of BALB/c mice with an SFV DNA replicon vector expressing the *Clostridium botulinum* neurotoxin A elicited antibody and lymphoproliferative responses [74]. Moreover, a single intranasal inoculation of a replication-deficient Ad vector expressing the heavy chain C-fragment of the *C. botulinum* neurotoxin C (BoNT/C) elicited high levels of BoNT/C-specific antibodies and protected against challenges with BoNT/C [75]. Related to malaria, SFV particles expressing the

Plasmodium falciparum Pf332 antigen elicited strong immune responses and immunological memory [76]. Moreover, Ad5- and Ad35-based expression of the *P. falciparum* circumsporozoite surface protein (CSP) elicited both cellular and serologic CSP antigen-specific responses in mice and induced strong malaria-specific immunity [77]. In another study, recombinant SIN particles were applied for the expression of the *P. voelii* circumsporozoite protein (CS), which induced a strong epitope-specific T cell response and provided a high degree of protection against malaria infection in mice immunized with 1×10^8 pfu SIN particles [78]. SIN DNA replicons have also been utilized for the expression of *Mycobacterium tuberculosis* antigen 85A (Ag85A), which provided long-term protection against *M. tuberculosis* in mice immunized with 5 µg SIN DNA [79]. Similarly, immunization of Swiss Webster mice with 1×10^7 pfu SIN particles expressing the protective antigen (PA) for *Bacillus antracis* elicited specific and neutralizing antibodies resulting in partial protection against *B. antracis* challenges [80].

3. Viral Vaccines for Cancer

A large number of cancer vaccine studies have been conducted with various viral vectors, as presented by examples below and in Table 2. For instance, glioblastomas have been targeted by SFV particles expressing endostatin [81]. In comparison to SFV-Lac Z particles and RV-based endostatin delivery, SFV-Endostatin showed superior inhibition of tumor growth and reduced intratumoral vascularization in a mouse B16 glioblastoma model. In another study, mice carrying B16 brain tumors were intratumorally administered DCs transduced with SFV-IL-18 particles in combination with IL-12 protein, which enhanced T helper type 1 responses from tumor specific CD4$^+$ and CD8$^+$ T cells and natural killers and antitumor immunity [82]. In a gene silencing approach, the miRT124 micro-RNA sequences targeting neurons were introduced into the replication-competent SFV4 vector, changing its tropism to mouse glioblastoma cells and following a single intraperitoneal injection into C57BL/6 mice with implanted CT-2A orthotopic gliomas resulted in significant inhibition of tumor growth and prolonged survival [83]. The chimeric VSVΔG-CHIKV vector, where the VSV G protein was replaced by the CHIKV envelope proteins (E3-E2-6K-E1), showed selective infection and elimination of tumor cells with an extended survival of mice with implanted CT-2A tumors from 40 to 100 days [84]. Oncolytic MV vectors expressing green fluorescence protein (GFP), carcinoembryonic antigen (CEA) and sodium iodide symporter (NIS) have demonstrated viral replication and cytopathic effects in glioblastoma cell lines [85]. Moreover, significant antitumor activity was detected in vivo. In a comparative study, Ad5/35 and HSV-1 both demonstrated 70% transduction efficiency in glioma cells [86]. However, in a glioblastoma mouse model where the MV fusogenic membrane glycoprotein (FMG) was expressed from both vectors, HSV-1-based treatment was superior to Ad5/35 therapy. Moreover, the better packaging capacity of HSV-1 favors its future use. In the case of clinical trials, a phase I, dose-escalation study was conducted with the Ad vector DNX-2401 (Delta-24-RGD) in 37 patients with recurrent high-grade glioma (HGG), which resulted in 20% of patients surviving more than 3 years [87]. Additionally, a more than 95% reduction in the tumor size was detected in three patients resulting in over 3 years of progression-free survival.

Related to breast cancer, an Ad vector was engineered with an E2F-1 promoter and the human interleukin-15 (IL-15) gene [88]. The novel SG400-E2F/IL-15 vector selectively killed tumor cells and IL-15 exhibited an immunomodulatory effect, which was confirmed in MDA-MB-231 breast cancer cells. Moreover, strong tumor growth inhibition was observed in BALB/c mice with implanted MDA-MB-231 tumors. Another approach relates to the utilization of adeno-associated virus (AAV) vectors for the delivery of short hairpin RNA (shRNA) targeting basal-like breast cancer (BLBC) [89]. It was demonstrated that the rAAV-PSMA2-shRNA vector efficiently transduced the BLBC cell lines, MDA-MB-468 and HCC1954, resulting in significantly decreased cell viability and induced apoptosis. Moreover, administration of rAAV-PSMA2-shRNA to a BLBC xenograft mouse model resulted in reduced tumor growth. In another AAV-based strategy, delivery of heart-specific miRNA sequences (miRT-1d) supported tumor-specific transgene expression and almost complete elimination in heart tissue [90]. Furthermore, insertion of the therapeutic suicide gene HSV-TK showed significant inhibition of tumor growth in polyoma middle T transgenic mice with multifocal breast tumors. In the context of breast cancer, Ad particles and a SIN DNA replicon expressing the rat HER2/neu gene showed inhibition of A2L2 tumor growth in pre-immunized BALB/c mice but not when the vaccination took place two days after the tumor challenge [91]. A prime-boost regimen with SIN DNA and Ad particles resulted in significant prolongation of survival rates. Moreover, it was demonstrated that intradermal immunization with SIN-HER2/neu DNA replicons elicited strong antibody responses in BALB/c mice [92]. Tumor protection was achieved with 80% less replicon DNA compared to conventional DNA plasmid vectors. In another approach the coxsackievirus A21 (CVA21) was applied for the expression of intercellular adhesion molecule-1 (ICAM-1) and decay-accelerating factor (DAF) [93]. Intravenous injection of CVA21-ICAM-1-DAF combined with intraperitoneal administration of doxorubicin hydrochloride resulted in significantly enhanced tumor regression in mice with MDA-MB-231 breast tumors. Related to clinical trials, six patients with recurrent breast cancer were included in a phase I dose-escalation study with an oncolytic HSV HF10 vector [94]. The outcome was no serious adverse events, and some therapeutic efficacy was registered.

In the case of cervical cancer, alphaviruses have been frequently used for preclinical immunization studies. For instance, VEE particles expressing the human papilloma virus-16 (HPV-16) E7 protein elicited CD8$^+$ T cell responses and prevented tumor development in immunized C57BL76 mice [95]. Moreover, when the HPV E6-E7 fusion was expressed from an SFV vector containing the translation enhancer signal from the SFV capsid gene, immunization of mice with SFVenh-HPV E6-E7 particles provided tumor regression and complete eradication of established tumors [96]. In another study, the combination of intradermal administration of SFV-HPV E6-E7 DNA replicons and electroporation resulted in 85% of immunized mice becoming tumor-free [97]. Remarkably, the therapeutic efficacy was achieved with a 200-fold lower dose, equivalent to 0.05 μg of SFV DNA, compared to conventional DNA plasmid vectors. Recently, GMP-grade production of SFV-HPV E6-E7 (Vvax001) has been produced for use in clinical trials [98]. A number of clinical trials have been conducted on HPV vaccines [99]. For instance, a vaccinia virus vector expressing HPV-16/18 E6/7 induced HPV-specific CTL immune responses in 28% and two out of eight patients showed tumor-free condition at 15 and 21 months, respectively, in a phase I/II trial [100]. In a phase III study in patients with HPV-induced anogenital intraepithelial neoplasia (AGIN), immunization with a recombinant MVA encoding the E2 protein from bovine papilloma virus (BPV) resulted in 90% lesion clearance in treated females and in 100% in male patients [101].

In the case of colon cancer, CT26 colon tumor models have been frequently evaluated. For instance, the non-cytopathic KUN vector expressing the granulocyte macrophage-stimulating factor (GM-CSF)—when administered intratumorally to BALB/c mice with CT26 xenografts—induced CD8$^+$ T cell responses, resulted in tumor regression and in cure of more than 50% of immunized animals [102]. In another study SFV particles expressing the vascular endothelial growth factor receptor-2 (VEGFR-2) was used for the immunization of BALB/c mice resulting in inhibition of tumor growth, reduction in tumor angiogenesis and prevention of metastatic spread [103]. Combination therapy with SFV-VEGFR-2 and SFV-IL-12 particles showed lower immune responses and inferior tumor growth inhibition compared to SFV-VEGFR-2 and SFV-IL-4 co-administration, which enhanced VEGFR-2-specific antibody responses and resulted in prolonged survival of immunized mice. Furthermore, immunization of mice with SFV-LacZ RNA replicons elicited antigen-specific and CD8$^+$ T cell responses after a single injection of 0.1 μg RNA [104]. Protection against tumor challenges was also achieved and tumor regression was observed in mice with pre-existing tumors. The vaccinia virus cowpox virus (CPVX) was engineered for improved tumor selectivity and oncolytic activity by the introduction of the fusion suicide gene-1 (FCU1), which converts the non-toxic prodrug 5-fluorocytosine (5-FC) into cytotoxic 5-fluorouracil (5-FU) and 5-fluorouridine-5′-monophosphate (5-FUMP) [105]. Systemic administration of the modified CPVX vector showed low accumulation in normal tissues but high tumor selectivity, which induced relevant inhibition of tumor growth. Moreover, co-administration of 5-FC enhanced the anti-tumor effect. Intratumoral CPVX administration induced relevant tumor growth inhibition in a LoVo colon cancer model. An Ad vector expressing CEA was administered to a mouse MC-38 colon cancer model [106]. Immunization of Ad-CEA in combination with the anti-PD-1 antibody showed enhanced anti-tumor activity and immune responses. Related to clinical trials, an oncolytic vaccinia virus was subjected to a phase I study in 11 patients with refractory advanced colorectal or other solid cancers [107]. No dose-related toxicity or treatment-related severe adverse events were detected and a strong inflammatory and Th1 cytokine induction support the potential immunity against cancer. The Newcastle disease virus (NDV) was subjected to immunotherapy in a phase III trial in 335 colorectal cancer patients [108]. The study indicated that NDV vaccinations provided prolonged survival and short-term improvement in quality of life.

Lung cancer has been targeted by alphavirus vectors and SFV-EGFP particles induced cell death in human H358a non-small cell lung cancer (NSCLC) cells and inhibited growth of H358a spheroids [109]. Intratumoral administration of SFV-EGFP to nu/nu mice with H358a xenografts induced apoptosis, which generated complete tumor regression in three out of seven mice. In another study, replication-competent SFV (VA7)-EGFP particles were compared to a conditionally replicating Ad vector (Ad5-Delta24TK-GFP) in nude mice implanted with A549 adenocarcinoma lung cells, which resulted in superior survival of SFV-immunized mice [110]. In contrast, systemic administration did not generate significant immune responses. In another study, immunization with SIN-LacZ particles elicited long-lasting memory T cell responses and provided protection against tumor challenges in mice [111]. Moreover, nude mice immunized with H2009 and A549 lung tumors showed reduced tumor growth after intratumoral administration of VSV-IFNβ [112]. Additionally, intratumoral injection of VSV-IFNβ resulted in tumor regression, extended survival, and the cure of 30% of mice with syngeneic LM2 lung tumors. In another approach, the Edmonston strain of MV expressing CEA showed potent killing of lung cancer cell lines and tumor regression in immunized mice [113]. Related to clinical trials, 78 NSCLC patients were treated with the TG4010 vaccine based on the MVA strain expressing human mucin-1 (MUC-1) and IL-2 [114]. It was discovered that improvement in survival correlated with the development of T cell responses against MUC-1.

Viral vector-based melanoma vaccine research has been intense with numerous preclinical studies and clinical trials conducted. In addition to the parallel study on colon cancer and melanoma for KUN-GM-CSF described above [102], yellow fever virus (YFV)—expressing the CTL epitope SIINFEKL of chicken ovalbumin—elicited SIINFEKL-specific CD8$^+$ lymphocytes and protected mice against challenges with malignant melanoma cells [115]. Alphaviruses have been frequently employed for melanoma treatment, where VEE vectors expressing the tyrosine-related protein-2 (TRP-2) demonstrated humoral immune responses, strong antitumor activity, and prolonged survival in a B16 mouse melanoma model [116]. Combination therapy with anti-CTL antigen-4 (CTLA-4) and anti-glucocorticoid-induced tumor necrosis factor receptor (GITR) monoclonal antibodies (mAbs) resulted in complete tumor regression in 50% and 90% of mice, respectively [117]. In another approach, co-administration of SFV-based expression of VEGFR-2 and IL-12 from one DNA replicon and survivin and β-hCG antigens from another DNA replicon was evaluated in a B16 mouse melanoma model [118]. Superior tumor growth inhibition and prolonged survival was achieved by combination therapy in comparison to immunization with either SFV DNA replicon alone. Moreover, the MV Leningrad-16 (L-16) strain showed statistically significant inhibition of tumor growth in a mel Z mouse melanoma model [119]. In another study, the VSV-GP vector pseudotyped with the non-neurotropic lymphocytic choriomeningitis virus (LCMV) provided prolonged survival in mice A375 xenograft and B16-OVA syngeneic mouse models [120]. Application of NDV vectors expressing IL-15 or IL-12 for intratumoral immunization of B16F10 melanoma tumor-bearing mice effectively suppressed tumor growth [121]. The 120-day survival rate for mice treated with rNDV-IL15 was 12.5% higher than for rNDV-IL12. Moreover, tumor re-challenge experiments indicated that the survival rate was 26.7% higher for rNDV-IL15 compared to rNDV-IL12. Furthermore, replication-competent CVA21 expressing ICAM-1/DAF resulted in rapid suppression of subcutaneous SK-Mel-28 melanoma xenografts in NOD-SCID mice [122]. Several clinical trials have been conducted for viral-based melanoma vaccines [123]. In this context, HSV-1 has been subjected to clinical phase I/IIb and phase III trials [124,125]. In the former, a 50% objective response rate was obtained in patients and a durable response lasting for more than 6 months was seen in 44% of patients [124]. In the phase III trial, the durable response rate improved, and longer median survival rates were obtained in patients with non-surgically resectable melanoma [125]. Moreover, a phase II/IIIb study resulted in significant clinical benefits and superior overall survival in stage III and IV melanoma patients [126]. The replication-competent reovirus, a dsRNA virus, was subjected to a phase II trial in patients with metastatic melanoma [127]. The treatment was well tolerated and reovirus replication was demonstrated in patient biopsies. In the case of CVA21-based clinical trials, stable disease was observed in 26.7% of patients in a phase Ib study [128] and durable responses in melanoma metastases were detected in a phase II trial [129,130]. A 15-year follow-up of an NDV-based clinical phase II trial demonstrated that NDV oncolysates were associated with prolonged survival in patients with lymph node-positive malignant melanoma [131].

Several types of vectors have been applied for vaccine development against ovarian cancer. The pseudotyped VSV-LCMV-GP demonstrated oncolytic activity in several ovarian cancer cell lines and in vivo in an ovarian A2780 tumor mouse model [132]. Superior reduction in tumor size was observed in combination with the JAK1/2 inhibitor ruxolitinib in both subcutaneous and orthotopic xenograft mouse models. Moreover, an MV containing a single-chain antibody (scFv) specific for the alpha-folate receptor (αFR), provided tumor specific targeting with no background infectivity of normal cells [133]. Mice with SKOV3ip.1 xenografts were intratumorally injected with MV-GFP and MV-αFR, which resulted in tumor volume reduction and increase in overall survival. Studies involving alphaviruses have been conducted on ovarian cancer such as combination therapy of SIN-IL-12 particles and the CPT-11 topoisomerase inhibitor irinotecan, which provided long-term survival in SCID mice implanted with aggressively growing human ovarian ES2 tumors [134]. Additionally, a prime-boost regimen of SFV-OVA and VV-OVA resulted in enhanced OVA-specific CD8+ T cell immune responses and enhanced anti-tumor activity in immunized C57BL/6 mice with implanted murine ovarian surface epithelial carcinoma (MOSEC) [135]. Related to clinical trials, a phase I study in patients with stage II-IV ovarian epithelial, fallopian tube, or primary peritoneal cavity cancer with the poxvirus ALVAC is in progress [136]. In a similar phase I trial, the safety and tolerability of the ALVAC vaccine was determined [137]. Furthermore, a phase II trial with fowlpox vaccinia virus in patients with epithelial ovarian, fallopian tube, or primary peritoneal carcinoma and whose tumors expressed the NY-ESO-1 or LAGE-1 antigen, evaluated the maintenance of remission at 12 months, time to failure of vaccine therapy, and cellular and humoral immunity [138].

In the case of pancreatic cancer, AAV2 expressing endostatin was administered intramuscularly or intravenously (portal vein) into Syrian golden hamsters previously inoculated into the pancreas with PGHAM-1 pancreatic cells [139]. The transplanted PGHAM-1 cells rapidly metastasized to the liver. After intramuscular injection the endostatin levels showed a modest increase and the numbers of metastases decreased. Intraportal administration resulted in significantly increased levels of endostatin and the size and number of metastases decreased substantially. Overall, intraportal injection was more efficient as an anti-angiogenic therapy. Oncolytic Ad vectors engineered with cell-targeting ligand SYENFSA (SYE) have demonstrated specific targeting of pancreatic cancer cells and efficient oncolysis of pancreatic ductal adenocarcinoma (PDAC) cells [140]. Moreover, VSV-GFP showed superior oncolytic activity in PDAC cell lines and in vivo compared to a conditionally replicative Ad vector (CRAd), Sendai virus and respiratory syncytial virus (RSV) [140]. However, the pancreatic HPAF-II cell line and a mouse HPAF-II model were resistant to VSV infections, which could be reduced by combination therapy with DEAE-dextran and ruxolitinib [141]. In another approach, SCID mice implanted with KLM1 and Capan-2 xenografts were immunized with MV vectors expressing SLAMblind showing significant suppression of tumor growth [142]. Furthermore, the chimeric orthopoxvirus CF33 efficiently killed six pancreatic cancer cell lines and caused regression in PANC-1 pancreatic xenografts after a single intratumoral injection of a low dose of 10^3 pfu [143]. CF33 was shown to preferentially replicate in tumors and non-injected distant xenografts were also affected. Related to clinical trials, eight patients with nonresectable pancreatic cancer were immunized intratumorally with an oncolytic HSV HF10 vaccine in a phase I dose-escalation study [94]. No serious adverse events occurred, and therapeutic efficacy was registered. In another phase I study, patients with nonresectable locally advanced pancreatic cancer showed only HSV HF10-unrelated adverse events after intratumoral administration [144]. Three patients showed partial responses (PR), stable disease (SD) was observed in four patients, and nine patients had progressive disease (PD). VEE-CEA vectors were administered intramuscularly in a phase I trial in pancreatic cancer patients [145]. Repeated VEE-CEA administration induced clinically relevant CEA-specific T cell antibody responses.

In the context of prostate cancer, intratumoral immunization with MV-CEA vectors resulted in a significant delay of tumor growth and prolonged survival in a prostate PC-3 mouse model [146]. Application of alphaviruses has demonstrated strong specific immune responses against prostate-specific membrane antigen (PSMA) [147] and six-transmembrane epithelial antigen of the prostate (STEAP) [148] after immunization of mice with VEE-PSMA and VEE-STEAP, respectively. Transgenic adenocarcinoma mouse prostate (TRAMP) mice showed long-term survival of 90% at 12 months after immunization with VEE particles expressing the prostate stem cell antigen (PSCA) [149]. VSV-LCMV-GP expressing luciferase (Luc) efficiently infected prostate cancer cell lines and showed long-term remission in intratumorally immunized Du145 and 22Rv1 mouse prostate cancer models [150]. In another approach, the combination therapy of oncolytic MV and mumps virus (MuV) vectors showed greater antitumor activity and prolonged survival in a PC-3 human prostate cancer model in comparison to MV and MuV vectors alone [151]. Related to clinical trials, in a phase I study, patients with castration resistant metastatic prostate cancer (CRPC) were immunized with either 0.9×10^7 or 3.6×10^7 IU of VEE-PSMA [152]. The treatment was well tolerated, but induced only weak PSMA-specific immune responses, which will require dose optimization to enhance the efficacy. In a phase I trial in 32 patients with hormone refractory metastatic prostate cancer vaccination with Ad5 expressing prostate-specific antigen (PSA), anti-PSA antibodies were elicited in 34% of patients, 68% showed anti-PSA responses, 48% had a longer PSA doubling time, and the survival time was prolonged in 55% of the patients [153]. POSTVAC (TRICOM) is a poxvirus vaccine candidate based on an attenuated recombinant VV prime vector and a fowlpox virus booster vector expressing B7-1, lymphocyte function associated antigen-3 (LFA-3) and ICAM-1 [154]. In a phase II trial, 125 minimally symptomatic CRPC patients were immunized with PROSTVAC, which showed an increase in the median overall survival but not progression free survival [155]. Similar findings were obtained in another phase II trial in 32 CRPC patients treated with PROSTVAC and GM-CSF [156]. Furthermore, in a phase III study in GRPC patients no differences were found in overall survival between patients treated with PROSTVAC, PROSTVAC + GM-CSF or placebo [157]. These findings indicate that other combination therapies with DNA vaccines and chemotherapies need to be explored [158] (Table 2).

Table 2. Examples of preclinical and clinical cancer vaccine studies.

Target	Antigen	Vector	Response	Ref
Brain				
GBM	Endostatin	SFV	Tumor regression, prolonged survival in mice	[81]
	IL-18 + IL-12	DC-SFV-IL-18	Enhanced antitumor immunity	[82]
CT-2A	miR124	SFV4	SFV replication in tumors, tumor regression	[83]
	Chimeric VLPs	VSVΔG-CHIKV	Tumor targeting, prolonged survival in mice	[84]
GBM	CEA	MV-CEA/GFP	MV replication in tumors	[85]
	MV FMG	Ad5/35	Transduction of glioma cells	[86]
	MV FMG	HSV-1	Superior to Ad in vitro and in vivo	[86]
HGG	oAd	DNX-2401	Long-term survival (>3 years) in phase I	[87]
Breast				
MDA-MB231	Ad	Ad-EF2/IL-15	Tumor growth inhibition in vitro, in mice	[88]
BLBC	PSMA2 shRNA	AAV	Reduced tumor growth in mouse model	[89]
MFB	miRT-1d, HSV-tk	AAV	Significant tumor growth inhibition in mice	[90]
A2L2	HER2/neu	Ad/SIN DNA	Tumor growth inhibition in mice	[91]
	HER2/neu	SIN DNA + Ad	Prolongation of survival in mice	[91]

Table 2. *Cont.*

Target	Antigen	Vector	Response	Ref
	HER2/neu	SIN DNA	Tumor protection with 80% less DNA	[92]
MDA-MB231	ICAM-1/DAF	CVA21	Strongly enhanced tumor regression in mice	[93]
Recurrent BC	oHSV	HSV HF10	Safety confirmed in phase I trial	[94]
Cervical				
C3	HPV E7	VEE	T cell responses, prevention of tumors	[95]
	HPV E6-E7	SFVenh	Complete eradication of established tumors	[96]
TC-1	HPV E6-E7	SFV DNA	85% tumor-free, 200-fold lower DNA dose	[97]
Adv CC	HPV-16/18 E6/7	VV	CTL in 28% of pts, 2 pts tumor-free in phase I	[100]
AGIN	BPV E2	MVA	90–100% lesion clearance in phase III	[101]
Colon				
CT26	GM-CSF	KUN	Tumor regression, cure of >50% of mice	[102]
	VEGFR-2	SFV	Inhibition of tumor growth and metastases	[103]
	VEGFR-2/IL-4	SFV	Prolonged survival in mice	[103]
	LacZ	SFV RNA	T cell responses, protection against tumors	[104]
LoVo	FCU1	CPVX	Tumor selectivity, tumor regression in mice	[105]
MC-38	CEA + anti-PD-1	Ad	Enhanced immune and anti-tumor responses	[106]
Phase I	CD	vvDD	Strongly induced immune responses in pts	[107]
Phase III	NDV 73-T	NDV	Prolonged survival in colon cancer patients	[108]
Lung				
NSCLC	EGFP	SFV	Complete tumor regression in 3 out of 7 mice	[109]
A549	EGFP	SFV vs. Ad	Superior survival of SFV over Ad therapy	[110]
CT26.CL25	EGFP	SIN	Protection against tumor challenges	[111]
A549, LM2	IFNβ	VSV	Tumor regression, cure of 30% of mice	[112]
A549, H2009	CEA	MV	Tumor regression in mice	[113]
Phase II	MUC-1, IL-2	MVA	T cell responses, improved survival of pts	[114]
Melanoma				
B16-OVA	GM-CSF	KUN	T cell responses, tumor regression in mice	[102]
B16-OVA	SIINFEKL	YFV	Protection against malignant melanoma	[115]
B16	TRP-2	VEE	Prolonged survival in mice	[116]
B16	TRP-2 + mAbs*	VEE	Complete tumor regression in 50–90% of mice	[117]
B16	VEGFR-2/IL-12 + Survivin/β-hCG	SFV DNA	Superior tumor growth inhibition after combination therapy	[118]
mel Z	MV L-16	MV	Inhibition of tumor growth in mice	[119]
A549, B16	GFP, Luc	VSV-LCMV GP	Prolonged survival in mice	[120]
B16F10	IL-15/IL-12	NDV	Efficient suppression of tumor growth	[121]
SK-Mel-28	ICAM-1/DAF	CVA21	Suppression of tumor growth in mice	[122]
Phase I/IIb	GM-CSF	HSV-1 T-VEC	50% objective response rate lasting > 6 months	[124]
Phase III	GM-CSF	HSV-1 T-VEC	Improved response, longer median survival	[125]
Phase II/IIIb	GM-CSF	HSV-1 T-VEC	Superior overall survival at stage III/IV	[126]
Phase II	Reolysin	Reovirus	Well tolerated, reovirus replication in biopsies	[127]
Phase 1b	CAVATAK	CVA21	Stable disease in 26.7% of patients	[128]
Phase II	CAVATAK	CVA21	Durable responses in metastatic melanoma	[129]
Phase II	NDV oncolysate	NDV	Prolonged survival in melanoma patients	[131]
Ovarian				
A2780	Luc + Rux	VSV-LCMV GP	Reduction in tumor growth	[132]
SKOV3ip.1	GFP, αFR	MV	Reduced tumor volume, prolonged survival	[133]
ES2	IL-12, CPT-11	SIN + CPT-11	Long-term survival in SCID mice	[134]
MOSEC	OVA	SFV	Enhanced anti-tumor activity in mice	[135]
Phase I	ALVAC	VV	Safety and tolerability studies	[136,137]
Phase II	Fowlpox	VV	Safety, maintenance of remission	[138]
Pancreatic				
PGHAM-1	Endostatin	AAV2	Tumor and metastases regression in hamsters	[139]
PADC	SYE	Ad	Efficient oncolysis of PDAC cells	[140]
PANC-1	GFP	VSV	Oncolytic activity in cell lines and in mice	[141]
Su86.86	GFP	VSV	Oncolytic activity in cell lines and in mice	[141]
KLM1,	SLAM	MV	Suppression of tumor growth in mice	[142]
Capan-2	SLAM	MV	Suppression of tumor growth in mice	[142]
PANC-1	Chimeric OPV	CF33	Replication in tumor cells, tumor regression	[143]
Phase I	oHSV	HSV HF10	Safety, therapeutic efficacy	[94]
Phase I	oHSV	HSV HF10	PR and SD in some patients	[144]
Phase I	CEA	VEE	T cell antibody responses	[145]
Prostate				
LNCaP	CEA	MV	Prolonged survival in mice	[146]
TRAMP-C	PSMA	VEE	Strong immune response in mice	[147]
TRAMP	STEAP	VEE	Prolonged survival in mice	[148]

Table 2. *Cont.*

Target	Antigen	Vector	Response	Ref
TRAMP-PSA	PSCA	VEE	90% survival rate in mice	[149]
Du145, 22Rv1	Luc	VSV-LCMV-GP	Long-term remission in mice	[150]
PC-3	MV, MuV	MV + MuV	Prolonged survival in mice	[151]
Phase I	PSMA	VEE	Modest neutralizing antibodies against PSMA	[152]
Phase I	PSA	Ad5	Antibody responses, prolonged survival	[153]
Phase II	Tricom	PROSTVAC	Prolonged median OS, not PFS	[155]
Phase III	Tricom + GM-CSF	PROSTVAC	Safe, no effect on OS	[157]

AAV, adeno-associated virus; Ad5, adenovirus type 5: Adv CC, advanced cervical cancer; AGIN, anogenital intraepithelial neoplasia; αFR, alpha folate receptor; BLBC, basal-like breast cancer; BPV, bovine papilloma virus; CEA, carcinoembryonic antigen; CD, yeast cytosine deaminase; CVA21, coxsackievirus A21; DAF, decay-accelerating factor; DC, dendritic cell; FCU1, fusion suicide gene 1; GBM, Glioblastoma multiforme; HGG, high-grade glioma; HPV, human papilloma virus; HSV-tk, herpes simplex virus-thymidine kinase; ICAM-1, intercellular adhesion molecule-1; Luc, luciferase; mAbs*, monoclonal antibodies against anti-CTL antigen-4 (CTLA-4) and anti-glucocorticoid-induced tumor necrosis factor receptor (GITR); MFB, multi-focal breast tumor; MOSEC, murine ovarian surface epithelial carcinoma; MV, measles virus; miRT-1d, micro-RNA targeting heart tissue; MVA, modified vaccinia virus Ankara; MuV, mumps virus; NDV, Newcastle disease virus; NSCLC, non-small cell lung cancer; oAd, oncolytic adenovirus; oHSV, oncolytic herpes simplex virus; OVA, ovalbumin; PFS, progression free survival; PROSTVAC, poxvirus vaccine consisting of VV and fowlpox virus; PSA, prostate-specific antigen; PSCA, prostate stem cell antigen; PSMA, prostate specific membrane antigen; pts, patients; Rux, ruxolitinib; SFV, Semliki Forest virus; SFVenh, SFV vector with translation enhancement signal from the SFV capsid gene; shRNA, short hairpin RNA; SIINFEKL, chicken ovalbumin epitope; SIN, Sindbis virus; SLAM, signaling lymphocyte activating molecule; STEAP, six transmembrane epithelial antigen of the prostate; TRAMP, transgenic adenocarcinoma of the mouse prostate; TRICOM, B71, LFA-3 and ICAM-1 expressed from PROSTVAC; VEE, Venezuelan equine encephalitis; VSV, vesicular stomatitis virus, VV, vaccinia virus; vvDD, oncolytic vaccinia virus vector expressing CD; YFV, yellow fever virus.

4. Vaccines against COVID-19

Naturally, vaccine development against the severe acute respiratory syndrome-coronavirus-2 (SARS-CoV-2) causing the COVID-19 pandemic has overshadowed any other vaccine initiative [159]. The impressive number of 155 vaccine candidates in preclinical and 47 candidates in clinical trials are based on inactivated and live attenuated vaccines, protein subunit and peptide vaccines, nucleic acids and viral vectors [160]. The focus here is uniquely on viral vector-based vaccines (Table 3).

The chimpanzee Ad vector ChAdOx1 nCoV-19 was engineered to express the SARS-CoV-2 S protein and when subjected to immunization of mice and rhesus macaques induced strong humoral and cellular immune responses and prevented pneumonia in macaques [161,162]. Similarly, Ad5-SARS-COV-2 S elicited strong S-specific antibody and cell-mediated immune responses in mice and rhesus macaques. Moreover, a single intramuscular or intranasal immunization with Ad5-S-nb2 provided protection against challenges with SARS-CoV-2 in macaques [163]. Preclinical studies in hamsters demonstrated that a single immunization with an Ad26 vector expressing SARS-CoV-2 S elicited neutralizing antibodies and protected immunized animals against pneumonia and death [164]. Immunization of rhesus macaques elicited strong neutralizing antibody responses and protected primates against SARS-CoV-2 [165]. In another preclinical approach, the full-length SARS-CoV-2 S gene was inserted into two positions of the MV genome [166]. Administration of the vaccine candidates to mice demonstrated efficient Th1-biased antibody and T cell responses after two immunizations. Considering that the lung is a vital organ for SARS-CoV-2 infection, MVA poxviruses have been suggested as potential candidates for COVID-19 vaccine development [167]. In this context, a novel vaccine platform was developed for MVA, where a unique three-plasmid system can efficiently generate recombinant MVA vectors from chemically synthesized DNA [168]. Using this technology, mice were immunized with fully synthetic MVA (sMVA) vectors co-expressing SARS-CoV-2 S and nucleocapsid, which elicited robust SARS-CoV-2 antigen-specific humoral and cellular immune responses including potent neutralizing antibodies.

Positive results from preclinical studies on COVID-19 vaccine candidates have supported the launch of several clinical trials. The first-in-human phase I dose-escalation, non-randomized clinical trial was conducted with three doses (5×10^{10}, 1×10^{11} and 1.5×10^{11}) of Ad5-SARS-CoV-2 S particles in 108 healthy volunteers [169]. The safety and tolerability of the treatment was good with only some minor pain reactions to the vaccination. Rapid SARS-CoV-2-specific T cell responses were detected 14 days after vaccination and humoral responses against SARS-CoV-2 reached peak levels at day 28 post-immunization. The Ad5-SARS-CoV-2 S vaccine candidate has now been subjected to a randomized, double-blind, placebo-controlled phase II trial in 603 healthy volunteers [170]. The two doses (1×10^{11} and 5×10^{10} virus particles) elicited significant neutralizing antibodies. Severe adverse reactions were observed in 24 (9%) of vaccinees, but no serious adverse reactions were reported. Overall, the immunization was safe and significant immune responses were induced in the majority of vaccinees after a single vaccination. Moreover, the recruitment of healthy adults 18 years of age and older is in progress for a global double-blind, placebo-controlled phase III trial with an immunization schedule of one intramuscular dose of Ad5-SARS-CoV-2 S [171]. Recruitment is in progress for a similar phase III trial for 18 to 85 years old volunteers for a single intramuscular administration of Ad5-SARS-CoV-2 S [172]. In another Ad based approach, the Ad26.COV2-S vaccine candidate was subjected to a randomized, double blind, placebo-controlled phase I/II study in 1045 healthy volunteers in Belgium and the USA [173]. Interim results demonstrated a good safety profile and immunogenicity after a single immunization [174]. A randomized, double-blind, placebo-controlled phase III study enrolling 60,000 participants is in progress [175].

The Ad-based Sputnik V vaccine developed at the Gamaleya Research Institute of Epidemiology and Microbiology in Russia caused some controversy due to its premature approval prior to the completion of any clinical phase III trials and even before the publication of findings from any preclinical or clinical studies with only a preliminary evaluation in 76 volunteers [176]. The rAd26-S/rAd5-S vaccine regimen is based on a prime vaccination with the Ad26-based SARS-CoV-2 S, followed by a booster vaccination with Ad5-SARS-Cov-2 S. Several weeks after the approval, the results from a phase I/II trial were published [177]. The results indicated a good safety profile with only mild and no serious adverse events. The intramuscular administration elicited strong SARS-CoV-2-specific antibodies in all vaccinated individuals. Despite being approved weeks earlier, the following statement was made in the publication: "further investigation is needed of the effectiveness of this vaccine for prevention of COVID-19" [177]. Recently, recruiting for two randomized, double-blind, placebo-controlled, multi-center phase III clinical trials in adult volunteers has started [178,179]. The simian ChAdOx1 nCoV-19 vaccine candidate showed promising preliminary results in a phase I/II trial [180]. The safety was good with no serious adverse events registered after a single intramuscular injection. The immune response was also promising with 32 out 35 vaccinees generating SARS-CoV-2-specific neutralizing antibodies. After a booster immunization, both humoral and cellular immune responses were detected in all vaccinees. The ChAdOx1 nCoV-19 vaccine candidate entered a randomized, double-blind, placebo-controlled multicenter phase III trial in 30,000 adults in August 2020 [181]. However, due to some suspect adverse events in patients, the phase III trial was put on hold in early September [182]. After an investigation into the issue, the trial resumed in the UK, but it remained on hold in the US until the FDA authorized the restart on 23 October 2020 [183].

Recently, the first-in-human phase I clinical trial with the MVA-SARS-2-S vaccine candidate in healthy volunteers was approved [184]. The study aims at assessing the safety and tolerability of the vaccine candidate and the enrolment of patients is in progress. A LV vector vaccine candidate based on minigenes of multiple conserved regions of SARS-CoV-2 is planned for a phase I/II clinical trial in 100 healthy volunteers [185]. Subcutaneous administration of 5×10^6 dendritic cells (DCs) transduced with the LV vector (LV-DC) in combination with intravenously injected 1×10^8 antigen-specific CTLs will be evaluated for safety and immunogenicity. Very recently, the MV-SARS-CoV-2 vaccine candidate TMV-083 was subjected to a randomized, placebo-controlled, two-center phase I clinical trial to evaluate the safety, tolerability and immunogenicity in 90 volunteers [186]. As it has been previously demonstrated that the replication-competent VSV-based SARS-CoV-2 S vaccine candidate (V590) can protect mice from SARS-CoV-2 pathogenesis [187], a phase I trial on the safety and tolerability is planned for 252 participants [188]. In another approach, a replication-competent VSV-ΔG vaccine, where the VSV G protein was replaced by SARS-CoV-2 S, resulted in potent SARS-CoV-2-specific neutralizing antibody responses in immunized golden Syrian hamsters [189]. Moreover, a single dose of 5×106 pfu of VSV-ΔG vaccine provided protection of hamsters against challenges with lethal doses of SARS-CoV-2. Additionally, the lung damage in immunized animals was minor and no viral load was detected. Next, the VSV- ΔG vaccine will be evaluated in humans in two phases [190]. In a phase I dose-escalation study, 18–55 years old volunteers will receive a single dose of 5×10^5, 5×10^6 and 5×10^7 pfu, respectively. In phase II, elderly subjects will receive a single dose as used in phase I or two immunization with 5×10^5 pfu 28 days apart. Finally, intranasal SARS-CoV-2 vaccine delivery is a potential option [191]. For instance, intranasal administration of an Ad5-based vector expressing the SARS-CoV-2 S receptor binding domain (RBD) elicited strong neutralizing antibody responses [192] (Table 3).

Table 3. Viral vector-based COVID-19 vaccine candidates.

Viral Vector	Stage	Response	Ref
Adenovirus			
ChAdOx1 nCoV-19	Preclinical	Strong immune response in mice and macaques	[161]
ChAdOx1 nCoV-19	Preclinical	Prevention of pneumonia in macaques	[162]
ChAdOx1 nCov-19	Phase I/II	Humoral and cellular responses in all vaccinees	[180]
ChAdOx1 nCoV-19	Phase III	Trial on hold because of suspect adverse events	[181]
Ad5-S-nb2	Preclinical	Strong immune response, SARS-CoV-2 protection	[163]
Ad5-S-nb2	Phase I	Humoral and T cell responses in volunteers	[169]
Ad5-S-nb2	Phase II	Significant immune responses in volunteers	[170]
Ad5-S-nb2	Phase III	Recruitment in progress	[171]
Ad5-S-nb2	Phase III	Recruitment in progress	[172]
Ad26.COV2.S	Preclinical	Protection against pneumonia in hamsters	[164]
Ad26.COV2.S	Preclinical	Protection against SARS-CoV-2 in macaques	[165]
Ad26.COV2.S	Phase I/II	Good safety and immunogenicity in volunteers	[173,174]
Ad26.COV2.S	Phase III	Recruitment in progress	[175]
rAd26-S/rAd5-S	Phase I/II	Good safety, humoral and cellular response	[177]
rAd26-S/rAd5-S	Phase III	Recruitment in progress	[178]
rAd26-S/rAd5-S	Phase III	Recruitment in progress	[179]
Ad5-CoV-2 S RBD	Preclinical	Neutralizing antibodies after nasal administration	[192]
Measles virus			
MV-SARS-CoV-2 S	Preclinical	Neutralizing and T cell antibody responses in mice	[166]
MV-SARS-CoV-2 S	Phase I	Recruiting in progress	[186]
Poxviruses			
sMVA	Preclinical	Potent neutralizing SARS-CoV-2 antibodies in mice	[168]
MVA-SARS-S	Phase I	Recruitment of participants in progress	[184]
Lentiviruses			
LV-DCs + CTL Ag	Phase I/II	Safety and immunogenicity evaluations in progress	[185]
Rhabdoviruses			
VSV-SARS-CoV2-S	Preclinical	Protection against SARS-CoV-2 pathogenesis in mice	[187]
VSV-SARS-CoV2-S	Phase I	Planned phase I trials on safety and tolerability	[188]
VSV-ΔG	Preclinical	Protection of hamsters against SARS-CoV-2	[189]
VSV-ΔG	Phase I/II	Recruitment in progress	[190]

Ad, adenovirus; Ag, antigen; ChAdOx1-S, simian adenovirus expressing SARS-CoV-2 S protein; CTLs, cytotoxic T lymphocytes; LV-DCs, lentivirus-transduced dendritic cells; MV, measles virus; MVA, modified vaccinia virus Ankara; RBD, receptor binding domain; sMVA, synthetic modified vaccinia virus Ankara; VSV, vesicular stomatitis virus.

5. Conclusions

The progress on viral vector-based vaccine development has been steady, targeting both infectious diseases and different types of cancers. Proof-of-concept has been demonstrated in numerous animal models resulting in robust antibody responses and protection against challenges with pathogens and tumor cells. Moreover, findings from vaccine trials have been encouraging. For instance, several vaccine candidates, based on VSV vectors, have provided protection in phase III trials [37,38]. Moreover, the EBOV vaccine based in the VSV-ZEBOV vector was approved in December 2019 under the brand name Ervebo by the FDA [193]. In the case of cancer vaccines, clinical data have confirmed robust immune responses previously shown in preclinical animal tumor models. Moreover, partial responses, stable disease, and prolonged overall survival have been demonstrated in clinical trials. For example, talimogene laherparepvec (TVEC), the oncolytic HSV-1 vector expressing GM-CSF, was approved for treatment of advanced melanoma by the FDA in October 2015 [194].

As presented in this review, there are many viral vectors to choose between for vaccine development. Clearly, not a single vector system can be declared superior. Although packaging capacity of foreign genes can be of importance, both viral antigens and tumor-associated antigens can be easily accommodated in almost any viral vector. Efficient packaging cell line systems have been engineered for many vector systems such as Ad, AAV, flaviviruses and lentiviruses, which has facilitated rapid and efficient large-scale production of vaccine candidates eligible for clinical applications. Self-replicating RNA virus vectors based on alphaviruses, flaviviruses, measles viruses and rhabdoviruses provide highly efficient cytoplasmic RNA amplification, a substantially favorable feature for generation of enhanced immune responses with reduced vaccine doses. In any case, it is not possible to recommend any universal vector system and each case needs to be evaluated based on the vaccine target, the handling of viral vectors and the preferred route of administration. Obviously, dealing with viral vectors requires a special attention related to safety. Since the advent of application of viral vectors, we have come a long way in engineering replication-deficient and oncolytic versions, which have proven safe for administration to humans. In comparison to conventional vaccines, viral vector-based vaccines have proven competitive related to costs and efficacy. In particular, alphavirus-based vaccines delivered as DNA or RNA replicons have been demonstrated to provide similar immune responses in preclinical animal models at 100- to 1000-fold lower concentrations compared to conventional DNA or RNA vaccines [16,54]. Similarly, protection against lethal challenges was obtained with 1×10^6–10^7 pfu of self-replicating RNA virus particles [20,22] compared to at least 1×10^{10} pfu Ad particles required [27,35]. Furthermore, approval of Ervebo and TVEC by the FDA presents strong evidence of the feasibility of additional viral vector-based vaccines reaching the market. However, further optimization related to vector engineering, delivery and dosing is required.

Finally, the COVID-19 pandemic has surely demonstrated how accelerated vaccine development can be realized. Today, 151 vaccine candidates have been subjected to preclinical studies and 42 vaccines have reached clinical trials. Although other approaches such as live-attenuated, peptide-, protein subunit-, DNA- and RNA-based vaccines have been taken, at least two Ad-based vaccine candidates are currently in phase III and one Ad-based vaccine has been approved, although only in Russia so far. COVID-19 vaccine development presents a good example on several levels. It demonstrates that in a time of a global crisis it is possible for academic institutions and commercial entities to work together efficiently. Moreover, the pandemic has demonstrated that it is appropriate and feasible to develop vaccine candidates based on different strategies including various types of viral vectors to achieve the goal as quickly as possible. Based on the current findings from both preclinical studies and clinical trials it is most likely that one type of COVID-19 vaccine will not be sufficient to overcome the pandemic. Therefore, it is of greatest importance that vaccine development can continue on all fronts with innovation and scientific approval as the cornerstone of all activities.

Funding: The authoring of this review received no external funding.

Conflicts of Interest: The author declares no conflict of interest.

References

1. Delrue, I.; Verzele, D.; Madder, A.; Nauwynck, H.J. Inactivated virus vaccines from chemistry to prophylaxis: Merits, risks and challenges. *Expert Rev. Vaccines* **2012**, *11*, 695–719. [CrossRef]
2. Chen, W.H.; Du, L.; Chag, S.M.; Ma, C.; Tricoche, N.; Tao, X.; Seid, C.A.; Hudspeth, E.M.; Lustigman, S.; Tseng, C.-T.; et al. Yeast-expressed recombinant protein of the receptor-binding domain in SARS-CoV spike protein with deglycosylated forms as a SARS vaccine candidate. *Hum. Vaccin. Immunother.* **2014**, *10*, 648–658. [CrossRef]
3. Wold, W.S.M.; Toth, K. Adenovirus vectors for gene therapy, vaccines and cancer gene therapy. *Curr. Gene Ther.* **2013**, *13*, 421–433. [CrossRef]
4. Lundstrom, K. RNA viruses as tools in gene therapy and vaccine development. *Genes* **2019**, *10*, 189. [CrossRef] [PubMed]

5. Schiedner, G.; Morral, N.; Parks, R.S.; Wu, Y.; Koopmans, S.C.; Langston, C.; Graham, F.L.; Beaudet, A.L.; Kochanek, S. Genomic DNA transfer with a high-capacity adenovirus vector results in improved in vivo gene expression and decreased toxicity. *Nat. Genet.* **1998**, *18*, 180–183. [CrossRef] [PubMed]

6. Strauss, J.H.; Strauss, E.G. The alphaviruses: Gene expression, replication and evolution. *Microbiol. Rev.* **1994**, *58*, 491–562. [CrossRef] [PubMed]

7. Pijlman, G.P.; Suhrbier, A.; Khromykh, A.A. Kunjin virus replicons: An RNA-based, non-cytopathic viral vector system for protein production, vaccine and gene therapy applications. *Exp. Opin. Biol. Ther.* **2006**, *6*, 134–145. [CrossRef]

8. Radecke, F.; Spielhofer, P.; Schneider, H.; Kaelin, K.; Huber, M.; Dötsch, C.; Christiansen, G.; Billeter, M.A. Rescue of measles viruses from cloned DNA. *EMBO J.* **1995**, *14*, 5773–5784. [CrossRef]

9. Osakada, F.; Callaway, E.M. Design and generation of recombinant rabies virus vectors. *Nat. Protoc.* **2013**, *8*, 1583–1601. [CrossRef]

10. Cone, R.D.; Mulligan, R.C. High-efficiency gene transfer into mammalian cells: Generation of helper-free recombinant retrovirus with broad mammalian host range. *Proc. Natl. Acad. Sci. USA* **1984**, *81*, 6349–6353. [CrossRef]

11. Fischer, A.; Hacein-Bey-Abina, S. Gene therapy for severe combined immunodeficiencies and beyond. *J. Exp. Med.* **2020**, *217*, e20190607. [CrossRef] [PubMed]

12. Vigna, E.; Naldini, L. Lentiviral vectors: Excellent tools for experimental gene transfer and promising candidates for gene therapy. *J. Gen. Med.* **2000**, *2*, 308–316. [CrossRef]

13. Torres, R.; Garcia, A.; Jimenez, M.; Rodriguez, S.; Ramirez, J.C. An integration-defective lentivirus-based resource for site-specific targeting of an edited safe-harbour locus in the genome. *Gene Ther.* **2014**, *21*, 343–352. [CrossRef] [PubMed]

14. Kwak, H.; Honig, H.; Kaufmann, H.L. Poxviruses as vectors for cancer immunotherapy. *Curr. Opin. Drug Discov. Devel.* **2003**, *6*, 161–168. [PubMed]

15. Bradley, S.; Jakes, A.D.; Harrington, K.; Pandha, H.; Melcher, A.; Errington-Mais, F. Applications of coxsackievirus A21 in oncology. *Oncolytic Virother.* **2014**, *3*, 47–55. [CrossRef]

16. Lundstrom, K. Self-amplifying RNA viruses as RNA vaccines. *Int. J. Mol. Sci.* **2020**, *21*, 5130. [CrossRef]

17. Kelvin, A.A. Outbreak of Chikungunya in the Republic of Congo and the global picture. *J. Infect. Dev. Ctries.* **2011**, *5*, 441–444. [CrossRef]

18. Jansen, K.A. The 2005–2007 Chikungunya epidemic in Reunion: Ambiguous etiologies, memories, and meaning-making. *Med. Anthropol.* **2013**, *32*, 174–189. [CrossRef]

19. Chattopadhyay, A.; Aquilar, P.V.; Bopp, N.E.; Yarovinsky, T.O.; Weaver, S.C.; Rose, J.K. A recombinant virus vaccine that protects both against Chikungunya and Zika virus infections. *Vaccine* **2018**, *36*, 3894–3900. [CrossRef]

20. Williams, A.J.; O'Brien, L.M.; Phillpots, R.J.; Perkins, S.D. Improved efficacy of gene optimized adenovirus-based vaccine for Venezuelan equine encephalitis virus. *Virol. J.* **2009**, *6*, 118. [CrossRef]

21. Reed, D.S.; Glass, P.J.; Bakken, R.R.; Barth, J.F.; Lind, C.M.; da Silva, L.; Hart, M.K.; Rayner, J.; Alterson, K.; Custer, M.; et al. Combined alphavirus replicon particle vaccine induces durable and cross-protective immune responses against equine encephalitis virus. *J. Virol.* **2014**, *88*, 12077–12086. [CrossRef] [PubMed]

22. Kamrud, K.I.; Custer, M.; Dudek, J.M.; Owens, G.; Alterson, K.D.; Lee, J.S.; Groebner, J.L.; Smith, J.F. Alphavirus replicon approach to promoterless analysis of IRES elements. *Virology* **2007**, *360*, 376–387. [CrossRef]

23. Tretyakova, I.; Tibbens, A.; Jokinen, J.D.; Johnson, D.M.; Lukashevich, I.S.; Pushko, P. Novel DNA-launched Venezuelan equine encephalitis virus vaccine with rearranged genome. *Vaccine* **2019**, *37*, 3317–3325. [CrossRef] [PubMed]

24. Tretyakova, I.; Plante, K.S.; Rossi, S.L.; Lawrence, W.S.; Peel, J.E.; Gudjohnsen, S.; Wang, E.; Mirchandani, D.; Tibbens, A.; Lamichhane, T.N. Venezuelan equine encephalitis vaccine with rearranged genome resists reversion and protects non-human primates from viremia after aerosol challenge. *Vaccine* **2020**, *38*, 3378–3386. [CrossRef] [PubMed]

25. Safronetz, D.; Mire, C.; Rosenke, K.; Feldmann, F.; Haddock, E.; Geissbert, T.; Feldmann, H. A recombinant vesicular stomatitis virus-based Lassa fever vaccine protects guinea pigs and macaques against challenge with geographically and genetically distinct Lassa viruses. *PLoS Negl. Trop. Dis.* **2015**, *9*, e0003736. [CrossRef] [PubMed]

26. Kainulainen, M.H.; Spengler, J.R.; Welch, S.R.; Coleman-McCray, J.D.; Harmon, J.R.; Klena, J.D.; Nichol, S.T.; Albarino, C.G.; Spiropoulou, C.F. Use of a scalable replicon-particle vaccine to protect against lethal Lassa virus infection in the guinea pig model. *J. Infect. Dis.* **2018**, *217*, 1957–1966. [CrossRef]

27. Maruyama, J.; Mateer, E.J.; Manning, J.T.; Sattler, R.; Seregin, A.V.; Bukreyeva, N.; Jones, F.R.; Balint, J.P.; Gabitzsch, E.S.; Huang, C.; et al. Adenoviral vector-based vaccine is fully protective against lethal Lassa fever challenge in Hartley guinea pigs. *Vaccine* **2019**, *37*, 6824–6831. [CrossRef] [PubMed]

28. Mateo, M.; Reynard, S.; Carnec, X.; Journeaux, A.; Baillet, N.; Schaeffer, J.; Picard, C.; Legras-Lachuer, C.; Allan, R.; Perthame, E.; et al. Vaccines inducing immunity to Lassa fever glycoprotein and nucleoprotein protect macaques after a single shot. *Sci. Transl. Med.* **2019**, *11*, eaaw3163. [CrossRef]

29. Inc., K.N. A Trial to Evaluate the Optimal Dose of MV-LASV. *Case Med. Res.* **2019**. [CrossRef]

30. Johnson, D.M.; Jokinen, J.D.; Wang, M.; Pfeiffer, T.; Tretyakova, I.; Carrion, R., Jr.; Griffiths, A.; Pushko, P.; Lukashevich, I.S. Bivalent Junin and Machupo experimental vaccine based on alphavirus RNA replicon vector. *Vaccine* **2020**, *38*, 2949–2959. [CrossRef]

31. Subissi, L.; Keita, M.; Mesfin, S.; Rezza, G.; Diallo, B.; Van Gucht, S.; Musa, E.O.; Yoti, Z.; Keita, S.; Djingarey, M.H.; et al. Ebola virus transmission caused by persistently infected survivors of the 2014-2016 outbreak in West Africa. *J. Infect. Dis.* **2018**, *218*, S287–S291. [CrossRef] [PubMed]

32. Pyankov, O.V.; Bodnev, S.A.; Pyankova, O.G.; Solodkyi, V.V.; Pyankov, S.A.; Setoh, Y.X.; Volchokova, V.A.; Suhrbier, A.; Volchikov, V.V.; Agafonov, A.A.; et al. A Kunjin replicon virus-like vaccine provides protection against Ebola virus infection in nonhuman primates. *J. Infect. Dis.* **2015**, *212* (Suppl. S2), S368–S371. [CrossRef] [PubMed]

33. Marzi, A.; Robertson, S.J.; Haddock, E.; Feldmann, F.; Hanley, P.W.; Scott, D.-P.; Strong, J.E.; Kobinger, G.; Best, S.M.; Feldmann, H. Ebola vaccine. VSV-EBOV rapidly protects macaques against infection with the 2014/2015 Ebola virus outbreak strain. *Science* **2015**, *349*, 739–742. [CrossRef] [PubMed]

34. Geisbert, T.W.; Feldmann, H. Recombinant vesicular stomatitis virus-based vaccines against Ebola and Marburg infections. *J. Infect. Dis.* **2011**, *204* (Suppl. S3), S1075–S1081. [CrossRef] [PubMed]

35. Sullivan, N.J.; Geisbert, T.W.; Geisbert, J.B.; Shedlock, D.J.; Xu, L.; Lamoreaux, L.; Custers, J.H.H.V.; Popernack, P.M.; Yang, Z.-Y.; Pau, M.G.; et al. Immune protection of nonhuman primates against Ebola virus with single low-dose adenovirus vectors encoding modified GPs. *PLoS Med.* **2006**, *3*, e177. [CrossRef]

36. Bukreyev, A.; Marzi, A.; Feldmann, F.; Zhang, L.; Yang, L.; Ward, J.M.; Dorward, D.W.; Pickles, R.J.; Murphy, B.R.; Feldmann, H.; et al. Chimeric human parainfluenza virus bearing the Ebola virus glycoprotein as the sole surface protein is immunogenic and highly protective against Ebola virus challenge. *Virology* **2009**, *383*, 348–361. [CrossRef] [PubMed]

37. Henao-Restrepo, A.M.; Longini, I.M.; Egger, M.; Dean, N.E.; Edmunds, W.J.; Camacho, A.; Carroll, M.W.; Doumbia, M.; Draguez, B.; Duraffour, S. Efficacy and effectiveness of an rVSV-vectored vaccine expressing Ebola surface glycoprotein: Interim results from the Guinea ring vaccination cluster-randomised trial. *Lancet* **2015**, *386*, 857–866. [CrossRef]

38. Henao-Restrepo, A.M.; Camacho, A.; Longini, I.M.; Watson, C.H.; Edmunds, W.J.; Egger, M.; Carroll, M.W.; Dean, N.E.; Diatta, I.; Doumbia, M.; et al. Efficacy and effectiveness of an rVSV-vectored vaccine in preventing Ebola virus disease: Final results from the Guinea ring vaccination, open-label, cluster-randomised trial (Ebola Ca Suffit!). *Lancet* **2017**, *389*, 505–518. [CrossRef]

39. Herbert, A.S.; Kuehne, A.I.; Barth, J.F.; Ortiz, R.A.; Nichols, D.K.; Zak, S.E.; Stonier, S.W.; Muhammad, M.A.; Bakken, R.R.; Prugar, L.I.; et al. Venezuelan equine encephalitis virus replicon particle vaccine protects nonhuman primates from intramuscular and aerosol challenge with ebolavirus. *J. Virol.* **2013**, *87*, 4852–4964. [CrossRef] [PubMed]

40. Khalil, S.M.; Tonkin, D.R.; Mattocks, M.D.; Snead, A.T.; Johnston, R.E.; White, L.J. A tetravalent alphavirus-vector based dengue vaccine provides effective immunity in an early life mouse model. *Vaccine* **2014**, *32*, 4068–4074. [CrossRef] [PubMed]

41. Hu, H.M.; Chen, H.W.; Hsiao, Y.; Wu, S.H.; Chung, H.H.; Hsieh, C.H.; Chong, P.; Leng, C.H.; Pan, C.H. The successful induction of T-cell and antibody responses by a recombinant measles virus-vectored tetravalent dengue vaccine provides partial protection against dengue-2 infection. *Hum. Vaccin. Immunother.* **2016**, *12*, 1678–1689. [CrossRef] [PubMed]

42. Torresi, J.; Ebert, G.; Pellegrini, M. Vaccines licensed and in clinical trials for the prevention of dengue. *Hum. Vaccin. Imunother.* **2017**, *13*, 1059–1072. [CrossRef]

43. Erasmus, J.H.; Khandhar, A.P.; Guderian, J.; Granger, B.; Archer, J.; Archer, M.; Cage, E.; Fuerte-Stone, J.; Larson, E.; Lin, S.; et al. A nanostructured lipid carrier for delivery of a replicating viral RNA provides single, low-dose protection against Zika. *Mol. Ther.* **2018**, *26*, 2507–2522. [CrossRef]

44. Poland, G.A.; Ovsyannikova, I.G.; Kennedy, R.B. Zika vaccine development: Current status. *Them. Rev. Vaccines* **2019**, *94*, 2572–2586. [CrossRef]

45. Durbin, A.P.; Karron, R.A.; Sun, W.; Vaughn, D.W.; Reynolds, M.J.; Perreault, J.R.; Thumar, B.; Men, R.; Lai, C.J.; Elkins, W.R.; et al. Attenuation and immunogenicity in humans of a live dengue virus type-4 vaccine candidate with a 30 nucleotide deletion in its 3′-untranslated region. *Am. J. Trop. Med. Hyg.* **2001**, *65*, 405–413. [CrossRef] [PubMed]

46. Ye, W.W.; Mason, B.B.; Chengalvala, M.; Cheng, S.M.; Zandle, G.; Lubeck, M.D.; Lee, S.G.; Mizitani, S.; Davis, A.R.; Hung, P.P. Co-expression of hepatitis B antigens by a non-defective adenovirus vaccine vector. *Arch. Virol.* **1991**, *118*, 11–27. [CrossRef] [PubMed]

47. Del Valle, J.R.; Devaux, P.; Hodge, G.; Wegner, N.J.; McChesney, M.B.; Cattaneo, R. A vectored measles virus induces hepatitis B surface antigen antibodies while protecting macaques against virus challenge. *J. Virol.* **2007**, *81*, 10597–10605. [CrossRef] [PubMed]

48. Reynolds, T.D.; Buonocore, L.; Rose, N.F.; Rose, J.K.; Robek, M.D. Virus-like vesicle-based therapeutic vaccine vectors for chronic hepatis B virus infection. *J. Virol.* **2015**, *89*, 10407–10415. [CrossRef] [PubMed]

49. Li, J.; Bao, M.; Ge, J.; Ren, S.; Zhou, T.; Qi, F.; Pu, X.; Dou, J. Research progress of therapeutic vaccines for treating chronic hepatitis B. *Hum. Vaccin. Immunother.* **2017**, *13*, 986–997. [CrossRef] [PubMed]

50. Zoulim, F.; Fournier, C.; Habersetzer, F.; Sprinzl, M.; Pol, S.; Coffin, C.S.; Leroy, V.; Ma, M.; Wedemeyer, H.; Lohse, A.W.; et al. Safety and immunogenicity of the therapeutic vaccine TG1050 in chronic hepatitis B patients: A phase 1b placebo-controlled trial. *Hum. Vaccines Immunother.* **2020**, *16*, 388–399. [CrossRef]

51. Gao, W.; Soloff, A.C.; Lu, X.; Montecalvo, A.; Nguyen, D.C.; Matsuoka, Y.; Robbins, P.D.; Swayne, D.E.; Donis, R.O.; Katz, J.M.; et al. Protection of mice and poultry from lethal H5N1 avian influenza virus through adenovirus-based immunization. *J. Virol.* **2006**, *80*, 1959–1964. [CrossRef] [PubMed]

52. Schultz-Cherry, S.; Dybing, J.K.; Davis, N.L.; Williamson, C.; Suarez, D.L.; Johnston, R.; Perdue, M.L. Influenza virus (A/HK/156/97) hemagglutinin expressed by an alphavirus replicon system protects against lethal infection with Hong Kong-origin H5N1 viruses. *Virology* **2000**, *278*, 55–59. [CrossRef] [PubMed]

53. Fleeton, M.N.; Chen, M.; Berglund, P.; Rhodes, G.; Parker, S.E.; Murphy, M.; Atkins, G.J.; Liljestrom, P. Self-replicative RNA vaccines elicit protection against influenza A virus, respiratory syncytial virus, and a tickborne encephalitis virus. *J. Infect. Dis.* **2001**, *183*, 1395–1398. [CrossRef] [PubMed]

54. Vogel, A.B.; Lambert, L.; Kinnear, E.; Busse, D.; Erbar, S.; Reufer, K.C.; Wicke, L.; Perkovic, M.; Beissert, T.; Haas, H.; et al. Self-amplifying RNA Vaccines Give Equivalent Protection against Influenza to mRNA Vaccines but at Much Lower Doses. *Mol. Ther.* **2018**, *26*, 446–455. [CrossRef]

55. Kreijtz, J.H.; Suezer, Y.; van Amerongen, G.; de Mutsert, G.; Schnierle, B.S.; Wood, J.M.; Kuiken, T.; Fouchier, R.A.; Lower, J.; Osterhaus, A.D.; et al. Recombinant modified vaccinia virus Ankara-based vaccine induces protective immunity in mice against infection with influenza virus H5N1. *J. Infect. Dis.* **2007**, *195*, 1598–1606. [CrossRef]

56. Kreijtz, J.H.; Goeijenbier, M.; Moesker, F.M.; van den Dries, L.; Goeijenbier, S.; De Gruyter, H.L.; Lehmann, M.H.; Mutsert, G.; van de Vijver, D.A.; Volz, A.; et al. Safety and immunogenicity of a modified-vaccinia- virus-Ankara-based influenza A H5N1 vaccine: A randomised double-blind phase 1/2a clinical trial. *Lancet Infect. Dis.* **2014**, *14*, 1196–1207. [CrossRef]

57. Liu, J.; Jaijyan, D.K.; Tang, Q.; Zhu, H. Promising Cytomegalovirus-based vaccine vector induces robust CD8(+) T-cell response. *Int. J. Mol. Sci.* **2019**, *20*, 4457. [CrossRef]

58. Abad-Fernandez, M.; Goonetilleke, N. Human cytomegalovirus-vectored vaccines against HIV. *Curr. Opin. HIV AIDS* **2019**, *14*, 137–142. [CrossRef]

59. Casimiro, D.R.; Tang, A.; Chen, L.; Fu, T.M.; Evans, R.K.; Davies, M.E.; Freed, D.C.; Hurni, W.; Aste-Amezaga, J.M.; Guan, L.; et al. Vaccine-induced immunity in baboons by using DNA and replication-incompetent adenovirus type 5 vectors expressing a human immunodeficiency virus type 1 gag gene. *J. Virol.* **2003**, *77*, 7663–7768. [CrossRef]

60. Guerbois, M.; Moris, A.; Combredet, C.; Najburg, V.; Ruffié, C.; Février, M.; Cayet, N.; Brandler, S.; Schwartz, O.; Tangy, F. Live attenuated measles vaccine expressing HIV-1 Gag virus like particles covered with gp160DeltaV1V2 is strongly immunogenic. *Virology* **2009**, *388*, 191–203. [CrossRef]

61. Brand, D.; Lemiale, F.; Turbica, I.; Buzelay, L.; Brunet, S.; Barin, F. Comparative analysis of humoral immune responses to HIV type 1 envelope glycoproteins in mice immunized with a DNA vaccine, recombinant Semliki Forest virus RNA, or recombinant Semliki Forest virus particles. *AIDS Res. Hum. Retrovir.* **1998**, *14*, 1369–1377. [CrossRef] [PubMed]

62. Giraud, A.; Ataman-Onal, Y.; Battail, N. Generation of monoclonal antibodies to native human immunodeficiency virus type 1 envelope glycoprotein by immunization of mice with naked RNA. *J. Virol. Methods* **1999**, *79*, 75–84. [CrossRef]

63. Ajbani, S.P.; Velhal, S.M.; Kadam, R.B.; Patel, V.V.; Lundstrom, K.; Bandivdekar, A.H. Immunogenicity of virus-like Semliki Forest virus replicon particles expressing Indian HIV-1C gag, env and pol RT genes. *Immunol. Lett.* **2017**, *190*, 221–232. [CrossRef] [PubMed]

64. Knudsen, M.L.; Ljungberg, K.; Tatoud, R.; Weber, J.; Esteban, M.; Liljestrom, P. Alphavirus replicon DNA expressing HIV antigens is an excellent prime for boosting with recombinant modified vaccinia Ankara (MVA) or with HIV gp140 protein antigen. *PLoS ONE* **2015**, *10*, e0117042. [CrossRef]

65. Bogers, W.M.; Oostermeijer, H.; Mooij, P.; Koopman, G.; Verschoor, E.J.; Davis, D.; Ulmer, J.B.; Brito, L.A.; Cu, Y.; Bannerjee, K.; et al. Potent immune responses in rhesus macaques induced by nonviral delivery of self-amplifying RNA vaccine expressing HIV type 1 envelope with a cationic emulsion. *J. Infect. Dis.* **2015**, *211*, 947–955. [CrossRef]

66. Melo, M.; Porter, E.; Zhang, Y.; Silva, M.; Li, N.; Dobosh, B.; Liquori, A.; Skog, P.; Landais, E.; Menis, S. Immunogenicity of RNA Replicons Encoding HIV Env Immunogens Designed for Self-Assembly into Nanoparticles. *Mol. Ther.* **2019**, *27*, 2080–2090. [CrossRef]

67. Altfeld, M.; Goulder, P.J. The STEP study provides a hint that vaccine induction of the right CD8+ T cell responses can facilitate immune control of HIV. *J. Infect. Dis.* **2011**, *203*, 753–755. [CrossRef]

68. Sekaly, R.-P. The failed HIV Merck vaccine study: A step back or a launching point for future vaccine development? *J. Exp. Med.* **2008**, *205*, 7–12. [CrossRef]

69. Gómez, C.E.; Nájera, J.L.; Sánchez, R.; Jiménez, V.; Esteban, M. Multimeric soluble CD40 ligand (sCD40L) efficiently enhances HIV specific cellular immune responses during DNA prime and boost with attenuated poxvirus vectors MVA and NYVAC expressing HIV antigens. *Vaccine* **2009**, *27*, 3165–3174. [CrossRef]

70. Rerks-Ngarm, S.; Pitisuttihum, P.; Nitayaphan, S.; Kaewkungwal, J.; Chiu, J.; Paris, R.; Premsri, N.; Namwat, C.; de Souza, M.; Adams, E.; et al. Vaccination with ALVAC and AIDSVAX to prevent HIV-1 infection in Thailand. *N. Engl. J. Med.* **2009**, *361*, 2209–2220. [CrossRef]

71. Available online: https://www.pharmaceutical-technology.com/news/niaid-hvtn-702-hiv-vaccine-south-africa/ (accessed on 6 October 2020).

72. Lemiale, F.; Korokhov, N. Lentiviral vectors for HIV disease prevention and treatment. *Vaccine* **2009**, *27*, 3443–3449. [CrossRef] [PubMed]

73. Norton, T.; Zhen, A.; Tada, T.; Kim, J.; Kitchen, S.; Landau, N.R. Lentiviral-based Dendritic Cell Vaccine Suppresses HIV Replication in Humanized Mice. *Mol. Ther.* **2019**, *27*, 960–973. [CrossRef]

74. Li, N.; Yu, Y.Z.; Yu, W.Y.; Sun, Z.W. Enhancement of the immunogenicity of DNA replicon vaccine of Clostridium botulinum neurotoxin serotype A by GM-CSF gene adjuvant. *Immunopharmacol. Immunotoxicol.* **2011**, *33*, 211–219. [CrossRef] [PubMed]

75. Xu, Q.; Pichichero, M.E.; Simpson, L.L.; Elias, M.; Smith, L.A.; Zeng, M. An adenoviral vector-based mucosal vaccine is effective in protection against botulism. *Gene Ther.* **2009**, *16*, 367–375. [CrossRef] [PubMed]

76. Andersson, C.; Vasconcelos, N.M.; Sievertzon, M.; Haddad, D.; Liljeqvist, S. Comparative immunization study using RNA and DNA constructs encoding a part of the Plasmodium falciparum antigen Pf332. *Scand. J. Immunol.* **2001**, *54*, 117–124. [CrossRef]

77. Shott, J.P.; McGrath, S.M.; Grazia Pau, M.; Custers, J.H.V.; Ophorst, O.; Demoitié, M.-A.; Dubois, M.-C.; Komisar, J.; Cobb, M.; Kester, K.E.; et al. Adenovirus 5 and 35 vectors expressing Plasmodium falciparum circumsporozoite surface protein elicit potent antigen-specific cellular IFN-gamma and antibody responses in mice. *Vaccine* **2008**, *26*, 2818–2823. [CrossRef]

78. Tsuji, M.; Bergmann, C.C.; Takita-Sonoda, Y.; Murata, K.; Rodrigues, E.G.; Nussenzweig, R.S.; Zavala, F. Recombinant Sindbis viruses expressing a cytotoxic T-lymphocyte epitope of a malaria parasite or of influenza virus elicit protection against the corresponding pathogen in mice. *J. Virol.* **1998**, *72*, 6907–6910. [CrossRef]

79. Kirman, J.R.; Turon, T.; Su, H.; Li, A.; Kraus, C.; Polo, J.M.; Belisle, J.; Morris, S.; Seder, R.A. Enhanced Immunogenicity to Mycobacterium tuberculosis by Vaccination with an Alphavirus Plasmid Replicon Expressing Antigen 85A. *Infect. Immun.* **2003**, *71*, 575–579. [CrossRef]

80. Thomas, J.M.; Moen, S.T.; Gnade, B.T.; Vargas-Inchaustegui, D.A.; Foltz, S.M.; Suarez, G.; Heidner, H.W.; König, R.; Chopra, A.K.; Peterson, J.W. Recombinant Sindbis virus vectors designed to express protective antigen of Bacillus anthracis protect animals from anthrax and display synergy with ciprofloxacin. *Clin. Vaccine Immunol.* **2009**, *16*, 1696–1699. [CrossRef]

81. Yamanaka, R.; Zullo, S.A.; Ramsey, J.; Onodera, M.; Tanaka, R.; Blaese, M. Induction of therapeutic antitumor antiangiogenesis by intratumoral injection of genetically engineered endostatin-producing Semliki Forest virus. *Cancer Gene Ther.* **2001**, *8*, 796–802. [CrossRef]

82. Yamanaka, R.; Tsuchiya, N.; Yajima, N.; Honma, J.; Hasegawa, H.; Tanaka, R.; Ramsey, J.; Blasé, R.M.; Xanthopoulos, K.G. Induction of an antitumor immunological response by an intratumoral injection of dendritic cells pulsed with genetically engineered Semliki Forest virus to produce interleukin-18 combined with the systemic administration of interleukin-12. *J. Neurosurg.* **2003**, *99*, 746–753. [CrossRef] [PubMed]

83. Martikainen, M.; Niittykoski, M.; von und zu Frauenberg, M.; Immonen, A.; Koponen, S. MicroRNA-attenuated clone of virulent Semliki Forest virus overcomes antiviral type I interferon in resistant mouse CT-2A glioma. *J. Virol.* **2015**, *89*, 10637–10647. [CrossRef] [PubMed]

84. Zhang, X.; Mao, G.; van den Pol, A.N. Chikungunya-vesicular stomatitis chimeric virus targets and eliminates brain tumors. *Virology* **2018**, *522*, 244–259. [CrossRef] [PubMed]

85. Allen, C.; Opyrchal, M.; Aderca, I.; Schroeder, M.A.; Sarkaria, J.N.; Domingo, E.; Federspiel, M.J.; Galanis, E. Oncolytic measles virus strains have a significant antitumor activity against glioma stem cells. *Gene Ther.* **2013**, *2*, 444–449. [CrossRef] [PubMed]

86. Hoffmann, D.; Wildner, O. Comparison of herpes simplex virus- and conditionally replicative adenovirus-based vectors for glioblastoma treatment. *Cancer Gene Ther.* **2007**, *14*, 627–639. [CrossRef]

87. Lang, F.F.; Conrad, C.; Gomez-Manzano, C.; Yung, W.K.A.; Sawaya, R.; Weinberg, J.S.; Prabhu, S.S.; Rao, G.; Fuller, G.N.; Aldape, K.D.; et al. Phase I study of DNX-2401 (Delta-24-RGD) oncolytic adenovirus; Replication and immunotherapeutic effects in recurrent malignant glioma. *J. Clin. Oncol.* **2018**, *36*, 1419–1427. [CrossRef]

88. Yan, Y.; Xu, H.; Wang, J.; Wu, X.; Wen, W.; Liang, Y.; Wang, L.; Liu, F.; Du, X. Inhibition of breast cancer cells by targeting E2F-1 gene and expressing IL-15 oncolytic adenovirus. *Biosci. Rep.* **2019**, *39*, BSR20190384. [CrossRef]

89. Pinto, C.; Silva, G.; Ribeiro, A.S.; Oliveira, M.; Garrido, M.; Bandeira, V.S.; Nascimento, A.; Coroadinha, A.S.; Peixoto, C.; Barbas, A.; et al. Evaluation of AAV-mediated delivery of shRNA to target basal-like breast cancer genetic vulnerabilities. *J. Biotechnol.* **2019**, *300*, 70–77. [CrossRef]

90. Trepel, M.; Körbelin, J.; Spies, E.; Heckmann, M.B.; Hunger, A.; Fehse, B.; Katus, H.A.; Kleinschmidt, J.A.; Müller, O.J.; Michelfelder, S. Treatment of multifocal breast cancer by systemic delivery of dual-targeted adeno-associated viral vectors. *Gene Ther.* **2015**, *22*, 840–847. [CrossRef]

91. Wang, X.; Wang, J.P.; Rao, X.M.; Price, J.E.; Zhou, H.S.; Lachman, L.B. Prime-boost vaccination with plasmid and adenovirus gene vaccines control HER2/neu+ metastatic breast cancer in mice. *Breast Cancer Res.* **2005**, *7*, R580–R588. [CrossRef]

92. Lachman, L.B.; Rao, X.M.; Kremer, R.H.; Ozpolat, B.; Kirjakova, G.; Price, J.E. DNA vaccination against neu reduces breast cancer incidence and metastasis in mice. *Cancer Gene Ther.* **2001**, *8*, 259–268. [CrossRef]

93. Skelding, K.A.; Barry, R.D.; Shafren, D.R. Enhanced oncolysis mediated by Coxsackievirus A21 in combination with doxorubicin hydrochloride. *Investig. New Drugs* **2012**, *30*, 568–581. [CrossRef]

94. Kasuya, H.; Kodera, Y.; Nakao, A.; Yamamura, K.; Gewen, T.; Zhiwen, W.; Hotta, Y.; Yamada, S.; Fujii, T.; Fukuda, S.; et al. Phase I dose-escalation clinical trial of HF10 oncolytic herpes virus in 17 Japanese patients with advanced cancer. *Hepatogastroenterology* **2014**, *61*, 599–605.

95. Velders, M.P.; McElhiney, S.; Cassetti, M.C.; Eiben, G.L.; Higgins, T.; Kovacs, G.R. Eradication of established tumors by vaccination with Venezuelan equine encephalitis virus replicon particles delivering human papillomavirus 16 E7 RNA. *Cancer Res.* **2001**, *61*, 7861–7867.

96. Daemen, T.; Riezebos-Brilman, A.; Bungener, L.; Regts, J.; Dontje, B.; Wilschut, J. Eradication of established HPV16-transformed tumours after immunisation with recombinant Semliki Forest virus expressing a fusion protein of E6 and E7. *Vaccine* **2000**, *21*, 1082–1088. [CrossRef]

97. Van de Wall, S.; Ljungberg, K.; Ip, P.P.; Boerma, A.; Knudsen, M.L.; Nijman, H.W.; Liljeström, P.; Daemen, T. Potent therapeutic efficacy of an alphavirus replicon DNA vaccine expressing human papilloma virus E6 and E7 antigens. *Oncoimmunology* **2018**, *7*, e1487913. [CrossRef] [PubMed]

98. Jorritsma-Smit, A.; van Zanten, C.J.; Schoemaker, J.; Meulenberg, J.J.M.; Touw, D.J.; Kosterink, J.G.W.; Nijman, H.W.; Daemen, T.; Allersma, D.P. GMP manufacturing of Vvax001, a therapeutic anti-HPV vaccine based on recombinant viral particles. *Eur. J. Pharm. Sci.* **2020**, *143*, 105096. [CrossRef] [PubMed]

99. Yang, A.; Farmer, E.; Wu, T.C.; Hung, C.-F. Perspectives for therapeutic HPV vaccine development. *J. Biomed. Sci.* **2016**, *75*. [CrossRef] [PubMed]

100. Borysiewicz, L.K.; Fiander, A.; Nimako, M.; Man, S.; Wilkinson, G.W.; Westmoreland, D.; Evans, A.S.; Adams, M.; Stacey, S.N.; Boursnell, M.E.; et al. A recombinant vaccinia virus encoding human papillomavirus types 16 and 18, E6 and E7 proteins as immunotherapy for cervical cancer. *Lancet* **1996**, *347*, 1523–1527. [CrossRef]

101. Rosales, R.; Lopez-Contreras, M.; Rosales, C.; Magallanes-Molina, J.R.; Gonzalez Vergara, R.; Arroyo-Cazarez, J.M.; Ricardez-Arenas, A.; Del Follo-Valencia, A.; Padilla-Arriaga, S.; Guerrero, M.V.; et al. Regression of human papillomavirus intraepithelial lesions is induced by MVA E2 therapeutic vaccine. *Hum. Gene Ther.* **2014**, *25*, 1035–1049. [CrossRef]

102. Hoang-Le, D.; Smeenk, L.; Anraku, I.; Pijlman, G.P.; Wang, X.P.; de Vrij, J. A Kunjin replicon vector encoding granulocyte macrophage colony-stimulating factor for intra-tumoral gene therapy. *Gene Ther.* **2009**, *16*, 190–199. [CrossRef] [PubMed]

103. Lyons, J.A.; Sheahan, B.J.; Galbraith, S.E. Inhibition of angiogenesis by a Semliki Forest virus vector expressing VEGFR-2 reduces tumour growth and metastasis in mice. *Gene Ther.* **2007**, *14*, 503–513. [CrossRef] [PubMed]

104. Ying, H.; Zaks, T.Z.; Wang, R.-F.; Irvine, K.R.; Kammula, U.S.; Marincola, F.M. Cancer therapy using a self-replicating RNA vaccine. *Nat. Med.* **1999**, *5*, 823–827. [CrossRef] [PubMed]

105. Ricordel, M.; Foloppe, J.; Pichon, C.; Sfrontato, N.; Antoine, D.; Tosch, C.; Cochin, S.; Cordier, P.; Quemeneur, E.; Camus-Bouclainville, C.; et al. Cowpox virus: A new and armed oncolytic poxvirus. *Mol. Ther. Oncolytics* **2017**, *7*, 1–11. [CrossRef] [PubMed]

106. Sun, Y.; Wang, S.; Yang, H.; Wu, J.; Li, S.; Qiao, G.; Wang, S.; Wang, X.; Zhou, X.; Osada, T.; et al. Impact of synchronized anti-PD-1 with Ad-CEA vaccination on inhibition of colon cancer growth. *Immunotherapy* **2019**, *11*, 953–966. [CrossRef]

107. Downs-Canner, S.; Guo, Z.S.; Ravindranathan, R.; Breitbach, C.J.; O'Malley, M.E.; Jones, H.L.; Moon, A.; McCart, J.A.; Shuai, Y.; Zeh, H.J.; et al. Phase I study of intravenous oncolytic poxvirus (vvDD) in patients with advanced solid cancers. *Mol. Ther.* **2016**, *24*, 1492–1501. [CrossRef]

108. Liang, W.; Wang, H.; Sun, T.M.; Yao, W.Q.; Chen, L.L.; Jin, Y.; Li, C.L.; Meng, F.J. Application of autologous tumor cell vaccine and NDV vaccine in treatment of tumors of digestive tract. *World J. Gastroenterol.* **2003**, *9*, 495–498. [CrossRef]

109. Murphy, A.M.; Morris-Downes, M.M.; Sheahan, B.J.; Atkins, G.J. Inhibition of human lung carcinoma cell growth by apoptosis induction using Semliki Forest virus recombinant particles. *Gene Ther.* **2000**, *7*, 1477–1482. [CrossRef]

110. Määttä, A.M.; Mäkinen, K.; Ketola, A.; Liimatainen, T.; Yongabi, F.N.; Vähä-Koskela, M. Replication competent Semliki Forest virus prolongs survival in experimental lung cancer. *Int. J. Cancer* **2008**, *123*, 1704–1711. [CrossRef]

111. Granot, T.; Yamanashi, Y.; Meruelo, D. Sindbis viral vectors transiently deliver tumor-associated antigens to lymph nodes and elicit diversified antitumor CD8+ T-cell immunity. *Mol. Ther.* **2014**, *22*, 112–122. [CrossRef]

112. Patel, M.R.; Jacobson, B.A.; Ji, Y.; Drees, J.; Tang, S.; Xiong, K. Vesicular stomatitis virus expressing interferon-β is oncolytic and promotes antitumor immune responses in a syngeneic murine model of non-small cell lung cancer. *Oncotarget* **2015**, *6*, 33165–33177. [CrossRef] [PubMed]

113. Patel, M.R.; Jacobson, B.A.; Belgum, H.; Raza, A.; Sadiq, A.; Drees, J.; Wang, H.; Jay-Dixon, J.; Etchison, R.; Federspiel, M.J.; et al. Measles vaccine strains for virotherapy of non-small cell lung carcinoma. *J. Thorac. Oncol.* **2014**, *9*, 1101–1110. [CrossRef]

114. Tosch, C.; Bastien, B.; Barraud, L.; Grellier, B.; Nourtier, V.; Gantzer, M.; Limacher, J.M.; Quemeneur, E.; Bendjama, K.; Préville, X. Viral based vaccine TG4010 induces broadening of specific immune response and improves outcome in advanced NSCLC. *J. Immunother. Cancer* **2017**, *5*, 70. [CrossRef] [PubMed]

115. McAllister, A.; Arbetman, A.E.; Mandl, S.; Pena-Rossi, C.; Andino, R. Recombinant yellow fever viruses are effective therapeutic vaccines for treatment of murine solid tumors and pulmonary metastases. *J. Virol.* **2000**, *74*, 9197–9205. [CrossRef]

116. Avogadri, F.; Merghoub, T.; Maughan, M.F.; Hirschhorn-Cymerman, D.; Morris, J.; Ritter, E. Alphavirus replicon particles expressing TRP-2 provide potent therapeutic effect on melanoma through activation of humoral and cellular immunity. *PLoS ONE* **2010**, *5*, e12670. [CrossRef]

117. Avogadri, F.; Zappasodi, R.; Yang, A.; Budhu, S.; Malandro, N.; Hisrchhorn-Cymerman, D. Combination of alphavirus replicon particle-based vaccination with immunomodulatory antibodies: Therapeutic activity in the B16 melanoma mouse model and immune correlates. *Cancer Immunol. Res.* **2014**, *2*, 448–458. [CrossRef]

118. Yin, X.; Wang, W.; Zhu, X.; Wang, Y.; Wu, S.; Wang, Z. Synergistic antitumor efficacy of combined DNA vaccines targeting tumor cells and angiogenesis. *Biochem. Biophys. Res. Comm.* **2015**, *465*, 239–244. [CrossRef]

119. Ammour, Y.; Ryabaya, O.; Shchetinina, Y.; Prokofeva, E.; Gavrilova, M.; Khochenkov, D.; Vorobyev, D.; Faizuloev, E.; Shohin, I.; Zverev, V.V.; et al. The Susceptibility of Human Melanoma Cells to Infection with the Leningrad-16 Vaccine Strain of Measles Virus. *Viruses* **2020**, *12*, 173. [CrossRef]

120. Kimpel, J.; Urbiola, C.; Koske, I.; Tober, R.; Banki, Z.; Wollmann, G. The Oncolytic virus VSV-GP is effective against malignant melanoma. *Viruses* **2018**, *10*, 108. [CrossRef]

121. Niu, Z.; Bai, F.; Sun, T.; Tian, H.; Yu, D.; Yin, J.; Li, S.; Li, T.; Cao, H.; Yu, Q.; et al. Recombinant newcastle disease virus expressing IL15 demonstrates promising antitumor efficiency in melanoma model. *Technol. Cancer Res. Treat.* **2015**, *14*, 607–615. [CrossRef] [PubMed]

122. Shafren, D.R.; Au, G.G.; Nguyen, T.; Newcombe, N.G.; Haley, E.S.; Beagley, L.; Johansson, E.S.; Hersey, P.; Barry, R.D. Systemic therapy of malignant human melanoma tumors by a common cold-producing enterovirus, Coxsackievirus A21. *Clin. Cancer Res.* **2004**, *10*, 53–60. [CrossRef] [PubMed]

123. Hromic-Jahjefendic, A.; Lundstrom, K. Viral vector-based melanoma gene therapy. *Biomedicines* **2020**, *8*, 60. [CrossRef] [PubMed]

124. Puzanov, I.; Milhem, M.M.; Minor, D.; Hamid, O.; Li, A.; Chen, L.; Chastain, M.; Gorski, K.S.; Anderson, A.; Chou, J.; et al. Talimogene laherparepvec in combination with ipilimumab in previously untreated, unresectable stage IIIB-IV melanoma. *J. Clin. Oncol.* **2016**, *34*, 2619–2626. [CrossRef] [PubMed]

125. Andtbacka, R.H.I.; Ross, M.; Puzanov, I.; Milhem, M.; Collichio, F.; Delman, K.A.; Amatruda, T.; Zager, J.S.; Cranmer, L.; Hsueh, E.; et al. Patterns of clinical response with talimogene laherparepvec (T-VEC) in patients with melanoma treated in the OPTiM phase III clinical trial. *Ann. Surg Oncol.* **2016**, *23*, 4169–4177. [CrossRef] [PubMed]

126. Bommareddy, P.K.; Patel, A.; Hossain, S.; Kaufman, H.L. Talimogene laherparepvec (T-VEC) and other oncolytic viruses for the treatment of melanoma. *Am. J. Clin. Dermatol.* **2017**, *18*, 1–15. [CrossRef] [PubMed]

127. Galanis, E.; Markovic, S.N.; Suman, V.J.; Nuovo, G.J.; Vile, R.G.; Kottke, T.J.; Nevala, W.K.; Thompson, M.A.; Lewis, J.E.; Rumilla, K.M.; et al. Phase II trial of intravenous administration of Reolysin® (Reovirus Serotype-3-dearing Strain) in patients with metastatic melanoma. *Mol. Ther.* **2012**, *20*, 1998. [CrossRef] [PubMed]

128. Silk, A.W.; Kaufman, H.; Gabrail, N.; Mehnert, J.; Bryan, J.; Norrell, J.; Medina, D.; Bommareddy, P.; Shafren, D.; Grose, M.; et al. Abstract CT026: Phase 1b study of intratumoral Coxsackievirus A21 (CVA21) and systemic pembrolizumab in advanced melanoma patients: Interim results of the CAPRA clinical trial. *Cancer Res.* **2017**, *77*, CT026.

129. Andtbacka, R.H.; Curti, B.D.; Hallmeyer, S.; Feng, Z.; Paustian, C.; Bifulco, C.; Fox, B.; Grose, M.; Shafren, D. Phase II calm extension study: Coxsackievirus A21 delivered intratumorally to patients with advanced melanoma induces immune-cell infiltration in the tumor microenvironment. *J. Immunother. Cancer* **2015**, *3*, P343. [CrossRef]

130. Andtbacka, R.H.I.; Curti, B.D.; Kaufman, H.; Daniels, G.A.; Nemunaitis, J.J.; Hallmeyer, L.E.S.; Lutzky, J.; Schultz, S.M.; Whitman, E.D.; Zhou, K.; et al. Final data from CALM: A phase II study of Coxsackievirus A21 (CVA21) oncolytic virus immunotherapy in patients with advanced melanoma. *J. Clin. Oncol.* **2015**, *33*, P9030. [CrossRef]

131. Batliwalla, F.M.; Bateman, B.A.; Serrano, D.; Murray, D.; Macphail, S.; Maino, V.C.; Ansel, J.C.; Gregersen, P.K.; Armstrong, C.A. A 15-year follow-up of AJJC stage III malignant melanoma patients treated postsurgically with Newcastle disease virus (NDV) oncolysate and determination of alterations in the CD8 T cell repertoire. *Mol. Med.* **1998**, *4*, 783–794. [CrossRef]

132. Dold, C.; Rodriguez Urbiola, C.; Wollmann, G.; Egerer, L.; Muik, A.; Bellmann, L.; Fiegl, H.; Marth, C.; Kimpel, J.; von Laer, D. Application of interferon modulators to overcome partial resistance to ovarian cancers to VSV-GP oncolytic viral therapy. *Mol. Ther Oncolytics* **2016**, *3*, 16021. [CrossRef] [PubMed]

133. Hasegawa, K.; Nakamura, T.; Harvey, M.; Ikeda, Y.; Oberg, A.; Figini, M.; Canevari, S.; Hartmann, L.C.; Peng, K.W. The use of a tropism-modified measles virus in folate receptor-targeted virotherapy of ovarian cancer. *Clin. Cancer Res.* **2006**, *12*, 6170–6178. [CrossRef] [PubMed]

134. Granot, T.; Meruelo, D. The role of natural killer cells in combinatorial anti-cancer therapy using Sindbis viral vector and irinotecan. *Cancer Gene Ther.* **2012**, *19*, 588–591. [CrossRef] [PubMed]

135. Zhang, Y.Q.; Tsai, Y.C.; Monie, A.; Wu, T.C.; Hung, C.F. Enhancing the therapeutic effect against ovarian cancer through a combination of viral oncolysis and antigen-specific immunotherapy. *Mol. Ther.* **2010**, *18*, 692–699. [CrossRef] [PubMed]

136. Sirolimus and Vaccine Therapy in Treating Patients with Stage II-IV Ovarian Epithelial, Fallopian Tube, or Primary Peritoneal Cavity Cancer. Available online: https://clinicaltrials.gov/ct2/show/NCT02833506 (accessed on 12 September 2020).

137. Vaccine Therapy in Stage II, III, or IV Epithelial Ovarian, Fallopian Tube, or Primary Peritoneal Cancers. Available online: https://clinicaltrials.gov/ct2/show/NCT00803569 (accessed on 12 September 2020).

138. Vaccine Therapy in Patients with Stage II, III, or IV Epithelial Ovarian, Fallopian Tube, or Peritoneal Cancer. Available online: https://clinicaltrials.gov/ct2/show/NCT00112957 (accessed on 12 September 2020).

139. Noro, T.; Miyake, K.; Suzuki-Miyake, N.; Igarashi, T.; Uchida, E.; Misawa, T.; Yamazaki, Y.; Shimada, T. Adeno-associated viral vector-mediated expression of endostatin inhibits tumor growth and metastasis in an orthotropic pancreatic cancer model in hamsters. *Cancer Res.* **2004**, *64*, 7486–7490. [CrossRef]

140. Nagasato, M.; Rin, Y.; Yamamoto, Y.; Henmi, M.; Ino, Y.; Yachida, S.; Ohki, R.; Hiraoka, N.; Tagawa, M.; Aoki, K. A tumor-targeting adenovirus with high gene-transduction efficiency for primary pancreatic cancer and ascites cells. *Anticancer Res.* **2017**, *37*, 3599–3605. [CrossRef]

141. Murphy, A.M.; Besmer, D.M.; Moerdyk-Schauwecker, M.; Moestl, N.; Ornelles, D.A.; Mukherjee, P. Vesicular stomatitis virus as an oncolytic agent against pancreatic ductal adenocarcinoma. *J. Virol.* **2012**, *86*, 3073–3087. [CrossRef]

142. Awano, M.; Fuijyki, T.; Shoji, K.; Amagai, Y.; Murakami, Y.; Furukawa, Y.; Sato, H.; Yoneda, M.; Kai, C. Measles virus selectively blind to signaling lymphocyte activity molecule has oncolytic efficacy against nectin-4 expressing pancreatic cells. *Cancer Sci.* **2016**, *107*, 1647–1652. [CrossRef]

143. O'Leary, M.P.; Choi, A.H.; Kim, S.I.; Chaurasiya, S.; Lu, J.; Park, A.K.; Woo, Y.; Warner, S.G.; Fong, Y.; Chen, N.G. Novel oncolytic chimeric orthopoxvirus causes regression of pancreatic cancer xenografts and exhibits abscopal effect at a single low dose. *J. Transl. Med.* **2018**, *16*, 110. [CrossRef]

144. Hirooka, Y.; Kasuya, H.; Ishikawa, T.; Kawashima, H.; Ohno, E.; Villalobos, I.; Naoe, Y.; Ichinose, T.; Koyama, N.; Tanaka, M.; et al. A phase I clinical trial of EUS-guided intratumoral injection of the oncolytic virus, HF10 for unresectable locally advanced pancreatic cancer. *BMC Cancer* **2018**, *18*, 596. [CrossRef]

145. Morse, M.A.; Hobelka, A.C.; Osada, T.; Berglund, P.; Hubby, B.; Negri, S. An alphavirus vector overcomes the presence of neutralizing antibodies and elevated numbers of Tregs to induce immune responses in humans with advanced cancer. *J. Clin. Investig.* **2010**, *120*, 3234–3241. [CrossRef] [PubMed]

146. Msaouel, P.; Iankov, I.D.; Allen, C.; Morris, J.C.; von Messling, V.; Cattaneo, R. Engineered measles virus as a novel oncolytic therapy against prostate cancer. *Prostate* **2009**, *69*, 82–91. [CrossRef] [PubMed]

147. Durso, R.J.; Andjelic, S.; Gardner, J.P.; Margitich, D.J.; Donovan, G.P.; Arrigale, R.R. A novel alphavirus vaccine encoding prostate-specific membrane antigen elicits potent cellular and humoral immune responses. *Clin. Cancer Res.* **2007**, *13*, 3999–4008. [CrossRef] [PubMed]

148. Garcia-Hernandez, M.L.; Gray, A.; Hubby, B.; Kast, W.M. In vivo effects of vaccination with six-transmembrane epithelial antigen of the prostate: A candidate antigen for treating prostate cancer. *Cancer Res.* **2007**, *67*, 1344–1351. [CrossRef]

149. Garcia-Hernandez, M.L.; Gray, A.; Hubby, B.; Klinger, O.J.; Kast, W.M. Prostate stem cell antigen vaccination induces a long-term protective immune response against prostate cancer in the absence of autoimmunity. *Cancer Res.* **2008**, *68*, 861–869. [CrossRef] [PubMed]

150. Urbiola, C.; Santer, F.R.; Petersson, M.; van der Pluijm, G.; Horninger, W.; Erlmann, P. Oncolytic activity of the rhabdovirus VSV-GP against prostate cancer. *Int. J. Cancer* **2018**, *143*, 1786–1796. [CrossRef] [PubMed]

151. Son, H.A.; Zhang, L.; Cuong, B.K.; Van Tong, H.; Cuong, L.D.; Hang, N.T.; Nhung, H.T.M.; Yamamoto, N.; Toan, N.L. Combination of Vaccine-Strain Measles and Mumps Viruses Enhances Oncolytic Activity against Human Solid Malignancies. *Cancer Investig.* **2018**, *7*, 106–117. [CrossRef]

152. Slovin, S.F.; Kehoe, M.; Durso, R.; Fernandez, C.; Olson, W.; Gao, J.P. A phase I dose escalation trial of vaccine replicon particles (VRP) expressing prostate-specific membrane antigen (PSMA) in subjects with prostate cancer. *Vaccine* **2013**, *31*, 943–949. [CrossRef]

153. Lubaroff, D.M.; Konety, B.R.; Link, B.; Gerstbrein, J.; Madsen, T.; Shannon, M.; Howard, J.; Paisley, J.; Boeglin, D.; Ratliff, T.L.; et al. Phase I clinical trial of an adenovirus/prostate-specific antigen vaccine for prostate cancer: Safety and immunologic results. *Clin. Cancer Res.* **2009**, *15*, 7375–7380. [CrossRef]

154. Madan, R.A.; Bilusic, M.; Heery, C.; Schlom, J.; Gulley, J.L. Clinical evaluation of TRICOM vector therapeutic cancer vaccines. *Semin Oncol.* **2012**, *39*, 296–304. [CrossRef]

155. Kantoff, P.W.; Schuetz, T.J.; Blumenstein, B.A.; Glode, L.M.; Bilhartz, D.L.; Wyand, M.; Manson, K.; Panicali, D.L.; Laus, R.; Schlom, J.; et al. Overall Survival Analysis of a Phase II Randomized Controlled Trial of a Poxviral-Based PSA-Targeted Immunotherapy in Metastatic Castration-Resistant Prostate Cancer. *J. Clin. Oncol.* **2010**, *28*, 1099–1105. [CrossRef] [PubMed]

156. Gulley, J.L.; Arlen, P.M.; Madan, R.A.; Tsang, K.-Y.; Pazdur, M.P.; Skarupa, L.; Jones, J.L.; Poole, D.J.; Higgins, J.P.; Hodge, J.W.; et al. Immunologic and prognostic factors associated with overall survival employing a poxviral-based PSA vaccine in metastatic castrate resistant prostate cancer. *Cancer Immunol, Immunother.* **2010**, *59*, 663–674. [CrossRef] [PubMed]

157. Gulley, J.L.; Borre, M.; Vogelzang, N.J.; Ng, S.; Agarwal, N.; Parker, C.C.; Pook, D.W.; Rathenborg, P.; Flaig, T.W.; Carles, J.; et al. Phase III Trial of PROSTVAC in asymptomatic or minimally symptomatic metastatic castration-resistant prostate cancer. *J. Clin. Oncol.* **2019**, *37*, 1051–1061. [CrossRef] [PubMed]

158. Boettcher, A.N.; Usman, A.; Morgans, A.; Vander Weele, D.J.; Sosman, J.; Wu, J.D. Past, current, and future of immunotherapies for prostate cancer. *Front. Oncol.* **2019**, *9*, 884. [CrossRef]

159. Lundstrom, K. The current status of COVID-19 vaccines. *Front. Genome Ed.* **2020**. [CrossRef]

160. Available online: https://www.who.int/publications/m/item/draft-landscape-of-covid-19-candidate-vaccines (accessed on 3 November 2020).

161. Folegatti, P.M.; Bellamy, D.; Roberts, R.; Powlson, J.; Edwards, N.J.; Mair, C.F.; Bowyer, G.; Poulton, I.; Mitton, C.H.; Green, N.; et al. Safety and immunogenicity of a novel recombinant simian Adenovirus ChAdOx2 as a vectored vaccine. *Vaccines* **2019**, *7*, 40. [CrossRef]

162. van Doremalen, N.; Lambe, T.; Spencer, A.; Belij-Rammerstorfer, S.; Purushotham, J.N.; Port, J.R.; Avanzato, V.A.; Bushmaker, T.; Flaxman, A.; Ulaszewska, M.; et al. ChAdOx1 nCov-19 vaccination prevents SARS-CoV-2 pneumonia in rhesus macaques. *Nature* **2020**, *586*, 578–582. [CrossRef]

163. Feng, L.; Wang, Q.; Shan, C.; Yang, C.; Feng, Y.; Wu, J.; Liu, X.; Zhou, Y.; Jian, R.; Hu, P.; et al. An adenovirus-vectored COVID-19 vaccine confers protection from SARS-CoV-2 challenge in rhesus macaques. *Nat. Commun.* **2020**, *11*, 4207. [CrossRef]

164. Tostanoski, L.H.; Wegmann, F.; Martinot, A.J.; Loos, C.; McMahan, K.; Mercado, N.B.; Yu, J.; Chan, C.N.; Bondoc, S.; Starke, C.E.; et al. Ad26 vaccine protects against SARS-CoV-2 severe clinical disease in hamsters. *Nat. Med.* **2020**. [CrossRef]

165. Mercado, N.N.B.; Zahn, R.; Wegmann, F.; Loos, C.; Chandrashekar, A.; Yu, J.; Liu, J.; Peter, L.; McMahan, K.; Tostanoski, H.; et al. Single-shot Ad26 vaccine protects against SARS-CoV-2 in rhesus macaques. *Nature* **2020**. [CrossRef]

166. Hörner, C.; Schürmann, C.; Auste, A.; Ebenig, A.; Muraleedharan, S.; Herrmann, M.; Schnierle, B.S.; Mühlebach, M.D. A Highly Immunogenic Measles Virus-based Th1-biased COVID-19 Vaccine. *bioRxiv* **2020**. [CrossRef]

167. Förster, R.; Fleige, H.; Sutter, G. Combating COVID-19: MVA vector vaccines applied to the respiratory tract as promising toward protective immunity in the lung. *Front. Immunol.* **2020**, *11*, 1959. [CrossRef] [PubMed]

168. Chiuppesi, F.; d'Alincourt Salazar, M.; Contreras, H.; Nguyen, V.H.; Martinez, J.; Park, S.; Nguyen, J.; Kha, M.; Iniguez, A.; Zhou, Q.; et al. Development of a synthetic poxvirus-based SARS-CoV-2 vaccine. *bioRxiv* **2020**. Preprint. [CrossRef]

169. Zhu, F.C.; Li, Y.-H.; Guan, X.-H.; Hou, L.H.; Wang, W.J.; Li, J.X.; Wu, S.P.; Wang, B.S.; Wang, Z.; Wang, L.; et al. Safety, tolerability, and immunogenicity of a recombinant adenovirus type-5 vectored COVID-19 vaccine: A dose-escalation, open label, non-randomised, first-in-human trial. *Lancet* **2020**, *395*, 1845–1854. [CrossRef]

170. Zhu, F.C.; Guan, X.H.; Li, Y.H.; Huang, J.Y.; Jiang, T.; Hou, L.H.; Li, J.X.; Yang, B.F.; Wang, L.; Wang, W.J. Immunogenicity and safety of a recombinant adenovirus type-5-vectored COVID-19 vaccine in healthy adults aged 18 years and older: A randomised, double-blind, placebo-controlled, phase 2 trial. *Lancet* **2020**, *396*, 479–488. [CrossRef]

171. Phase III Trial of A COVID-19 Vaccine of Adenovirus Vector in Adults 18 Years Old. Available online: https://clinicaltrials.gov/ct2/show/NCT04526990 (accessed on 5 November 2020).

172. Clinical Trial of Recombinant Novel Coronavirus Vaccine (Adenovirus Type 5 Vector) Against COVID-19. Available online: https://clinicaltrials.gov/ct2/show/NCT04540419 (accessed on 5 November 2020).

173. A Study of Ad26.COV2.S in Adults (COVID-19). Available online: https://clinicaltrials.gov/ct2/show/NCT04436276 (accessed on 5 November 2020).

174. Sadoff, J.; Le Gars, M.; Shukarev, G.; Heerwegh, D.; Truyers, C.; de Marit Groot, A.; Stoop, J.; Tete, S.; Van Damme, W.; Leroux-Roels, I.; et al. Safety and immunogenicity of the Ad26.COV.S COVID-19 vaccine candidate: Interim results of a phase 1/2a, double-blind, randomized, placebo-controlled trial. *medRxiv* **2020**. [CrossRef]

175. A Study of Ad26.COV2.S for the Prevention of SARS-CoV-2-Mediated COVID-19 in Adult Participants (ENSEMBLE). Available online: https://clinicaltrials.gov/ct2/show/NCT04505722 (accessed on 5 November 2020).

176. Callaway, E. Russia's fast-track coronavirus vaccine draws outrage over safety. *Nature* **2020**, *584*, 334–335. [CrossRef]

177. Logunov, D.Y.; Dolzhikova, I.V.; Zubkova, O.V.; Tukhvatullin, A.I.; Shcheblyakov, D.V.; Dzharullaeva, A.S.; Grousova, D.M.; Erokhova, A.S.; Kovyrshina, A.V.; Botikov, A.G.; et al. Safety and immunogenicity of an rAd26 and rAd5 vector-based heterologous prime-boost COVID-19 vaccine in two formulations: Two open, non-randomised phase 1/2 studies from Russia. *Lancet* **2020**, *396*, 887–897. [CrossRef]

178. Clinical Trial of Efficacy, Safety, and Immunogenicity of Gam-COVID-Vac Vaccine against COVID-19 (RESIST). Available online: https://clinicaltrials.gov/ct2/show/NCT04530396 (accessed on 5 November 2020).

179. Clinical Trial of Efficacy, Safety, and Immunogenicity of Gam-COVID-Vac Vaccine against COVID-19 in Belarus. Available online: https://clinicaltrials.gov/ct2/show/NCT04564716 (accessed on 5 November 2020).

180. Folegatti, P.M.; Ewer, K.J.; Aley, P.K.; Angus, B.; Becker, S.; Belij-Rammerstorfer, S.; Bellamy, D.; Bibi, S.; Bittaye, M.; Clutterbuck, E.A.; et al. Safety and immunogenicity of the ChAsOx1 nCoV-19 vaccine against SARS-CoV-2: A preliminary report of a phase 1/2 single-blind, randomised controlled trial. *Lancet* **2020**, *396*, 467–478. [CrossRef]

181. Phase III Double-blind, Placebo-controlled Study of AZD1222 for the Prevention of COVID-19 in Adults. Available online: https://clinicaltrials.gov/ct2/show/NCT04516746 (accessed on 5 November 2020).

182. Phillips, N.; Cyranoski, D.; Mallapaty, S. A leading coronavirus vaccine trial is on hold: Scientists react. *Nature* **2020**. [CrossRef]

183. Global Clinical Trials of COVID-19 Vaccine Resume. Available online: https://www.ox.ac.uk/news/2020-11-23-global-clinical-trials-covid-19-vaccine-resume (accessed on 5 November 2020).

184. Safety, Tolerability and Immunogenicity of the Candidate Vaccine MVA-SARS-2-S against COVID-19. Available online: https://clinicaltrials.gov/ct2/show/NCT04569383 (accessed on 5 November 2020).

185. Immunity and Safety of Covid-19 Synthetic Minigene Vaccine. Available online: https://clinicaltrials.gov/ct2/show/NCT04276896 (accessed on 5 November 2020).

186. Clinical Trial to Evaluate the Safety and Immunogenicity of the COVID-19 Vaccine (COVID-19-101). Available online: https://clinicaltrials.gov/ct2/show/NCT04497298 (accessed on 5 November 2020).

187. Brett, J.B.; Rothlauf, P.W.; Chen, R.E.; Kafai, N.M.; Fox, J.M.; Smith, B.K.; Shrihari, S.; McCune, B.T.; Harvey, I.B.; Keeler, S.P.; et al. Replication-Competent Vesicular Stomatitis Virus Vaccine Vector Protects against SARS-CoV-2-Mediated Pathogenesis in Mice. *Cell Host Microbe* **2020**, *28*, 465–474. [CrossRef]

188. Dose Ranging Trial to Assess Safety and Immunogenicity of V590 (COVID-19 Vaccine) in Healthy Adults (V590-001). Available online: https://clinicaltrials.gov/ct2/show/NCT04569786 (accessed on 5 November 2020).

189. Yahalom-Ronen, Y.; Tamir, H.; Melamed, S.; Politi, B.; Shifman, O.; Achdout, H.; Vitner, E.B.; Israeli, O.; Milrot, E.; Stein, D.; et al. A single dose of recombinant VSV-ΔG-spike provides protection against SARS-CoV-2 challenge. *bioRxiv* **2020**. [CrossRef]

190. Evaluate the Safety, Immunogenicity and Potential Efficacy of an rVSV-SARS-CoV-2-S Vaccine. Available online: https://clinicaltrials.gov/ct2/show/NCT04608305 (accessed on 5 November 2020).

191. Higgins, T.S.; Wu, A.W.; Illing, E.A.; Sokoloski, K.J.; Weaver, B.A.; Anthony, B.P.; Hughes, N.; Ting, J.Y. Intranasal Antiviral Drug Delivery and Coronavirus Disease 2019 (COVID-19): A State of the Art Review. *Otolaryngol. -Head Neck Surg.* **2020**, *163*, 682–694. [CrossRef] [PubMed]

192. King, R.; Silva-Sanchez, A.; Peel, J.N.; Botta, D.; Meza-Perez, S.; Allie, R.; Schultz, M.D.; Liu, M.; Bradley, J.E.; Qiu, S.; et al. Single-dose intranasal administration of AdCOVID elicits systemic and mucosal immunity against SARS-CoV-2 in mice. *bioRxiv* **2020**. [CrossRef]
193. Ollmann Saphire, E. A vaccine against Ebola virus. *Cell* **2020**, *181*, 6. [CrossRef]
194. Conry, R.M.; Westbrook, B.; McKee, S.; Norwood, T.G. Talimogene laherparepvec: First in class oncolytic virotherapy. *Hum. Vaccin. Immunother.* **2018**, *14*, 839–846. [CrossRef]

Publisher's Note: MDPI stays neutral with regard to jurisdictional claims in published maps and institutional affiliations.

 *viruses*

Article

JSRV Intragenic Enhancer Element Increases Expression from a Heterologous Promoter and Promotes High Level AAV-Mediated Transgene Expression in the Lung and Liver of Mice

Darrick L. Yu, Natalie Chow and Sarah K. Wootton *

Department of Pathobiology, Ontario Veterinary College, University of Guelph, Guelph, ON N1G 2W1, Canada;
darrickyu@gmail.com (D.L.Y.); nataliesmchow@gmail.com (N.C.)
* Correspondence: kwootton@uoguelph.ca

Received: 30 September 2020; Accepted: 4 November 2020; Published: 6 November 2020

Abstract: Jaagsiekte sheep retrovirus (JSRV) induces tumors in the distal airways of sheep and goats. A putative intragenic enhancer, termed JE, localized to the 3′ end of the JSRV *env* gene, has been previously described. Herein we provide further evidence that the JE functions as a transcriptional enhancer, as it was able to enhance gene expression when placed in either forward or reverse orientation when combined with a heterologous chicken beta actin promoter. We then generated novel composite promoters designed to improve transgene expression from adeno-associated virus (AAV) gene therapy vectors. A hybrid promoter consisting of the shortest JE sequence examined (JE71), the U3 region of the JSRV long terminal repeat (LTR), and the chicken beta actin promoter, demonstrated robust expression in vitro and in vivo, when in the context of AAV vectors. AAV-mediated transgene expression in vivo from the hybrid promoter was marginally lower than that observed for AAV vectors encoding the strong CAG promoter, but greatly reduced in the heart, making this promoter/enhancer combination attractive for non-cardiac applications, particularly respiratory tract or liver directed therapies. Replacement of the murine leukemia virus intron present in the original vector construct with a modified SV40 intron reduced the promoter/enhancer/intron cassette size to 719 bp, leaving an additional ~4 kb of coding capacity when packaged within an AAV vector. Taken together, we have developed a novel, compact promoter that is capable of directing high level transgene expression from AAV vectors in both the liver and lung with diminished transgene expression in the heart.

Keywords: adeno-associated virus (AAV) vector; jaagsiekte sheep retrovirus (JSRV); LTR; enhancer; transduction

1. Introduction

Adeno-associated virus (AAV) is a small, non-pathogenic, non-enveloped, single stranded DNA virus belonging to the *Parvoviridae* family [1,2]. In 2017, Luxturna (voretigene neparvovec-rzyl), a recombinant vector based on AAV was approved by the U.S. Food and Drug Administration for the treatment of biallelic RPE65 gene mutation-associated retinal dystrophy, a rare form of inherited vision loss that may result [3]. In 2019, a second AAV gene therapy for the treatment of children less than two years of age with spinal muscular atrophy (SMA), termed Zolgensma, was the second directly administered gene therapy approved in the U.S., demonstrating the growing promise of AAV based gene therapies for the treatment of a wide variety of conditions and inherited disorders.

AAV offers many advantages over other gene delivery vectors such as adenovirus vectors due to its superior transducing efficiency in vivo, its ability to promoter sustained transgene expression, its low immunogenicity and the fact that it can and is being used in a wide range of clinical applications [4].

However, the limited packaging capacity of AAV vectors (~4.7 kb) necessitates the selection of promoter/enhancer elements that are as small as possible, yet retain a high degree of expression, particularly in scenarios where the transgene is of a considerable size, such as in the case of cystic fibrosis transmembrane conductance regulator (CFTR) (4.4 kb). The optimal promoter/enhancer combination for AAV vectored gene therapy applications would be one that has high activity in the target cell population, but minimal to low activity in non-target cells.

Jaagsiekte sheep retrovirus (JSRV) is a simple betaretrovirus that is capable of inducing a form of lung cancer in sheep known as ovine pulmonary adenocarcinoma [5]. Viral gene expression is primarily governed by promoter and enhancer elements located within the long terminal repeat (LTR) sequences that are found on the 5′ and 3′ terminal ends of the integrated provirus [6]. More recently, a putative enhancer sequence, known as JE, has been located outside of the 3′ LTR within the env gene, just prior to the beginning of the 3′ LTR sequence. Adeno-associated virus (AAV) vectors bearing the putative JE enhancer sequence in conjunction with the JSRV long terminal repeat (LTR) demonstrated enhanced, tissue specific expression. Augmentation of the JSRV LTR with the JE resulted in a >4-fold enhancement in lung directed transgene expression and a ~2-fold improvement in liver when vectors were administered to mice [7].

We sought to further delineate the manner in which JE functions and provide further evidence that it is able to function as a transcriptional enhancer. In addition, we hypothesized that promoter/enhancer cassettes based on the JE and the JSRV LTR sequences would be highly effective for in vivo gene delivery purposes, owing to their ability to promote high level protein expression and their relatively compact size.

2. Materials and Methods

2.1. Cell Culture

Human embryonic kidney (HEK 293, ATCC CRL-1573) cells, HEK 293T cells, rat fibroblast (208F) cells, and HTX cells, a pseudodiploid subclone of HT-1080 fibrosarcoma cells [8], were maintained in Dulbecco's modified Eagle's medium (DMEM; Thermo Fisher Scientific, Ottawa Canada) supplemented with 10% fetal bovine serum (FBS) (GIBCO, Invitrogen), 100 units/mL penicillin, 100 µg/mL streptomycin and 2 mM L-glutamine in a humidified 5% CO_2 atmosphere at 37 °C.

2.2. Molecular Cloning

PCR was used to amplify JE or LTR sequences from the molecular clone of JSRV, pCMV-JS21 [9]. Vectors encoding human alkaline phosphatase reporter gene (hPLAP) were derived from an AAV vector plasmid, AEEE1AP, as described previously [7]. The vector was modified to replace the Enzootic-Nasal Tumor Virus-1 (ENTV-1) enhancer/promoter component with enhancer elements derived from Jaagsiekte Sheep Retrovirus (JSRV), acting on the chicken beta actin promoter to drive expression of hPLAP. Splicing of the hPLAP encoding transcript was promoted by the presence of a murine leukemia virus env intron [10] found between the enhancer/promoter and the hPLAP gene. Following the hPLAP gene was the SV40 polyA tail for the polyadenylation of transcripts. JE and LTR sequences were cloned into XbaI and BglII sites in the AAV vector plasmid, AEEE1AP [7] containing a murine leukemia virus retrovirus intron and hPLAP reporter gene. The chicken beta actin promoter (CBA) was cloned downstream of the JE or JSRV LTR sequences into BglII and KpnI sites. The sequence for the putative enhancers is as follows: JE71: ACATATGAAATATAGAAATATGTTACAGCACCAACATCTTATGG AGCTTTTAAAAAATAAAGAGAGGGGAG; JE184:ACCCTGATTGGTGTAGGAATACTTGTGTTTAT TATAATTGTCGTAATCCTTATATTTCCTTGCCTTGTTCGTGGCATGGTTCGCGATTTTCTAAAGATG AGAGTTGAAATGCTGCATATGAAATATAGAAATATGTTACAGCACCAACATCTTATGGAGCTTTT AAAAAATAAAGAGAGGGGAG; and JE324: CGTTAGACCTTTTACAACTGCATAATGAGATTCTTG ATATTGAAAATTCGCCGAAGGCTACACTAAATATAGCCGATACTGTTGATAATTTCTTGCAAAA TTTATTCTCTAATTTTCCTAGTCTCCATTCGCTGTGGAAAACCCTGATTGGTGTAGGAATACTT

GTGTTTATTATAATTGTCGTAATCCTTATATTTCCTTGCCTTGTTCGTGGCATGGTTCGCGATTTTCT
AAAGATGAGAGTTGAAATGCTGCATATGAAATATAGAAATATGTTACAGCACCAACATCTTATGG
AGCTTTTAAAAAATAAAGAGAGGGGAG.

2.3. Transfection of Mammalian Cells

Approximately 5×10^6 cells were seeded onto three 10-cm tissue culture dishes for each construct 24 h prior to transfection. Cells were transfected with 10 μg each of pCMV-βgal and the construct of interest using polyethylenimine (Polysciences Inc., Warrington, PA, USA) for 208F rat fibroblast cells and calcium phosphate for human HEK 293, HEK 293T and HTX cells. Polyethylenimine transfection was conducted according to manufacturer's directions. Transfection using the calcium phosphate method was conducted as described previously [11].

2.4. Preparation of Cell Lysates

After 48 h, cells were washed with 10 mL of ice cold PBS and scraped off the plate using a rubber policeman into 1 mL of PBS. Cells and PBS were spun down at 4000 rpm for 2 min to recover cells. 500 μL of TMNC lysis buffer (50 mM Tris-HCl (pH 7.5), 5 mM $MgCl_2$, 100 mM NaCl, and 4% (wt/vol) 3-[(3-cholamidopropyl)-dimethylammonio]-1-propanesulfonate (CHAPS) was added to each cell pellet before pipetting up and down to suspend the cells. Cells were allowed to lyse for 15 min on ice. Cell debris was removed by centrifuging the sample at 14,000 rpm and the supernatant was recovered for use in subsequent assays. A Bradford assay was performed on cell lysates according to the method of Sambrook and Russell to determine total protein concentration [11].

2.5. Beta-Galactosidase Assay

Beta-galactosidase assays were performed by the method of J. Miller [12]. The following solutions were mixed together prior to performing the assay: $100\times$ Mg^{2+} solution containing 0.1 M $MgCl_2$ and 4.5 M β-mercaptoethanol, $1\times$ ONPG solution containing 4 mg/mL o-nitrophenyl-β-D-Galactoside (ONPG) dissolved in 0.1 M dibasic sodium phosphate buffer pH (7.5). 3 μL of $100\times$ Mg^{2+}, 66 μL of $1\times$ ONPG, 30 μL of cell lysate, and 201 μL of 0.1 M sodium phosphate were mixed together to initiate the reaction. Reactions were incubated at 37 °C until a faint yellow color developed. Reactions were stopped by adding 500 μL of 1 M Na_2CO_3. To determine beta-galactosidase activity, absorbance was read at 420 nm using a BioTeK Powerwave XS2 plate reader.

2.6. In Vivo Administration of AAV Vectors

Mouse experiments were performed in accordance with the guidelines set forth by the Canadian Council on Animal Care (CCAC). Eight-week old C57BL6/J mice were obtained from Charles River Laboratories (Saint-Constant, QC). AAV vectors were produced by cotransfection of HEK 293 cells with genome and packaging plasmids as described previously [13]. AAV vector titers were determined by quantitative polymerase chain reaction (qPCR) analysis as described elsewhere [14]. AAV vectors were administered via three different routes of administration: intravenous, intraperitoneal, and intranasal to determine relative promoter activity. For intravenous delivery, a phosphate buffered saline (PBS) solution containing 2×10^{10} vector genomes of AAV vectors was injected in a 100 μL volume into the tail vein. For intraperitoneal, 8×10^{10} vector genomes were injected into the intraperitoneal space in a 500 μL total volume containing the AAV vector plus PBS. For intranasal delivery, 1×10^{10} vector genomes were delivered in two aliquots of 40 μL each in order to maximize the chances that tissues deep in the lung would be transduced. A modified method of intranasal delivery was used so as to ensure vector delivery to the distal lung [15]. Mice were euthanized 4 weeks post vector administration, and lungs were perfused through the heart with 20 mL of PBS and then separated into individual lobes. For consistency, the same lung lobe from each mouse was either flash frozen in liquid nitrogen or fixed in 2% paraformaldehyde-PBS for 16 h at 22 °C. Half of other major organs, including the liver, spleen, pancreas, nose, heart, and kidney were fixed for 24 h at 22 °C, with the other half placed into liquid nitrogen for a subsequent enzymatic assay of hPLAP activity. Tissues were stained for vector-encoded heat-stable hPLAP as described previously [7]. Gross pictures of stained tissues were taken using a Zeiss dissecting scope (Zeiss Canada, Toronto, ON, Canada).

2.7. Determination of Alkaline Phosphatase Activity

For in vitro studies, cell lysates were heated at 65 °C for 1 h to inactivate endogenous alkaline phosphatases. For 208F cells, 30 µg of total protein as determined by Bradford assay was loaded into each well of a 96 well plate for each 10 cm dish. For HTX, HEK 293, and HEK 293T cells, 100 µg of total protein as determined by Bradford assay were loaded into each well of a 96 well plate. Mouse tissues were harvested 4 weeks after vector administration, snap-frozen in liquid nitrogen and stored at −80 °C until assayed. A small piece (approximately 0.5 cm in diameter; ~30 mg) of the tissue to be analyzed was homogenized in TMNC lysis buffer (50 mM Tris HCl pH 7.5, 5 mM $MgCl_2$, 100 mM NaCl, 4% (wt/vol) CHAPS) using a Precellys 24 homogenizer (Bertin Technologies, Montigny-le-Bretonneux, France), with ~200 µL of TMNC buffer in a FastPrep™ Lysing Matrix A tube (MP Bio, Santa Ana, CA, USA). Tissue homogenates were placed in a water bath at 65 °C for 1 h to inactivate endogenous heat-labile AP activity and subsequently clarified by centrifugation at 17,900× *g* for 15 min at 4 °C to remove cell debris. The protein content of each sample was determined by the method of Bradford, and the AP activity in tissue lysates was determined, in triplicate, by a fluorometric assay using the 4-methylumbelliferyl phosphate (MUP) (Sigma, St. Louis, MO, USA) substrate, as described previously [7]. The mean and standard deviation were calculated for each of the different cell lines and vector constructs, as well as for lung, liver, heart, pancreas and spleen for each of the diffesrent vector treatment groups. To correct for differences in transfection efficiency for each of the plasmid constructs described in Figure 1, the mean of AP activity was divided by the mean of β-gal activity.

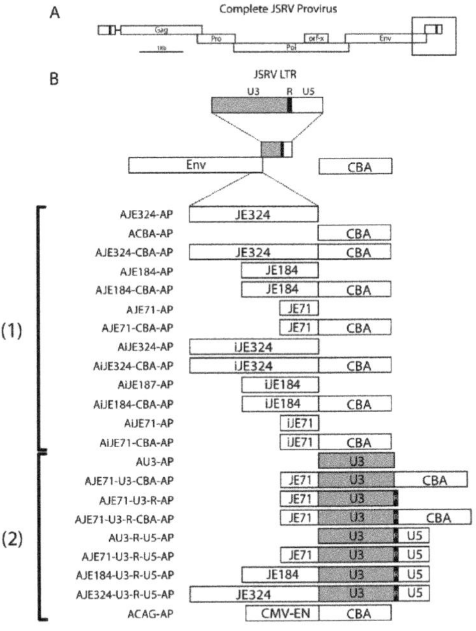

Figure 1. Schematic Diagram of Promoter/Enhancer Constructs. (**A**) The structure of an integrated Jaagsiekte sheep retrovirus (JSRV) provirus. A box is drawn around the 3′ end of the *env* gene and the 3′ long terminal repeat (LTR) where the putative enhancer lies. (**B**) Promoter/enhancer constructs used to drive human placental alkaline phosphatase reporter gene expression and determine if JE functions as a transcriptional enhancer. JE71, JE184, and JE324 refer to three different lengths of the putative JE enhancer region, 71, 184 and 324 bp in length, respectively. The JE sequences are taken from the 3′ end of the *env* gene just prior to the start of the 3′ LTR. In some constructs, the JE was inverted, denoted with an "i". JE and inverted JE were placed in front of the chicken beta actin (CBA) promoter to determine if they could function upon a heterologous promoter. (2) A number of constructs were also evaluated for their utility as gene expression vectors. Constructs bearing the U3 region of the JSRV LTR, the U3 and R region of the LTR, or the full length JSRV LTR (U3, R and U5 regions), in combination with the chicken beta actin promoter were also tested. A subset of these was evaluated in vivo: ACBA-AP, AJE71-CBA-AP, AJE71-U3-CBA-AP, ACAG-AP (a construct bearing the CAG promoter, which combines the CMV immediate early enhancer, chicken beta actin promoter and rabbit β-globin intron). AP denotes the presence of a human placental alkaline phosphatase reporter gene.

3. Results

3.1. JE on Its Own Functions Poorly as a Promoter

Constructs bearing short, medium, or long length JE sequences (pAJE71-AP, pAJE184-AP, and pAJE324-AP, respectively) on their own demonstrated little or no expression when transfected into human embryonic kidney (HEK 293), human fibrosarcoma (HTX), or rat fibroblast (208F) cells (Figure 2). Similarly, constructs bearing inverted versions of short, medium, or long length JE sequences (pAiJE71-AP, pAiJE184-AP, and pAiJE324-AP, respectively) also demonstrated low or no expression in these cell lines. The lack of expression for the JE or inverted JE sequences on their own suggest that the JE is unable to function as a promoter in either forward or reverse orientations in these cell lines. Interestingly, there was a low level of activity in HEK 293T cells, but not 293 cells for forward and reverse JE constructs, suggesting that the SV40 large T antigen was able to promote transcription from the AAV2 inverted terminal repeats (ITR) present on the 5′ flanking end of the JE. No SV40 origin of replication could be found within the plasmid backbone, suggesting that any increase in expression within 293T cells was independent of plasmid replication by the T antigen.

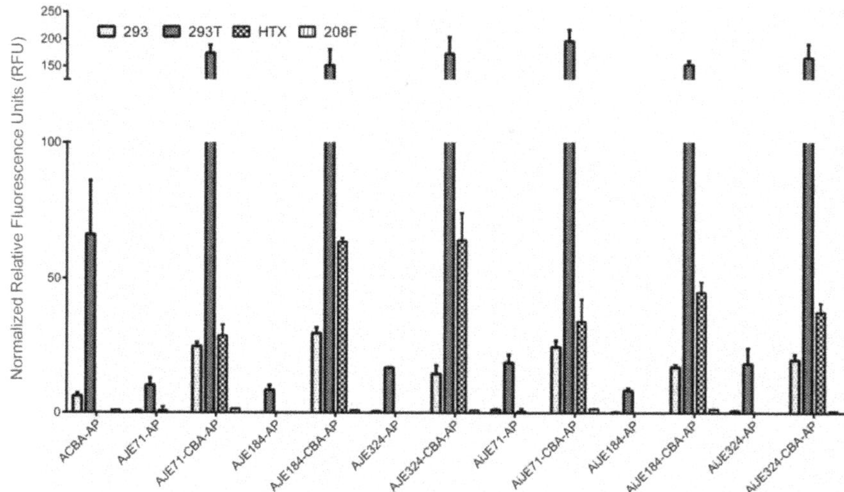

Figure 2. Transfection of four different cell lines (human embryonic kidney (HEK) 293, HEK 293T, rat fibroblast (208F), and human fibrosarcoma (HTX)) with various constructs incorporating the normal orientation JE or the inverted JE. The presence of the inverted JE is denoted by an "i". Three different sizes were compared corresponding to JE sequences 71, 184, or 324 bp in length. These inverted and non-inverted JE sequences were either placed on their own, or in front of the chicken beta actin promoter to determine if the JE could function as a promoter, or in conjunction with a heterologous promoter (chicken beta actin promoter) as an enhancer. The chicken beta actin promoter was also transfected to indicate a baseline level of expression. Normalized Relative Fluorescence Units (RFU) are reported, where AP activity was normalized to β-galactosidase activity. Experiments were conducted in triplicate with three biological replicates.

3.2. JE Enhances Expression from a Heterologous Promoter, in Both Forward and Reverse Orientations, and Extending JE from the 5′ End Does Not Appear to Further Increase Expression

Inclusion of the chicken beta actin promoter (CBA) markedly increased expression from constructs bearing the JE in all cell lines tested (Figure 2). Comparing pAJE71-AP (possessing the short JE, but lacking the chicken beta actin promoter) to pAJE71-CBA-AP (possessing both short JE and the chicken beta actin promoter), a nearly 20-fold increase was observed in 293T cells and an equivalent or greater increase was observed in 293 and HTX cell lines. In a similar manner, comparing pAJE184-AP (possessing medium length JE alone) to pAJE184-CBA-AP (possessing medium length JE plus the chicken beta actin promoter) and pAJE324-AP (possessing the long version of JE alone) to pAJE324-CBA-AP (long JE in addition to the chicken beta actin promoter), there was a ~20-fold increase in expression in 293T cells and a further increase in other cell lines. Extension of the JE from 71 to 184 and 324 base pairs did not appear to confer a corresponding increase in expression, suggesting that the putative enhancer element is located within the original 71 bp region and no additional enhancer elements are located within the 184 or 324 bp regions. A 2-fold or greater increase in transgene expression was also observed when comparing short, medium, or long JE sequences in conjunction with the chicken beta actin promoter relative to the chicken beta actin promoter on its own, demonstrating that it is not merely the chicken beta actin promoter that was able to confer higher expression.

For another series of constructs, the putative JE sequences were placed in an inverted orientation relative to their orientation in the JSRV provirus. Inverted JE sequences corresponding to short, medium and long lengths placed in front of the chicken beta actin promoter (pAiJE72-CBA-AP, pAIJE184-CBA-AP, and pAiJE324-CBA-AP, respectively) all demonstrated a similar improvement as their non-inverted counterparts in 293, HTX, 293T and 208F cells, while constructs containing the inverted JE (pAiJE72-AP, pAiJE184-AP, and pAiJE324-AP) of varying lengths but lacking the chicken beta actin promoter conferred low expression in all cell lines but 293T.

3.3. A Hybrid Promoter Consisting of JE, the U3 and R Regions of the LTR, and the Chicken Beta Actin Promoter Is Highly Active in a Variety of Cell Lines

The relatively low activity of the wild-type JSRV promoter is dwarfed by the high activity brought about by the combination of the JE, the U3 and R regions of the JSRV LTR placed upstream of the chicken beta actin promoter (pAJE71-U3R-CBA-AP) in 293, 293T, 208F, and HTX cells. In fact, expression was so high in 293T and HTX cells that it matched or exceeded expression by the CAG promoter (pACAGAP), currently one of the best promoters for constitutive expression from a variety of cell lines (Figure 3). A vector consisting of the JE71, U3, but not the R region of the JSRV LTR (pAJE71-U3-CBA-AP) expressed to high levels in the same cell lines, reaching levels comparable to the CAG promoter (Figure 3). Both pAJE71-U3-CBA-AP and pAJE71-U3R-CBA-AP demonstrate expression that was much higher than a vector encoding a promoter consisting of the CBA promoter alone, demonstrating the ability of the JE and JSRV LTR elements to enhance expression from a heterologous promoter.

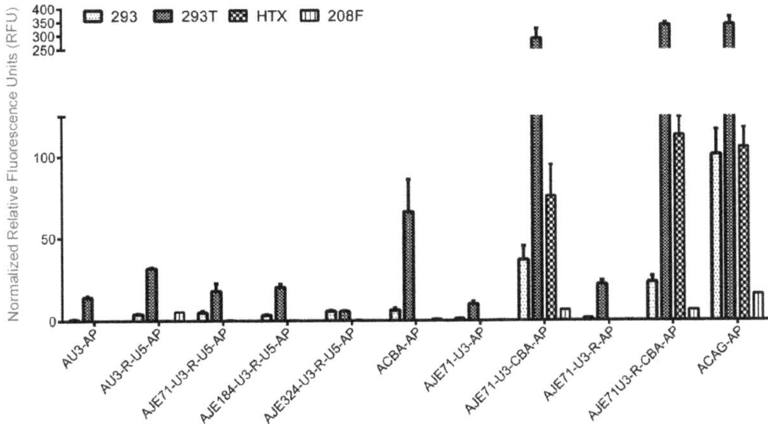

Figure 3. Transfection of four different cell lines (HEK 293, HEK 293T, 208F, and HTX) with various constructs incorporating JE, the chicken beta actin promoter (CBA), and components of the JSRV LTR to determine their suitability for in vivo gene delivery experiments. The JE was investigated to see if it would work with its own homologous promoter, the JSRV LTR, consisting of U3-R-U5 regions, or in conjunction with just the U3/U3 and R regions to enhance expression from the chicken beta actin promoter. These hybrid promoter/enhancer cassettes were compared to the strong CAG promoter and the CBA promoter on its own. Normalized Relative Fluorescence Units (RFU) are reported, where AP activity was normalized to β-galactosidase activity. Experiments were conducted in triplicate with three biological replicates.

Constructs possessing components of the JE or JSRV LTR but lacking the chicken beta actin promoter demonstrated low activity in 293, 293T, 208F and HTX cell lines (Figure 3). pAU3-AP, pAU3-R-U5-AP, pAJE72-U3-R-U5-AP, pAJE184-U3-R-U5-AP, pAJE324-U3-R-U5-AP, pAJE71-U3-AP, and pAJE71-U3-R-AP each exhibited poor expression in 293, 293T, 208F and HTX cells. This may be due to the high specificity of JSRV LTR components for tissues of the respiratory tract, the primary target for JSRV infection.

3.4. The JE/JSRV LTR/CBA Hybrid Promoter Is Active in the Lungs, Nose, Trachea, Liver, Heart, Spleen, and Pancreas of Mice

Expression of the JSRV promoter is primarily limited to the respiratory tract (lung, nose, trachea) and liver when mice are transduced via an AAV vector [8]. Inclusion of the JE enhancer element appeared to improve expression in these tissues, as well as enhance expression in the spleen and pancreas in mice transduced with AAV vectors via the intranasal, intraperitoneal and intravenous route (Figure 4). A hybrid promoter consisting of the JE71, U3 region of the JSRV LTR, and CBA promoter (AJE71-U3-CBA-AP) was highly active in a variety of tissues when transduced into mice via AAV vectors. High levels of expression could be seen in the tissues of the respiratory tract, including the lung, nose and trachea, as well as the liver, spleen and pancreas both grossly (Figure 4) and histologically (Figure 5). Unlike the JSRV LTR, expression was also observed in the heart, but not to the same extent as the CAG promoter construct, which demonstrated an extremely high level of expression (Figure 4). Note that since the vectors were all packaged into the same AAV capsid, in this case AAV serotype 6 (AAV6), the differences in reporter gene expression observed in vivo are due to the tissue specificity of the promoter and not the capsid. Moreover, the use of a heat stable placental alkaline phosphatase reporter gene (hPLAP) allows for heat inactivation of all endogenous alkaline phosphatase while retaining hPLAP enzymatic activity. As such, alkaline phosphatase staining, as evidence by the purple color, is only observed if the tissue has been transduced by the AAV vector expressing hPLAP. This can be observed in the mock infected tissues, where no purple staining was detected, either grossly or histologically.

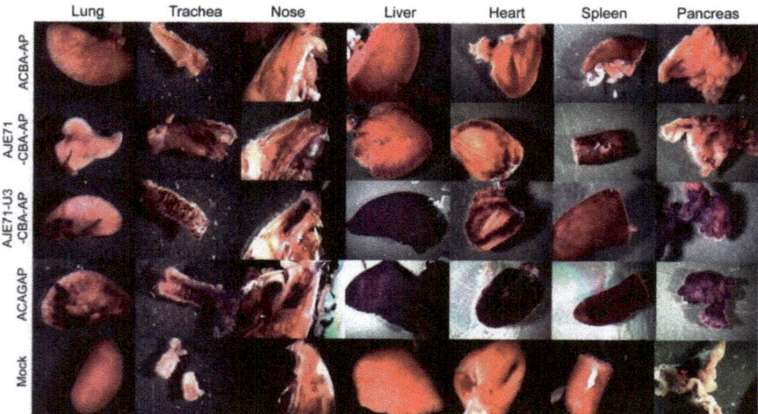

Figure 4. Representative gross images of tissues stained to indicate the presence of human placental alkaline phosphatase reporter gene expression from mice transduced with various JE/CBA constructs. AAV vectors were administered to mice by three different routes of administration concurrently: intravenous (2×10^{10} vg), intraperitoneal (8×10^{10} vg), and intranasal (1×10^{10} vg) to determine relative promoter activity. Mice were euthanized 4 weeks post vector administration and tissues stained for human alkaline phosphatase reporter gene (hPLAP) expression.

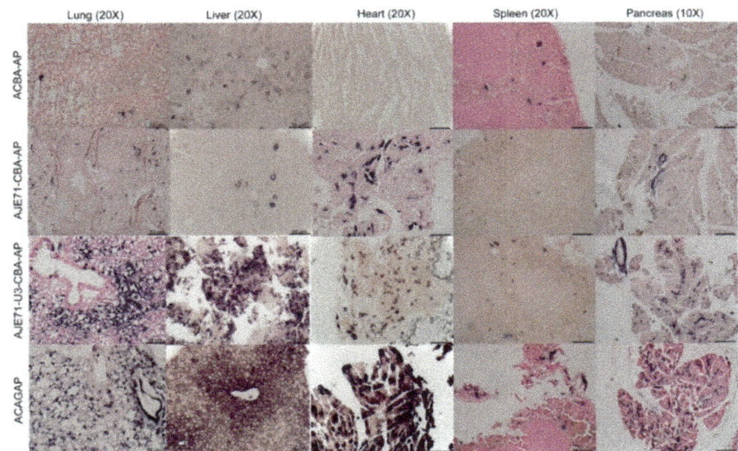

Figure 5. Representative histological alkaline phosphatase staining of tissue sections from mice transduced simultaneously via the intravenous (2×10^{10} vg), intraperitoneal (8×10^{10} vg), and intranasal (1×10^{10} vg) routes with various AAV vectors bearing JE/CBA promoter/enhancer sequences and imaged 4 weeks post-transduction. Scale bar represents 50 µM.

In histological sections, strong AP expression could be observed in lung alveolar cells and liver hepatocytes of mice transduced with a vector that bore the short JE and the CBA (AJE71-CBA-AP), similar to that of ACAGAP (Figure 5). Histological sections of heart transduced with ACAGAP demonstrate very dark staining of individual cells transduced with this vector, compared to the AJE71-U3-CBA-AP, which exhibits a fainter degree of staining (Figure 5). Efficient expression was also observed in pancreatic cells for AJE71-U3-CBA-AP and ACAGAP. Within the spleen, it appeared that no vector was particularly effective.

Quantification of hPLAP enzymatic activity in tissues harvested from the vector transduced mice revealed that in all tissues evaluated, expression from AJE71-U3-CBA-AP was reduced relative to ACAG-AP, but was much higher than ACBA-AP. AJE71-CBA-AP was marginally higher than ACBA-AP but far lower than AJE71-U3-CBA-AP, demonstrating the utility of the JSRV U3 region (Figure 6).

3.5. A Shorter Intron Can Be Used in Conjunction with AJE71-U3-CBA-AP without Reducing Transgene Expression

The MLV env intron (581 bp) was exchanged for a shorter sequence based on an optimized SV40 intron (93 bp) modified to include consensus splice sites. Transfection of HEK 293 cells with the SV40 intron containing construct compared to the parental plasmid containing the MLV intron demonstrated no visible difference in hPLAP expression (Figure 7).

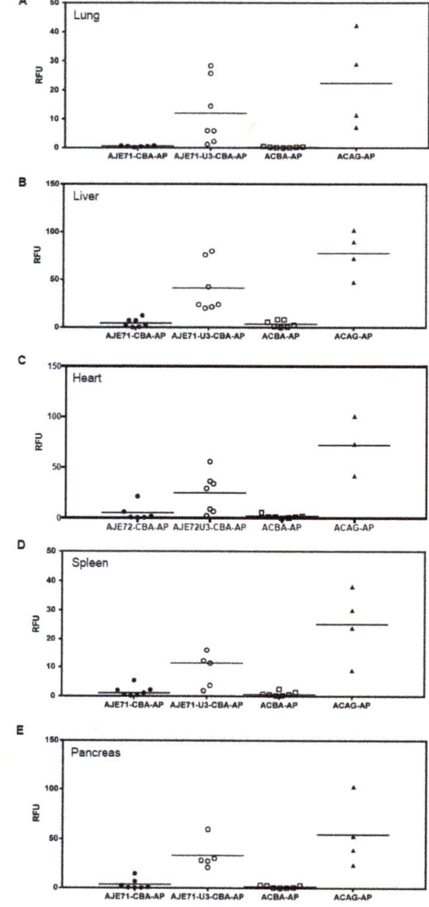

Figure 6. Quantification of alkaline phosphatase activity within homogenized (**A**) lung, (**B**) liver, (**C**) heart, (**D**) spleen and (**E**) pancreas from mice transduced via the intravenous (2×10^{10} vg), intraperitoneal (8×10^{10} vg), and intranasal (1×10^{10} vg) routes with various AAV vectors and harvested for hPLAP enzyme activity analysis 4 weeks post-transduction.

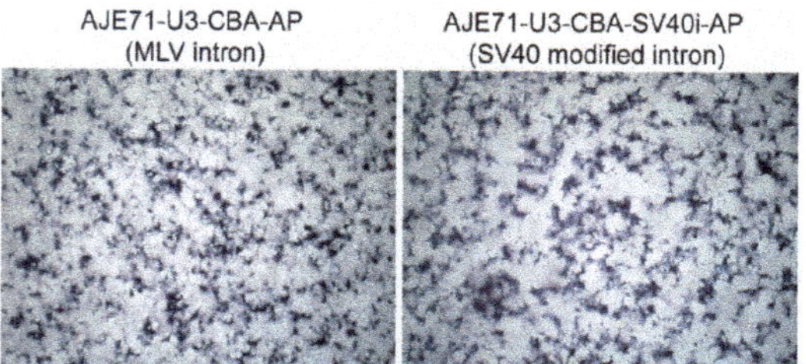

Figure 7. Transfection of two different plasmids in HEK 293 cells, utilizing different introns, employing the JE71-U3-CBA promoter. In the image on the left, an intron derived from MLV is used. In the image on the right, a modified SV40 intron was utilized, reducing the size of the promoter–enhancer–intron combination by 488 bp while maintaining a similar amount of expression. Experiments were conducted in triplicate with three biological replicates. Images were taken at 100× magnification.

4. Discussion

In its native context, within an integrated JSRV provirus, JE may function to enhance expression from the 5′ LTR sequence to drive viral gene expression, or may participate in the process of tumor development by dysregulating the expression of cellular proto-oncogenes.

A lack of expression was observed in constructs bearing only the JE, compared to constructs that possessed the JE and a functional heterologous promoter, the chicken beta actin promoter. Furthermore, inclusion of the JE in any of its varying lengths described here and in either forward or reverse orientations proved to show an increase in protein expression when combined with the chicken beta actin promoter. These characteristics strongly suggest that the JE is able to function as an enhancer sequence [16]. Extension of the JE sequence did not appear to further enhance expression in the cell lines tested, suggesting that no additional enhancer sequences are localized to the area immediately upstream of the 3′ LTR, or at least ones that are active in the cell lines tested. In addition, lengthening of the JE increased the distance between the promoter and inverted terminal repeat of the AAV vector, and no further increase in expression was observed, suggesting that the distance of the promoter from the ITR was not a factor in increasing expression, as previously hypothesized [7]. The particular region where the JE exists overlaps with an RNA export element termed the SPRE or RejRE, which functions in conjunction with a region overlapping the signal peptide of the JSRV envelope to facilitate export of unspliced genomic RNA [17]. However, it is unlikely that the JE functions as an RNA export element in this context as the JSRV envelope is not present in any of the experiments described. Furthermore, there should be no deficit in the ability of transcripts to be exported from the nucleus as they all contain well characterized introns: either the rabbit β-globin intron for the ACAGAP construct or the Moloney murine leukemia virus intron for the other constructs assayed. In addition, the JE sequence should not be transcribed at a high level as it preceded the sequence encoding the promoter in all of the constructs where it was present. Previous attempts at 5′ Rapid Amplification of cDNA Ends (RACE) and RT-PCR were not able to detect the presence of any transcripts except those originating from the R region, when JE71 was placed in front of the JSRV LTR [7].

Low levels of expression observed for the constructs containing JE alone, in either forward or inverted orientation (pAJE71-AP, pAJE184-AP, pAJE324-AP, pAiJE71-AP, pAiJE184AP, and pAiJE324-AP) indicate that JE itself is not able to function effectively as a promoter. This rules out the possibility of JE functioning as a promoter for a transactivating non coding RNA, as is the case in Moloney murine leukemia virus and feline leukemia virus, which both express noncoding RNAs from their 3′ LTRs that are able to transactivate signaling pathways involved in cancer [18–20].

Evidence for a possible expression enhancing region or putative enhancer sequence located in regions flanking the JSRV env gene came to light in a paper by Sinn et al., wherein greatly increased JSRV envelope pseudotyped lentiviral vector titers were observed when env flanking sequences were included in the envelope expression cassette [21]. Previous work with the Prague and Schmidt-Ruppin strains of Rous sarcoma virus has demonstrated that sequences immediately preceding the 3′ LTR are able to enhance expression in reporter gene studies [22,23]. Furthermore, precedence exists for the presence of a region upstream of the 3′ LTR determining the spectrum of disease observed in other retroviruses, such as the exogenous virus-specific region (XSR) sequence of the Prague strain of Rous sarcoma virus, which was demonstrated to be a determinant of oncogenicity [24] and a similar region known as the E region in avian leukosis virus (subgroup J) that has been shown to contribute to oncogenicity in certain chicken breeds [25].

The identity of the transcription factor that binds to the JE has yet to be elucidated; however, one promising candidate is AP-2, for which there are four predicted sites within the JE71 sequence (Figure 8). A wide variety of other transcription factors are also predicted to bind; these include glucocorticoid receptor (GR), for which there are five sites, pituitary-specific positive transcription factor 1 (Pit-1A), for which there are five sites, and c-ETS-2, for which there are four predicted sites. Activation of these pathways through overexpression of transcription factors and/or using pharmacological agents could help to identify which one is responsible for the observed enhancer activity.

Figure 8. Transcription Factors Predicted to Bind to JE-71. A number of different transcription factors have been predicted to bind to the JE-71 sequence using the EMBOSS 6.5.7 tool tfscan and graphically presented by Geneious. Particularly promising putative transcription factors with multiple predicted binding sites include AP-2 (4 sites), glucocorticoid receptor (GR/GR beta, 5 sites), pituitary-specific positive transcription factor 1 (Pit-1A, 5 sites), and c-ETS-2 (4 sites).

A high level of expression could be observed for AJE71-U3-CBA-AP in vivo, nearly matching ACAGAP for some individual mice (Figures 4–6). However, on an aggregate level, expression was generally lower for all of the organs assayed. Amongst the biggest difference was in the heart, where expression was >2-fold higher in ACAGAP compared to AJE71-U3-CBA-AP. This property might make the JE71-U3-CBA promoter particularly useful when widespread, constitutive expression in a variety of organs is called for, but expression in the heart is to be minimized. The high level of expression observed in the respiratory tract and liver may make this promoter effective for respiratory tract diseases such as lung, paranasal sinus, and nasal cavity cancer, or a monogenic disorder such as cystic fibrosis. At approximately 650 bp in size, JE71-U3-CBA is not particularly large for a promoter, and may be combined with a short intron such as an optimized SV40 intron (93 bp) for a total enhancer–promoter–intron size of approximately 743 bp, leaving ~4 kb for the transgene and polyA signal when used in the context of an AAV vector, which has a coding capacity of ~4.7 kb. This might be attractive for the design of an AAV vector based therapeutic employing the truncated CFTR fragment previously shown to be able to rescue the processing of endogenous FΔ508 CFTR in vivo [25]. JE71-U3-CBA and JE71-U3R-CBA enhancer/promoter combinations functioned as effectively as the CAG promoter in the cell lines tested, highlighting the potential of these sequences as alternatives to the commonly used CAG promoter (1621 bp) for in vitro protein expression purposes in mammalian cells.

In summary, we demonstrate the presence of an intragenic transcriptional enhancer element, found within the 3′ end of the env gene. This JSRV-based enhancer element demonstrated strong expression in a variety of tissues, particularly respiratory and hepatic, and promoter/enhancer/intron cassettes derived from these elements can be pared down to a sufficiently small size (<630 bp in length), suitable for genome constraints imposed by AAV vector systems.

Author Contributions: Conceptualization, D.L.Y. and S.K.W.; methodology, D.L.Y. and N.C.; formal analysis, D.L.Y. and S.K.W.; writing—original draft preparation, D.L.Y.; writing—review and editing, S.K.W.; supervision, S.K.W.; project administration, D.L.Y.; funding acquisition, S.K.W. All authors have read and agreed to the published version of the manuscript.

Funding: This research was funded by NSERC Discovery Grant number RGPIN-2018-04737, Cystic Fibrosis Canada grant ID 3017, and the Mason Research Fund.

Acknowledgments: We would like to acknowledge all the veterinary technicians at the Animal Isolation Unit, University of Guelph, for their involvement in the care and maintenance of the animals used in this study.

References

1. Gonçalves, M.A. Adeno-associated virus: From defective virus to effective vector. *Virol. J.* **2005**, *2*, 43. [CrossRef] [PubMed]
2. Grieger, J.C.; Samulski, R.J. Adeno-associated virus vectorology, manufacturing, and clinical applications. *Methods Enzymol.* **2012**, *507*, 229–254. [CrossRef] [PubMed]
3. Bainbridge, J.W.; Mehat, M.S.; Sundaram, V.; Robbie, S.J.; Barker, S.E.; Ripamonti, C.; Georgiadis, A.; Mowat, F.M.; Beattie, S.G.; Gardner, P.J.; et al. Long-term effect of gene therapy on Leber's congenital amaurosis. *N. Engl. J. Med.* **2015**, *372*, 1887–1897. [CrossRef] [PubMed]
4. Naso, M.F.; Tomkowicz, B.; Perry, W.L.; Strohl, W.R. Adeno-Associated Virus (AAV) as a Vector for Gene Therapy. *BioDrugs* **2017**, *31*, 317–334. [CrossRef]
5. Palmarini, M.; Fan, H. Molecular biology of jaagsiekte sheep retrovirus. *Curr. Top. Microbiol. Immunol.* **2003**, *275*, 81–115.
6. Rosales Gerpe, M.C.; van Lieshout, L.P.; Domm, J.M.; Ingrao, J.C.; Datu, J.; Walsh, S.R.; Yu, D.L.; Jong, J.; Krell, P.J.; Wootton, S.K. The U3 and Env Proteins of Jaagsiekte Sheep Retrovirus and Enzootic Nasal Tumor Virus Both Contribute to Tissue Tropism. *Viruses* **2019**, *11*, 1061. [CrossRef]
7. Yu, D.L.; Linnerth-Petrik, N.M.; Halbert, C.L.; Walsh, S.R.; Miller, A.D.; Wootton, S.K. Jaagsiekte sheep retrovirus and enzootic nasal tumor virus promoters drive gene expression in all airway epithelial cells of mice but only induce tumors in the alveolar region of the lungs. *J. Virol.* **2011**, *85*, 7535–7545. [CrossRef]
8. Rasheed, S.; Nelson-Rees, W.A.; Toth, E.M.; Arnstein, P.; Gardner, M.B. Characterization of a newly derived human sarcoma cell line (HT-1080). *Cancer* **1974**, *33*, 1027–1033. [CrossRef]
9. Palmarini, M.; Sharp, J.M.; de las Heras, M.; Fan, H. Jaagsiekte sheep retrovirus is necessary and sufficient to induce a contagious lung cancer in sheep. *J. Virol.* **1999**, *73*, 6964–6972. [CrossRef]
10. Miller, A.D.; Chen, F. Retrovirus packaging cells based on 10A1 murine leukemia virus for production of vectors that use multiple receptors for cell entry. *J. Virol.* **1996**, *70*, 5564–5571. [CrossRef]
11. Sambrook, J.; Russell, D. *Molecular Cloning: A Laboratory Manual*, 3rd ed.; Cold Spring Harbor Lab. Press: Plainview, NY, USA, 2001.
12. Miller, J.H. *Experiments in Molecular Genetics*; Cold Spring Harb. Lab. Press: Plainview, NY, USA, 1972.
13. van Lieshout, L.P.; Domm, J.M.; Wootton, S.K. AAV-Mediated Gene Delivery to the Lung. *Methods Mol. Biol.* **2019**, *1950*, 361–372. [CrossRef]
14. Aurnhammer, C.; Haase, M.; Muether, N.; Hausl, M.; Rauschhuber, C.; Huber, I.; Nitschko, H.; Busch, U.; Sing, A.; Ehrhardt, A.; et al. Universal real-time PCR for the detection and quantification of adeno-associated virus serotype 2-derived inverted terminal repeat sequences. *Hum. Gene Ther. Methods* **2012**, *23*, 18–28. [CrossRef]
15. Santry, L.A.; Ingrao, J.C.; Yu, D.L.; de Jong, J.G.; van Lieshout, L.P.; Wood, G.A.; Wootton, S.K. AAV vector distribution in the mouse respiratory tract following four different methods of administration. *BMC Biotechnol.* **2017**, *17*, 43. [CrossRef]
16. Banerji, J.; Rusconi, S.; Schaffner, W. Expression of a beta-globin gene is enhanced by remote SV40 DNA sequences. *Cell* **1981**, *27*, 299–308. [CrossRef]
17. Caporale, M.; Arnaud, F.; Mura, M.; Golder, M.; Murgia, C.; Palmarini, M. The signal peptide of a simple retrovirus envelope functions as a posttranscriptional regulator of viral gene expression. *J. Virol.* **2009**, *83*, 4591–4604. [CrossRef] [PubMed]
18. Choi, S.; Faller, D. The long terminal repeats of a murine retrovirus encode a trans-activator for cellular genes. *J. Biol. Chem.* **1994**, *269*, 19691–19694. [PubMed]
19. Choi, S.; Faller, D. A transcript from the long terminal repeats of a murine retrovirus associated with trans activation of cellular genes. *J. Virol.* **1995**, *69*, 7054–7060. [CrossRef]
20. Forman, L.; Pal-Ghosh, R.; Spanjaard, R.; Faller, D.; Ghosh, S. Identification of LTR-specific small non-coding RNA in FeLV infected cells. *FEBS Lett.* **2009**, *583*, 1386–1390. [CrossRef] [PubMed]
21. Sinn, P.; Burnight, E.; Shen, H.; Fan, H.; McCray, P.J. Inclusion of Jaagsiekte sheep retrovirus proviral elements markedly increases lentivirus vector pseudotyping efficiency. *Mol. Ther.* **2005**, *11*, 460–469. [CrossRef]
22. Luciw, P.; Bishop, J.; Varmus, H.; Capecchi, M. Location and function of retroviral and SV40 sequences that enhance biochemical transformation after microinjection of DNA. *Cell* **1983**, *33*, 705–716. [CrossRef]
23. Laimins, L.; Tsichlis, P.; Khoury, G. Multiple enhancer domains in the 3′ terminus of the Prague strain of Rous sarcoma virus. *Nucleic Acids Res.* **1984**, *12*, 6427–6442. [CrossRef]

24. Tsichlis, P.; Donehower, L.; Hager, G.; Zeller, N.; Malavarca, R.; Astrin, S.; Skalka, A. Sequence comparison in the crossover region of an oncogenic avian retrovirus recombinant and its nononcogenic parent: Genetic regions that control growth rate and oncogenic potential. *Mol. Cell. Biol.* **1982**, *2*, 1331–1338. [CrossRef] [PubMed]

25. Chesters, P.; Smith, L.; Nair, V. E (XSR) element contributes to the oncogenicity of Avian leukosis virus (subgroup J). *J. Gen. Virol.* **2006**, *87*, 2685–2692. [CrossRef] [PubMed]

Publisher's Note: MDPI stays neutral with regard to jurisdictional claims in published maps and institutional affiliations.

Review

Viral Related Tools against SARS-CoV-2

Laura Fernandez-Garcia [1,2,†], Olga Pacios [1,2,†], Mónica González-Bardanca [1,2], Lucia Blasco [1,2], Inés Bleriot [1,2], Antón Ambroa [1,2], María López [1,2], German Bou [1,2,3] and Maria Tomás [1,2,3,*]

[1] Microbiology Department-Research Institute Biomedical A Coruña (INIBIC), Hospital A Coruña (CHUAC), University of A Coruña (UDC), 15006 A Coruña, Spain; laugemis@gmail.com (L.F.-G.); olgapacios776@gmail.com (O.P.); monica.gonzalez.bardanca@sergas.es (M.G.-B.); luciablasco@gmail.com (L.B.); bleriot.ines@gmail.com (I.B.); anton17@mundo-r.com (A.A.); maria.lopez.diaz@sergas.es (M.L.); German.Bou.Arevalo@sergas.es (G.B.)

[2] Study Group on Mechanisms of Action and Resistance to Antimicrobials (GEMARA) of Spanish Society of Infectious Diseases and Clinical Microbiology (SEIMC), 28003 Madrid, Spain

[3] Spanish Network for the Research in Infectious Diseases (REIPI), 41071 Sevilla, Spain

* Correspondence: MA.del.Mar.Tomas.Carmona@sergas.es; Tel.: +34-981-176-399; Fax: +34-981-178-273

† These authors contributed equally to this work.

Received: 18 September 2020; Accepted: 15 October 2020; Published: 16 October 2020

Abstract: At the end of 2019, a new disease appeared and spread all over the world, the COVID-19, produced by the coronavirus SARS-CoV-2. As a consequence of this worldwide health crisis, the scientific community began to redirect their knowledge and resources to fight against it. Here we summarize the recent research on viruses employed as therapy and diagnostic of COVID-19: (i) viral-vector vaccines both in clinical trials and pre-clinical phases; (ii) the use of bacteriophages to find antibodies specific to this virus and some studies of how to use the bacteriophages themselves as a treatment against viral diseases; and finally, (iii) the use of CRISPR-Cas technology both to obtain a fast precise diagnose of the patient and also the possible use of this technology as a cure.

Keywords: SARS-CoV-2; COVID-19; phages; CRISPR; viruses; prevention; diagnosis; treatment

---

1. Introduction

At the end of 2019, a new virus appeared in the city of Wuhan (China) and quickly spread throughout the world, causing a global pandemic. The virus is closely related to the SARS-CoV (Severe Acute Respiratory Syndrome Coronavirus), thus named SARS-CoV-2 [1,2]. It is a β-coronavirus, carrying single-stranded, positive-sense RNA genome and four main structural proteins: spike (with two subunits, S1 and S2), envelope, membrane and nucleocapsid (N) [3]. SARS-CoV-2, as SARS-CoV, enters the cell through the receptor-binding domain (RBD) of the S1, which recognizes the angiotensin-converting protein 2 (ACE2), present in the surface of host cells [4]. SARS-CoV-2 provokes COVID-19, a new disease that produces a wide range of symptoms ranging from an asymptomatic carrier state to respiratory distress syndrome and even acute heart injury with the risk of secondary infections [5]. The rapid spread of the virus and the absence of treatment for this new disease have led researchers all over the world to join forces in the search for a solution by using all available resources.

Since the beginning of this pandemic, all the medical resources were focused on two main points: diagnostic and treatment of the disease. The diagnosis of SARS-CoV-2 was firstly based on molecular approaches [6], the real-time RT-PCR assay has become the election method to detect the presence of the virus as it is a specific and sensitive method to disclose viral RNA from respiratory tract samples [7]. In order to establish the presence of the virus, following the WHO's indications, clinical laboratories from all around the world are using various primer pairs: the spike gene, the RNA-dependent RNA polymerase gene (RdRp), the nucleocapsid gene and the envelope gene [8]. Besides, some serologic analyses have been used to diagnose an active or past infection by quantifying the presence of IgM and IgG in the patient serum [8]. An interesting systematic review and meta-analysis has been carried out concerning the serological assays [9]. Authors concluded that the sensitivity of this technique was higher three weeks after the symptom onset, compared with the first week, and that heterogeneity was found in all analyses. Among the advantages of serological assays, we find that they are cheaper and easier to implement at the point of care, but, above all, they can identify asymptomatic individuals previously infected by SARS-CoV-2. Moreover, serological tests could be deployed as surveillance tools to better understand the epidemiology of SARS-CoV-2. Many serological tests for Covid-19 have become available in a very short period, and this is precisely where their main disadvantage resides: the pace of development of serological tests has been so fast that it has exceeded that of rigorous evaluation. Therefore, uncertainty about the accuracy of serological assays remains important [9].

Concerning the treatment of this disease, and due to its rapid development, finding an effective treatment against it was imperative. Thus, researchers and medical doctors began to test existing medicines and repurposing them as COVID-19 treatments, highlighting: (i) nucleoside analogs, as favipiravir (used for influenza virus, Ebola, chikungunya, yellow fever, enterovirus and norovirus treatment) [10,11], ribavirin (used for treating the respiratory syncytial virus, hepatitis C virus and also against SARS and MERS) [12], remdesivir (used for HIV treatment) [11,13] or galidesivir [14]; (ii) antiparasitics as chloroquine (used against malaria, with positive in vitro results against SARS, MERS, Ebola, HIV, Nipah and Hendra viruses, although no protection was found in vivo against these viruses) [15–18]; (iii) protease inhibitors (lopinavir and ritonavir used as HIV treatments) [19]; (iv) indole-derivate molecules as arbidol (used against hepatitis viruses) [20]; and finally, (v) convalescent plasma therapy from patients who recovered from the infection [21].

Paradoxically, an efficient prevention strategy to combat SARS-CoV-2 could come from different human viruses, e.g., in the form of a vaccine vector. A virus is known as an extremely small infective particle, which can only replicate inside a host. Since their discovery, they have been identified as the cause of a great number of diseases, but more recently, they have also been considered a solution for some of them [22]. Viruses can be genetically modified to express antigens of interest, turning them into efficient vectors that deliver immunogenic particles inside the human body [23]. The usefulness of the viral vectors is based on: (i) their high specificity for their targets, (ii) their ability for gene transduction and (iii) their capacity to generate strong cellular and humoral immune responses without an adjuvant [22]. Besides, all the viruses used as vectors are genetically modified to eliminate their replicative capacity and to decrease or eradicate their pathogenicity. However, a potential problem with viral vectors is the pre-existing immunity, due to previous viral exposure [22].

Indeed, there are viruses able to specifically infect bacteria as well. These are called bacteriophages, and they can also represent an interesting tool useful in the analysis of SARS-CoV-2, in the diagnostic of the disease and in its treatment. Bacteriophages (also known as "phages") are the natural predators of bacteria, highly specific: They recognize the bacterial receptors on the surface of the prokaryotic cells and strongly attach to them [24]. Since the discovery of bacteriophages in 1915 [25], they have been used as an alternative treatment for critical bacterial infections, on some occasions even life-threatening [26,27]. In the last decade, i.e., in the post-antibiotic era, the therapy based on lytic phages (phage therapy) or phage derived proteins (enzybiotics) such as, for instance, phage-encoded endolysins [28], has gained popularity, being one of the few options currently available for infections caused by multi-drug resistant (MDR) bacteria [26,29]. Phages have demonstrated their innocuousness for humans, although some concerns still need to be investigated such as the purity of the preparation [26]. However, this is far from being the only use for phages; they might be a good option to isolate neutralizing antibodies against other infectious diseases, caused by parasites [30] or viruses [31], using the phage display technique.

Highly related to bacteriophages are the Cluster Regulatory Interspaced Palindromic Repeats (CRISPR), discovered in 1993 and firstly named as short regulatory repeats (SRSRs) [32]. It was years later when their function as a bacterial immunity system against bacteriophages was reported [33,34].

CRISPR fragments are phage-derived sequences harbored by bacteria in their chromosomes that act as an acquired immunity system in prokaryotes: when a bacterium that has been infected by a bacteriophage is re-infected by the same type of phage, CRISPR-Cas system recognizes the viral DNA/RNA repeated sequences and digests the spacer segments between the repetitions, using the endonuclease activity of Cas (CRISPR associated) proteins [35]. This system has been extensively studied by many scientific researchers from all over the world and belonging to very different domains [36]. The importance of this technology has been increasing in the last decade, and nowadays, it is even possible to replace one DNA fragment by another; therefore, CRISPR-Cas system is currently considered one of the most important tools to genetic edition, treatment of diseases and genetic modification of mammalian cells, among others [36].

Throughout this work, we have analyzed innovative methods of diagnostic and treatment of this new disease, the COVID-19, based on the use of human viruses, bacterial viruses (bacteriophages), or virus-related tools (CRISPR). Due to the novelty of the topic here discussed and the amount of information available, in this review, some articles that have not been peer-reviewed are cited.

2. Human Viruses as Prevention

Nowadays, there are several types of viral vectors depending on the type of virus used: retrovirus [37], lentivirus [38], Sendai virus [39], cytomegalovirus [40], poxvirus [41], adenovirus [42], adeno-associated virus (AAV) [43], among others. These vectors have been used against several diseases such as HIV [44–48], hepatitis [49], tuberculosis [50,51], influenza [52,53] and even cancer [54,55]. The most common viruses used for the development of vaccines against human infectious diseases are poxvirus and adenovirus. Poxviruses were the first viruses ever used as vaccine and so the best known with a safety and efficacy widely demonstrated; on the other hand, adenoviruses have been deeply analyzed especially due to its easy production, great transduction efficiency, a broad spectrum of tropism and their transgene expression [22]. The following studies and/or clinical trials measured an elicited humoral response (quantified by ELISA or Western blot) and a neutralizing response (by neutralization assays using either the live virus or a pseudovirus). Neutralizing antibodies can, as their name implies, neutralize the biological effects of the antigen and interfere with their infectivity without a need for immune cells. Currently developed SARS-CoV- and MERS-CoV-specific neutralizing antibodies include monoclonal antibodies (mAbs), their functional antigen-binding fragment (Fab), the single-chain variable region fragment (scFv), or single-domain antibodies. They target S1-RBD, S1-NTD, or the S2 region, blocking the binding of RBDs to their respective receptors and interfering with S2-mediated membrane fusion or entry into the host cell, thus inhibiting viral infections [56].

One of the best-known poxvirus vectors is the Modified Vaccinia Ankara (MVA), unable to replicate in most mammalian cells, thus becoming a safe vector that expresses antigens which elicit an immune response [57]. MVA has been recently modified by Chiuppesi et al. to co-express SARS-CoV-2 spike (S) and nucleocapsid (N) antigens with the aim of testing its immunogenicity and developing a candidate vaccine against COVID-19. In the study, the authors challenged several mice with two MVA vectors, sMVA-S, and sMVA-N vectors, expressing the S and N antigen, respectively. Both vectors were evaluated in a murine model by co-immunization at different doses, and they observed similar SARS-CoV-2 antigen-specific humoral and cellular immune responses in vaccine groups receiving sMVA-S and sMVA-N alone or in combination. Authors claimed that both vectors expressing the S and N antigens can stimulate potent SARS-CoV-2-specific humoral and cellular immune responses in mice, either expressed isolated or in combination. For neutralizing experiments, they used SARS-CoV-2 pseudovirus and detected neutralizing antibodies in all vaccine groups receiving the S antigen. The authors claimed that these neutralizing responses increased after the booster immunization [57].

The adenoviral vector most commonly used for clinical trials and experimental gene therapy applications is HAdV-C5, abbreviated as Ad5 [58]. The research group of Zhu et al. performed a phase-2 trial using a replication-defective Ad5 expressing the spike glycoprotein of SARS-CoV-2, to assess its level of safety, tolerability, and immunogenicity in a group of healthy adults. This trial did not report serious adverse events within 28 days post-vaccination. They found a peak in specific T-cells at day 14 post-vaccination, whereas the peak in neutralizing antibodies anti-spike occurred at day 28 post-vaccination, detected through both live SARS-CoV-2 virus neutralization and pseudovirus neutralization tests [59].

Nowadays, this vaccine is being tested in humans in a phase-3 clinical trial (Table 1) [60]. Similarly, at the University of Oxford, scientists have designed a chimpanzee adenovirus (ChAdOx1) vectored vaccine encoding a codon-optimized full-length spike protein of SARS-CoV-2 [61]. The authors reported that a single vaccination with ChAdOx1 nCoV-19 was effective in preventing damage to the lungs upon high dose, indicating that vaccination prevents virus replication in the lower respiratory tract, but no reduction in viral shedding from the nose was observed. The biggest limitation of this study was that animals were challenged with a high dose of virus via multiple routes, which does not simulate a realistic human exposure [61]. These researchers performed a phase-1/2 randomized trial in healthy adults and observed that those vaccinated with the ChAdOx1 nCoV-19 (5×10^{10} viral particles) experimented a few mild/moderate secondary effects during the first days after the vaccination. Nevertheless, authors demonstrated that their vaccine is effective with a single-dose, without several adverse reactions, and detected the presence of high levels of neutralizing antibodies as well as spike-specific antibodies 28 days after vaccination [62]. Currently, this vaccine is in a phase-3 clinical trial in different countries (Table 1) [63].

Table 1. Viral-vector vaccine candidates and their current state of development according to the WHO.

Developer Institution	Country/s	Type of Viral-Vector	Current State
University of Oxford/ AstraZeneca	United Kingdom	ChAdOx1-S	Clinical trial Phase 3
Beijing Institute of Biotechnology/ CanSino Biological Inc.	China	Ad5	Clinical trial Phase 2
Janssen Pharmaceutical Companies	Belgium	Ad26	Clinical trial Phase $\frac{1}{2}$
Gamaleya Research Institute	Russia	Adenovirus	Clinical trial Phase 1
ReiThera/LEUKOCARE/Uncercells	Italy/Germany/Belgium	Adenovirus	Clinical trial Phase 1
Institute Pasteur/Themis/Univ. of Pittsburgh CVR/Merck Sharp & Dohme	France/United States	Measles	Clinical trial Phase 1
Medicago Inc.	Canada	Plant-derivated VLP	Clinical trial Phase 1
ID Pharma	Japan	Sendai virus	Pre-clinical
Ankara University	Turkey	Adenovirus	Pre-clinical
Massachusetts General Hospital/Massachusetts Eye and Ear/AveXis	United States	Adenovirus	Pre-clinical
GeoVax/BravoVax	United States/China	MVA	Pre-clinical
German center for infection Research/IDT Biologike GmbH	Germany	MVA	Pre-clinical
IDIBAPS-Hospital clinic	Spain	MVA	Pre-clinical
Altimmune	United States	Adenovirus	Pre-clinical
Erciyes University	Turkey	Ad5	Pre-clinical
ImmunityBio Inc/NantKwest Inc.	United States	Ad5	Pre-clinical
Greffex	United States	Ad5	Pre-clinical
Stabilitech Biopharma Ltd.	United Kingdom	Ad5	Pre-clinical
Valo Therapeutics Ltd.	United Kingdom	Adenovirus	Pre-clinical
Vaxart	United States	Ad5	Pre-clinical
National Biotechnology Center (CNB-CSIC)	Spain	MVA	Pre-clinical
University of Georgia/University of Iowa	United States	Parainfluenza virus	Pre-clinical

Table 1. *Cont.*

Bharat Biotech/Thomas Jefferson University	India/United States	Rabies virus	Pre-clinical
National Research Centre	Egypt	Influenza A	Pre-clinical
National Center for Genetic Engineering and Biotechnology (BIOTEC)/ GPO	Thailand	Flu virus	Pre-clinical
KU Leuven	Belgium	YF17D	Pre-clinical
Cadila Healthcare Limited	India	Measles	Pre-clinical
FBRI SRC VB Vector/Rospotrebnadzor	Russia	Measles	Pre-clinical
German center for infection Research/CanVirex AG	Germany	Measles	Pre-clinical
Tonix Pharma/Southern Research	United States	Horsepox	Pre-clinical
BiOCAD/ IEM	Russia	Influenza	Pre-clinical
FBRI SRC VB Vector/Rospotrebnadzor	Russia	Influenza A	Pre-clinical
Fundação Oswaldo Cruz/ Instituto Buntantan	Brazil	Influenza	Pre-clinical
University of Hong Kong	China	Influenza	Pre-clinical
IAVI/ Merk	Italy/United States	VSV	Pre-clinical
University of Manitoba	Canada	VSV	Pre-clinical
University of Western Ontario	United States	VSV	Pre-clinical
Aurobindo Pharma	India	VSV	Pre-clinical
FBRI SRC VB Vector/Rospotrebnadzor	Russia	VSV	Pre-clinical
Israel Institute for Biological Research/Weizman Institute of Science	Israel	VSV	Pre-clinical
UW-Madison/FluGen/Bharat Biotech	United States	Influenza	Pre-clinical
Intravacc/Wageningen Bioveterinary Research/Utrecht University	The Netherlands	Newcastle disease virus	Pre-clinical
The Lancaster University	United Kingdom	Avian paramyxovirus	Pre-clinical
University of Manitoba	Canada	VLP	Pre-clinical
Bezmialem Vakif University	Turkey	VLP	Pre-clinical
Middle East Technical University	Turkey	VLP	Pre-clinical
VBI Vaccines Inc.	United States	VLP	Pre-clinical
IrsiCaixa AIDS Research/IRTA-CReSA/Barcelona Supercomputing Centre/Grifols	Spain	VLP	Pre-clinical
Mahidol University/The Government Pharmaceutical Organization (GPO)/Siriraj Hospital	Thailand	VLP	Pre-clinical
Navarrabiomed, Oncoinmunology group	Spain	VLP	Pre-clinical
Saiba GmbH	Switzerland	VLP	Pre-clinical
Imophoron Ltd. and Bristol University's Max Planck Centre	United Kingdom	VLP	Pre-clinical
Doherty Institute	Australia	VLP	Pre-clinical
OSIVAX	France	VLP	Pre-clinical
ARTES Biotechnology	Germany	VLP	Pre-clinical
University of Sao Paulo	Brazil	VLP	Pre-clinical

VLP—Virus-like particle; VSV—vesicular stomatitis virus.

Consistently with their preliminary results, Mercado et al. immunized several rhesus macaques with another adenoviral vector (Ad26) expressing also the spike protein. However, the immunogen that they used was the full-length membrane-bound S protein with a mutation of the furin cleavage site and two proline stabilizing mutations [64]. They reported a robust immune response based on neutralizing antibodies, obtaining complete protection against the SARS-CoV-2 challenge in 5 out of 6 animals [65]. Based on the previous results, they are now performing a phase-1/2 trial in healthy adults in which they are going to administrate two intramuscular doses of the vaccine (Ad26COVS1) [65,66].

In the same context, the Gamaleya Research Institute of Russia had performed two phase-1 clinical trials with adeno-based vaccines. In these clinical trials, they are going to test the safety of two different vaccines, one based in Ad26 and the other in Ad5, both containing the Spike protein of the SARS-CoV-2 and the lyophilizate of the two mentioned above, for the preparation of a solution for intramuscular injection [67,68]. The combination of Ad26 and Ad5 expressing the spike protein is now in phase-3 [67,68].

Moreover, companies from Italy, Germany, and Belgium have joined forces to develop a simian adenoviral vector-based vaccine that expressed the S protein of the SARS-CoV-2, whose phase 1 clinical trial has begun in Italy this summer [69].

Moreover, the Pasteur Institute in collaboration with two companies and the University of Pittsburgh have developed a Measles-vector vaccine expressing a modified surface glycoprotein of the SARS-CoV-2. This vaccine candidate is nowadays in phase 1 clinical trial, to test the safety, tolerability and immunogenicity of a vaccine that is going to be administrated intramuscularly in two doses separated by 28 days in 90 healthy adults [70]. Furthermore, Medicago Inc. has developed another phase-1 trial testing a virus-like particle vaccine that will be injected into healthy adults with or without an adjuvant, trying different doses of the vaccine [71].

Xiamen University is developing an intranasal spray viral-vector vaccine, based on influenza A virus expressing the spike protein (Table 1). This is currently in a phase-1 clinical trial, and the spray is being nasally administered in one dose in 60 healthy adults [72].

Apart from all these clinical trials, according to the World Health Organization (WHO), nowadays there are 49 viral-vector candidates in pre-clinical evaluation: 9 using adenoviruses, 4 using MVA, 7 using influenza A virus, 3 using Measles virus, 5 using VSV (vesicular stomatitis virus), 7 using other viruses, and 12 using virus-like particles [73] (Table 1).

3. Bacteriophages

As detailed above, bacteriophages are the natural viruses of bacteria that have been used to treat diseases for a long time. Therefore, in 1988, de la Cruz et al. modified the filamentous phage F1 from *Escherichia coli* in order to express repetitive regions from the circumsporozoite protein of *Plasmodium falciparum* [74]. The resulting phages displayed the recombinant protein on their capsid surface and were found to act as carriers capable of producing immunological responses in rabbits. This is one example of how one of the oldest and most abundant entities on Earth has been turned into a powerful therapeutic weapon. Since then, researchers have been investigating the potential of phages in the fight against other infectious diseases. Phage display libraries are a remarkably useful tool that allows the identification of the best ligands for a given target [75], permitting the construction of large libraries consisting of numerous antibody genes [76]. This type of libraries has been used since 1992 to identify specific monoclonal antibodies (mAb) against certain bacteria or viruses [77,78], and a high percentage of human therapeutic antibodies have been developed by this technique [79]. In the past, there have been several examples of phage display libraries expressing viral peptides that have successfully inhibited infections, for instance, the ones caused by adenovirus type 2 [80], hepatitis B virus [81], hantavirus, sin nombre virus [82], and Andes virus [83]. This justifies the use of these libraries as a diagnostic and treatment tools of SARS-CoV-2 (Figure 1).

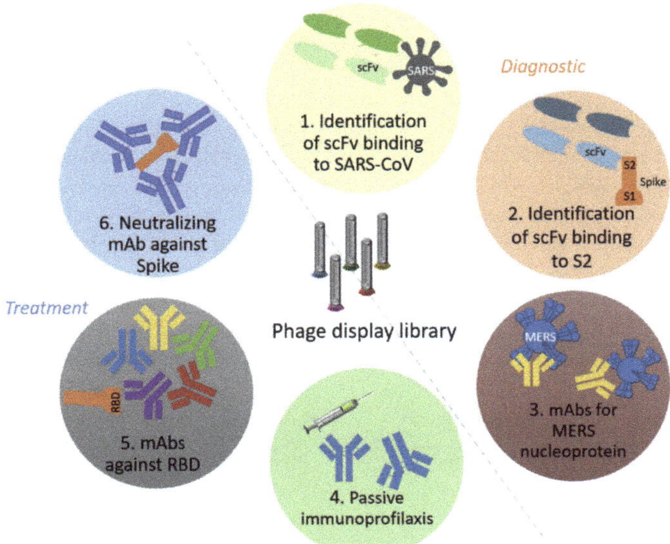

Figure 1. Uses of phage-display libraries in the diagnostic (1, 2 y 3) and treatment (4, 5, and 6) of SARS-CoV-2. scFv: single-chain variable fragment. S2: spike subunit 2. mAbs: monoclonal antibodies. MERS: Middle-East respiratory syndrome. RBD: receptor-binding protein.

3.1. Bacteriophages as Diagnostic Tools: Phage-Display Libraries

Phage display is a powerful technique for the identification and isolation of peptides or proteins [76]. This technique consists of expressing foreign peptides on the surface of bacteriophages, frequently filamentous bacteriophages isolated from *E. coli*, but not exclusively. Indeed, phagemids are the most commonly used vector in phage display technique: these filamentous-phage-derived vectors contain the replication origin of a plasmid, a selective marker, the intergenic region (usually containing the packing sequence), a gene of a phage coat protein, restriction enzyme recognition sites, a promoter and a DNA segment encoding a signal peptide [84]. Phagemids have small genomes, which makes them suitable to accommodate larger foreign DNA fragments. Moreover, they are more efficient in transformation, which allows for obtaining a phage display library with high diversity. A variety of restriction enzyme recognition sites are available in the genome of phagemids, which is convenient for DNA recombination and gene manipulation. Furthermore, the expression level of fusion proteins can be easily controlled and, finally, phagemids are usually genetically more stable than recombinant phages [84].

Nowadays, the most common phages used are the M13 (*Inoviridae*), T4 (*Myoviridae*), T7 (*Podoviridae*) and Lambda (*Syphoviridae*) [85]. Most of the proteins are displayed as fusion proteins with the N- or C-terminus of different phage surface proteins: coat proteins pIII or pVIII on M13 [86], capsid proteins HOC (highly antigenic outer capsid) or SOC (small outer capsid) on T4 [87], the capsid protein pX on T7 [88] and the head protein pD or the tail protein pV on Lambda phage [89]. To select a specific mAb from the library, phages must be subjected to a multiple-cycle process, and after each one, antibodies showing the highest affinity are chosen for the next round. During each cycle, the library is incubated with the target, previously immobilized on a solid support, washed, eluted, amplified and reselected [90]. In the case of phage display libraries expressing mAbs, these are often quantified by ELISA or similar techniques [91]. In the context of the current COVID-19 pandemic, it is worthy to consider that the use of this method has led to major discoveries concerning highly related coronavirus, like SARS or MERS: (i) identification of two single-chain fragment variable (scFv) antibodies that are highly specific for SARS-CoV [92] and (ii) proposal of a new method, called "Yin-Yang", for selecting mAbs by using crude antigens of MERS-CoV (Middle East Respiratory Syndrome Coronavirus), which led to the isolation of three mAb against the MERS-CoV nucleocapsid protein [93]. Besides, several researchers have suggested highly specific diagnostic methods that use phage display libraries for the S1 subunit of the SARS-CoV spike protein, with no cross-reaction with other coronaviruses [75].

Phage display libraries have led to several discoveries associated with coronaviruses that have caused serious human diseases in the past, such as SARS-CoV or MERS. One example is the identification of an scFv antibody, called B1, which binds the S2 of the spike protein of the SARS-CoV both in vitro and in vivo, exhibiting a potent neutralizing activity [94]. Moreover, this same technique allowed for the detection of a human Fab (Fragment antigen-binding) molecule against the spike protein of SARS-CoV, named M1A that could be used in passive immunoprophylaxis [95]. However, the Fc (Fragment crystallizable) region of the antibody is needed to enable the development of a proper immune response [96], so structural modifications (as the authors suggested) would be interesting in order to enhance its protective and neutralizing capacities. Following with these libraries, five types of mAb against the receptor-binding domain (RBD) of the SARS-CoV-2 spike protein have been identified [97]. Finally, phage display libraries lead to the identification of two important mAbs, one of which neutralizes the RBD of the SARS-CoV-2 [98] (Figure 1).

3.2. Bacteriophages as Treatments

In addition to the phage-libraries, other strategies employing phages have been developed as treatments. Regarding the coronavirus type, Ren et al. developed phages that bear specific gastroenteritic coronavirus peptides, which induced humoral and cell-mediated immunity in mice, suggesting that phage-based vaccines may be efficient heterologous antigens for initiating host humoral and cellular immune responses [99]. Furthermore, Lauster et al. modified the icosahedral capsid of Q-beta-phage (Qβ) to display sialic acid ligands that bind to the trimeric haemagglutinin (HA) of the influenza A virus (IAV). These researchers demonstrated that the Qβ-phage capsids can act as highly specific inhibitors of IAV, completely blocking its entry to cells by covering the whole envelope of the virus. However, this method is still undergoing preclinical development [100].

4. CRISPR-Cas

CRISPR-Cas is a bacterial adaptive immune system that was first demonstrated employing a nuclease enzyme (Cas9) that came in 2007 from Barrangou et al. [34]. However, it was Marraffini et al. who proved, in 2008, that CRISPR did not work by RNA interference but by cutting DNA [101]. In parallel, Deveau and Horvath's groups realized that viral DNA was always digested at the same positions upon infection when the bacterium displayed its CRISPR-Cas immunity system [102,103]. Consequently, they claimed that Cas9 catalyzes the digestion of the DNA at precise positions, encoded by specific sequences of "programmable" RNA (CRISPR-RNA or crRNA), which opened the door to the revolution of CRISPR: a molecular tool that allows accurate site-directed digestion in the DNA. CRISPR can also provide a precise, sensitive diagnostic technique as well as an elegant therapeutic option, which has been applied to identify Zika virus [104], human papillomavirus [105], African Swine Fever virus [106], *Staphylococcus aureus* [107] and *Pseudomona aeruginosa* [108], among others.

4.1. CRISPR-Cas as a Molecular Tool of Diagnostic of COVID-19

In the last few months, several projects related to CRISPR have appeared or have been modified in response to the current crisis caused by the COVID-19 pandemic [109]. All these techniques use mainly the Cas13 and Cas12 proteins because of their capacity to cut single-strands of either DNA or RNA [110]. Most of CRISPR based techniques have been developed to use LAMP or RT_LAMP (Reverse transcription loop-mediated isothermal amplification). This technique was developed to simplify the PCR process, with shorter reaction times and no need for specific equipment [111]. Besides, these methods can be developed without high technology or difficulties, allowing the technicians to perform the diagnostic of the disease directly in the sample collection points. Among this research, we highlight six main diagnostic tests using CRISPR technology (Table 2):

(i)   SHERLOCK: **S**pecific **H**igh-sensitivity **E**nzymatic **R**eporter un**LOCK**ing. This technique uses the RNAse activity of the CRISPR-Cas13a protein, which needs only a small specific RNA guide [112]. The system was adapted to a simple test against SARS-CoV-2, called STOPCovid (**S**HERLOCK **T**esting in **O**ne **P**ot), which counts nowadays with two versions: STOPCovid.v1 and STOPCovid.v2 [113]. Both of them use LAMP technique for RNA amplification and can detect up to 100 viral genome copies per reaction in 45–60 min. STOPCovid.v2 uses magnetic beads to simplify the RNA extraction and reduce its duration [113]. Researchers have developed a simple test format that can be performed without complex instrumentation and can detect the virus in saliva samples [114]. This method has been clinically validated by a different research group, who have decreased the limit of detection, thus increasing its sensitivity [115].

(ii)   DETECTR: **D**NA **E**ndonuclease **T**arg**E**ted **C**RISPR **T**rans **R**eporter. This system uses the CRISPR-Cas12a protein to detect SARS-CoV-2 through its nucleoprotein and envelope genes, based on the method of RT-LAMP, which includes a simultaneous retrotranscription process. This technique allows the detection of the virus in naso- and oropharyngeal samples within 30–40 min. The limit of detection is 10 copies per microliter [116].

(iii)   CARMEN: **C**ombinatorial **A**rrayed **R**eactions for **M**ultiplexed **E**valuation of **N**ucleic-acids. This method combines SHERLOCK with microfluidic technology, enabling the analysis of numerous types of samples from patients. The system was developed to detect 169 human-associated viruses, including SARS-CoV-2. Moreover, it can be used for viral detection in several types of samples, ranging from plasma to nasal swab samples [117].

(iv)   AIOD-CRISPR: **A**ll **I**n **O**ne **D**ual CRISPR-Cas12a. This system uses the Cas12a protein in a fast, specific, simple method for the visual detection of SARS-CoV-2 and HIV viruses by the naked eye. This method can also be performed at a single temperature, thus avoiding the need for techniques such as LAMP. It detected 1.3 copies of a plasmid expressing the nucleocapsid protein of SARS-CoV-2, although it has not yet been tested with clinical samples [118].

(v)   CONAN: **C**as3-**O**perated **N**ucleic **A**cid detectio**N**. This CRISPR-based tool employs mainly Cas3 endonuclease, in combination with Cas5, 6, 7, 8, and 11, which mediates targeted DNA cleavage. When combined with isothermal amplification methods, CONAN provides a rapid and sensitive method to detect SARS-CoV-2, with a reliability of 90% [119].

(vi)   CRISPR-COVID: A few months ago, another CRISPR-based tool suitable for the diagnostic of SARS-CoV-2 infection was developed, also based on the Cas13a endonuclease. Scientists claimed that this technique was extremely sensitive and specific, with almost a single-copy sensitivity, as they were able to identify as low as 7.5 copies of viral RNA per reaction in some cases. Furthermore, they did not detect any false positives and the time needed per reaction was only 40 min [120].

Table 2. Comparison of the main characteristics of some novel diagnostic methods for SARS-CoV-2 and the gold standard COVID-19 RT-PCR assay.

	COVID-19 RT-PCR	STOP-Covid [a] (SHERLOCK)	DETECTR	CARMEN	AIOD-CRISPR	CONAN	CRISPR-COVID
Gene Target	Spike protein RdRp Nucleocapsid	Spike ORF1ab Nucleocapsid	Envelop Nucleocapsid	ORF1ab	Nucleocapsid	Nucleocapsid	ORF1ab Nucleocapsid
Sample type	RNA	RNA	DNA	RNA	DNA	DNA	RNA
Assay reaction time	120 min	60 min	30–40 min	~30 min	40 min	30–40 min	40 min
N° of samples/ reaction	1	1	1	1000	1	1	1
Results	Quantitative	Semi-quantitative	Qualitative	Quantitative	Quantitative	Quantitative	Qualitative
Detection limit	>10 viral copies	42 viral copies	10 viral copies	10 viral copies	1.3 copies of SARS-CoV-2 Nucleocapsid gene plasmids	100 viral copies	7.5 viral copies
FDA Approval	Yes	Yes	In process	In process	-	-	-

[a] This section includes both versions, STOPCovid.v1 and STOPCovid.v2. -: without information.

Recently, Fozouni and collaborators developed an innovative technique based on the direct detection of SARS-CoV-2 from nasal swab RNA extracts using an amplification-free CRISPR-Cas13a-based mobile phone assay. The sensitivity of the technique was around 100 copies/µL and the duration under 30 min, being able to detect a set of positive clinical samples in under 5 min [121].

4.2. CRISPR-Cas as a Treatment

CRISPR technology has also been proposed as a treatment for IAV and COVID-19 by using the PAC-MAN method (**P**rophylactic **A**ntiviral **C**RISPR in hu**MAN** cells). This system uses the Cas13d protein, which has RNAse activity, to destroy the highly conserved genomic RNA regions of the coronavirus. Cas13d enzyme effectively inhibited and degraded SARS-CoV-2 viral RNA in respiratory epithelial cells. The authors suggest several possible delivery forms for the Cas13d protein and its RNA guides, such as nanoparticles, a DNA-based liposomal strategy and a ribonucleoprotein complex. Furthermore, Cas13d is capable of processing its RNA guides so that multiple RNAs with different targets can be delivered at the same time, thus increasing the chances of complete viral eradication. Although this approach has produced promising results in the laboratory, it is still at the pre-clinical trial stage and must be tested in animal models before being tested in humans [122].

Other authors have suggested AAV as a suitable delivery vehicle for the Cas13d, as each viral particle can pack more than three RNA guides. Moreover, AAVs are excellent, safe delivery systems, and they also have specific lung cell serotypes, enabling administration via the respiratory route. Nevertheless, this delivery method, like the treatment based on CRISPR-Cas13, is still at the pre-clinical trial stage [123].

5. Discussion

Among this review, we have revised all the viral-based vaccines and viral-related techniques (bacteriophages and CRISPR) that are currently been used in the diagnostic and treatment of SARS-CoV-2. Here, we have summarized all the vaccines that are currently under study (according to WHO), which use viruses as vectors. Moreover, we described the use of phage-display libraries to select monoclonal antibodies specifically against SARS-CoV-2 and how human viruses are used as vectors in vaccines. Finally, in addition to the present techniques, we have reported the new tools that have been developed as new CRISPR diagnosis and treatment methods.

The use of virus-based vaccines has been studied for many years, although until now only one viral-vector vaccine has been approved for use in humans, the rVSV-ZEBOV-GP indicated against Ebola [124]. Nevertheless, some of them are in the final steps of the clinical trials [125]. Despite all the advantages that this kind of vaccines have, they still have some disadvantages such as the pre-existing immunity that can be found against the most common viral-vectors (poxvirus and adenovirus), which might decrease the efficacy of the vaccine [126], or the lack of proper animal models [127]. Nevertheless, in the last months, new murine models have been developed by adding the human ACE-2 receptor to mice by knock-in [128] or transducing the mouse using an adenovirus that expressed the hACE-2 [129]. The viral-vector vaccines are an adequate option in the fight against the COVID-19 disease, being five of the twenty-seven candidate vaccines in a clinical trial to date [73].

However, prophylaxis is not the only way to defeat disease, it is as important to have reliable diagnostic methods and proper treatments. Here we described several new specific diagnostic methods based both in the use of phage-display libraries and CRISPR-Cas. The phage-display libraries are one of the most effective ways to generate a great number of peptides, proteins, or antibodies in a small period [90]. They have been proved useful in the analysis of several autoimmune diseases [130,131] and to produce human antibody therapeutics [132], such as *Helicobacter pylori* [133], *P. aeruginosa* [134], *S. aureus* [135], *Leishmania* [136], Citomegalovirus [137] and Rabies virus [138], among others. However, this technology has a few limitations, like the diversity of the peptides and their quality, which depends on the origin and diversity of the library, as well as on the process employed to evaluate the antibodies [139].

Most of the tools exposed in this review are still under analysis or waiting for their approval, except for the CRISPR-Cas diagnostic systems, some of which are currently accepted with clinical validation. These tools have solved most of the problems that this diagnostic technology had, such as the need for PAM (Protospacer adjacent motif) sequences, quantification of the sample, need to pre-treat the sample or the detection of more than one target per reaction [140]. Interestingly, Fozouni et al. used crRNAs targeting SARS-CoV-2 RNA quantifying viral load using enzyme kinetics, which allows for improvements in the sensitivity and specificity of the diagnosis of COVID19. This innovative assay in combination with mobile phone-based quantification can provide rapid, low-cost, point-of-care screening to aid in the control of SARS-CoV-2 [121]. However, the use of CRISPR technology to treat the disease has to face the main problem of the delivery of the system to the target cells, being proposed several options as phagemids [141] or viruses [142]. Moreover, another problem is the presence of undesirable secondary mutations: although the CRISPR system has a very low frequency of secondary mutations [143], some studies have demonstrated the unnecessary perfect match for the function of the system [144,145].

6. Conclusions

This work reviews the ultimate tools already developed and in process for the diagnosis and treatment of the new disease COVID-19 using human viruses, bacteriophages, and the bacterial immune system CRISPR-Cas. These methods are the next step in the development of more specific and precise diagnostic tools as well as a new point of view in the treatment of this pandemic, but also useful for many other diseases. Despite the rapid outcome of all the studies presented here, their results leave no doubt about their usefulness against the SARS-CoV-2. They represent an extraordinary opportunity to defeat this disease as well as an incredible example of a common effort of the scientific community all around the world.

Author Contributions: Review and writing by L.F.-G. and O.P. (both authors contributed equally to the work); review and correcting by L.B.; review by I.B., A.A., M.L., M.G.-B. and G.B.; supervision, analysis, writing and funding by M.T. All authors have read and agreed to the published version of the manuscript.

Funding: This study was funded by grants PI16/01163 and PI19/00878 awarded to M. Tomás within the State Plan for R+D+I 2013–2016 (National Plan for Scientific Research, Technological Development and Innovation 2008-2011) and co-financed by the ISCIII-Deputy General Directorate for Evaluation and Promotion of Research—European

Regional Development Fund "A way of Making Europe" and Instituto de Salud Carlos III FEDER, Spanish Network for the Research in Infectious Diseases (REIPI, RD16/0016/0006) and by the Study Group on Mechanisms of Action and Resistance to Antimicrobials, GEMARA (SEIMC, http://www.seimc.org/).

Acknowledgments: Spanish Network for the Research in Infectious Diseases (REIPI, RD16/0016/0006) and by the Study Group on Mechanisms of Action and Resistance to Antimicrobials, GEMARA (SEIMC, http://www.seimc.org/).

Conflicts of Interest: All authors declare no conflict of interest.

References

1. Drosten, C.; Günther, S.; Preiser, W.; van der Werf, S.; Brodt, H.R.; Becker, S.; Rabenau, H.; Panning, M.; Kolesnikova, L.; Fouchier, R.A.; et al. Identification of a novel coronavirus in patients with severe acute respiratory syndrome. *N. Engl. J. Med.* **2003**, *348*, 1967–1976. [CrossRef] [PubMed]
2. Ksiazek, T.G.; Erdman, D.; Goldsmith, C.S.; Zaki, S.R.; Peret, T.; Emery, S.; Tong, S.; Urbani, C.; Comer, J.A.; Lim, W.; et al. A novel coronavirus associated with severe acute respiratory syndrome. *N. Engl. J. Med.* **2003**, *348*, 1953–1966. [CrossRef] [PubMed]
3. Cui, J.; Li, F.; Shi, Z.L. Origin and evolution of pathogenic coronaviruses. *Nat. Rev. Microbiol.* **2019**, *17*, 181–192. [CrossRef] [PubMed]
4. Zhou, P.; Yang, X.L.; Wang, X.G.; Hu, B.; Zhang, L.; Zhang, W.; Si, H.R.; Zhu, Y.; Li, B.; Huang, C.L.; et al. A pneumonia outbreak associated with a new coronavirus of probable bat origin. *Nature* **2020**, *579*, 270–273. [CrossRef]
5. Khan, M.N.; Sarker, M. A review of Coronavirus 2019 COVID-19 a life threating disease all over the world. *World Cancer Res. J.* **2020**, *7*, e1586. [CrossRef]
6. Ahn, D.G.; Shin, H.J.; Kim, M.H.; Lee, S.; Kim, H.S.; Myoung, J.; Kim, B.T.; Kim, S.J. Current Status of Epidemiology, Diagnosis, Therapeutics, and Vaccines for Novel Coronavirus Disease 2019 (COVID-19). *J. Microbiol. Biotechnol.* **2020**, *30*, 313–324. [CrossRef]
7. Wang, C.; Horby, P.W.; Hayden, F.G.; Gao, G.F. A novel coronavirus outbreak of global health concern. *Lancet* **2020**, *395*, 470–473. [CrossRef]
8. World Health Organization. Laboratory testing of 2019 novel coronavirus (2019-nCoV) in suspected human cases. In *Interim Guidance*; World Health Organization: Geneva, Switzerland, 2020.
9. Lisboa Bastos, M.; Tavaziva, G.; Abidi, S.K.; Campbell, J.R.; Haraoui, L.P.; Johnston, J.C.; Lan, Z.; Law, S.; MacLean, E.; Trajman, A.; et al. Diagnostic accuracy of serological tests for covid-19: Systematic review and meta-analysis. *BMJ* **2020**, *370*, m2516. [CrossRef]
10. De Clercq, E. New Nucleoside Analogues for the Treatment of Hemorrhagic Fever Virus Infections. *Chem. Asian J.* **2019**, *14*, 3962–3968. [CrossRef]
11. Wang, M.; Cao, R.; Zhang, L.; Yang, X.; Liu, J.; Xu, M.; Shi, Z.; Hu, Z.; Zhong, W.; Xiao, G. Remdesivir and chloroquine effectively inhibit the recently emerged novel coronavirus (2019-nCoV) in vitro. *Cell Res.* **2020**, *30*, 269–271. [CrossRef]
12. Zumla, A.; Chan, J.F.; Azhar, E.I.; Hui, D.S.; Yuen, K.Y. Coronaviruses—Drug discovery and therapeutic options. *Nat. Rev. Drug Discov.* **2016**, *15*, 327–347. [CrossRef] [PubMed]
13. Sheahan, T.P.; Sims, A.C.; Graham, R.L.; Menachery, V.D.; Gralinski, L.E.; Case, J.B.; Leist, S.R.; Pyrc, K.; Feng, J.Y.; Trantcheva, I.; et al. Broad-spectrum antiviral GS-5734 inhibits both epidemic and zoonotic coronaviruses. *Sci. Transl. Med.* **2017**, *9*. [CrossRef]
14. Costanzo, M.; De Giglio, M.A.R.; Roviello, G.N. SARS-CoV-2: Recent Reports on Antiviral Therapies Based on Lopinavir/Ritonavir, Darunavir/Umifenovir, Hydroxychloroquine, Remdesivir, Favipiravir and Other Drugs for the Treatment of the New Coronavirus. *Curr. Med. Chem.* **2020**, *27*. [CrossRef] [PubMed]
15. Neely, M.; Kalyesubula, I.; Bagenda, D.; Myers, C.; Olness, K. Effect of chloroquine on human immunodeficiency virus (HIV) vertical transmission. *Afr. Health Sci.* **2003**, *3*, 61–67. [PubMed]
16. Vincent, M.J.; Bergeron, E.; Benjannet, S.; Erickson, B.R.; Rollin, P.E.; Ksiazek, T.G.; Seidah, N.G.; Nichol, S.T. Chloroquine is a potent inhibitor of SARS coronavirus infection and spread. *Virol. J.* **2005**, *2*, 69. [CrossRef] [PubMed]

17. Freiberg, A.N.; Worthy, M.N.; Lee, B.; Holbrook, M.R. Combined chloroquine and ribavirin treatment does not prevent death in a hamster model of Nipah and Hendra virus infection. *J. Gen. Virol.* **2010**, *91*, 765–772. [CrossRef] [PubMed]

18. Dowall, S.D.; Bosworth, A.; Watson, R.; Bewley, K.; Taylor, I.; Rayner, E.; Hunter, L.; Pearson, G.; Easterbrook, L.; Pitman, J.; et al. Chloroquine inhibited Ebola virus replication in vitro but failed to protect against infection and disease in the in vivo guinea pig model. *J. Gen. Virol.* **2015**, *96*, 3484–3492. [CrossRef]

19. Chu, C.M.; Cheng, V.C.; Hung, I.F.; Wong, M.M.; Chan, K.H.; Chan, K.S.; Kao, R.Y.; Poon, L.L.; Wong, C.L.; Guan, Y.; et al. Role of lopinavir/ritonavir in the treatment of SARS: Initial virological and clinical findings. *Thorax* **2004**, *59*, 252–256. [CrossRef]

20. Boriskin, Y.S.; Leneva, I.A.; Pecheur, E.I.; Polyak, S.J. Arbidol: A broad-spectrum antiviral compound that blocks viral fusion. *Curr. Med. Chem.* **2008**, *15*, 997–1005. [CrossRef]

21. Duan, K.; Liu, B.; Li, C.; Zhang, H.; Yu, T.; Qu, J.; Zhou, M.; Chen, L.; Meng, S.; Hu, Y.; et al. Effectiveness of convalescent plasma therapy in severe COVID-19 patients. *Proc. Natl. Acad. Sci. USA* **2020**, *117*, 9490–9496. [CrossRef]

22. Ura, T.; Okuda, K.; Shimada, M. Developments in Viral Vector-Based Vaccines. *Vaccines* **2014**, *2*, 624–641. [CrossRef] [PubMed]

23. Chen, Y.H.; Keiser, M.S.; Davidson, B.L. Viral Vectors for Gene Transfer. *Curr. Protoc. Mouse Biol.* **2018**, *8*, e58. [CrossRef]

24. Garretto, A.; Miller-Ensminger, T.; Wolfe, A.J.; Putonti, C. Bacteriophages of the lower urinary tract. *Nat. Rev. Urol.* **2019**, *16*, 422–432. [CrossRef] [PubMed]

25. Twort, F.W. An investigation on the nature of ultra-microscopic viruses. *Lancet* **1915**, *186*, 1241–1243. [CrossRef]

26. Furfaro, L.L.; Payne, M.S.; Chang, B.J. Bacteriophage Therapy: Clinical Trials and Regulatory Hurdles. *Front. Cell Infect. Microbiol.* **2018**, *8*, 376. [CrossRef] [PubMed]

27. Schooley, R.T.; Biswas, B.; Gill, J.J.; Hernandez-Morales, A.; Lancaster, J.; Lessor, L.; Barr, J.J.; Reed, S.L.; Rohwer, F.; Benler, S.; et al. Development and use of Personalized Bacteriophage-Based Therapeutic Cocktails to Treat a Patient with a Disseminated Resistant Acinetobacter baumannii Infection. *Antimicrob. Agents Chemother.* **2017**, *61*. [CrossRef]

28. Blasco, L.; Ambroa, A.; Trastoy, R.; Bleriot, I.; Moscoso, M.; Fernández-Garcia, L.; Perez-Nadales, E.; Fernández-Cuenca, F.; Torre-Cisneros, J.; Oteo-Iglesias, J.; et al. In vitro and in vivo efficacy of combinations of colistin and different endolysins against clinical strains of multi-drug resistant pathogens. *Sci. Rep.* **2020**, *10*, 7163. [CrossRef]

29. Pacios, O.; Blasco, L.; Bleriot, I.; Fernandez-Garcia, L.; Gonzalez Bardanca, M.; Ambroa, A.; Lopez, M.; Bou, G.; Tomas, M. Strategies to Combat Multidrug-Resistant and Persistent Infectious Diseases. *Antibiotics* **2020**, *9*. [CrossRef]

30. Goulart, L.R.; da, S.R.V.; Costa-Cruz, J.M. Anti-parasitic Antibodies from Phage Display. *Adv. Exp. Med. Biol.* **2017**, *1053*, 155–171. [CrossRef]

31. Barbas, C.F.; Burton, D.R. Selection and evolution of high-affinity human anti-viral antibodies. *Trends Biotechnol.* **1996**, *14*, 230–234. [CrossRef]

32. Mojica, F.J.; Juez, G.; Rodriguez-Valera, F. Transcription at different salinities of Haloferax mediterranei sequences adjacent to partially modified PstI sites. *Mol. Microbiol.* **1993**, *9*, 613–621. [CrossRef] [PubMed]

33. Mojica, F.J.; Diez-Villasenor, C.; Garcia-Martinez, J.; Soria, E. Intervening sequences of regularly spaced prokaryotic repeats derive from foreign genetic elements. *J. Mol. Evol.* **2005**, *60*, 174–182. [CrossRef] [PubMed]

34. Barrangou, R.; Fremaux, C.; Deveau, H.; Richards, M.; Boyaval, P.; Moineau, S.; Romero, D.A.; Horvath, P. CRISPR provides acquired resistance against viruses in prokaryotes. *Science* **2007**, *315*, 1709–1712. [CrossRef] [PubMed]

35. Barrangou, R.; Marraffini, L.A. CRISPR-Cas systems: Prokaryotes upgrade to adaptive immunity. *Mol. Cell* **2014**, *54*, 234–244. [CrossRef]

36. Lander, E.S. The Heroes of CRISPR. *Cell* **2016**, *164*, 18–28. [CrossRef]

37. Cavazzana-Calvo, M.; Hacein-Bey, S.; de Saint Basile, G.; Gross, F.; Yvon, E.; Nusbaum, P.; Selz, F.; Hue, C.; Certain, S.; Casanova, J.L.; et al. Gene therapy of human severe combined immunodeficiency (SCID)-X1 disease. *Science* **2000**, *288*, 669–672. [CrossRef]

38. Tebas, P.; Stein, D.; Binder-Scholl, G.; Mukherjee, R.; Brady, T.; Rebello, T.; Humeau, L.; Kalos, M.; Papasavvas, E.; Montaner, L.J.; et al. Antiviral effects of autologous CD4 T cells genetically modified with a conditionally replicating lentiviral vector expressing long antisense to HIV. *Blood* **2013**, *121*, 1524–1533. [CrossRef]

39. Slobod, K.S.; Shenep, J.L.; Lujan-Zilbermann, J.; Allison, K.; Brown, B.; Scroggs, R.A.; Portner, A.; Coleclough, C.; Hurwitz, J.L. Safety and immunogenicity of intranasal murine parainfluenza virus type 1 (Sendai virus) in healthy human adults. *Vaccine* **2004**, *22*, 3182–3186. [CrossRef]

40. Hansen, S.G.; Ford, J.C.; Lewis, M.S.; Ventura, A.B.; Hughes, C.M.; Coyne-Johnson, L.; Whizin, N.; Oswald, K.; Shoemaker, R.; Swanson, T.; et al. Profound early control of highly pathogenic SIV by an effector memory T-cell vaccine. *Nature* **2011**, *473*, 523–527. [CrossRef]

41. Rerks-Ngarm, S.; Pitisuttithum, P.; Nitayaphan, S.; Kaewkungwal, J.; Chiu, J.; Paris, R.; Premsri, N.; Namwat, C.; de Souza, M.; Adams, E.; et al. Vaccination with ALVAC and AIDSVAX to prevent HIV-1 infection in Thailand. *N. Engl. J. Med.* **2009**, *361*, 2209–2220. [CrossRef]

42. Coughlan, L.; Bradshaw, A.C.; Parker, A.L.; Robinson, H.; White, K.; Custers, J.; Goudsmit, J.; Van Roijen, N.; Barouch, D.H.; Nicklin, S.A.; et al. Ad5:Ad48 hexon hypervariable region substitutions lead to toxicity and increased inflammatory responses following intravenous delivery. *Mol. Ther.* **2012**, *20*, 2268–2281. [CrossRef] [PubMed]

43. Ferreira, V.; Petry, H.; Salmon, F. Immune Responses to AAV-Vectors, the Glybera Example from Bench to Bedside. *Front. Immunol.* **2014**, *5*, 82. [CrossRef] [PubMed]

44. Xin, K.Q.; Mizukami, H.; Urabe, M.; Toda, Y.; Shinoda, K.; Yoshida, A.; Oomura, K.; Kojima, Y.; Ichino, M.; Klinman, D.; et al. Induction of robust immune responses against human immunodeficiency virus is supported by the inherent tropism of adeno-associated virus type 5 for dendritic cells. *J. Virol.* **2006**, *80*, 11899–11910. [CrossRef]

45. Perreau, M.; Pantaleo, G.; Kremer, E.J. Activation of a dendritic cell-T cell axis by Ad5 immune complexes creates an improved environment for replication of HIV in T cells. *J. Exp. Med.* **2008**, *205*, 2717–2725. [CrossRef]

46. Gomez, C.E.; Najera, J.L.; Perdiguero, B.; Garcia-Arriaza, J.; Sorzano, C.O.; Jimenez, V.; Gonzalez-Sanz, R.; Jimenez, J.L.; Munoz-Fernandez, M.A.; Lopez Bernaldo de Quiros, J.C.; et al. The HIV/AIDS vaccine candidate MVA-B administered as a single immunogen in humans triggers robust, polyfunctional, and selective effector memory T cell responses to HIV-1 antigens. *J. Virol.* **2011**, *85*, 11468–11478. [CrossRef]

47. Chiuppesi, F.; Vannucci, L.; De Luca, A.; Lai, M.; Matteoli, B.; Freer, G.; Manservigi, R.; Ceccherini-Nelli, L.; Maggi, F.; Bendinelli, M.; et al. A lentiviral vector-based, herpes simplex virus 1 (HSV-1) glycoprotein B vaccine affords cross-protection against HSV-1 and HSV-2 genital infections. *J. Virol.* **2012**, *86*, 6563–6574. [CrossRef] [PubMed]

48. Hansen, S.G.; Sacha, J.B.; Hughes, C.M.; Ford, J.C.; Burwitz, B.J.; Scholz, I.; Gilbride, R.M.; Lewis, M.S.; Gilliam, A.N.; Ventura, A.B.; et al. Cytomegalovirus vectors violate CD8 + T cell epitope recognition paradigms. *Science* **2013**, *340*, 1237874. [CrossRef]

49. Cavenaugh, J.S.; Awi, D.; Mendy, M.; Hill, A.V.; Whittle, H.; McConkey, S.J. Partially randomized, non-blinded trial of DNA and MVA therapeutic vaccines based on hepatitis B virus surface protein for chronic HBV infection. *PLoS ONE* **2011**, *6*, e14626. [CrossRef]

50. Tameris, M.D.; Hatherill, M.; Landry, B.S.; Scriba, T.J.; Snowden, M.A.; Lockhart, S.; Shea, J.E.; McClain, J.B.; Hussey, G.D.; Hanekom, W.A.; et al. Safety and efficacy of MVA85A, a new tuberculosis vaccine, in infants previously vaccinated with BCG: A randomised, placebo-controlled phase 2b trial. *Lancet* **2013**, *381*, 1021–1028. [CrossRef]

51. Smaill, F.; Jeyanathan, M.; Smieja, M.; Medina, M.F.; Thanthrige-Don, N.; Zganiacz, A.; Yin, C.; Heriazon, A.; Damjanovic, D.; Puri, L.; et al. A human type 5 adenovirus-based tuberculosis vaccine induces robust T cell responses in humans despite preexisting anti-adenovirus immunity. *Sci. Transl. Med.* **2013**, *5*, 205ra134. [CrossRef]

52. Lin, J.; Calcedo, R.; Vandenberghe, L.H.; Bell, P.; Somanathan, S.; Wilson, J.M. A new genetic vaccine platform based on an adeno-associated virus isolated from a rhesus macaque. *J. Virol.* **2009**, *83*, 12738–12750. [CrossRef] [PubMed]

53. Berthoud, T.K.; Hamill, M.; Lillie, P.J.; Hwenda, L.; Collins, K.A.; Ewer, K.J.; Milicic, A.; Poyntz, H.C.; Lambe, T.; Fletcher, H.A.; et al. Potent CD8 + T-cell immunogenicity in humans of a novel heterosubtypic influenza A vaccine, MVA-NP+M1. *Clin. Infect. Dis.* **2011**, *52*, 1–7. [CrossRef]

54. Carter, B.J. Adeno-associated virus vectors in clinical trials. *Hum. Gene Ther.* **2005**, *16*, 541–550. [CrossRef]

55. Chan, W.M.; Rahman, M.M.; McFadden, G. Oncolytic myxoma virus: The path to clinic. *Vaccine* **2013**, *31*, 4252–4258. [CrossRef]

56. Jiang, S.; Hillyer, C.; Du, L. Neutralizing Antibodies against SARS-CoV-2 and Other Human Coronaviruses. *Trends Immunol.* **2020**, *41*, 355–359. [CrossRef]

57. Chiuppesi, F.; Salazar, M.D.; Contreras, H.; Nguyen, V.H.; Martinez, J.; Park, S.; Nguyen, J.; Kha, M.; Iniguez, A.; Zhou, Q.; et al. Development of a Synthetic Poxvirus-Based SARS-CoV-2 Vaccine. *bioRxiv* **2020**. [CrossRef]

58. Wold, W.S.; Toth, K. Adenovirus vectors for gene therapy, vaccination and cancer gene therapy. *Curr. Gene* **2013**, *13*, 421–433. [CrossRef]

59. Zhu, F.C.; Guan, X.H.; Li, Y.H.; Huang, J.Y.; Jiang, T.; Hou, L.H.; Li, J.X.; Yang, B.F.; Wang, L.; Wang, W.J.; et al. Immunogenicity and safety of a recombinant adenovirus type-5-vectored COVID-19 vaccine in healthy adults aged 18 years or older: A randomised, double-blind, placebo-controlled, phase 2 trial. *Lancet* **2020**, *396*, 479–488. [CrossRef]

60. US National Library of Medicine. *Phase III Trial of A COVID-19 Vaccine of Adenovirus Vector in Adults 18 Years Old and above, on NIH*; USA National Library of Medicine: Bethesda, MD, USA, 2020.

61. Van Doremalen, N.; Lambe, T.; Spencer, A.; Belij-Rammerstorfer, S.; Purushotham, J.N.; Port, J.R.; Avanzato, V.A.; Bushmaker, T.; Flaxman, A.; Ulaszewska, M.; et al. ChAdOx1 nCoV-19 vaccination prevents SARS-CoV-2 pneumonia in rhesus macaques. *Nature* **2020**. [CrossRef]

62. Folegatti, P.M.; Ewer, K.J.; Aley, P.K.; Angus, B.; Becker, S.; Belij-Rammerstorfer, S.; Bellamy, D.; Bibi, S.; Bittaye, M.; Clutterbuck, E.A.; et al. Safety and immunogenicity of the ChAdOx1 nCoV-19 vaccine against SARS-CoV-2: A preliminary report of a phase 1/2, single-blind, randomised controlled trial. *Lancet* **2020**, *396*, 467–478. [CrossRef]

63. US National Library of Medicine. *Phase III Doubled-Blind, Placebo-Controlled Study of AZD1222 for the Prevention of COVID-19 in Adults, on NIH*; USA National Library of Medicine: Bethesda, MD, USA, 2020.

64. Mercado, N.B.; Zahn, R.; Wegmann, F.; Loos, C.; Chandrashekar, A.; Yu, J.; Liu, J.; Peter, L.; McMahan, K.; Tostanoski, L.H.; et al. Single-shot Ad26 vaccine protects against SARS-CoV-2 in rhesus macaques. *Nature* **2020**. [CrossRef]

65. Janssen Pharmaceutical Companies. A Study of Ad26COVS1 in Adults (COVID-19). Available online: https://clinicaltrials.gov/ct2/show/record/NCT04436276?term=NCT04436276&draw=2&rank=1 (accessed on 4 August 2020).

66. Companies, J.P. *A Study of Ad26.COV2.S for the Prevention of SARS-CoV-2-Mediated COVID-19 in Adult Participants (ENSEMBLE), on NIH*; US National Library of Medicine: Bethesda, MD, USA, 2020.

67. Gamaleya Research Institute. An Open Study of the Safety, Tolerability and Immunogenicity of the Drug "Gam-COVID-Vac" Vaccine Against COVID-19. Available online: https://clinicaltrials.gov/ct2/show/ NCT04436471?term=vaccine&cond=covid-19&draw=4 (accessed on 4 August 2020).

68. Gamaleya Research Institute. An Open Study of the Safety, Tolerability and Immunogenicity of "Gam-COVID-Vac Lyo" Vaccine Against COVID-19. Available online: https://clinicaltrials.gov/ct2/show/ record/NCT04437875?term=vaccine&cond=covid-19&draw=4 (accessed on 4 August 2020).

69. ReiThera. European Consortium for the Fast-Track Development of a Single-Dose Adenovirus-Based COVID-19 Vaccine. Available online: https://www.reithera.com/2020/04/23/reithera-leukocare-and-univercells-announce-pan-european-consortium-for-the-fast-track-development-of-a-single-dose-adenovirus-based-covid-19-vaccine/ (accessed on 17 August 2020).

70. Institute Pasteur. Clinical Trial to Evaluate the Safety and Immunogenicitiy of the COVID-19 Vaccine (COVID-19-101). Available online: https://clinicaltrials.gov/ct2/show/record/NCT04497298?term=vaccine& cond=covid-19&draw=2&rank=1 (accessed on 17 August 2020).

71. Medicago Inc. Safety, Tolerability and Immunogenicinity of a Coronavirus-Like Particle COVID-19 Vaccine in Adults Aged 18–55 Years. Available online: https://clinicaltrials.gov/ct2/show/record/NCT04450004?term=vaccine&cond=covid-19&draw=2 (accessed on 4 August 2020).

72. University Xiamen. *A Phase I Clinical Trial of Influenza Virus Vector COVID-19 Vaccine for Intranasal Spray*; Chinese Clinical Trial Registry: Hong Kong, China, 2020.

73. World Health Organization. Draft Landscape of COVID-19 Candidate Vaccines. 9 September 2020. Available online: https://www.who.int/publications/m/item/draft-landscape-of-covid-19-candidate-vaccines (accessed on 17 September 2020).

74. De la Cruz, V.F.; Lal, A.A.; McCutchan, T.F. Immunogenicity and epitope mapping of foreign sequences via genetically engineered filamentous phage. *J. Biol. Chem.* **1988**, *263*, 4318–4322.

75. Wang, C.; Sun, X.; Suo, S.; Ren, Y.; Li, X.; Herrler, G.; Thiel, V.; Ren, X. Phages bearing affinity peptides to severe acute respiratory syndromes-associated coronavirus differentiate this virus from other viruses. *J. Clin. Virol.* **2013**, *57*, 305–310. [CrossRef] [PubMed]

76. Ebrahimizadeh, W.; Rajabibazl, M. Bacteriophage vehicles for phage display: Biology, mechanism, and application. *Curr. Microbiol.* **2014**, *69*, 109–120. [CrossRef] [PubMed]

77. Marks, J.D.; Hoogenboom, H.R.; Griffiths, A.D.; Winter, G. Molecular evolution of proteins on filamentous phage. Mimicking the strategy of the immune system. *J. Biol. Chem.* **1992**, *267*, 16007–16010. [PubMed]

78. Christensen, D.J.; Gottlin, E.B.; Benson, R.E.; Hamilton, P.T. Phage display for target-based antibacterial drug discovery. *Drug Discov. Today* **2001**, *6*, 721–727. [CrossRef]

79. Kretzschmar, T.; von Ruden, T. Antibody discovery: Phage display. *Curr. Opin. Biotechnol.* **2002**, *13*, 598–602. [CrossRef]

80. Hong, S.S.; Boulanger, P. Protein ligands of the human adenovirus type 2 outer capsid identified by biopanning of a phage-displayed peptide library on separate domains of wild-type and mutant penton capsomers. *EMBO J.* **1995**, *14*, 4714–4727. [CrossRef]

81. Dyson, M.R.; Murray, K. Selection of peptide inhibitors of interactions involved in complex protein assemblies: Association of the core and surface antigens of hepatitis B virus. *Proc. Natl. Acad. Sci. USA* **1995**, *92*, 2194–2198. [CrossRef]

82. Larson, R.S.; Brown, D.C.; Ye, C.; Hjelle, B. Peptide antagonists that inhibit Sin Nombre virus and hantaan virus entry through the beta3-integrin receptor. *J. Virol.* **2005**, *79*, 7319–7326. [CrossRef]

83. Hall, P.R.; Hjelle, B.; Njus, H.; Ye, C.; Bondu-Hawkins, V.; Brown, D.C.; Kilpatrick, K.A.; Larson, R.S. Phage display selection of cyclic peptides that inhibit Andes virus infection. *J. Virol.* **2009**, *83*, 8965–8969. [CrossRef] [PubMed]

84. Qi, H.; Lu, H.; Qiu, H.J.; Petrenko, V.; Liu, A. Phagemid vectors for phage display: Properties, characteristics and construction. *J. Mol. Biol.* **2012**, *417*, 129–143. [CrossRef] [PubMed]

85. Marintcheva, B. Chapter 5—Phage Display. In *Harnessing the Power of Viruses*; Elsevier: Amsterdam, The Netherlands, 2018; Volume 2018, pp. 133–160.

86. Hess, G.T.; Cragnolini, J.J.; Popp, M.W.; Allen, M.A.; Dougan, S.K.; Spooner, E.; Ploegh, H.L.; Belcher, A.M.; Guimaraes, C.P. M13 bacteriophage display framework that allows sortase-mediated modification of surface-accessible phage proteins. *Bioconjug. Chem.* **2012**, *23*, 1478–1487. [CrossRef] [PubMed]

87. Wu, J.; Tu, C.; Yu, X.; Zhang, M.; Zhang, N.; Zhao, M.; Nie, W.; Ren, Z. Bacteriophage T4 nanoparticle capsid surface SOC and HOC bipartite display with enhanced classical swine fever virus immunogenicity: A powerful immunological approach. *J. Virol. Methods* **2007**, *139*, 50–60. [CrossRef]

88. Krumpe, L.R.; Atkinson, A.J.; Smythers, G.W.; Kandel, A.; Schumacher, K.M.; McMahon, J.B.; Makowski, L.; Mori, T. T7 lytic phage-displayed peptide libraries exhibit less sequence bias than M13 filamentous phage-displayed peptide libraries. *Proteomics* **2006**, *6*, 4210–4222. [CrossRef]

89. Cicchini, C.; Ansuini, H.; Amicone, L.; Alonzi, T.; Nicosia, A.; Cortese, R.; Tripodi, M.; Luzzago, A. Searching for DNA-protein interactions by lambda phage display. *J. Mol. Biol.* **2002**, *322*, 697–706. [CrossRef]

90. Bazan, J.; Calkosinski, I.; Gamian, A. Phage display—A powerful technique for immunotherapy: 1. Introduction and potential of therapeutic applications. *Hum. Vaccin. Immunother.* **2012**, *8*, 1817–1828. [CrossRef]

91. Chakravarthy, B.; Ménard, M.; Brown, L.; Atkinson, T.; Whitfield, J. Identification of protein kinase C inhibitory activity associated with a polypeptide isolated from a phage display system with homology to PCM-1, the pericentriolar material-1 protein. *Biochem. Biophys. Res. Commun.* **2012**, *424*, 147–151. [CrossRef]

92. Liu, Z.X.; Yi, G.H.; Qi, Y.P.; Liu, Y.L.; Yan, J.P.; Qian, J.; Du, E.Q.; Ling, W.F. Identification of single-chain antibody fragments specific against SARS-associated coronavirus from phage-displayed antibody library. *Biochem. Biophys. Res. Commun* **2005**, *329*, 437–444. [CrossRef]

93. Lim, C.C.; Woo, P.C.Y.; Lim, T.S. Development of a Phage Display Panning Strategy Utilizing Crude Antigens: Isolation of MERS-CoV Nucleoprotein human antibodies. *Sci. Rep.* **2019**, *9*, 6088. [CrossRef]

94. Duan, J.; Yan, X.; Guo, X.; Cao, W.; Han, W.; Qi, C.; Feng, J.; Yang, D.; Gao, G.; Jin, G. A human SARS-CoV neutralizing antibody against epitope on S2 protein. *Biochem. Biophys. Res. Commun.* **2005**, *333*, 186–193. [CrossRef] [PubMed]

95. Kang, X.; Yang, B.A.; Hu, Y.; Zhao, H.; Xiong, W.; Yang, Y.; Si, B.; Zhu, Q. Human neutralizing Fab molecules against severe acute respiratory syndrome coronavirus generated by phage display. *Clin. Vaccine Immunol.* **2006**, *13*, 953–957. [CrossRef] [PubMed]

96. Begum, N.; Horiuchi, S.; Tanaka, Y.; Yamamoto, N.; Ichiyama, K.; Yamamoto, N. New approach for generation of neutralizing antibody against human T-cell leukaemia virus type-I (HTLV-I) using phage clones. *Vaccine* **2002**, *20*, 1281–1289. [CrossRef]

97. Wu, Y.; Li, C.; Xia, S.; Tian, X.; Kong, Y.; Wang, Z.; Gu, C.; Zhang, R.; Tu, C.; Xie, Y.; et al. Identification of Human Single-Domain Antibodies against SARS-CoV-2. *Cell Host Microbe* **2020**, *27*, 891–898.e895. [CrossRef] [PubMed]

98. Zeng, X.; Li, L.; Lin, J.; Li, X.; Liu, B.; Kong, Y.; Zeng, S.; Du, J.; Xiao, H.; Zhang, T.; et al. Blocking antibodies against SARS-CoV-2 RBD isolated from a phage display antibody library using a competitive biopanning strategy. *bioRxiv* **2020**. [CrossRef]

99. Ren, X.; Liu, B.; Yin, J.; Zhang, H.; Li, G. Phage displayed peptides recognizing porcine aminopeptidase N inhibit transmissible gastroenteritis coronavirus infection in vitro. *Virology* **2011**, *410*, 299–306. [CrossRef] [PubMed]

100. Lauster, D.; Klenk, S.; Ludwig, K.; Nojoumi, S.; Behren, S.; Adam, L.; Stadtmüller, M.; Saenger, S.; Zimmler, S.; Hönzke, K.; et al. Phage capsid nanoparticles with defined ligand arrangement block influenza virus entry. *Nat. Nanotechnol.* **2020**, *15*, 373–379. [CrossRef] [PubMed]

101. Marraffini, L.A.; Sontheimer, E.J. CRISPR interference limits horizontal gene transfer in staphylococci by targeting DNA. *Science* **2008**, *322*, 1843–1845. [CrossRef]

102. Deveau, H.; Barrangou, R.; Garneau, J.E.; Labonte, J.; Fremaux, C.; Boyaval, P.; Romero, D.A.; Horvath, P.; Moineau, S. Phage response to CRISPR-encoded resistance in Streptococcus thermophilus. *J. Bacteriol.* **2008**, *190*, 1390–1400. [CrossRef]

103. Horvath, P.; Romero, D.A.; Coute-Monvoisin, A.C.; Richards, M.; Deveau, H.; Moineau, S.; Boyaval, P.; Fremaux, C.; Barrangou, R. Diversity, activity, and evolution of CRISPR loci in Streptococcus thermophilus. *J. Bacteriol.* **2008**, *190*, 1401–1412. [CrossRef]

104. Pardee, K.; Green, A.A.; Takahashi, M.K.; Braff, D.; Lambert, G.; Lee, J.W.; Ferrante, T.; Ma, D.; Donghia, N.; Fan, M.; et al. Rapid, Low-Cost Detection of Zika Virus Using Programmable Biomolecular Components. *Cell* **2016**, *165*, 1255–1266. [CrossRef]

105. Chen, J.S.; Ma, E.; Harrington, L.B.; Da Costa, M.; Tian, X.; Palefsky, J.M.; Doudna, J.A. CRISPR-Cas12a target binding unleashes indiscriminate single-stranded DNase activity. *Science* **2018**, *360*, 436–439. [CrossRef]

106. He, Q.; Yu, D.; Bao, M.; Korensky, G.; Chen, J.; Shin, M.; Kim, J.; Park, M.; Qin, P.; Du, K. High-throughput and all-solution phase African Swine Fever Virus (ASFV) detection using CRISPR-Cas12a and fluorescence based point-of-care system. *Biosens. Bioelectron.* **2020**, *154*, 112068. [CrossRef]

107. Guk, K.; Keem, J.O.; Hwang, S.G.; Kim, H.; Kang, T.; Lim, E.K.; Jung, J. A facile, rapid and sensitive detection of MRSA using a CRISPR-mediated DNA FISH method, antibody-like dCas9/sgRNA complex. *Biosens. Bioelectron.* **2017**, *95*, 67–71. [CrossRef] [PubMed]

108. Gootenberg, J.S.; Abudayyeh, O.O.; Kellner, M.J.; Joung, J.; Collins, J.J.; Zhang, F. Multiplexed and portable nucleic acid detection platform with Cas13, Cas12a, and Csm6. *Science* **2018**, *360*, 439–444. [CrossRef]

109. Davies, K.; Barrangou, R. COVID-19 and the CRISPR Community Response. *Cris. J.* **2020**, *3*, 66–67. [CrossRef] [PubMed]

110. Wang, X.; Shang, X.; Huang, X. Next-generation pathogen diagnosis with CRISPR/Cas-based detection methods. *Emerg. Microbes Infect.* **2020**, *9*, 1682–1691. [CrossRef]

111. Li, J.J.; Xiong, C.; Liu, Y.; Liang, J.S.; Zhou, X.W. Loop-Mediated Isothermal Amplification (LAMP): Emergence as an Alternative Technology for Herbal Medicine Identification. *Front. Plant Sci.* **2016**, *7*, 1956. [CrossRef] [PubMed]

112. Kellner, M.J.; Koob, J.G.; Gootenberg, J.S.; Abudayyeh, O.O.; Zhang, F. SHERLOCK: Nucleic acid detection with CRISPR nucleases. *Nat. Protoc.* **2019**, *14*, 2986–3012. [CrossRef]

113. Joung, J.; Ladha, A.; Saito, M.; Kim, N.G.; Woolley, A.E.; Segel, M.; Barretto, R.P.J.; Ranu, A.; Macrae, R.K.; Faure, G.; et al. Detection of SARS-CoV-2 with SHERLOCK One-Pot Testing. *N. Engl. J. Med.* **2020**. [CrossRef]

114. Joung, J.; Ladha, A.; Saito, M.; Segel, M.; Bruneau, R.; Huang, M.-l.W.; Kim, N.-G.; Yu, X.; Li, J.; Walker, B.D.; et al. Point-of-care testing for COVID-19 using SHERLOCK diagnostics. *MedRxiv* **2020**. [CrossRef]

115. Patchsung, M.; Jantarug, K.; Pattama, A.; Aphicho, K.; Suraritdechachai, S.; Meesawat, P.; Sappakhaw, K.; Leelahakorn, N.; Ruenkam, T.; Wongsatit, T.; et al. Clinical validation of a Cas13-based assay for the detection of SARS-CoV-2 RNA. *Nat. Biomed. Eng.* **2020**. [CrossRef] [PubMed]

116. Broughton, J.P.; Deng, X.; Yu, G.; Fasching, C.L.; Servellita, V.; Singh, J.; Miao, X.; Streithorst, J.A.; Granados, A.; Sotomayor-Gonzalez, A.; et al. CRISPR-Cas12-based detection of SARS-CoV-2. *Nat. Biotechnol.* **2020**, *38*, 870–874. [CrossRef]

117. Ackerman, C.M.; Myhrvold, C.; Thakku, S.G.; Freije, C.A.; Metsky, H.C.; Yang, D.K.; Ye, S.H.; Boehm, C.K.; Kosoko-Thoroddsen, T.-S.F.; Kehe, J.; et al. Massively multiplexed nucleic acid detection using Cas13. *Nature* **2020**, *582*, 277–282. [CrossRef]

118. Ding, X.; Yin, K.; Li, Z.; Liu, C. All-in-One Dual CRISPR-Cas12a (AIOD-CRISPR) Assay: A Case for Rapid, Ultrasensitive and Visual Detection of Novel Coronavirus SARS-CoV-2 and HIV virus. *bioRxiv* **2020**. [CrossRef]

119. Yoshimi, K.; Takeshita, K.; Yamayoshi, S.; Shibumura, S.; Yamauchi, Y.; Yamamoto, M.; Yotsuyanagi, H.; Kawaoka, Y.; Mashimo, T. Rapid and accurate detection of novel coronavirus SARS-CoV-2 using CRISPR-Cas3. *MedRxiv* **2020**. [CrossRef]

120. Hou, T.; Zeng, W.; Yang, M.; Chen, W.; Ren, L.; Ai, J.; Wu, J.; Liao, Y.; Gou, X.; Li, Y.; et al. Development and evaluation of a rapid CRISPR-based diagnostic for COVID-19. *PLoS Pathog.* **2020**, *16*, e1008705. [CrossRef]

121. Abbott, T.R.; Dhamdhere, G.; Liu, Y.; Lin, X.; Goudy, L.; Zeng, L.; Chemparathy, A.; Chmura, S.; Heaton, N.S.; Debs, R.; et al. Development of CRISPR as an Antiviral Strategy to Combat SARS-CoV-2 and Influenza. *Cell* **2020**, *181*, 865–876. [CrossRef]

122. Fozouni, P.; Son, S.; de León Derby, M.D.; Knott, G.J.; Gray, C.N.; D'Ambrosio, M.V.; Zhao, C.; Switz, N.A.; Kumar, G.R.; Stephens, S.I.; et al. Direct detection of SARS-CoV-2 using CRISPR-Cas13a and a mobile phone. *MedRxiv* **2020**. [CrossRef]

123. Nguyen, T.M.; Zhang, Y.; Pandolfi, P.P. Virus against virus: A potential treatment for 2019-nCov (SARS-CoV-2) and other RNA viruses. *Cell Res.* **2020**, *30*, 189–190. [CrossRef]

124. USA Food and Drugs Agency. First FDA-Approved Vaccine for the Prevention of Ebola Virus Disease, Marking a Critical Milestone in Public Health Preparedness and Response. Available online: https://www.fda.gov/news-events/press-announcements/first-fda-approved-vaccine-prevention-ebola-virus-disease-marking-critical-milestone-public-health (accessed on 13 October 2020).

125. Coughlan, L. Factors Which Contribute to the Immunogenicity of Non-replicating Adenoviral Vectored Vaccines. *Front. Immunol.* **2020**, *11*, 909. [CrossRef] [PubMed]

126. Cooney, E.L.; Collier, A.C.; Greenberg, P.D.; Coombs, R.W.; Zarling, J.; Arditti, D.E.; Hoffman, M.C.; Hu, S.L.; Corey, L. Safety of and immunological response to a recombinant vaccinia virus vaccine expressing HIV envelope glycoprotein. *Lancet* **1991**, *337*, 567–572. [CrossRef]

127. Lurie, N.; Saville, M.; Hatchett, R.; Halton, J. Developing Covid-19 Vaccines at Pandemic Speed. *N. Engl. J. Med.* **2020**, *382*, 1969–1973. [CrossRef] [PubMed]

128. Sun, S.H.; Chen, Q.; Gu, H.J.; Yang, G.; Wang, Y.X.; Huang, X.Y.; Liu, S.S.; Zhang, N.N.; Li, X.F.; Xiong, R.; et al. A Mouse Model of SARS-CoV-2 Infection and Pathogenesis. *Cell Host Microbe* **2020**, *28*, 124–133.e124. [CrossRef]

129. Sun, J.; Zhuang, Z.; Zheng, J.; Li, K.; Wong, R.L.; Liu, D.; Huang, J.; He, J.; Zhu, A.; Zhao, J.; et al. Generation of a Broadly Useful Model for COVID-19 Pathogenesis, Vaccination, and Treatment. *Cell* **2020**. [CrossRef]

130. Latrofa, F.; Pichurin, P.; Guo, J.; Rapoport, B.; McLachlan, S.M. Thyroglobulin-thyroperoxidase autoantibodies are polyreactive, not bispecific: Analysis using human monoclonal autoantibodies. *J. Clin. Endocrinol. Metab.* **2003**, *88*, 371–378. [CrossRef]

131. Payne, A.S.; Ishii, K.; Kacir, S.; Lin, C.; Li, H.; Hanakawa, Y.; Tsunoda, K.; Amagai, M.; Stanley, J.R.; Siegel, D.L. Genetic and functional characterization of human pemphigus vulgaris monoclonal autoantibodies isolated by phage display. *J. Clin. Investig.* **2005**, *115*, 888–899. [CrossRef]

132. Venkatesh, N.; Im, S.H.; Balass, M.; Fuchs, S.; Katchalski-Katzir, E. Prevention of passively transferred experimental autoimmune myasthenia gravis by a phage library-derived cyclic peptide. *Proc. Natl. Acad. Sci. USA* **2000**, *97*, 761–766. [CrossRef]

133. Houimel, M.; Corthesy-Theulaz, I.; Fisch, I.; Wong, C.; Corthesy, B.; Mach, J.; Finnern, R. Selection of human single chain Fv antibody fragments binding and inhibiting Helicobacter pylori urease. *Tumour. Biol.* **2001**, *22*, 36–44. [CrossRef]

134. Molina-Lopez, J.; Sanschagrin, F.; Levesque, R.C. A peptide inhibitor of MurA UDP-N-acetylglucosamine enolpyruvyl transferase: The first committed step in peptidoglycan biosynthesis. *Peptides* **2006**, *27*, 3115–3121. [CrossRef]

135. Yacoby, I.; Shamis, M.; Bar, H.; Shabat, D.; Benhar, I. Targeting antibacterial agents by using drug-carrying filamentous bacteriophages. *Antimicrob. Agents Chemother.* **2006**, *50*, 2087–2097. [CrossRef] [PubMed]

136. Coelho, E.A.; Chavez-Fumagalli, M.A.; Costa, L.E.; Tavares, C.A.; Soto, M.; Goulart, L.R. Theranostic applications of phage display to control leishmaniasis: Selection of biomarkers for serodiagnostics, vaccination, and immunotherapy. *Rev. Soc. Bras. Med. Trop.* **2015**, *48*, 370–379. [CrossRef] [PubMed]

137. Carlsson, F.; Trilling, M.; Perez, F.; Ohlin, M. A dimerized single-chain variable fragment system for the assessment of neutralizing activity of phage display-selected antibody fragments specific for cytomegalovirus. *J. Immunol. Methods* **2012**, *376*, 69–78. [CrossRef] [PubMed]

138. Zhao, X.L.; Yin, J.; Chen, W.Q.; Jiang, M.; Yang, G.; Yang, Z.H. Generation and characterization of human monoclonal antibodies to G5, a linear neutralization epitope on glycoprotein of rabies virus, by phage display technology. *Microbiol. Immunol.* **2008**, *52*, 89–93. [CrossRef]

139. Rahbarnia, L.; Farajnia, S.; Babaei, H.; Majidi, J.; Veisi, K.; Ahmadzadeh, V.; Akbari, B. Evolution of phage display technology: From discovery to application. *J Drug Target* **2017**, *25*, 216–224. [CrossRef]

140. Li, Y.; Li, S.; Wang, J.; Liu, G. CRISPR/Cas Systems towards Next-Generation Biosensing. *Trends Biotechnol.* **2019**, *37*, 730–743. [CrossRef]

141. Citorik, R.J.; Mimee, M.; Lu, T.K. Sequence-specific antimicrobials using efficiently delivered RNA-guided nucleases. *Nat. Biotechnol.* **2014**, *32*, 1141–1145. [CrossRef]

142. Xu, C.L.; Ruan, M.Z.C.; Mahajan, V.B.; Tsang, S.H. Viral Delivery Systems for CRISPR. *Viruses* **2019**, *11*. [CrossRef]

143. Jiang, W.; Marraffini, L.A. CRISPR-Cas: New Tools for Genetic Manipulations from Bacterial Immunity Systems. *Annu. Rev. Microbiol.* **2015**, *69*, 209–228. [CrossRef]

144. Jackson, R.N.; Golden, S.M.; van Erp, P.B.; Carter, J.; Westra, E.R.; Brouns, S.J.; van der Oost, J.; Terwilliger, T.C.; Read, R.J.; Wiedenheft, B. Structural biology. Crystal structure of the CRISPR RNA-guided surveillance complex from Escherichia coli. *Science* **2014**, *345*, 1473–1479. [CrossRef]

145. Gasiunas, G.; Barrangou, R.; Horvath, P.; Siksnys, V. Cas9-crRNA ribonucleoprotein complex mediates specific DNA cleavage for adaptive immunity in bacteria. *Proc. Natl. Acad. Sci. USA* **2012**, *109*, E2579–E2586. [CrossRef] [PubMed]

Publisher's Note: MDPI stays neutral with regard to jurisdictional claims in published maps and institutional affiliations.

Article

Parenterally Administered Porcine Epidemic Diarrhea Virus-Like Particle-Based Vaccine Formulated with CCL25/28 Chemokines Induces Systemic and Mucosal Immune Protectivity in Pigs

Chin-Wei Hsu [1], Ming-Hao Chang [2], Hui-Wen Chang [1], Tzong-Yuan Wu [2,3,*] and Yen-Chen Chang [1,*]

1 Graduate Institute of Molecular and Comparative Pathobiology, School of Veterinary Medicine, National Taiwan University, Taipei 106, Taiwan; robby951159@gmail.com (C.-W.H.); huiwenchang@ntu.edu.tw (H.-W.C.)

2 Department of Bioscience Technology, Chung Yuan Christian University, Taoyuan 320, Taiwan; jhon2003111333@gmail.com

3 Department of Medical Research, China Medical University Hospital, China Medical University, Taichung 406, Taiwan

* Correspondence: tywu@cycu.edu.tw (T.-Y.W.); yenchenchang@ntu.edu.tw (Y.-C.C.)

Received: 23 July 2020; Accepted: 30 September 2020; Published: 2 October 2020

Abstract: Generation of a safe, economical, and effective vaccine capable of inducing mucosal immunity is critical for the development of vaccines against enteric viral diseases. In the current study, virus-like particles (VLPs) containing the spike (S), membrane (M), and envelope (E) structural proteins of porcine epidemic diarrhea virus (PEDV) expressed by the novel polycistronic baculovirus expression vector were generated. The immunogenicity and protective efficacy of the PEDV VLPs formulated with or without mucosal adjuvants of CCL25 and CCL28 (CCL25/28) were evaluated in post-weaning pigs. While pigs intramuscularly immunized with VLPs alone were capable of eliciting systemic anti-PEDV S-specific IgG and cellular immunity, co-administration of PEDV VLPs with CCL25/28 could further modulate the immune responses by enhancing systemic anti-PEDV S-specific IgG, mucosal IgA, and cellular immunity. Upon challenge with PEDV, both VLP-immunized groups showed milder clinical signs with reduced fecal viral shedding as compared to the control group. Furthermore, pigs immunized with VLPs adjuvanted with CCL25/28 showed superior immune protection against PEDV. Our results suggest that VLPs formulated with CCL25/28 may serve as a potential PEDV vaccine candidate and the same strategy may serve as a platform for the development of other enteric viral vaccines.

Keywords: porcine epidemic diarrhea virus; VLP; chemokines; pig; vaccine

---

1. Introduction

Porcine epidemic diarrhea (PED), caused by the porcine epidemic diarrhea virus (PEDV), is a contagious enteric viral disease that occurs specifically in pigs. PEDV is classified in the order *Nidovirales*, family *Coronaviridae*, and genus *Alphacoronavirus*. It contains a positive-sense, single-stranded, ~28-kilobase RNA genome, incorporated with the nucleocapsid (N) protein and enveloped in a membranous outer coat comprising three structural proteins: membrane (M), envelope (E), and S (S) proteins [1]. The S protein is a multifunctional molecular apparatus responsible for specific host and tissue recognition and induction of protective humoral as well as cellular immunities for viral neutralization and elimination [2–4]. The M protein, the most abundant structural protein, is responsible for the generation of virus particles. The small E protein, which accounts for a minority of the envelope component, plays a critical role in the viral morphogenesis and the final step of the budding process [5,6]. PEDV has a tropism for enterocytes of villous tips and further leads to atrophic enteritis [7,8]. Pigs of all ages can be affected by the disease; however, the severity of the gastrointestinal clinical signs is age-dependent as a result of the slower turnover of enterocytes and incomplete innate immunity in neonatal piglets than in post-weaning pigs [9,10]. Historically, PED has had minimal effect on piglets in a sporadic to epidemic manner. However, since 2010, PEDV has evolved into highly virulent viruses, designated as genogroup 2 (G2) strains, which are different from previous G1 strains and result in high mortality in neonatal piglets and devastating global outbreaks [11,12]. The high death rates cause tremendous economic losses, thus highlighting the requirement for an effective vaccine strategy.

In the past, the available tools for the prevention of PEDV infection have included live-attenuated or inactivated vaccines derived from G1 PEDVs, such as strains CV777, DR13, P5-V, and SM98-1 [13,14]. The traditional vaccines only confer partial cross-protection against the novel highly virulent PEDV G2 strains despite only up to 10% difference in the amino acid sequence of the S proteins between the G1 and G2 strains [1,15]. To control the outbreaks of virulent G2 viruses, two conditionally licensed vaccines have been commercialized in the United States. One is a recombinant vaccine expressing the PEDV S protein using a replication-deficient Venezuelan equine encephalitis virus packaging system, while the other is an inactivated vaccine based on the whole virus of non-S INDEL PEDV strain. The third vaccine candidate, which is in commercial development by Vaccine and Infectious Disease Organization—International Vaccine Centre, is a subunit vaccine that uses mammalian HEK-293 T cell-expressed PEDV S1 proteins [16]. However, the efficacy of these commercial vaccines in stimulating solid lactogenic immunity against the disease in suckling piglets has been inconsistent [17,18]. Many other attempts have also been made to develop effective vaccines, but most of them have been incapable of inducing mucosal immunity [19–22]. Although the immunogenicity of the live-attenuated as well as inactivated vaccines is considered more effective [23], the live-attenuated approach has safety concerns and poses risks of genetic recombination or virulence restoration of the wild-type strains. Besides, both the approaches are time-consuming in vaccine development [24,25]. In the quest for novel strategies, the next-generation vaccine is expected to promptly deal with RNA viruses harboring a high mutation rate [1]. The subunit approach is considered to be a better strategy for the development of a safer vaccine; however, it falls short on the ability to elicit the optimal immunogenicity. With developments in biotechnology, the use of virus-like particles (VLPs) as an advanced subunit vaccine offers advantages of being less time-consuming to develop, being safer, and eliciting adequate immunogenicity. Thus, it is a new balanced approach that avoids the trade-offs between security and immunogenicity with regard to vaccine development to the maximum extent.

Virus-like particles, which exhibit the size and geometry of the viral structures and closely resemble the corresponding native virion but without the viral genomic nucleic acids [26], have been shown to improve the immunogenicity. The absence of the genetic material renders VLPs replication-incompetent. The nanometer dimensions in a range of supramolecular particulate antigens (20–200 nm) allow VLPs to not only freely drain into lymph nodes but also to be efficiently uptaken by antigen-presenting cells [27,28], which is conducive to T cell responses, including CD4$^+$ T helper cells and CD8$^+$ cytotoxic T cells [29]. The efficient cross-presentation is mainly mediated by the appropriate size of the VLPs to activate lymphoid dendritic cells, which only reside in lymph nodes and are essential to cellular immune responses [30–32]. Additionally, VLPs characterized by nanoparticles with a highly repetitive array of conformational epitopes can directly interact with B cells and facilitate the subsequent humoral immunity [26,33,34]. Therefore, VLP is an effective and safe tool to stimulate both cellular and humoral immunity in the absence of intracellular replication. Considering the high complexity of the enveloped PEDV VLPs, a polycistronic baculovirus expression vector system (P-BEVS), which involves co-expressing polycistronic genes via internal ribosome entry sites, has been adopted to produce VLPs comprising all of the envelope components [35,36]. Although a recent report has demonstrated that BEVS can successfully produce PEDV VLPs that are capable of inducing PEDV-specific humoral immunity in mice [37], the efficacy of PEDV VLPs against PEDV challenge in pigs still needs further evaluations.

Majorly targeting the superficial villous enterocytes, the PEDV is categorized as a type I enteropathogenic virus, and local mucosal and cellular immunities are important to control the viral infection [38]. The establishment of mucosal immunity requires a specific microenvironment to promote the development of the defense models, such as immunoglobulin class switching to IgA and J chain, and program surface homing ligands or receptors of immune cells [39]. Therefore, in the natural situation, the best immunization route is through the affected compartment. Many studies have proved that oral inoculation of virulent enteric viruses had better induction of mucosal IgA [35,40,41]. However, oral administration of equal vaccine dosage is technically difficult and labor-intensive in the swine industry. Intramuscular injection is a common immunization route in the field due to its operational practicability, but at the expense of protective mucosal immunity. To enhance the mucosal immune responses of parenteral administration, studies have used chemokines as molecular adjuvants to potentially drive the immune responses toward intestinal mucosal immunity. Among the various chemokines, small and large intestine-associated chemokines, CCL25/TECK (thymus-expressed chemokine) and CCL28 (mucosae-associated epithelial chemokine) [36], have been shown to up-regulate the localization of immune cells with CCR9 and CCR10, respectively, to the mucosal sites after systemic immunization [42–44]. Besides, the synergistic effects of CCL25 and CCL28 in trafficking IgA$^+$ cells into the intestines has been confirmed in mice [45]. Our previous work also proved that inactivated PEDV co-adjuvanted with porcine CCL25 and CCL28 is competent in enhancing mucosal and systemic antibody responses and protective efficacy in pigs [46]. Therefore, intramuscular injection of an immunogen in combination with CCL25 and CCL28 could be a potential strategy for developing vaccines against enteric diseases.

In the present study, a P-BEVS-derived PEDV VLP comprising the S, M, and E proteins of the highly virulent PEDV G2 strain was successfully generated. The immunogenicity of the VLP, including systemic PEDV S-specific IgG, mucosal IgA, and cellular immunity, was evaluated in a 4-week-old pig model, and accompanied with an assessment of the protection against a homogenous PEDV challenge at 11-week-old pigs. Furthermore, an advanced statistical method known as generalized estimating equations (GEE), which has high statistical power in a small sample size as well as the ability to examine the effects of multiple factors on an outcome, was used for data analysis.

2. Materials and Methods

2.1. Plasmid Construction

The S nucleotide sequence with the original signal peptide replaced by the honeybee melittin signal peptide was kindly provided by Dr. Yu-Chan Chao at Academia Sinica. The S, M, and E genes were derived from the Taiwan G2b PEDV-PT strain (Genbank accession no. KP276252). The S gene was codon-optimized to an insect cell system and synthesized by ProTech (ProTech, Taipei, Taiwan). The 2A-like sequence isolated from *Perina nuda* virus (PnV) and the M and E genes were inserted into the XbaI and NotI sites, respectively, in the pBac-mcsI-PnV339-eGFP-Rhir-mcsII vector [47]. Following that, the 2A-M-PnV339-eGFP-Rhir-E sequence, along with the honeybee melittin signal peptide, hexahistidine tag, and S gene, was included in the pFastBac1 plasmid (Invitrogen, Carlsbad, CA, USA) using the NEBuilder® HiFi DNA Assembly Kit (New England Biolabs, Ipswich, MA, USA) to generate pFastBac1-HM6H-PEDV-S-2A-M-PnV339-eGFP-Rhir-E (Figure 1). It was used as the recombinant baculovirus transfer vector to recombine with the bacmid DNA in *E. coli* (strain DH10Bac, Invitrogen). The recombinant bacmid containing PEDV S, M, and E genes was transfected into Sf21 cells using Cellfectin™ (Life Technologies, Carlsbad, CA, USA) to generate the recombinant baculovirus, SME-Bac.

pFastBac1-HM6H-PEDV-S-2A-M-Pnv339-eGFP-Rhir-E

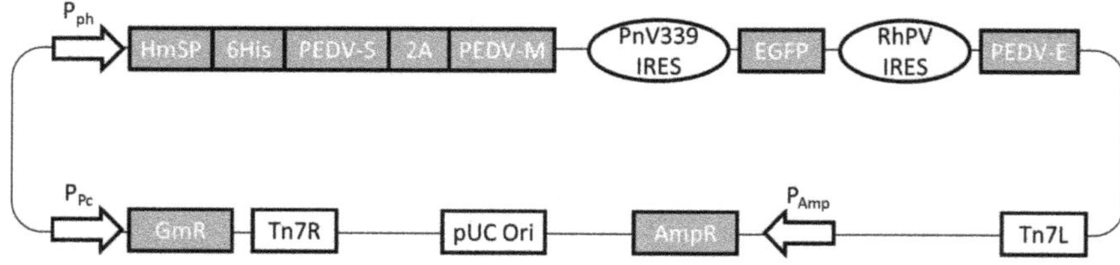

Figure 1. The construction map of the plasmid pFastBac1- HM6H- PEDV- S- 2A- M- Pnv339- eGFP- Rhir-E. The recombinant Taiwan G2b PEDV-PT strain spike (S) gene, with the honey bee melittin signal peptide and 6xHis-tag, and the membrane (M) gene linked by the 2A-like sequence were driven by the polyhedrin promoter. The envelope (E) gene was translated through the internal ribosome entry site (IRES) of Rhopalosiphum padi virus (RhPV). Enhanced green fluorescent protein (EGFP) gene was inserted into the plasmid and expressed by a truncated perina nuda picorna-like virus IRES (PnV339 IRES).

2.2. Generation of VLPs

Sf21 cells were passaged to reach a cell density of 1×10^7 in each T75 flask. After SME-Bac infection at a multiplicity of infection (MOI) of 1 in Sf21 cells for 5 days, the culture medium was collected, centrifuged at 600× g for 5 min to remove the cell debris, and passed through a 0.22 μm filter. VLPs in 10 mL of the supernatant were collected using sucrose cushion ultracentrifugation at 9000× g, 4 °C, for 90 min using a Beckman SW-41 rotor (Beckman Instruments, Spinco Division, Palo Alto, CA, USA). The precipitated VLPs were resuspended in 100 μL phosphate-buffered saline (PBS).

2.3. Indirect Fluorescent Antibody Test

Sf21 cells were passaged to reach a cell density of 2×10^5 cells/well of a 24-well plate. After the Sf21 cells were infected with 5 MOI of SME-Bac for 4 days, each well was washed three times with 200 μL of PBS supplemented with 0.1% tween-20 (PBST) and fixed using 200 μL of 4% paraformaldehyde on ice for 20 min. The paraformaldehyde was then removed and each well was blocked with 200 μL blocking buffer (3% bovine serum albumin) and incubated at room temperature (RT) for 1 h. Two hundred microliters of anti-PEDV S monoclonal antibody, P4B [48], diluted in blocking buffer (1:200 ratio) was added to each well and allowed to incubate at RT for 2 h. After three PBST washes and a 1 h incubation with Alexa Flour® 594-conjugated AffiniPure goat anti-mouse IgG (Jackson ImmunoResearch, West Grove, PA, USA) diluted 1:400 in blocking buffer, the wells were washed three times with PBST and observed under a fluorescence microscope.

2.4. Western Blotting

The precipitated VLPs were loaded onto an 8% sodium dodecyl sulfate-polyacrylamide electrophoresis gel. After electrophoresis, the proteins were transferred to a methanol-activated polyvinylidene fluoride membrane at 300 mA for 180 min. The recombinant S protein was detected using a rabbit anti-His-tag polyclonal antibody (1:2000 dilution, Rockland, NY, USA). Following this, a horseradish peroxidase (HRP)-conjugated goat anti-rabbit IgG (1:5000 dilution, Cell Signaling Technology, Massachusetts, USA) was used as the secondary antibody for signal detection. The membrane was developed using Immobilon™ Western ECL Substrate (Millipore, MA, USA).

To determine the expression of S protein on the VLPs, the PEDV-challenged and PEDV-free porcine sera were used to perform Western blot. The prepared VLPs and EGFP-Bac, which was the same plasmid without PEDV S, E, and M sequences (pFastBac1-HM6H-P-2A-PnV339-eGFP-Rhir) and acted as a negative control, were loaded and protein electrophoresis was conducted. The following steps were similar to those described above, but instead, the primary and secondary antibodies were replaced by PEDV-challenged porcine serum (1:1000 dilution in blocking buffer) and HRP-conjugated goat anti-pig IgG (1:1000 dilution; Kirkegaard & Perry Laboratories, MD, USA), respectively.

2.5. Characterization of VLPs Using Electron Microscopy

For the preparation of the microscopic grids, an aliquot of 10 µL of samples was added to the carbon-coated grid for 1 min and then removed using a filter paper. The grids were stained with 2% phosphotungstic acid (PTA) for 1 min. Following that, the excess PTA was drained and the grids were completely dried for 6 h before being examined under a Tecnai G^2 Spirit TWIN transmission electron microscope (FEI Company, OR, USA).

2.6. Expression and Purification of CC Chemokines

The CC chemokines, CCL25 and CCL28, were prepared as described in a previous study [46]. The aqueous formulations comprised the immunogen with/without CC chemokines for different groups (Table 1). The amounts of VLP and chemokines CCL25 or CCL28 were quantified using Western blot followed by ImageJ analysis and Pierce™ BCA Protein Assay Kit (Thermo Fisher Scientific, Waltham, MA, USA).

Table 1. Vaccine formulation and immunization program.

Group	Immunogen	Adjuvant	
		CC Chemokine	Freund's Adjuvant *
Control	None	None	Yes
VLP	1.8 mg of VLP (0.2 µg S protein)	None	Yes
VLP + CCL25/28	1.8 mg of VLP (0.2 µg S protein)	30 µg CCL25 and 30 µg CCL28	Yes

* 1st immunization: 0.5 mL of complete Freund's adjuvant; 2nd immunization: 0.5 mL of incomplete Freund's adjuvant; 3rd immunization: 0.5 mL of incomplete Freund's adjuvant.

2.7. Cell Lines and Viruses

The highly virulent viral stock of PEDV Pintung 52 passage 7 (PEDVPT-P7) was derived from PEDVPT-P5 (GenBank accession no. KY929405) as described in previous studies [19–21]. Vero C1008 cells (American Type Culture Collection no. CRL-1586) were used for viral preparation and the neutralizing assay. The culture medium was Dulbecco's modified Eagle medium (DMEM, Gibco, Grand Island, NY, USA) supplemented with 10% fetal bovine serum (GE Healthcare, Uppsala, Sweden), 250 ng/mL amphotericin B, 100 U/mL penicillin, and 100 µg/mL streptomycin. The viral titer of PEDVPT-P7 was 1.78×10^5 TCID$_{50}$/mL, as determined using the endpoint titration assay with a ten-fold serial dilution in triplicates.

2.8. Immunization Program of Pigs

Twenty-three-week-old castrated male Large White × Duroc crossbred pigs, which were PEDV-seronegative and had no PEDV fecal shedding, were selected from a conventional pig farm. These pigs were randomly separated into three groups, including the VLP group ($n = 7$), VLP+CCL ($n = 7$), and control ($n = 6$) groups. After acclimation for one week, pigs in each group were intramuscularly primed with the 0.5 mL regimen shown in Table 1 on day 0. In the control group, pigs were immunized with adjuvanted Dulbecco's PBS (DPBS, Gibco), containing Freund's complete adjuvant (Sigma-Aldrich, St. Louis, MO, USA). Pigs in VLP and VLP+CCL groups were injected with 1.8 mg VLP (containing 0.2 µg S protein) diluted in 0.5 mL adjuvanted DPBS with or without

30 µg CCL25 and 30 µg CCL28. When boosting on days 14 and 35, the formulations were identical to those used for priming, except that the Freund's complete adjuvant was replaced with Freund's incomplete adjuvant (Sigma-Aldrich). At 0, 14, 28, and 49 days post-prime immunization (DPPI), blood anti-coagulated using ethylenediamine tetraacetic acid (EDTA) was collected along with oral swabs for detecting IFN-γ-producing cells, systemic IgG and neutralizing antibody titers, and mucosal IgA titers. At 49 DPPI, all the pigs were orally challenged with 5 mL of 10^5 TCID$_{50}$/mL PEDVPT-P7 to evaluate the protective efficacy. Stool consistency was monitored daily along with the collection of fecal swabs for detecting viral shedding. The animal experimental procedure was reviewed and approved by the Institutional Animal Care and Use Committee of the National Taiwan University (Taiwan, China) with the approval no. NTU107EL-00105.

2.9. Evaluation of Systemic IgG and Mucosal IgA Levels

PEDV-specific antibodies in the plasma and saliva were detected using an in-house PEDV S-based indirect enzyme-linked immunosorbent assay, as described in the previous study [48]. Briefly, the 96-well flat-bottom microplates (Thermo Fisher Scientific) were coated with 2 µg/mL recombinant S protein diluted in coating buffer (KPL, Gaithersburg, MD, USA) at 4 °C overnight. Following that, each well was washed six times with 200 µL/well of washing buffer (KPL) using a microplate washer (BioTek Instruments, Inc., Winooski, VT, USA) and then blocked with 300 µL/well of blocking buffer (KPL) at RT for 1 h. After six times of washing, the plasma IgG was evaluated by adding 100 µL per well of 40-fold diluted plasma samples in blocking buffer (KPL) and incubating at RT for 1 h, while the salivary IgA titer was detected by adding 100 µL per well of two-fold diluted salivary supernatant in blocking buffer (KPL) and incubating at 4 °C overnight. Following washing at the end of incubation, 100 µL/well of HRP-conjugated goat anti-pig IgG (KPL) at a 1:1000 dilution and HRP-conjugated goat anti-pig IgA (Abcam, Cambridge, UK) at a 1:5000 dilution were used to detect porcine IgG and IgA, respectively. After incubation at RT prior to the wash step, 50 µL of ABST® Peroxidase Substrate System (KPL) was added to each well and the reaction was allowed to develop at RT for 5 and 45 min for IgG and IgA measurements, respectively. The reactions were stopped by adding 50 µL of stopping solution (KPL). The optical density (OD) values were read at 405 nm using the EMax® Plus Microplate Reader (Molecular Devices, Crawley, UK). The IgG and IgA titers have been expressed as sample-to-positive ratios (S/P ratios), defined as the difference between the OD values of the sample and the negative control divided by the difference between the OD values of the positive and negative controls. The positive control samples were plasma or salivary samples from pigs challenged with PEDV in previous experiments.

2.10. Neutralizing Antibody Assay

For the evaluation of neutralizing antibody titers, 100 µL of Vero cells were seeded into 96-well culture plates (Thermo Fisher Scientific) at a density of 3×10^5 cells/mL and incubated at 37 °C, 5% CO$_2$ overnight to reach 80–90% confluency. Plasma samples of the pigs were heated at 56 °C for 30 min to inactivate the complement. The ten-fold diluted, inactivated plasma samples were two-fold serially diluted in post-inoculation (PI) medium containing DMEM supplemented with 0.3% tryptose phosphate broth (Sigma-Aldrich), 0.02% yeast extract (Acumedia, Lansing, CA, USA), and 10 µg/mL trypsin (Gibco). Each well contained 50 µL of 100 TCID$_{50}$ PEDVPT-P5 and 50 µL of diluted plasma samples and was incubated at 37 °C, 5% CO$_2$, for 1 h. Subsequently, the mixture was added to 90%-confluent Vero cells following two washes with PI medium and allowed to incubate at 37 °C, 5% CO$_2$, for 1 h. The mixture was then removed and fresh PI medium was added and allowed to incubate at 37 °C, 5% CO$_2$, for one day. The cytopathic effects were then observed under an inverted light microscope (Nikon, Tokyo, Japan). The neutralizing titer was defined as the last dilution without cytopathic effects.

2.11. Isolation of Peripheral Blood Mononuclear Cells

For the functional assay of peripheral blood mononuclear cells (PBMCs), 10 mL of blood, containing 1 mL of 1% EDTA (Merck, Darmstadt, Germany) at pH 7.5–8.0, was collected and centrifuged at 1811× g, 4 °C, for 30 min. The buffy coat was harvested and diluted in 6 mL of RPMI-1640 medium (Gibco) for subsequent density gradient centrifugation. The diluted buffy coat was gently applied to an equal volume of Ficoll-Paque™ PLUS (GE Healthcare), and centrifuged at 1811× g, 20 °C, for 30 min. The isolated PBMCs, located at the interface of RPMI-1640 and Ficoll-Paque™ PLUS (GE Healthcare), were collected and mixed with three volumes of sterile ammonium chloride potassium (ACK) lysis buffer, containing 0.15 M NH_4Cl, 1.0 M $KHCO_3$, and 0.01 M EDTA at pH 7.2–7.4. After incubation at 4 °C for 5 min, the cells were centrifuged at 201× g, 20 °C, for 10 min to collect the erythrocyte-free pellets. The pellet was resuspended in RPMI-1640 medium and centrifuged at 129× g, 20 °C, for 10 min to get rid of the platelets. The platelet-free pellets were then diluted to a final concentration of 3×10^6 PBMCs/mL in CTL-Test™ medium (Cellular Technology, LLC, Cleveland, OH, USA) for subsequent use.

2.12. Enzyme-Linked Immunospot Assay of PEDV S-Specific IFN-γ

According to the manufacturer's instructions, the total PEDV S-specific IFN-γ secreting-cells were analyzed using enzyme-linked immunospot (ELISPOT) assay with anti-porcine IFN-γ pre-coated plates and detecting antibodies purchased from Cellular Technology. The freshly isolated PBMCs were seeded into the anti-porcine IFN-γ pre-coated plates at a density of 3×10^5 cells/well and incubated at 37 °C for 24 h with CTL-Test™ medium (mock) or CTL-Test™ medium containing 10 µg/mL of in-house full-length recombinant S (treatment) [48] or 0.1 µg/mL concanavalin A (Sigma-Aldrich) (positive control). After incubation for one day, IFN-γ detection and color development were performed according to the manufacturer's protocol. The scanning and counting were performed using CTL ImmunoSpot® analyzers and the results were analyzed using ImmunoSpot® software version 7.0.23.2.

2.13. Stool Consistency Scoring and Body Weight Measurement

The clinical signs for each pig were monitored and recorded daily. Based on previous studies [22,46], the severity of diarrhea was graded as 0: normal consistency; 1: loose consistency; 2: semi-fluid consistency; 3: liquid consistency. The body weight (BW) of each pig was measured weekly.

2.14. RNA Extraction, Complementary DNA Synthesis, and Probe-Based Quantitative Real-Time PCR

To detect fecal viral shedding after the viral challenge, feces collected from rectal swabs were resuspended in 900 µL of DPBS (Gibco) and mixed using a vortex. The resuspended samples were centrifuged at 13,793× g for 10 min. The viral RNA was extracted using the Cador® Pathogen 96 QIAcube® HT Kit according to the manufacturer's instructions (Qiagen, Hilden, Germany). Reverse transcription was performed using the QuantiNova™ Reverse Transcription Kit (Qiagen) to synthesize cDNA for subsequent quantitative real-time PCR, as described previously [49]. The detection limit of the assay was 4.7 $\log_{10}$ RNA copies per mL based on the standard curve of the in vitro transcribed PEDV RNA.

2.15. Statistical Analysis

The collected data were characterized by a small sample size, missing data, and independence between repeated measurements during the period of study with a non-normal distribution, as checked using the Shapiro–Wilk test. Traditional modeling techniques, such as repeated measures analysis of variance, reduce the statistical power and contribute to interpretation issues by list-wise deletion of missing data and data distortion after transformation [50,51]. The use of alternative statistical methods, such as GEE and linear mixed effects, for analysis of the longitudinal data overcomes the problems of valuable data reduction and inflexible correlation structure. Besides, several studies have proved that these advanced statistical methods are capable of enhancing power even with a small sample size and missing data as compared to the traditional models [52,53]. In the collected data set, GEE as an extension of the generalized linear model is preferable over linear mixed effects due to the robust standard errors, no limitation of normality, and more emphasis on the population-level trajectories rather than within-subject changes [54]. The aforementioned advantages render GEE increasingly popular in clinical trials.

A descriptive statistical analysis of all the experimental data was performed using a 95% confident interval for mean values to summarize the sample features. GEE with an exchangeable correlation structure and identity link function was used for further inferential statistics. The outcome variables (systemic IgG, oral IgA, neutralizing antibody titers, and fecal viral shedding) were modeled with a treatment factor (control; VLP; VLP+CCL), a repeated measure factor (pre-vaccination; 14, 28, and 49 DPPI), and a BW factor. Based on the significant interaction terms (group × BW; time × BW; group × time × BW), BW was considered as a covariate; therefore, the model was adjusted at a fixed BW before further statistical analysis. Given that the significant interaction term (group × time) was noted in all the data sets, the results were presented as post hoc comparisons of the simple main effects, which revealed the effect of the different treatments at different times. The data were analyzed using SPSS (SPSS for Mac, v. 24.0; IBM, Chicago, IL, USA). A p value of 0.05 was considered statistically significant. All the graphics were prepared using GraphPad Prism 6.0 (GraphPad software, San Diego, CA, USA). Results have been expressed as mean ± standard error of the mean (SEM).

3. Results

3.1. Preparation and Characterization of PEDV VLPs Expressed Using Recombinant Baculovirus, SME-Bac

After the transfer construct pFastBac1-HM6H-PEDV-S-2A-M-Pnv339-eGFP-Rhir-E in the recombinant baculovirus, SME-Bac was transduced into Sf21 cells, the recombinant S protein expression and VLP production were evaluated using Western blot. Using a previously characterized PEDV S-displaying baculovirus (S-Bac) [22] as the positive control, the His-tagged S protein was successfully detected in the culture medium. The amounts of S protein in the VLP-containing supernatants after three–five days of infection with 1, 3, and 5 MOI of SME-Bac are shown in Figure 2A. The S proteins were slightly larger than 170 kDa in size. The optimal VLP production condition was found to be five days after infection with 1 MOI of SME-Bac. To determine the recombinant S protein being displayed on the surface of the VLPs, the infected Sf21 cells were probed using an indirect immunofluorescence assay with the PEDV S monoclonal antibody and Alexa Flour® 594-conjugated secondary antibody. Compared to the non-infected cells, there were strong fluorescence signals on the plasma membrane of Sf21 cells transduced with SME-Bac (Figure 3I), indicating that the S proteins were successfully being expressed on the surface of SME-Bac-infected Sf21 cells. Furthermore, to identify the S protein expressed on the VLPs, the Western blotting detected by PEDV-challenged porcine serum was performed. The result is demonstrated in Figure 2B. The size of protein bands pointed out by the arrow heads was approximately 200 kDa, indicative of the S protein.

To evaluate the stability of P-BEVS that express PEDV VLP, we compared the infectivity and S protein expression level of passages 4, 11, 14, or 15 of SME-Bac. In the SME-Bac, EGFP, translated under the control of PnV339 IRES, could be used to monitor the infectivity of the recombinant baculovirus during VLP preparation after multiple passages. As shown in Figure 4A, the green fluorescence was similar between Sf21 cells infected with SME-Bac passages 4, 11, 14, and 15. However, the EGFP still can be translated via the cap-dependent mechanism when the S-2A-M sequences are lost during passaging. To further confirm the results, we monitored the S proteins expression in the passages 4, 11, 14, and 15. It showed that the expression level of S protein in both passages were consistent and similar (Figure 4B). Thus, the stability of SME-Bac could sustain at least 15 passages in the present study.

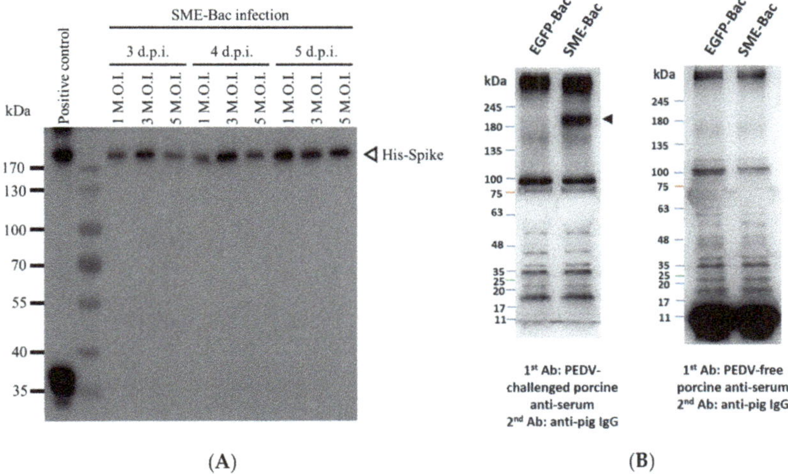

(A)   (B)

Figure 2. The detection of PEDV spike (S) proteins in the infected Sf21 supernatant in various virus-like particles (VLP) production conditions by Western blot. (**A**) The samples were purified by conducting sucrose cushion and detected by anti-His tag antibodies. Positive control was Sf21 cells infected with S-Bac. (**B**) The Western blots of the VLPs stained with PEDV-challenged porcine serum were conducted. The protein bands of PEDV S protein are indicated by arrow heads. The EGFP-Bac was the sample collected from Sf21 cells infected with pFastBac1-HM6H-P-2A-PnV339-eGFP-Rhir and acted as a negative control.

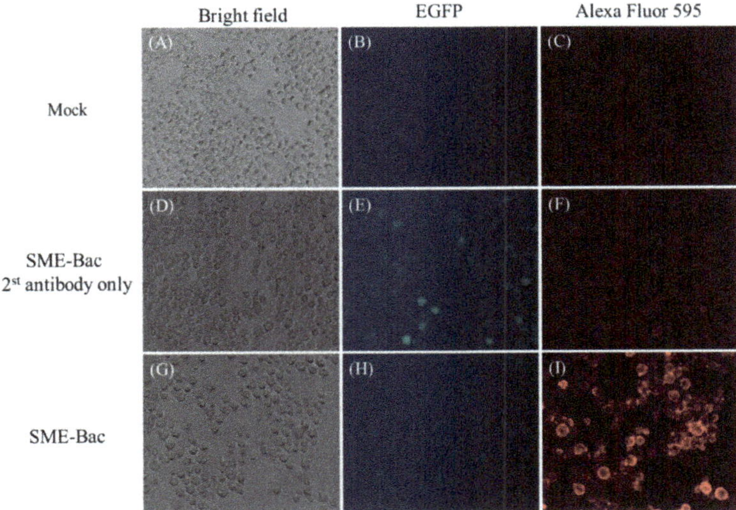

Figure 3. The detection of recombinant spike (S) proteins in SME-Bac infected cells. Sf21 cells in a total number of 2×10^5 cells were infected with 5 MOI of SME-Bac for 4 days. (**A,D,G**) The morphologies of the Sf21 cells with different treatments under bright field. (**B,E,H**) The Sf21 cells successfully infected by SME-Bac showed green fluorescence under fluorescent microscope. (**C,F,I**) In the indirect fluorescent assay, Sf21 cells expressing PEDV S protein displayed red fluorescent signals under fluorescent microscope. The mock-infected cells and SME-Bac-infected cells stained with secondary antibody were used as negative controls.

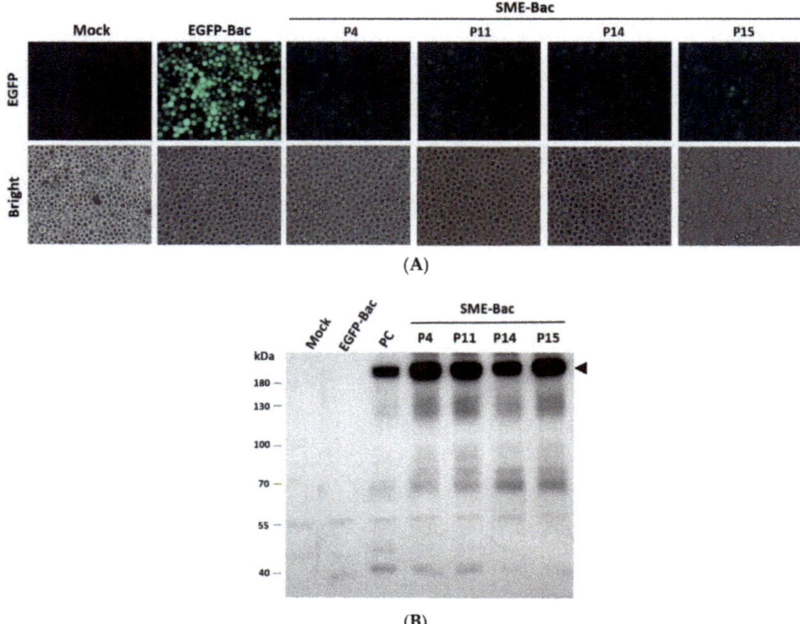

Figure 4. Detection of PEDV spike protein expression in Sf21 cells infected with different passages of SME-Bac. (**A**) The expression level of EGFP of Sf21 cells infected with 1 MOI of EGFP-Bac or SME-Bac passages 4, 11, 14, or 15 (P4, P11, P14, or P15) for 4 days were observed by fluorescence microscopy. (**B**) The PEDV S protein expression in the cell lysates after SME-Bac passages 4, 11, 14, or 15 infection was detected by Western blotting. The arrow head demonstrates the amount of S protein in different passages.

3.2. Negative Staining Electron Microscopy of PEDV VLPs

To investigate whether the co-expressed S, M, and E proteins can successfully assemble into VLPs, the sample was collected and purified from the supernatant of SME-Bac-infected Sf21 cells, and then examined using transmission electron microscopy (TEM). The TEM image has been shown in Figure 5. There were numerous VLPs, approximately 100 nm in diameter, displaying similar morphology to coronavirus (black arrows and inset figure) and some rod-shaped virions, approximately 200 nm in size, resembling baculovirus (white arrows). Although a small number of baculoviruses were precipitated together with the VLPs, further purification of VLP for removing the baculovirus was not performed. Since baculovirus itself can elicit innate immune responses by regulating cytokines and promote B cell and T cell activation [55–57], the remaining baculoviruses are used as an adjuvant to enhance the effect of the VLPs.

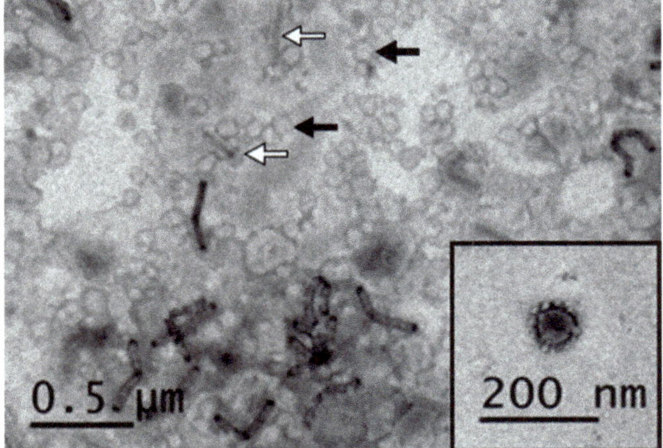

Figure 5. PEDV-like particles released in the culture supernatant of 1 MOI SME-Bac-infected Sf21 cells after 5-day infection. The electron micrograph demonstrates the morphology of VLPs (black arrows and inset figure) and baculovirus virions (white arrows).

3.3. Changes in Body Weight

The BW of each pig was measured weekly during the experiment. The BW showed a linear increase in all the groups. However, there was no significant difference in the mean BW among the three groups (Figure 6).

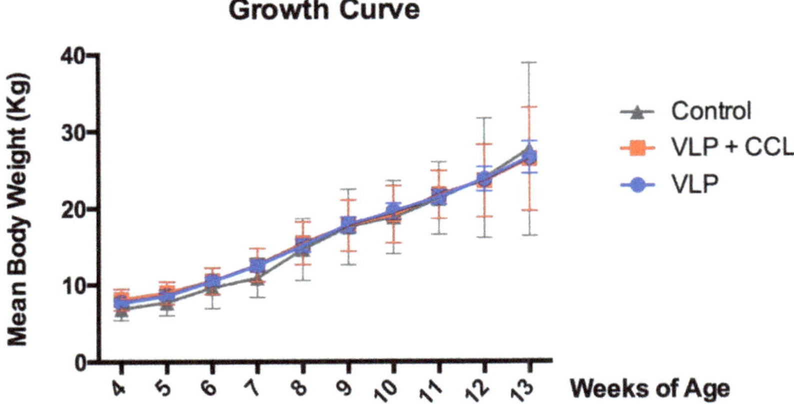

Figure 6. Changes of the weekly body weight. The body weights of all pigs in each group were measured every week. The weekly averaged body weights in each group are demonstrated as mean ± SEM. The mean values in control, VLP, and VLP+CCL groups are presented as gray, blue, and red lines, respectively.

3.4. Detection of Systemic and Mucosal S-Specific Antibody Titers

Compared to the control group, elevated IgG titers in the plasma were observed in both the VLP and VLP+CCL groups post-immunization. At 49 DPPI, the titers of control, VLP, and VLP+CCL groups were 0.06 ± 0.01, 0.41 ± 0.12, and 0.69 ± 0.13, respectively (Figure 7). The statistical method of GEE was used to assess the systemic IgG and revealed the main effects of time (Wald chi-square = 42.504, $p < 0.001$), treatment (Wald chi-square = 116.400, $p < 0.001$), and BW (Wald chi-square = 5.896,

$p = 0.015$). Significant interactions of treatment × time (Wald chi-square = 724.532, $p < 0.001$), treatment × BW (Wald chi-square = 27.901, $p < 0.001$), time × BW (Wald chi-square = 16.578, $p = 0.001$), and treatment × time × BW (Wald chi-square = 136.937, $p < 0.001$) were presented. After the interaction effects of BW were removed, post hoc comparisons of the simple main effects revealed that systemic IgG levels were significantly higher in the VLP and VLP+CCL groups than in the control group at 14 DPPI, and significantly higher in the VLP+CCL group than in the control and VLP groups at 28 and 49 DPPI.

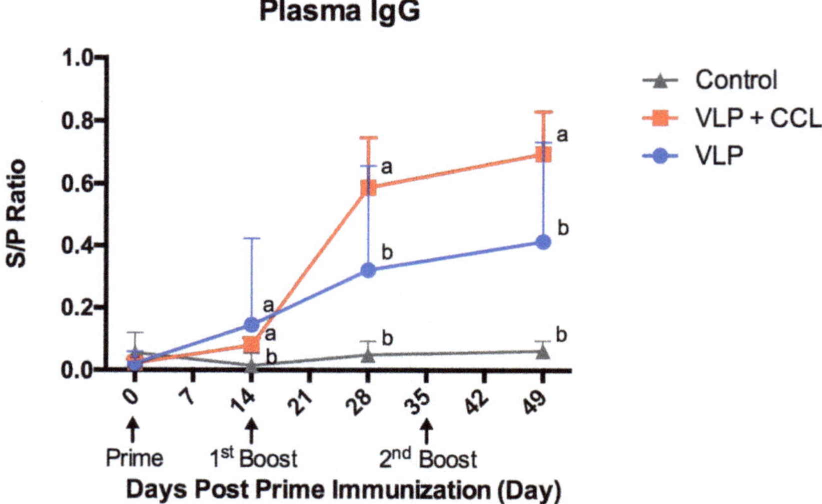

Figure 7. Systemic PEDV S-specific IgG titers following the prime, 1st boost, and 2nd boost. Pigs were immunized with different regimens at 0, 14, and 35 DPPI. Blood was collected at 0 (pre-priming), 14, 28, and 49 DPPI for evaluating PEDV-specific IgG titer by the PEDV S-based ELISA. The results are shown as averaged values of sample-to-positive ratios (S/P ratio) with error bars representing the SEM. The values in control, VLP, and VLP+CCL groups are demonstrated by gray, blue, and red lines, respectively. Different alphabets indicate significant differences among different groups ($p < 0.05$).

Compared to the control group, the oral IgA S/P titers elicited in the VLP and VLP+CCL groups at 49 DPPI were 0.16 ± 0.05 and 0.15 ± 0.05, respectively (Figure 8). Statistical analysis of oral IgA revealed the main effects of time (Wald chi-square = 21.983, $p < 0.001$), treatment (Wald chi-square = 11.703, $p = 0.020$), and BW (Wald chi-square = 5.674, $p = 0.017$). Significant interactions of treatment × time (Wald chi-square = 297.607, $p < 0.001$), treatment x BW (Wald chi-square = 19.334, $p = 0.001$), time × BW (Wald chi-square = 19.783, $p = 0.001$), and treatment × time × BW (Wald chi-square = 157.940, $p < 0.001$) were presented. Following the removal of the interaction effects of BW, post hoc comparisons of the simple main effects indicated that the mucosal IgA levels were significantly higher in the VLP+CCL group than in the control group at 49 DPPI.

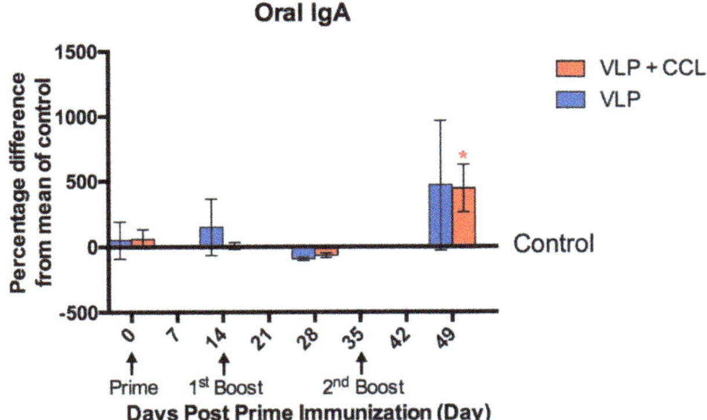

Figure 8. Oral PEDV S-specific IgA titers after the prime, 1st boost, and 2nd boost immunizations. Pigs were immunized at 0, 14, and 35 DPPI. Oral swabs were collected at day 0, 14, 28, and 49 DPPI to evaluate PEDV-specific IgA in the saliva by the PEDV S-based ELISA. The data are displayed as the percentage difference from the mean of control, which is defined as the percentage difference between treatment and control group divided by the mean of the control group, with SEM at different time points. The results of control, VLP, and VLP+CCL groups are presented as gray, blue, and red bars, respectively. The asterisk indicates the significant statistical difference between treatment and control groups ($p < 0.05$).

3.5. Evaluation of Neutralizing Antibody Titers in the Blood

Titers of neutralizing antibodies in the blood were elevated in both the VLP and VLP+CCL groups at 28 DPPI (mean ± SEM) but slightly decreased at 49 DPPI (mean ± SEM) (Figure 9). The main effects of time (Wald chi-square = 6.073, $p = 0.048$), treatment (Wald chi-square = 13.107, $p = 0.001$), and BW (Wald chi-square = 1.100, $p = 0.294$) were identified using statistical analysis. Interactions of treatment × time (Wald chi-square = 28.809, $p < 0.001$), treatment × BW (Wald chi-square = 8.196, $p = 0.017$), and treatment × time × BW (Wald chi-square = 21.322, $p < 0.001$) were significant, while time × BW (Wald chi-square = 0.621, $p = 0.733$) was non-significant. After the BW was adjusted, post hoc comparisons of the simple main effects revealed significant differences in the neutralizing antibody levels in the VLP and VLP+CCL groups at 28 DPPI, and in the VLP+CCL group at 49 DPPI, compared to the control group.

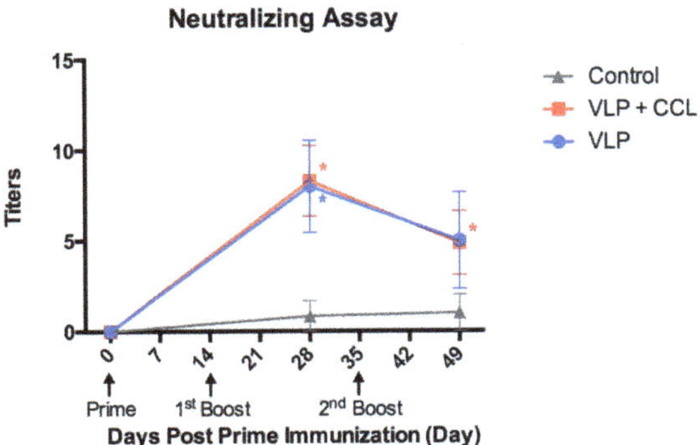

Figure 9. The titers of plasma neutralizing antibodies against PEDV following the prime, 1st boost, and 2nd boost immunization. The neutralizing activity against PEDVPT-P5 was performed. The values are displayed as mean ± SEM and presented as gray, blue, and red lines of control, VLP, and VLP+CCL groups, respectively. The asterisk indicates the significant statistical difference between treatment and control groups ($p < 0.05$).

3.6. Assessment of S-Specific Interferon-γ-Secreting Cells in the PBMCs

To evaluate the specific cellular immunity against PEDV, we quantified the endpoint PEDV-S specific IFN-γ-secreting T-cells in PBMCs using the ELISPOT assay. Although the mean values of the VLP and VLP+CCL groups were 31.29 ± 8.59 and 36.14 ± 12.72 spot counts per well, which were higher than those in the control group (16.60 ± 7.44 spot counts per well) (Figure 10), there was no significant difference among the different groups.

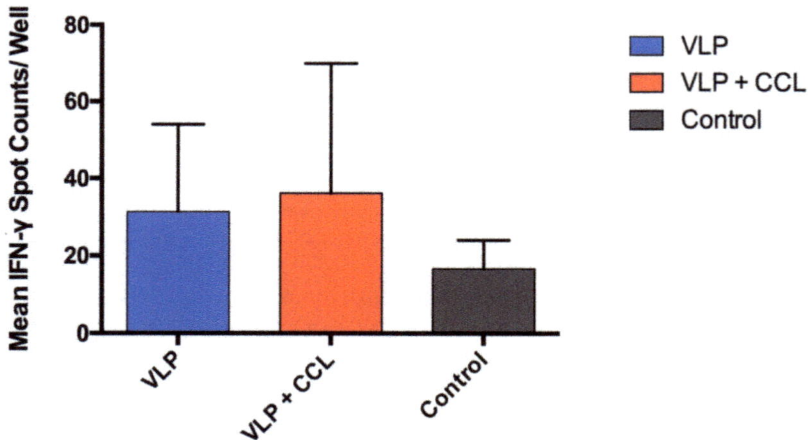

Figure 10. The result of PEDV S-specific Interferon-γ-secreting cell count in the peripheral blood mononuclear cells (PBMCs). The PBMCs were prepared from the peripheral blood of pigs at 49 DPPI. Interferon-γ-secreting cells were enumerated by the ELISPOT assay. The data are shown as mean ± SEM and the results of control, VLP, and VLP+CCL groups are, respectively, illustrated as gray, blue, and red bars.

3.7. Evaluation of the Protection Provided by the VLP Adjuvant with/without CCL25 and CCL28 against Virulent PEDV Challenge

To evaluate the protection offered by the different regimens, all the pigs were orally challenged with PEDVPT-P7. The onset time of diarrhea in the pigs was variable and ranged from three to six days post-challenge (DPC). Upon determination of the peak fecal score in the control group (six DPC, $n = 4$), two pigs showed watery diarrhea (score 3) but the other two pigs presented normal feces (score 0). Comparatively, when the peak fecal score was determined in the VLP group (six DPC, $n = 5$), moderate diarrhea (score 2) was observed in two pigs, mild diarrhea (score 1) in one pig, and normal feces (score 0) in two pigs, while only three pigs in the VLP+CCL group ($n = 5$) showed disconnected mild diarrhea (score 1) over three–eight DPC. The total scores of the control and VLP groups gradually decreased over six–nine DPC and no clinical signs were observed in any of the groups over 10–13 DPC. Overall, pigs immunized with VLP and VLP+CCL showed milder diarrhea symptoms than those in the control group (Figure 11A).

For quantification of PEDV loads in the feces, a PEDV N-based real-time RT-PCR was performed. In the control group, viral shedding started with 1.73 ± 3.46 $\log_{10}$ copies/mL at three DPC, reached the peak of 4.26 ± 4.92 $\log_{10}$ copies/mL at five DPC, and then declined after six DPC. In the VLP group, viral shedding was detected as 2.27 ± 3.18 $\log_{10}$ copies/mL at four DPC and fluctuated over four–eight DPC with a peak of 2.66 ± 3.65 $\log_{10}$ copies/mL at five DPC. Pigs in the VLP+CCL group exhibited average fecal viral shedding of 2.28 ± 3.20 $\log_{10}$ copies/mL at three DPC and lasted for six days with a peak of 2.75 ± 3.79 $\log_{10}$ copies/mL at four DPC (Figure 11B). Statistical analysis revealed the main effects of time (Wald chi-square = 225.571, $p < 0.001$), treatment (Wald chi-square = 10.095, $p = 0.039$), and BW (Wald chi-square = 0.097,

$p = 0.755$). Significant interactions of treatment × time (Wald chi-square = 6.345×10^{11}, $p < 0.001$), treatment × BW (Wald chi-square = 11.397, $p = 0.022$), time × BW (Wald chi-square = 224.419, $p < 0.001$), and treatment × time × BW (Wald chi-square = 55716955.8, $p < 0.001$) were presented. After the covariates of BW values in the model were fixed, although there were no significant differences in the viral shedding among all the groups, the pigs in the control group exhibited higher peak viral shedding than the other two groups.

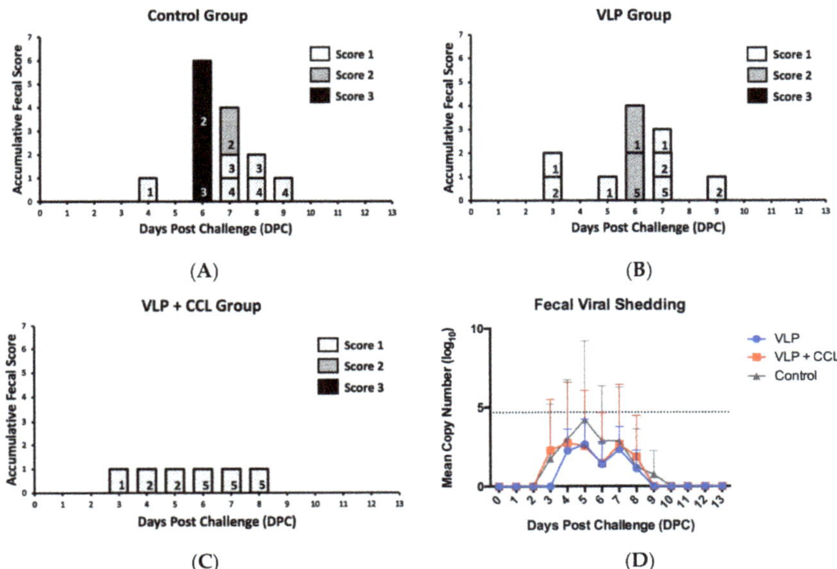

Figure 11. Evaluation of the protective efficacy of different treatments. Pigs in all groups were challenged by the highly virulent porcine epidemic diarrhea virus Pintung 52 (PEDV-PT) strain passage 7. The post-challenge pigs were monitored for stool consistency and fecal viral loads for 13 days. (**A–C**) The result of stool consistency scoring in each group. According to the stool consistency, the stool was graded as 0 for normal; 1 for loose feces; 2 for semi-fluid feces; and 3 for watery feces. The total number of pigs in the control, VLP, and VLP + CCL groups was 4, 5, and 5 pigs, respectively. The Arabic numerals labeled in the bar indicate the individuals in each group. (**D**) The result of fecal viral shedding in each group detected by probe-based quantitative reverse transcription PCR (RT-qPCR) targeting the PEDV N gene. The results were presented as mean value of $\log_{10}$ RNA copies/mL ± SEM. The detection limit of RT-qPCR was 4.7 $\log_{10}$ (copies/mL) marked as a dotted line.

4. Discussion

In this study, PEDV VLPs were generated and characterized for developing a safe and potent immunogen against highly virulent strains in pigs. To induce effective protection via a convenient parenteral route, we incorporated CCL25 and CCL28, which have been shown to be effective in our previous study [46], as mucosal adjuvants in this strategy. Our results demonstrated that the regimen is capable of eliciting not only systemic PEDV S-specific IgG and IFN-γ-producing cells in PBMCs but also mucosal PEDV S-specific IgA. Compared to the control group, the clinical signs in pigs of both the VLP and VLP+CCL groups were markedly palliated accompanied by lower viral shedding without watery diarrhea. Therefore, PEDV VLP formulated with CCL25 and CCL28 may be a potential PEDV vaccine candidate and the strategy might serve as a platform for the development of other enteric viral vaccines.

In this study, the S gene in the P-BEVS vector was in the same ORF as the M protein flanking with the 2A-like peptide sequence derived from PnV viruses [58]. Thus, the S and M proteins were translated by the same ribosome with the same yield. The E protein was controlled by the RhPV IRES, which would mediate the cap-independent translation through the same mRNA carrying the coding sequences of S and M proteins. Thus, the VLP of PEDV generated by the P-BEVS system should express the S, M, and E proteins simultaneously in the SME-Bac-infected Sf21 cells. In the present study, the detection of S, M, and E proteins by using PEDV-challenged porcine serum was performed and only the S protein was successfully detected in the SME-Bac VLP. We speculate that the failure to detect the M and E proteins using PEDV hyperimmune porcine serum by Western blotting might be due to two possibilities. First, the E and M proteins in VLPs derived from the P-BEVS system may not produce glycoproteins to generate complex M and E proteins, as the insect cell-produced glycoproteins have clearly different N-glycans from those produced by mammalian cells [59,60]. The protein structure and immunogenicity of the E and M proteins of the SME-Bac VLP might exhibit some differences from those of PEDV virions. Second, poor immunogenicity of the PEDV E protein has also been previously demonstrated [61]. Due to these detection limits, the Western blot and IFA were performed to confirm the expression of the S protein and TEM was performed to demonstrate the formation of the VLPs.

Humoral and cellular immunity play an indispensable role in the generation of an effective vaccine [62]. It is well known that lactogenic passive immune-transferring pathogen-specific IgA is one of the effective strategies to protect newborn piglets, which have immature immune systems, against G2 PEDV infection [63]. However, the crucial role of memory T cell responses in PEDV has also been proposed to protect pigs from reinfection by displaying undetected fecal viral shedding and absence of systemic and mucosal antibody responses [64]. Therefore, vaccines that can elicit both humoral and cellular immune responses might be potent in preventing the disease. Herein, after three intramuscular injections, serological and IFN-γ-secreting cell measurements revealed that the VLP and VPL+CCL groups were able to stimulate both immune responses in pigs. Both humoral and cellular immunities are measured by the interaction with in-house recombinant S protein, which is well-established regarding the confirmations of its biological function and immunogenicity in our previous studies [21,22,51,65,66] and the result could more correlate with the clinical protectivity than measured by the interaction with inactivated virions. The successful induction of both humoral and cellular immunities in the condition of using a lower amount of S protein (0.2 μg/dose) in our VLP regimen than that in other subunit vaccine studies [21,67,68] suggests that VLP is an effective strategy to induce potent humoral and cellular immune responses [69–71].

Secretory IgA, which is the first line of mucosal immunity, was observed to be significantly elevated after the second boost in the VLP+CCL group as compared to the control group. The statistical result seems contradictory to the raw data, which represented similar mean IgA S/P ratios between the VLP and VLP+CCL groups at 49 DPPI. Such a contradiction in the results could have arisen depending on whether the effects of BW are taken into consideration or not. The observation of enhanced IgG, IgA, and neutralizing titers in the presence of CCL25 and CCL28 is comparable to many related published reports [42,44,45,72,73] and serves as evidence that CCL25 and CCL28 are involved in chemotaxis and immunostimulation [74,75]. In addition, a milder clinical sign was also observed in pigs of the VLP+CCL group as compared to the VLP group. However, when compared to the animals immunized using the inactivated virus formulated with CCL25/28 in a previous study [46], pigs in the VLP+CCL group showed relatively less protection against the PEDV challenge. It might be due to the PEDV exposure age at 11 weeks old and failure to elicit the optimal immune response, as it can be observed that the mean PEDV S-specific IgG titers were relatively low, with mean S/P ratios of around 0.4 and 0.7 in the VLP and VLP-CCL groups, respectively. However, based on the results of our previous work, the titer of systemic IgG or neutralizing antibodies might play a minor role in the protection against PEDV [21]. In the context of the pigs injected with the same dose of CCL25/28 as in the previous study [46], the fair effect of the immunization might be caused by the suboptimal antigen concentration in the VLPs. On the other hand, the use of the soluble chemokines may also contribute to the ineffective stimulation of immune responses. Several studies have indicated that chemokine-incorporated VLPs can stimulate robust antigen-specific immune responses, while modest immune responses are noted in VLPs with soluble chemokines [76,77], highlighting the influence of co-delivery of an antigen and chemokine on promoting effective immune stimulation. Hence, the adequate regimens of VLP and CCL or even co-delivery of all the components still need further optimization and should also be applied to sows to evaluate the protection for litters via lactogenic immunity.

The immune responses elicited by immunization are affected by multiple factors, such as intrinsic host factors, perinatal host factors, and nutritional factors [78]. To evaluate the immunogenicity and potential protectivity of the novel VLP immunogen, an appropriate animal model is important for the preclinical investigation. In the present study, PEDV-seronegative post-weaning pigs were used for the VLP immunization and viral challenging experiments. This animal model has been well-established in our previous studies for preliminarily evaluating the immunogenicity of potential PEDV vaccine candidates [21,46]. After confirming the VLP regimen is capable of eliciting PEDV-specific humoral and cellular responses, considering that vulnerable suckling piglets should be protected by colostrum antibody transferred from immunized sows [63], the VLP in combination with chemokines strategy should be applied to gilts and sows to evaluate the immunogenicity and protective efficacy of lactogenic immunity in neonatal piglets against PEDVs.

To induce immunity and protectivity in pigs against the emerging G2b PEDV strains, we have successfully generated the G2b PEDV-based VLP vaccine and demonstrated the efficacy against a homologous G2b PEDV challenge in pigs. It has been reported that memory $CD4^+$ T cells against human cold coronaviruses, such as human coronavirus (HCoV) OC43, HCoV-229E, HCoV-NL63, and HCoV-HKU1, and monoclonal IgA against severe acute respiratory syndrome coronavirus (SARS-CoV) can provide cross-reactivity against severe acute respiratory syndrome coronavirus 2 (SARS-CoV-2) [79,80]. As for the PEDVs, the nucleotide sequences of the S protein among G1 and G2 strains differ within 10%, and antiserum of G2 PEDVs has been proved to be able to provide partial cross-reactivity against G1 PEDVs and vice versa [15,81,82]. Accordingly, the vaccines derived from G2 strains might cross-protect pigs against G1 PEDV infection. The efficacy of our G2b-based PEDV VLP vaccine against G1 PEDVs should also be evaluated in the future.

In the present study, although the regimen used for immunization of the VLP+CCL group still needs to be modified, the strategy was able to induce mucosal and systemic immune responses via intramuscular administration. In addition, it also provided partial protection by resulting in palliated clinical signs and reduced viral shedding following challenge with the highly virulent PEDV strain. Additionally, VLP derived from P-BEVS serves as a potent immunogen that is capable of inducing humoral and cellular immunities in pigs. Of note, this study also points out the importance of integrating BW as a covariate into evaluation when investigating vaccine efficacy. In summary, VLP in combination with CC chemokines could be a promising candidate for mucosal vaccines against other enteric or mucosal pathogens.

Author Contributions: Conceptualization, H.-W.C. and T.-Y.W.; methodology, T.-Y.W., C.-W.H., and M.-H.C.; software, C.-W.H.; validation, H.-W.C., Y.-C.C., and T.-Y.W.; formal analysis, C.-W.H.; investigation, C.-W.H.; resources, H.-W.C., Y.-C.C., and T.-Y.W.; data curation, C.-W.H. and M.-H.C.; writing—original draft preparation, C.-W.H. and M.-H.C.; writing—review and editing, H.-W.C., Y.-C.C., and T.-Y.W.; visualization, H.-W.C., Y.-C.C., and T.-Y.W.; supervision, H.-W.C., Y.-C.C., and T.-Y.W.; project administration, H.-W.C., Y.-C.C., and T.-Y.W.; funding acquisition, H.-W.C., Y.-C.C., and T.-Y.W. All authors have read and agreed to the published version of the manuscript.

Funding: This research was funded by Ministry of Science and Technology, Taiwan, China and the grant number are MOST 109-2321-B-033-001, 109-2313-B-002-016-MY3 and 109-2313-B-002-052.

Conflicts of Interest: The authors declare no conflict of interest.

References

1. Lee, C. Porcine Epidemic Diarrhea Virus: An Emerging and Re-Emerging Epizootic Swine Virus. *Virol. J.* **2015**, *12*. [CrossRef]
2. Li, W.; van Kuppeveld, F.J.M.; He, Q.; Rottier, P.J.M.; Bosch, B.J. Cellular Entry of the Porcine Epidemic Diarrhea Virus. *Virus Res.* **2016**, *226*. [CrossRef] [PubMed]

3. Nam, E.; Lee, C. Contribution of the Porcine Aminopeptidase N (CD13) Receptor Density to Porcine Epidemic Diarrhea Virus Infection. *Vet. Microbiol.* **2010**, *144*. [CrossRef] [PubMed]

4. Costantini, V.; Lewis, P.; Alsop, J.; Templeton, C.; Saif, L.J. Respiratory and Fecal Shedding of Porcine Respiratory Coronavirus (PRCV) in Sentinel Weaned Pigs and Sequence of the Partial S-Gene of the PRCV Isolates. *Arch. Virol.* **2004**, *149*. [CrossRef] [PubMed]

5. Masters, P.S. The Molecular Biology of Coronaviruses. *Adv. Virus Res.* **2006**, *66*. [CrossRef]

6. Venkatagopalan, P.; Daskalova, S.M.; Lopez, L.A.; Dolezal, K.A.; Hogue, B.G. Coronavirus Envelope (E) Protein Remains at the Site of Assembly. *Virology* **2015**, *478*. [CrossRef] [PubMed]

7. Debouck, P.; Pensaert, M.; Coussement, W. The Pathogenesis of an Enteric Infection in Pigs, Experimentally Induced by the Coronavirus-like Agent, CV 777. *Vet. Microbiol.* **1981**, *6*. [CrossRef]

8. Li, B.X.; Ge, J.W.; Li, Y.J. Porcine Aminopeptidase N Is a Functional Receptor for the PEDV Coronavirus. *Virology* **2007**, *365*. [CrossRef]

9. Jung, K.; Saif, L.J. Porcine Epidemic Diarrhea Virus Infection: Etiology, Epidemiology, Pathogenesis and Immunoprophylaxis. *Vet. J.* **2015**, *204*. [CrossRef]

10. Shibata, I.; Tsuda, T.; Mori, M.; Ono, M.; Sueyoshi, M.; Uruno, K. Isolation of Porcine Epidemic Diarrhea Virus in Porcine Cell Cultures and Experimental Infection of Pigs of Different Ages. *Vet. Microbiol.* **2000**, *72*. [CrossRef]

11. Stevenson, G.W.; Hoang, H.; Schwartz, K.J.; Burrough, E.R.; Sun, D.; Madson, D.; Cooper, V.L.; Pillatzki, A.; Gauger, P.; Schmitt, B.J.; et al. Emergence of Porcine Epidemic Diarrhea Virus in the United States: Clinical Signs, Lesions, and Viral Genomic Sequences. *J. Vet. Diagn. Investig.* **2013**, *25*. [CrossRef] [PubMed]

12. Li, W.; Li, H.; Liu, Y.; Pan, Y.; Deng, F.; Song, Y.; Tang, X.; He, Q. New Variants of Porcine Epidemic Diarrhea Virus, China, 2011. *Emerg. Infect. Dis.* **2012**, *18*. [CrossRef]

13. Song, D.; Park, B. Porcine Epidemic Diarrhoea Virus: A Comprehensive Review of Molecular Epidemiology, Diagnosis, and Vaccines. *Virus Genes* **2012**, *44*. [CrossRef] [PubMed]

14. Song, D.; Moon, H.; Kang, B. Porcine Epidemic Diarrhea: A Review of Current Epidemiology and Available Vaccines. *Clin. Exp. Vaccine Res.* **2015**, *4*. [CrossRef] [PubMed]

15. Wang, X.; Chen, J.; Shi, D.; Shi, H.; Zhang, X.; Yuan, J.; Jiang, S.; Feng, L. Immunogenicity and Antigenic Relationships among Spike Proteins of Porcine Epidemic Diarrhea Virus Subtypes G1 and G2. *Arch. Virol.* **2016**, *161*. [CrossRef]

16. Makadiya, N.; Brownlie, R.; Van Den Hurk, J.; Berube, N.; Allan, B.; Gerdts, V.; Zakhartchouk, A. S1 Domain of the Porcine Epidemic Diarrhea Virus Spike Protein as a Vaccine Antigen. *Virol. J.* **2016**, *13*. [CrossRef]

17. Khamis, Z.; Menassa, R. Porcine Epidemic Diarrhea Virus. In *Prospects of Plant-Based Vaccines in Veterinary Medicine*; Springer International Publishing: Berlin/Heidelberg, Germany, 2018; pp. 255–266, ISBN 9783319901374.

18. Crawford, K.; Lager, K.M.; Kulshreshtha, V.; Miller, L.C.; Faaberg, K.S. Status of Vaccines for Porcine Epidemic Diarrhea Virus in the United States and Canada. *Virus Res.* **2016**, *226*. [CrossRef]

19. Kao, C.F.; Chiou, H.Y.; Chang, Y.C.; Hsueh, C.S.; Jeng, C.R.; Tsai, P.S.; Cheng, I.C.; Pang, V.F.; Chang, H.W. The Characterization of Immunoprotection Induced by a Cdna Clone Derived from the Attenuated Taiwan Porcine Epidemic Diarrhea Virus Pintung 52 Strain. *Viruses* **2018**, *10*, 543. [CrossRef]

20. Chang, Y.C.; Kao, C.F.; Chang, C.Y.; Jeng, C.R.; Tsai, P.S.; Pang, V.F.; Chiou, H.Y.; Peng, J.Y.; Cheng, I.C.; Chang, H.W. Evaluation and Comparison of the Pathogenicity and Host Immune Responses Induced by a G2b Taiwan Porcine Epidemic Diarrhea Virus (Strain Pintung 52) and Its Highly Cell-Culture Passaged Strain in Conventional 5-Week-Old Pigs. *Viruses* **2017**, *9*, 121. [CrossRef]

21. Chang, Y.C.; Chang, C.Y.; Tsai, P.S.; Chiou, H.Y.; Jeng, C.R.; Pang, V.F.; Chang, H.W. Efficacy of Heat-Labile Enterotoxin B Subunit-Adjuvanted Parenteral Porcine Epidemic Diarrhea Virus Trimeric Spike Subunit Vaccine in Piglets. *Appl. Microbiol. Biotechnol.* **2018**, *102*. [CrossRef]

22. Chang, C.Y.; Hsu, W.T.; Chao, Y.C.; Chang, H.W. Display of Porcine Epidemic Diarrhea Virus Spike Protein on Baculovirus to Improve Immunogenicity and Protective Efficacy. *Viruses* **2018**, *10*, 346. [CrossRef] [PubMed]

23. Amanna, I.J.; Slifka, M.K. Wanted, Dead or Alive: New Viral Vaccines. *Antiviral Res.* **2009**, *84*. [CrossRef] [PubMed]

24. Gerdts, V.; Zakhartchouk, A. Vaccines for Porcine Epidemic Diarrhea Virus and Other Swine Coronaviruses. *Vet. Microbiol.* **2017**, *206*. [CrossRef] [PubMed]

25. Lai, M.M.C. Recombination in Large RNA Viruses: Coronaviruses. *Semin. Virol.* **1996**, *7*. [CrossRef]

26. Bachmann, M.F.; Jennings, G.T. Vaccine Delivery: A Matter of Size, Geometry, Kinetics and Molecular Patterns. *Nat. Rev. Immunol.* **2010**, *10*. [CrossRef]

27. Mohsen, M.O.; Gomes, A.C.; Cabral-Miranda, G.; Krueger, C.C.; Leoratti, F.M.; Stein, J.V.; Bachmann, M.F. Delivering Adjuvants and Antigens in Separate Nanoparticles Eliminates the Need of Physical Linkage for Effective Vaccination. *J. Control. Release* **2017**, *251*. [CrossRef]

28. Ahsan, F.; Rivas, I.P.; Khan, M.A.; Torres Suárez, A.I. Targeting to Macrophages: Role of Physicochemical Properties of Particulate Carriers—Liposomes and Microspheres—On the Phagocytosis by Macrophages. *J. Control. Release* **2002**, *79*. [CrossRef]

29. Banchereau, J.; Steinman, R.M. Dendritic Cells and the Control of Immunity. *Nature* **1998**, *392*. [CrossRef]

30. Den Haan, J.M.M.; Lehar, S.M.; Bevan, M.J. CD8+ but Not CD8- Dendritic Cells Cross-Prime Cytotoxic T Cells in Vivo. *J. Exp. Med.* **2000**, *192*. [CrossRef]

31. Dudziak, D.; Trumpfheller, C.; Yamazaki, S.; Cheong, C.; Liu, K.; Lee, H. Differential Antigen Processing by Dendritic Cell Subsets in Vivo. *Science* **2007**. [CrossRef]

32. Ohteki, B.T.; Fukao, T.; Suzue, K.; Maki, C. Interleukin 12-Dependent Interferon Gamma Production by CD8alpha+ Lymphoid Dendritic Cells. *J. Exp. Med.* **1999**, *189*. [CrossRef] [PubMed]

33. Manolova, V.; Flace, A.; Bauer, M.; Schwarz, K.; Saudan, P.; Bachmann, M.F. Nanoparticles Target Distinct Dendritic Cell Populations According to Their Size. *Eur. J. Immunol.* **2008**, *38*. [CrossRef] [PubMed]

34. Lin, Y.-T.; Teng, C.-Y.; Villaflores, O.B.; Chen, Y.-J.; Liu, M.-K.; Chan, H.-L.; Jinn, T.-R.; Wu, T.-Y. Using Internal Ribosome Entry Sites to Facilitate Engineering of Insect Cells and Used in Secretion Proteins Production. *J. Taiwan Inst. Chem. Eng.* **2017**, *71*. [CrossRef]

35. Song, D.S.; Oh, J.S.; Kang, B.K.; Yang, J.S.; Moon, H.J.; Yoo, H.S.; Jang, Y.S.; Park, B.K. Oral Efficacy of Vero Cell Attenuated Porcine Epidemic Diarrhea Virus DR13 Strain. *Res. Vet. Sci.* **2007**, *82*. [CrossRef]

36. Meurens, F.; Berri, M.; Whale, J.; Dybvig, T.; Strom, S.; Thompson, D.; Brownlie, R.; Townsend, H.G.G.; Salmon, H.; Gerdts, V. Expression of TECK/CCL25 and MEC/CCL28 Chemokines and Their Respective Receptors CCR9 and CCR10 in Porcine Mucosal Tissues. *Vet. Immunol. Immunopathol.* **2006**, *113*. [CrossRef]

37. Wang, C.; Yan, F.; Zheng, X.; Wang, H.; Jin, H.; Wang, C.; Zhao, Y.; Feng, N.; Wang, T.; Gao, Y.; et al. Porcine Epidemic Diarrhea Virus Virus-like Particles Produced in Insect Cells Induce Specific Immune Responses in Mice. *Virus Genes* **2017**, *53*. [CrossRef]

38. Chattha, K.S.; Roth, J.A.; Saif, L.J. Strategies for Design and Application of Enteric Viral Vaccines. *Annu. Rev. Anim. Biosci.* **2015**, *3*. [CrossRef]

39. Brandtzaeg, P.; Johansen, F.E. Mucosal B Cells: Phenotypic Characteristics, Transcriptional Regulation, and Homing Properties. *Immunol. Rev.* **2005**, *206*. [CrossRef]

40. Zhang, L.; Wang, W.; Wang, S. Effect of Vaccine Administration Modality on Immunogenicity and Efficacy. *Expert Rev. Vaccines* **2015**, *14*. [CrossRef]

41. Yuan, L.; Kang, S.-Y.; Ward, L.A.; To, T.L.; Saif, L.J. Antibody-Secreting Cell Responses and Protective Immunity Assessed in Gnotobiotic Pigs Inoculated Orally or Intramuscularly with Inactivated Human Rotavirus. *J. Virol.* **1998**, *72*. [CrossRef]

42. Aldon, Y.; Kratochvil, S.; Shattock, R.J.; McKay, P.F. Chemokine-Adjuvanted Plasmid DNA Induces Homing of Antigen-Specific and Non–Antigen-Specific B and T Cells to the Intestinal and Genital Mucosae. *J. Immunol.* **2020**, *204*. [CrossRef] [PubMed]

43. Kutzler, M.A.; Wise, M.C.; Hutnick, N.A.; Moldoveanu, Z.; Hunter, M.; Reuter, M.A.; Yuan, S.; Yan, J.; Ginsberg, A.A.; Sylvester, A.; et al. Chemokine-Adjuvanted Electroporated DNA Vaccine Induces Substantial Protection from Simian Immunodeficiency Virus Vaginal Challenge. *Mucosal Immunol.* **2016**, *9*. [CrossRef] [PubMed]

44. Rainone, V.; Dubois, G.; Temchura, V.; Überla, K.; Clivio, A.; Nebuloni, M.; Lauri, E.; Trabattoni, D.; Veas, F.; Clerici, M. CCL28 Induces Mucosal Homing of HIV-1-Specific IgA-Secreting Plasma Cells in Mice Immunized with HIV-1 Virus-like Particles. *PLoS ONE* **2011**, *6*, e26979. [CrossRef] [PubMed]

45. Feng, N.; Jaimes, M.C.; Lazarus, N.H.; Monak, D.; Zhang, C.; Butcher, E.C.; Greenberg, H.B. Redundant Role of Chemokines CCL25/TECK and CCL28/MEC in IgA + Plasmablast Recruitment to the Intestinal Lamina Propria after Rotavirus Infection. *J. Immunol.* **2006**, *176*. [CrossRef] [PubMed]

46. Hsueh, F.C.; Chang, Y.C.; Kao, C.F.; Hsu, C.W.; Chang, H.W. Intramuscular Immunization with Chemokine-Adjuvanted Inactive Porcine Epidemic Diarrhea Virus Induces Substantial Protection in Pigs. *Vaccines* **2020**, *8*, 102. [CrossRef] [PubMed]

47. Lo, Y.-W.; Wu, T.-Y. *Development of Flu Vaccine by a Novel Polycistronic Baculovirus Expression Vector*; Chung Yuan Christian University: Taoyuan City, Taiwan, 2010.

48. Chang, C.-Y.; Peng, J.-Y.; Cheng, Y.-H.; Chang, Y.-C.; Wu, Y.-T.; Tsai, P.-S.; Chiou, H.-Y.; Jeng, C.-R.; Chang, H.-W. Development and Comparison of Enzyme-Linked Immunosorbent Assays Based on Recombinant Trimeric Full-Length and Truncated Spike Proteins for Detecting Antibodies against Porcine Epidemic Diarrhea Virus. *BMC Vet. Res.* **2019**, *15*. [CrossRef]

49. Jung, K.; Wang, Q.; Scheuer, K.A.; Lu, Z.; Zhang, Y.; Saif, L.J. Pathology of US Porcine Epidemic Diarrhea Virus Strain PC21A in Gnotobiotic Pigs. *Emerg. Infect. Dis.* **2014**, *20*. [CrossRef]

50. Muth, C.; Bales, K.L.; Hinde, K.; Maninger, N.; Mendoza, S.P.; Ferrer, E. Alternative Models for Small Samples in Psychological Research: Applying Linear Mixed Effects Models and Generalized Estimating Equations to Repeated Measures Data. *Educ. Psychol. Meas.* **2016**, *76*. [CrossRef]

51. Maxwell, S.E.; Delaney, H.D.; Kelley, K. *Designing Experiments and Analyzing Data: A Model Comparison Perspective*, 3rd ed.; Routledge: New York, NY, USA, 2018; ISBN 978-1-315-16978-1.

52. Ma, Y.; Mazumdar, M.; Memtsoudis, S.G. Beyond Repeated-Measures Analysis of Variance: Advanced Statistical Methods for the Analysis of Longitudinal Data in Anesthesia Research. *Reg. Anesth. Pain Med.* **2012**, *37*. [CrossRef]

53. Zhang, S.; Cao, J.; Ahn, C. A GEE Approach to Determine Sample Size for Pre- and Post-Intervention Experiments with Dropout. *Comput. Stat. Data Anal.* **2014**, *69*. [CrossRef]

54. Wang, M.; Kong, L.; Li, Z.; Zhang, L. Covariance Estimators for Generalized Estimating Equations (GEE) in Longitudinal Analysis with Small Samples. *Stat. Med.* **2016**, *35*. [CrossRef]

55. Abe, T.; Takahashi, H.; Hamazaki, H.; Miyano-Kurosaki, N.; Matsuura, Y.; Takaku, H. Baculovirus induces an innate immune response and confers protection from lethal influenza virus infection in mice. *J. Immunol.* **2003**, *171*, 1133–1139. [CrossRef] [PubMed]

56. Heinimäki, S.; Tamminen, K.; Malm, M.; Vesikari, T.; Blazevic, V. Live baculovirus acts as a strong B and T cell adjuvant for monomeric and oligomeric protein antigens. *Virology* **2017**, *511*, 114–122. [CrossRef] [PubMed]

57. Suzuki, T.; Chang, M.O.; Kitajima, M.; Takaku, H. Baculovirus activates murine dendritic cells and induces non-specific NK cell and T cell immune responses. *Cell. Immunol.* **2010**, *262*, 35–43. [CrossRef]

58. Wu, C.Y.; Lo, C.F.; Huang, C.J.; Yu, H.T.; Wang, C.H. The complete genome sequence of Perina nuda picorna-like virus, an insect-infecting RNA virus with a genome organization similar to that of the mammalian picornaviruses. *Virology* **2002**, *294*, 312–323. [CrossRef]

59. Altmann, F.; Staudacher, E.; Wilson, I.B.; Marz, L. Insect cells as hosts for the expression of recombinant glycoproteins. *Glycoconj. J.* **1999**, *16*, 109–123. [CrossRef]

60. Jarvis, D.L. Developing baculovirus-insect cell expression systems for humanized recombinant glycoprotein production. *Virology* **2003**, *310*, 1–7. [CrossRef]

61. Gimenez-Lirola, L.G.; Zhang, J.; Carrillo-Avila, J.A.; Chen, Q.; Magtoto, R.; Poonsuk, K.; Baum, D.H.; Pineyro, P.; Zimmerman, J. Reactivity of Porcine Epidemic Diarrhea Virus Structural Proteins to Antibodies against Porcine Enteric Coronaviruses: Diagnostic Implications. *J. Clin. Microbiol.* **2017**, *55*, 1426–1436. [CrossRef]

62. Slifka, M.K.; Amanna, I. How Advances in Immunology Provide Insight into Improving Vaccine Efficacy. *Vaccine* **2014**, *32*. [CrossRef]

63. Langel, S.N.; Paim, F.C.; Lager, K.M.; Vlasova, A.N.; Saif, L.J. Lactogenic Immunity and Vaccines for Porcine Epidemic Diarrhea Virus (PEDV): Historical and Current Concepts. *Virus Res.* **2016**, *226*. [CrossRef]

64. Id, V.D.K.; Kim, Y.; Yang, M.; Vannucci, F.; Molitor, T.; Cheeran, M.C. Immune Responses to Porcine Epidemic Diarrhea Virus (PEDV) in Swine and Protection against Subsequent Infection. *PLoS ONE* **2020**, *15*, e0231723. [CrossRef]

65. Chang, C.Y.; Cheng, I.C.; Chang, Y.C.; Tsai, P.S.; Lai, S.Y.; Huang, Y.L.; Jeng, C.R.; Pang, V.F.; Chang, H.W. Identification of Neutralizing Monoclonal Antibodies Targeting Novel Conformational Epitopes of the Porcine Epidemic Diarrhoea Virus Spike Protein. *Sci. Rep.* **2019**, *9*. [CrossRef] [PubMed]

66. Chang, C.Y.; Hsu, W.T.; Tsai, P.S.; Chen, C.M.; Cheng, I.C.; Chao, Y.C.; Chang, H.W. Oral Administration of Porcine Epidemic Diarrhea Virus Spike Protein Expressing in Silkworm Pupae Failed to Elicit Immune Responses in Pigs. *AMB Express* **2020**, *10*. [CrossRef] [PubMed]

67. Oh, J.; Lee, K.W.; Choi, H.W.; Lee, C. Immunogenicity and Protective Efficacy of Recombinant S1 Domain of the Porcine Epidemic Diarrhea Virus Spike Protein. *Arch. Virol.* **2014**, *159*. [CrossRef] [PubMed]

68. Subramaniam, S.; Yugo, D.M.; Heffron, C.L.; Rogers, A.J.; Sooryanarain, H.; LeRoith, T.; Overend, C.; Cao, D.; Meng, X.J. Vaccination of Sows with a Dendritic Cell-Targeted Porcine Epidemic Diarrhea Virus S1 Protein-Based Candidate Vaccine Reduced Viral Shedding but Exacerbated Gross Pathological Lesions in Suckling Neonatal Piglets. *J. Gen. Virol.* **2018**, *99*. [CrossRef] [PubMed]
69. Zeltins, A. Construction and Characterization of Virus-like Particles: A Review. *Mol. Biotechnol.* **2013**, *53*. [CrossRef] [PubMed]
70. Kushnir, N.; Streatfield, S.J.; Yusibov, V. Virus-like Particles as a Highly Efficient Vaccine Platform: Diversity of Targets and Production Systems and Advances in Clinical Development. *Vaccine* **2012**, *31*. [CrossRef]
71. Buonaguro, L.; Tagliamonte, M.; Tornesello, M.L.; Buonaguro, F.M. Developments in Virus-like Particle-Based Vaccines for Infectious Diseases and Cancer. *Expert Rev. Vaccines* **2011**, *10*. [CrossRef]
72. Hieshima, K.; Kawasaki, Y.; Hanamoto, H.; Nakayama, T.; Nagakubo, D.; Kanamaru, A.; Yoshie, O. CC Chemokine Ligands 25 and 28 Play Essential Roles in Intestinal Extravasation of IgA Antibody-Secreting Cells. *J. Immunol.* **2004**, *173*. [CrossRef]
73. Kathuria, N.; Kraynyak, K.A.; Carnathan, D.; Betts, M.; Weiner, D.B.; Kutzler, M.A. Generation of Antigen-Specific Immunity Following Systemic Immunization with DNA Vaccine Encoding CCL25 Chemokine Immunoadjuvant. *Hum. Vaccines Immunother.* **2012**, *8*. [CrossRef]
74. Mohan, T.; Deng, L.; Wang, B. CCL28 Chemokine: An Anchoring Point Bridging Innate and Adaptive Immunity. *Int. Immunopharmacol.* **2017**, *51*. [CrossRef] [PubMed]
75. Wang, C.; Liu, Z.; Xu, Z.; Wu, X.; Zhang, D.; Zhang, Z.; Wei, J. The Role of Chemokine Receptor 9/Chemokine Ligand 25 Signaling: From Immune Cells to Cancer Cells (Review). *Oncol. Lett.* **2018**, *16*. [CrossRef] [PubMed]
76. Sun, X.; Zhang, H.; Xu, S.; Shi, L.; Dong, J.; Gao, D.; Chen, Y.; Feng, H. Membrane-Anchored CCL20 Augments HIV Env-Specific Mucosal Immune Responses. *Virol. J.* **2017**, *14*. [CrossRef] [PubMed]
77. Mohan, T.; Berman, Z.; Luo, Y.; Wang, C.; Wang, S.; Compans, R.W.; Wang, B.Z. Chimeric Virus-like Particles Containing Influenza HA Antigen and GPI-CCL28 Induce Long-Lasting Mucosal Immunity against H3N2 Viruses. *Sci. Rep.* **2017**, *7*. [CrossRef]
78. Zimmermann, P.; Curtis, N. Factors That Influence the Immune Response to Vaccination. *Clin. Microbiol. Rev.* **2019**, *32*. [CrossRef]
79. Ejemel, M.; Li, Q.; Hou, S.; Schiller, Z.A.; Tree, J.A.; Wallace, A.; Amcheslavsky, A.; Kurt Yilmaz, N.; Buttigieg, K.R.; Elmore, M.J.; et al. A cross-reactive human IgA monoclonal antibody blocks SARS-CoV-2 spike-ACE2 interaction. *Nat. Commun.* **2020**, *11*, 4198. [CrossRef]
80. Mateus, J.; Grifoni, A.; Tarke, A.; Sidney, J.; Ramirez, S.I.; Dan, J.M.; Burger, Z.C.; Rawlings, S.A.; Smith, D.M.; Phillips, E.; et al. Selective and cross-reactive SARS-CoV-2 T cell epitopes in unexposed humans. *Science* **2020**, *370*. [CrossRef]
81. Choudhury, B.; Dastjerdi, A.; Doyle, N.; Frossard, J.P.; Steinbach, F. From the field to the lab—An European view on the global spread of PEDV. *Virus Res.* **2016**, *226*, 40–49. [CrossRef]
82. Lin, C.-M.; Gao, X.; Oka, T.; Vlasova, A.N.; Esseili, M.A.; Wang, Q.; Saif, L.J. Antigenic relationships among porcine epidemic diarrhea virus and transmissible gastroenteritis virus strains. *J. Virol.* **2015**, *89*, 3332–3342. [CrossRef]

Review

Exploring the Prospects of Engineered Newcastle Disease Virus in Modern Vaccinology

Muhammad Bashir Bello [1,2], Khatijah Yusoff [2,3], Aini Ideris [2,4], Mohd Hair-Bejo [2,5], Abdurrahman Hassan Jibril [6], Ben P. H. Peeters [7] and Abdul Rahman Omar [2,5,*]

1   Department of Veterinary Microbiology, Faculty of Veterinary Medicine, Usmanu Danfodiyo University PMB, Sokoto 2346, Nigeria; bbtambuwal@gmail.com
2   Laboratory of Vaccines and Immunotherapeutics, Institute of Bioscience, Universiti Putra Malaysia, Serdang, Selangor 43400, Malaysia; kyusoff@upm.edu.my (K.Y.); aiini@upm.edu.my (A.I.); mdhair@upm.edu.my (M.H.-B.)
3   Department of Microbiology, Faculty of Biotechnology and Biomolecular Sciences, Universiti Putra Malaysia, Serdang, Selangor 43400, Malaysia
4   Department of Veterinary Clinical Studies, Faculty of Veterinary Medicine, Universiti Putra Malaysia Serdang, Selangor 43400, Malaysia
5   Department of Veterinary Pathology and Microbiology, Faculty of Veterinary Medicine, Universiti Putra Malaysia Serdang, Selangor 43400, Malaysia
6   Department of Veterinary Public Health and Preventive Medicine, Faculty of Veterinary Medicine, Usmanu Danfodiyo University PMB, Sokoto 2346, Nigeria; jibrilah50@yahoo.com
7   Department of Virology, Wageningen Bioveterinary Research, POB 65, NL8200 Lelystad, The Netherlands; ben.peeters@wur.nl
*   Correspondence: aro@upm.edu.my; Tel.:+603-89472111

Received: 21 February 2020; Accepted: 15 March 2020; Published: 16 April 2020

Abstract: Many traditional vaccines have proven to be incapable of controlling newly emerging infectious diseases. They have also achieved limited success in the fight against a variety of human cancers. Thus, innovative vaccine strategies are highly needed to overcome the global burden of these diseases. Advances in molecular biology and reverse genetics have completely restructured the concept of vaccinology, leading to the emergence of state-of-the-art technologies for vaccine design, development and delivery. Among these modern vaccine technologies are the recombinant viral vectored vaccines, which are known for their incredible specificity in antigen delivery as well as the induction of robust immune responses in the vaccinated hosts. Although a number of viruses have been used as vaccine vectors, genetically engineered Newcastle disease virus (NDV) possesses some useful attributes that make it a preferable candidate for vectoring vaccine antigens. Here, we review the molecular biology of NDV and discuss the reverse genetics approaches used to engineer the virus into an efficient vaccine vector. We then discuss the prospects of the engineered virus as an efficient vehicle of vaccines against cancer and several infectious diseases of man and animals.

Keywords: Newcastle disease virus; reverse genetics; vaccines; infectious diseases; cancer

1. Introduction

Vaccines are undoubtedly among the most effective fighters of infectious diseases. They have historically been used to completely eradicate or at least substantially reduce the menace of many human and animal diseases [1–3]. Conventional vaccines can be broadly classified into two groups. The first group includes the live attenuated vaccines that are highly effective due to their ability to induce immune responses that are essentially similar to those due to natural infection [4,5]. Unfortunately these vaccines often retain the tendency of reversion back to virulence. The second group includes the inactivated vaccines, which are known to be incredibly safe as a result of their non-replicating nature, but are often poorly immunogenic and therefore, do not elicit a long lasting immunity [6]. Furthermore, both live attenuated and inactivated vaccines have failed to effectively curb the menace of a variety of major global pathogens. These limitations altogether quest for the need to develop novel vaccine strategies that could potentially overcome the weaknesses of the conventional vaccines. Interestingly, the recent advancements in molecular genetics and bioinformatics have paved way for the emergence of next generation vaccine technologies such as synthetic peptides, DNA vaccines, recombinant viral-vectored vaccines, and reverse genetics-based vaccines, which are all currently on the path of revolutionizing medical and veterinary vaccinology [7,8].

Newcastle disease virus is one of the most important avian viral pathogens that inflict huge economic losses in the global poultry industry [9]. The virus is highly genetically diverse, with currently more than 20 phylogenetically distinct genotypes based on the recently proposed NDV taxonomy criteria [10]. Following the recovery of the first strain of the virus by reverse genetics 20 years ago [11], tremendous progress has been recorded in the genetic manipulation of various strains of the virus. So far, NDV has been engineered to express rationally designed, safe and highly stable protective antigens against several human and animal pathogens (Table 1). The virus has also been genetically reprogrammed for improved oncolytic efficacy against a variety of human cancers (Table 2). Thus, the prospects of genetically engineered NDV in the era of modern vaccinology cannot be over emphasized. In this review, we start by describing the molecular biology of NDV and the various approaches used in the genetic manipulation of the virus for effective vaccine delivery. We then discuss the potentials of the engineered virus as a vaccine against cancer and other life threatening diseases in man and against economically important viruses in various domestic animals.

2. Newcastle Disease Virus

2.1. Architecture and Genome Organisation

Ultrastructurally, the particles of NDV are pleomorphic in shape with diameters ranging from 100–500nm. They essentially consist of the ribonucleoprotein (RNP) surrounded by the viral envelop with its surface glycoproteins that project as spikes. The RNP is made up of the RNA genome completely encapsidated by a protein called nucleocapsid protein (NP). Other proteins associated with the RNP include the large protein (L), which is the RNA-dependent RNA polymerase and its co-factor, the phosphoprotein (P) [12]. Together, they form a helical structure surrounded by a lipid bilayer envelop with surface projections of hemagglutinin-neuraminidase (HN) and fusion proteins (F). The matrix protein (M) is found just beneath the viral envelop and maintains the shape and structure of the virion [13].

The genome of NDV is either 15,198, 15,192 or 15,186 bp in size [14]. It is a single stranded, non-segmented negative sense RNA that consists of leader (55 nucleotides) and trailer (114 nucleotides) terminal sequences separated by six genes in the order 3′-NP-P-M-F-HN-L-5′. These terminal sequences are highly conserved across most of the paramyxoviruses [12,15] and house the regulatory signals for virus replication [16]. In addition, each gene in the NDV genome encodes a single protein and is characterized by having a coding sequence flanked by highly conserved gene start (GS) and a gene end (GE) transcriptional signals [17]. These features of NDV genome are shared by most of the paramyxoviruses, suggesting a common transcription and replication strategy.

2.2. Virulence Determinants

Using reverse genetics, RNP associated viral structural proteins (NP, P and L) have been shown to collectively contribute to NDV virulence. Experiments involving the swapping of NP between pathogenically different strains revealed a change in virulence of the generated recombinant virus only when NP along with its other replication proteins partners (P and L) were swapped together [18]. Notably, the NP encapsidates the entire genomic RNA while the L functions as an RNA-dependent

RNA polymerase that binds to P protein and forms a complex that recognizes the RNP for initiating the process of transcription and replication [19]. Recent evidence has indicated that the L protein can significantly contribute to NDV virulence through its role in virus replication [20]. However, specific regions within the L protein responsible for determining NDV virulence are yet to be discovered. The F protein is the major determinant of NDV virulence. It is synthesized in an inactive form, F_0, and then becomes activated following its proteolytic cleavage into F1 and F2. The enzyme involved in the cleavage is determined by the amino acid composition of the cleavage site, which in turn varies in virulent and avirulent NDV strains [21]. Avirulent pathotypes generally have a monobasic F cleavage site acted upon by extracellular trypsin like proteases. On the other hand, the virulent strains have multiple basic amino acid residues at their cleavage sites that are activated by intracellular proteases of the furin family [22]. In addition to the cleavage site's amino acid composition, the sites of post translational modifications, notably glycosylation sites, may play a significant role in the virus virulence. Recently, [23] observed an increase in virus virulence following deletion of N glycans from the heptad repeats of the NDV F protein. Furthermore, [24] showed a dramatic increase in NDV virulence when the cytoplasmic tail from a mesogenic strain was used to replace its counterparts in another lentogenic strain, suggesting the possible role of the F protein's cytoplasmic domain in determining the virulence of NDV. The HN protein is also believed to be a determinant of NDV virulence. Depending on the location of the termination codon, the HN protein has several subtypes of varying length, with most of the avirulent strains encoding the longest protein made up of 616 amino acids. In contrast, the virulent strains encode the shortest HN proteins of about 571 amino acids. Lentogenic strains such as LaSota and B1 encode an HN protein of 577 amino acids in length [25]. Other lengths of HN protein reported include 572, 580, 582, and 585 amino acids. To investigate whether the HN length diversity is directly associated with the virus virulence, several NDV chimeras with shortened or extended HN length were generated by reverse genetics and tested for pathogenicity using standard procedures. It was revealed that varying the lengths of HN directly affects the functions related to viral replication but not pathogenicity [26]. V protein, a non-structural protein encoded by the p gene, also plays a vital role in determining NDV virulence. The role of V protein is clearly manifested during viral pathogenesis as an interferon antagonist, by selectively targeting STAT-1 for degradation [27]. Recombinant NDVs that cannot express V protein often show no evidence of growth in 9–10 days chicken embryonated eggs and a very impaired growth in cell culture. Those recombinant viruses also manifest sensitivity to exogenous interferon administration, indicating that the V protein plays a role in NDV replication and virulence [27]. Collectively, virulence of NDV is determined by individual or collective roles of both structural and non- structural viral proteins.

2.3. Transcription and Replication

The infectious cycle of NDV follows the same pattern as other members of the family *Paramyxoviridae* obviously due to the common genomic features shared among these viruses [28,29]. The viral polymerase complex first recognizes a single promoter located in the 3′ leader sequence of the genome and then moves towards the 5′ end by responding to the conserved gene start (GS) and gene end (GE) signals located at the beginning and end of each gene respectively, to produce various mRNAs [30]. Upon arrival at the GE, the polymerase complex terminates transcription, scans through the intergenic region, and then starts transcription of the next gene and continues in that order until it transcribes the last gene, from where it finally dissociates from the RNA [31]. As the transcription progresses, the synthesized mRNAs are translated into viral proteins whose accumulation in the cell causes the viral polymerase to switch from transcription to replication of the entire viral genome. Further details of the NDV replication cycle have been discussed elsewhere [32]. Noteworthy, efficient replication of NDV can only occur when the genomic length of the virus size is in a multiple of six. This phenomenon is referred to "rule of six" and occurs because during replication, the NP of most paramyxoviruses is associated with exactly six nucleotides on the genomic RNA [33]. Therefore, in any experiment involving the genetic manipulation of NDV, the rule of six must be strictly obeyed [30].

3. Reverse Genetics System

Reverse genetics is the term used to describe the recovery of recombinant viruses from their cloned cDNA [34]. Viruses generated by reverse genetics can be engineered to either encode desired mutations in the indigenous viral genes or express heterologous antigens as additional proteins. Reverse genetics is therefore a state-of-the-art recombinant DNA technology with considerable impact in modern vaccine design, development and delivery. The first group of viruses known to be amenable to reverse genetics were the positive sense RNA viruses [35] whose genetic material has the same polarity as cellular mRNA and therefore can directly act as templates for protein synthesis. However, owing to peculiar complexities and challenges associated with the replication of negative sense RNA viruses, reverse genetics could not be immediately applied to manipulate those viruses [31] until in 1994 when rabies virus became the first negative sense RNA virus to be successfully rescued from its cloned cDNA [36]. Subsequently, reverse genetics system was established in other negative sense viruses and in 1999, the rescue of the first Newcastle disease virus (NDV) strain entirely from its cloned cDNA was accomplished [11].

3.1. Recovery of Recombinant NDV

The constructs needed for the recovery of recombinant NDV are helper plasmids and a full length cDNA clone. The helper plasmids are eukaryotic expression vectors that encode the minimum molecular machinery (NP, P and L genes) for the transcription and replication of NDV [37]. On the other hand, the full length cDNA clone represents the entire antigenome of NDV vectored by a transcription vector usually under the control of a T7 promoter. To rescue a recombinant NDV, the full length cDNA construct and the helper plasmids are co-delivered at an optimized ratio, into cells expressing T7 RNA polymerase for the initiation of virus infectious cycle [38]. Commonly used cells in NDV recovery are human epitheloid carcinoma (Hep-2) cells infected with modified vaccinia virus expressing T7 RNA polymerase and genetically engineered Baby Hamster kidney cells (BST-T7) that constitutively express the T7 polymerase. Immediately after transfection, viral RNA will be transcribed from the full length construct and the proteins expressed from helper plasmids will associate to produce the RNP template. Unfortunately, this particular step is largely inefficient and, therefore, the most rate-limiting step in NDV reverse genetics [32]. Once RNP is intracellularly organized, the virus infectious life cycle becomes activated. In a nut shell, the success of generating an infectious virus from a cloned full length cDNA of NDV requires the precise transcription of the full length construct into a viral RNA; optimal co-expression of the NP, P and L genes at a level sufficient to kick-start the replication process; as well as the expression of other viral genes for onward progression of the replication process and the release of the assembled viral particles (Figure 1).

3.2. Recent Improvement in NDV Rescue System

The difficulties associated with NDV reverse genetics are largely due to the need to co-transfect at least four plasmid constructs into the same cell. In order to enhance the transfection efficiency, [39] developed a two-plasmid reverse genetics system that dramatically improved the rescue efficiency of NDV. When compared with the conventional four plasmid system, the newly developed system appeared to be superior, enabling an earlier and increased amount of the recovered viruses. In some cases, viruses that could not be rescued using the traditional system were successfully recovered with the aid of this improved reverse genetics system. More recently, a single plasmid-based NDV reverse genetics system was developed [40]. In this approach, the full length construct under the control of T7 promoter was designed to have additional T7 promoter sequences upstream of the GS of P and L genes. Thus, when the plasmid was transfected into cells earlier infected with a fowl pox virus expressing T7 RNA polymerase, a full length viral RNA and two subgenomic RNAs were transcribed. Arguably, some of the subgenomic and full length viral RNAs might be capped and polyadenylated, respectively, at 5′ and 3′ ends by some of the fowl pox enzymes. The mRNAs are in turn translated by the cellular protein synthesis machinery into NP, P and L proteins. Therefore, mere delivery of this single plasmid into cells infected with FP-T7 leads to the assembly of RNP intracytoplasmically, and the eventual production of infectious virus particle. To substantiate the point that the fowl pox enzymes were responsible for the capping and polyadenylation of the subgenomic RNAs, the same plasmid construct was transfected into BSR cells constitutively expressing T7 polymerase and no virus was rescued. Thus, this technique provides an improvement in NDV rescue efficiency and may have important applications in future NDV reverse genetics experiments [40].

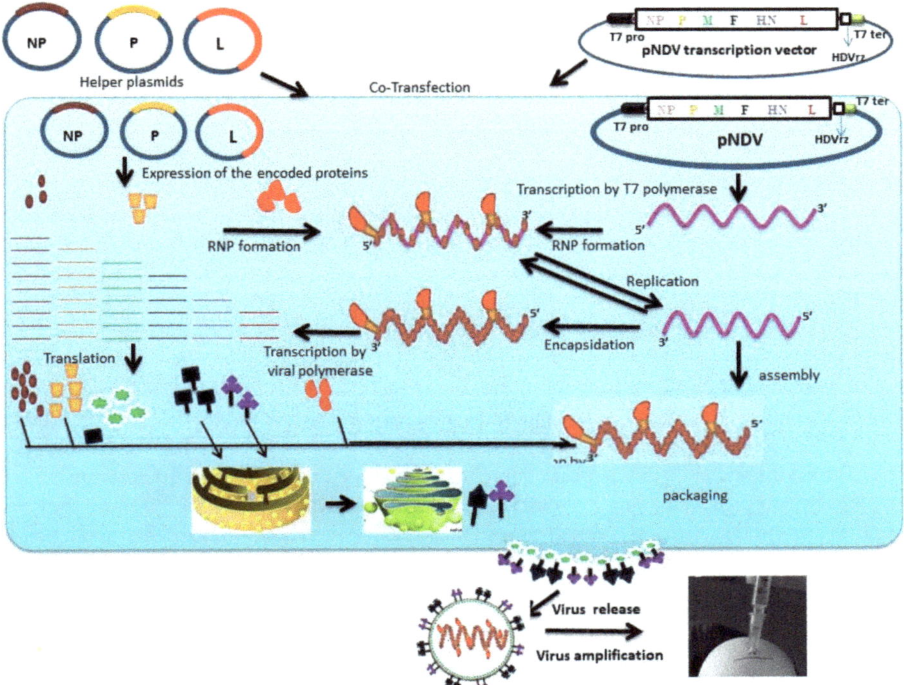

Figure 1. Reverse genetics approach for the generation of engineered Newcastle disease virus in BSR-T7 cells. Full length NDV antigenome flanked by T7 promoter (T7 pro), and autocatalytic hepatitis delta virus ribozyme (HDVrz) is cloned into the transcription vector to form pNDV. Helper plasmids (expression constructs for NP, P and L) are co-transfected with pNDV into BSR-T7 cells, which constitutively express T7 RNA polymerase. Within the cytoplasm, the T7 polymerase transcribes the viral RNA using the pNDV as a template. The helper plasmids express the encoded proteins for association with the transcribed RNA to form the ribonucleoprotein (RNP) template for onward replication cycles. The viral polymerase gradiently transcribes the viral genome into respective mRNAs, which are subsequently translated into proteins. NP, P and L proteins assemble with the newly synthesized negative sense viral RNA while the F and HN proteins are post-translationally modified in the endoplasmic reticulum and Golgi apparatus before being transported to the cell surface. The final event is the release of the recombinant virus by budding, which is then amplified in specific pathogen-free chicken embryonated eggs.

3.3. Strategies of Foreign Gene Expression Using NDV Vector

Similar to other paramyxoviruses, NDV has been shown to tolerate the insertion of one or more additional genes into its genome without compromising the biological features of the virus. Traditionally, those foreign genes (FG) are inserted as independent transcription units (ITU) made up of GE, IR, GS, and Kozak sequences followed by the coding region of the gene of interest (GOI) [41].

If the foreign antigens are required to be displayed on the surface of NDV, it is important that they are fused with the cytoplasmic and transmembrane domains of the NDV F protein [42]. Noteworthy, the expression level of the FG is considerably dependent on its genomic location in the NDV backbone. Recently, the optimal site for FG expression has been shown to be the P-M junction, which yields the strongest expression signals compared to all other locations in the NDV genome [43]. Contrastingly, the sequential transcription phenomenon states that the closer the gene is to the 3′ end of the genome, the higher the level of its expression [44]. Given that the P-M junction is not located at the extreme 3′ end of the NDV genome, it is logical to ask the question, what exactly is responsible for the strongest level of gene expression at this genomic site? Previous studies have shown that efficient replication of paramyxoviruses requires an optimal NP:P ratio [28]. Therefore, inserting additional gene at locations upstream of the P gene in the NDV genome might affect the relative abundance of the downstream proteins leading to the disruption of the NP:P ratio, which ultimately affects viral replication. On the other hand, when additional genes are expressed at gene borders downstream of the P gene, this NP:P ratio remains unaffected [43]. This probably explains the choice of P-M as the optimal site for FG expression.

Another means of FG expression is via an internal ribosome entry site (IRES), which allows the expression of two genes from a single mRNA transcript [45]. In this system, the IRES sequence is inserted immediately downstream of the stop codon of any gene in the NDV backbone followed by the coding region of the gene of interest (Figure 2). During transcription, the two genes separated by the IRES sequence are transcribed into a single mRNA such that the first gene is translated using the default cap-dependent translational machinery while the translation of the downstream gene is cap-independent [46]. The advantage of this system is that the level of gene expression can be regulated by taking advantage of the sequential transcription mechanism of NDV. Thus, when the foreign gene is an immunogen for which a high level of gene expression is needed, the IRES sequence can be inserted immediately downstream the NP gene, followed by the gene of interest. On the other hand, if the FG is a proinflammatory cytokine used in cancer immunotherapy, the IRES and the foreign gene can be located downstream of the M gene so that its expression can be moderate to avoid cytokine storms [47]. Using this strategy, [48] generated a chimeric NDV expressing the F and G proteins of avian paramyxovirus type C as a potential vaccine candidate for turkeys (Table 1).

Table 1. NDV as an effective vaccine vector in various animal species.

Host	Disease	Immunogen	NDV Backbone	Type of Immunity	Reference
Monkeys	Severe acute respiratory syndrome	SARS-CoV S	Beaudette C	SARS-CoV S-specific CD8+ T cells	[49]
Camel	Middle east respiratory syndrome	MERS-CoV S	LaSota	Neutralizing antibodies	[50]
Cattle	Rift valley fever	RVFV Gn	LaSota	Neutralizing antibodies	[51]
Cattle	Infectious bovine rhinotracheitis	BHV-1 gD	LaSota	Neutralizing IgG and IgA	[52]
Cattle	Bovine ephemeral fever	BEFV G	Lasota	Neutralizing antibodies	[53]
Chicken	Infectious bronchitis	IBV S1 or full S	Lasota	CD4+ and CD8+ cells; Neutralizing antibodies	[42,54]
Chicken	Infectious laryngotracheitis	ILTV gD	Lasota	Neutralizing antibodies	[55]
Chicken	Highly pathogenic avian influenza	H5, H7, H9	Lasota	Neutralizing antibodies	[56,57]
Chicken	Infectious bursal disease	IBDV VP2	Lasota	-	[58]
Dogs and cats	Rabies	RV G	Lasota	Neutralizing antibodies	[59]
Goose	Gosling plaque	VP3 of goose parvovirus	NA-1	Neutralizing antibodies	[60]
Guinea pigs	Acquired immune deficiency syndrome	HIV gp160	Lasota	Neutralizing antibodies	[61]
Horse	West Nile fever	WNV *PrM/E*	Lasota	CD4+ and CD8+ cells; Neutralizing IgG	[62]
Mice	Acute pneumonia	RSV F	Hitchner B1	CD8+ cells	[63]
Mice	Nipah encephalitis	Nipah virus G and F	Lasota	T and B cells; Neutralizing antibodies	[64]
Mice	Vesicular stomatitis	VSV G	LaSota	Neutralizing antibodies	[65]
Mice	Viral gastroenteritis	NV VP1	LaSota and Beaudette C	CD8+ cells; Neutralizing antibodies	[66]

Table 1. *Cont.*

Host	Disease	Immunogen	NDV Backbone	Type of Immunity	Reference
Minks	Canine distemper	CDV F and HN	LaSota	Neutralizing antibodies	[67]
Monkey	Ebola	EBOV GP	Beaudette C	CD8+ cells: virus specific IgA, and IgG	[68,69]
Pigs	Nipah encephalitis	Nipah virus G and F	LaSota	T and B cells; Neutralizing antibodies	[64]
Monkeys	Parainfluenza	HPIV3 HN	Beaudette C	Neutralizing antibodies	[70]
Turkey	Turkeys rhinotracheitis	F and G of AMPV type CG of AMPV type A and B	LaSota	Neutralizing antibodies	[48,71]

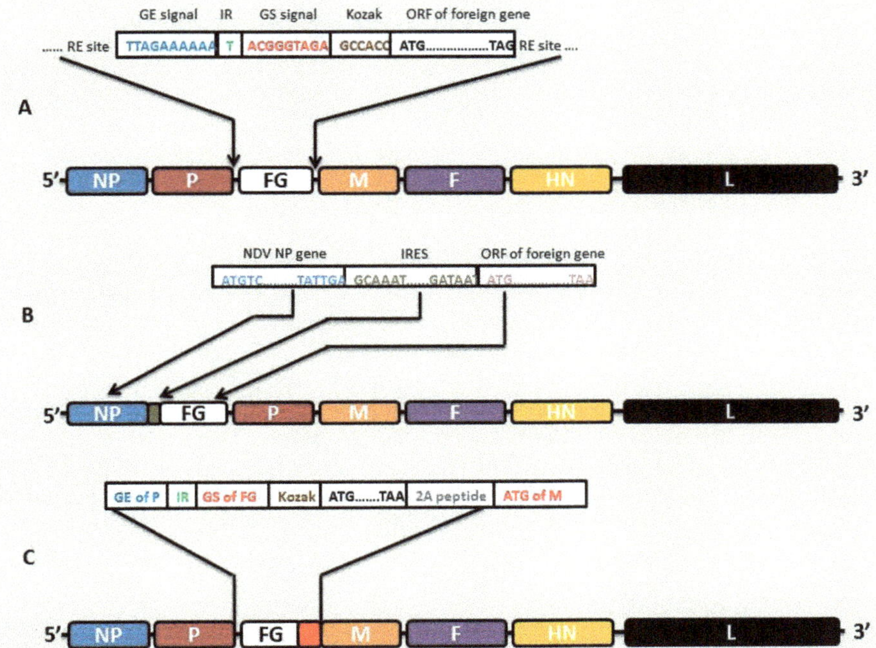

Figure 2. Strategies for foreign gene (FG) expression using NDV as a vector. (**A**) Expression of a foreign gene (FG) as an additional transcription unit. The FG along with NDV transcriptional signals and a Kozak sequence is cloned in a non-coding region, preferably between the P and M genes. (**B**) IRES-mediated expression of an FG. An IRES sequence is placed in between the coding regions of any NDV gene and the FG. (**C**) Expression of an FG via 2A peptide-mediated fusion with any NDV gene.

In another strategy, NDV has been engineered to express FG via a novel integrating transcriptional unit [72]. In this system, a fusion sequence made up of the foreign gene and 2AUbi is inserted immediately upstream of the translation start codon for any gene in the NDV backbone (Figure 2). The 2AUbi is actually a 94 amino acids sequence made up of the self-cleaving foot and mouth disease virus 2A peptide and ubiquitin coding sequences [73]. Therefore, the FG is expressed as a fusion protein with the downstream protein, which is post-translationally separated by the self-cleaving activity of 2A peptide in vivo. When compared with the system utilizing the independent transcriptional unit (ITU), this system was shown to have higher efficiency of expression. So far, expression of FG via an integrating transcriptional unit has only been demonstrated for NP, M and L genes in the NDV back bone. Preliminary results indicate that M gene yields the optimal result using this system of FG expression [72].

In summary, various approaches are available to express FG using the NDV backbone, providing room for a variety of choices depending on the level and manner of gene expression desired. The ITU system is by far the most popular and frequently utilized method of foreign gene expression using NDV vector. Other methods recently evolving include the IRES system as well the integrating transcriptional unit systems. While in the ITU and integrating transcriptional unit systems, the optimal site for foreign gene expression is the P-M junction, with the latter being immediately upstream of the M gene start codon; in the IRES mediated expression, the NP-P junction yields the highest level of gene expression. However, further research is needed to identify the level of foreign gene expression as a fusion protein particularly with the P, F and HN genes of NDV, to verify the claim that expression is highest when the foreign gene is located immediately upstream of the M initiation codon. This will definitely have practical implications for the design of recombinant NDV vectored vaccines.

4. Engineered NDV as a Vaccine Vector against Infectious Diseases

4.1. Unique Attributes of NDV as a Vaccine Vector

One of the most attractive applications of reverse genetics technology is in the manipulation of viral RNA genomes to express vaccine antigens against several human and animal pathogens [74]. While a number of viruses have been used as vaccine vectors, the efficiency of gene delivery by NDV in both human and animals is unparalleled, due to some unique architectural features of the viral genome. In the first place, the genome of NDV, unlike those of pox and herpes viruses, is simple and encodes only a few structural proteins, which reduces the number of proteins expressed along with the FG and, therefore, enhances the specific immune response against the expressed heterologous proteins [19]. Secondly, each gene has conserved GS and GE sequences, respectively, at its 3′ and 5′ ends, which implies that foreign antigens can be efficiently expressed in the same manner as indigenous NDV proteins, when flanked with those transcriptional signals [32]. In addition, the entire replication cycle of NDV takes place in the cytoplasm, thereby avoiding the risk of random integration of the viral genome into the host cell DNA [16]. These useful properties collectively make NDV an efficient vector for vaccine delivery in different host species.

4.2. Engineered NDV as a Bivalent Vaccine in Poultry

To use a recombinant NDV as a bivalent vaccine in poultry, it is imperative that the backbone virus is either a naturally attenuated strain such as LaSota and Hitchner B1 strains, or a genetically engineered NDV encoding some attenuation-inducing mutations particularly at the F protein cleavage site, which is the major determinant of NDV virulence. Although any lentogenic strain of NDV could serve as a bivalent vaccine in poultry, the use of NDV strain more closely related to the circulating field NDV strains stands to offer better efficacy especially in terms of reduction of virus shedding post challenge [8]. Thus, the current trend in ND control focuses on the use of reverse genetics tools to generate the so-called genotype-matched live attenuated NDV vaccines [75–77]. Such genotype-matched vaccines, when used as vaccine vectors for other poultry diseases, have the potential to excellently protect against both diseases. The following are examples of diseases in poultry against which NDV has been used to successfully deliver vaccine antigens.

Highly pathogenic avian Influenza is arguably the most fatal viral disease in the global poultry industry. It is caused by avian influenza virus (AIV) of the genus *Influenzavirus* A in the family orthomyxoviridae [78]. The virus has a negative-sense, segmented single-stranded RNA genome encoding at least 10 proteins: polymerase basic 1 (PB1), polymerase basic 1 (PB2), polymerase acid (PA), hemagglutinin (HA), nucleoprotein (NP), neuraminidase (NA), matrix 1 (M1), matrix 2 (M2), nonstructural 1 (NS1), and nonstructural 2 (NS2) [79]. Among these proteins, the HA and NA are the most important inducers of neutralizing immunity [80]. Despite the availability of conventional vaccines, outbreaks due H5NI and H7N7 AIV subtypes have dramatically increased especially in the last two decades. The inactivated AIV vaccines are not effective as they provide suboptimal protection against the disease. The live attenuated vaccines may be highly immunogenic but are not recommended by the OIE due to the potential for novel subtype emergence as a result of genetic reassortment [80]. Thus, alternative vaccine platforms must be explored to control the menace of this fatal avian pathogen. Recombinant viral vectors such as turkey herpesvirus, fowlpox virus, adenovirus, infectious laryngotracheitis (ILT) virus, and Marek's disease virus (MDV) have all demonstrated promising AIV vaccine delivery efficiency [81]. Similarly, recombinant NDV expressing H5 or H7 of AIV has severally been shown to demonstrate excellent protective efficacy against challenge with both virulent NDV and HPAI virus in chicken [56,82,83]. Details of the protective efficacy of NDV vectored AIV vaccines have recently been reviewed else-where [57].

Infectious bronchitis, an economically important disease of poultry caused by a rapidly evolving coronavirus known as avian infectious bronchitis virus (IBV), has been incriminated in crippling the productivity of the poultry industry all over the world [84]. Current control strategies for the disease rely on the use of traditional live and inactivated vaccines that may not cross protect against other heterologous serotypes [85], necessitating the need to develop vaccines based on the prevailing regionally important serotypes. Since field IBV strains are difficult to attenuate [86], the use of recombinant live vaccines could serve as an attractive alternative in the control of infectious bronchitis in chicken. Unfortunately, due to their large genomes, recovery of recombinant IBV strains by reverse genetics can be highly cumbersome [87]. However, since several studies have shown that most of the neutralizing epitopes against IBV are found in the S1 protein, recombinant viral vectors expressing the IBV S1 protein have the potential to induce protective immune responses in vaccinated chickens. Recently, recombinant NDV strain LaSota engineered to express LX4 type IBV S1 protein has been shown to induce strong cellular and humoral immunity, which provides complete protection against both virulent NDV and LX4 type IBV challenge [54]. Indeed, after vaccination with this bivalent vaccine and subsequent challenge with a virulent IBV field isolate, a marked reduction in the replication of the challenge virus in the trachea was observed especially after booster vaccination [54]. More recently, engineered NDV expressing IBV complete S, S1 or S2 proteins was shown to completely protect against both virulent NDV and IBV challenge, with the highest protective efficacy demonstrated by rNDV expressing complete IBV S protein [42]. Similarly, Abozeid, H.H. et al. ([88]) showed that recombinant NDV expressing codon-optimized S glycoprotein of the Egyptian IBV variant strain IBV/Ck/EG/CU/4/2014 provided excellent protection against both NDV and IBV challenge in chicken. Thus, rNDV is an efficient vector for the delivery of IBV-specific immunogens in chickens.

Another important avian pathogen threatening the global poultry production is infectious bursal disease virus (IBDV). The virus specifically replicates in the chicken's bursa of Fabricius, resulting in the destruction of the developing B lymphocytes, leading to immunosuppression and enhanced vulnerability to several infectious diseases [89]. Despite the availability of vaccines, control of IBD is still a problem owing to the emergence of antigenically distinct variants as a result of the use of the current IBD vaccines [90]. These variants evade neutralizing antibodies raised against the classical IBDV vaccines, making the control of the disease using the currently available vaccines unreliable. Therefore, to efficiently control this immunosuppressive disease of chicken, there is the need to rationally design vaccines that do not lead to the emergence of novel antigenic variants of the field virus [91]. Since several studies have identified the IBDV VP2 protein to contain the major immunoprotective epitopes against the virus [92], recombinant vaccines based on VP2 protein stand to outperform the currently used IBD vaccines. Interestingly, genetically modified NDV engineered to express IBDV VP2 antigen upstream of the NP gene in the NDV backbone was shown to induce 90% protection against both virulent ND and IBD challenges. More so, a booster vaccination using this bivalent vaccine led to 100% protection against both diseases [58]. This clearly demonstrates the potential of NDV as a promising vector for the delivery of IBDV immunogens in poultry.

Turkey rhinotracheitis affects the upper respiratory tract of young turkeys leading to high economic losses especially when superimposed with secondary bacterial infection. The disease is caused by avian metapneumovirus (AMPV) subtypes A, B, C, and D, with A and B subtypes being the most widely distributed [93]. Although effective live attenuated vaccines have been developed against this disease, their reversion back to virulence is of serious concern to poultry farmers all over the world. For instance, live vaccines based on AMPV subtypes A and B were shown to acquire a few mutations and regain virulence in turkeys, causing clinical disease similar in severity to the original field strain [94]. Thus, effective control of the disease begs the development of an effective and stable vaccine with no risk of reversion. The surface glycoproteins (F and G) of AMPV are known to participate not only in pathogenicity but also in the induction of neutralizing immunity in the vaccinated host. With the advent of recombinant DNA techniques, viruses vectoring AMPV F or G have been generated and shown to demonstrate some level of protection against turkey rhinotracheitis [71]. More recently, vaccination of turkeys with a recombinant NDV expressing the F and G proteins of AMPV type C (APMV-C) was shown to elicit both NDV-specific and AMPV-C specific humoral immune responses, which substantially protected against challenge with the two virulent viruses [48]. This further demonstrates the incredible ability of NDV to serve as an efficient vaccine vector in turkeys.

In the geese industry, gosling plaque (otherwise known as goose hepatitis) accounts for the highest rates of mortality and mobility. The disease is caused by a goose parvovirus that mainly causes gastrointestinal disturbances due to the predilection of the virus for the intestinal wall [95]. Live and inactivated vaccines are available to control this disease in areas where it is endemic. However, the currently available vaccines have many setbacks that necessitate the need to develop alternative vaccines. Apart from the risk of reversion of live vaccines back to virulence, the cost of vaccines that are prepared in SPF geese allantoic fluid is outrageously high, owing to the scarcity of SPF geese or duck embryonated eggs. Thus, improved and more cost-effective vaccines are needed to adequately control this economically important disease of geese. Among the structural proteins of GPV, the VP3 contains the highest number of neutralizing epitopes, a property that qualifies it to be used as a suitable vaccine antigen against the virus. Recombinant genotype VII-based NDV expressing the VP3 of GPV developed as a bivalent vaccine was shown to simultaneously induce strong neutralizing antibodies against NDV and GPV in China [60]. Since the prevailing NDV strains in China for the past two decades have been genotype VII isolates, this chimeric NDV and GPV vaccine stand to offer excellent protection against the field NDV strains (because it is genotype-matched) and at the same time adequately protect against GPV.

Other important avian pathogens against which NDV was used to deliver immunogens include infectious laryngotracheitis virus (ILTV). This pathogen primarily affects the upper respiratory system of chicken where it causes severe clinical signs such as dyspnoea, coughing and conjunctivitis, which inevitably lead to inappetence, reduced feed conversion efficiency, and a drastic decrease in egg production [96]. For a very long period of time, live attenuated vaccines have been widely used to control the disease. However, those vaccines have been shown to retain some level of residual virulence strong enough to cause clinical disease in chicken especially when administered as a spray [97]. Furthermore, they are associated with latent infection, which might later be reactivated especially in long-lived layers and breeders [98]. Thus, alternative vaccine platforms are required to optimally control this economically important disease of chicken. Deletion mutants of the virus devoid of glycoprotein J have been generated and shown to be highly protective in chicken [98]. Unfortunately, these vaccine candidates demonstrate poor in vitro replication kinetics. Attention therefore has shifted to recombinant fowl pox or HVT vectored vaccines that do not possess the risk of establishing latent infection. Again, these vaccines are only partially protective against ILTV [99]. Recently, a recombinant NDV LaSota strain engineered to express the glycoprotein D of the ILTV has been shown to induce a strong immune response that completely protected the vaccinated birds against the highly virulent ILTV challenge. Indeed, the level of neutralizing antibodies induced by the recombinant NDV vectored vaccine was even higher than that of commercial vaccines [55,100]. Given this superior humoral response and the stability of the virus, recombinant NDV expressing ILTV gD is a potential next generation vaccine candidate against ILTV infection in chicken.

4.3. Engineered NDV as a Vaccine Vector in Ruminant and Monogastric Animals

Middle East respiratory syndrome corona virus (MERS-COV) is the causative agent of a fatal human disease characterized by gastrointestinal and respiratory disorders [101]. The virus is harbored by bats and camels with the possibility of spill-over from these animals to man. Since camel–human contact in the Middle East is more frequent than bat–man interaction, the virus is more likely to be transmitted to man from camels, which were recently declared the major reservoirs of the virus in Saudi Arabia [102,103]. Hence, the most appropriate strategy to effectively control the disease is the development of a vaccine that will neutralize the virus in camel reservoirs. Recently, recombinant NDV Beaudette C strain was engineered to express MERS-CoV protein using reverse genetics approaches. When this chimeric virus was used to immunize camels in a prime-boost manner via the intramuscular route at the dose of 2×10^9 EID_{50} per camel, neutralizing antibodies against MERS-CoV were elicited [50]. Although a challenge experiment was not performed in the above study due to the unavailability of large animal containment facility, the rise in the neutralizing antibodies titre, as determined by ELISA and virus neutralization tests, especially after the booster vaccination, suggests the potential immunoprotective property of the vaccine [50]. More so, the fact that NDV can easily be prepared in large quantities in embryonated chicken eggs makes this potential bivalent vaccine less expensive and therefore easily affordable by the less privileged farmers. NDV is therefore a potential vaccine delivery agent against infectious diseases in camels.

Bovine enzootic fever, otherwise known as 3-day sickness, is an acute febrile disease of cattle and buffalo caused by an arthropod-borne pathogen called bovine ephemeral fever virus (BEFV). Although the disease is rarely fatal, it causes huge economic loses in the form of drastic reduction of milk yield among the dairy cattle [104]. Various vaccine preparations ranging from live, inactivated to subunit vaccines are utilized to control this disease in different countries. However, despite the effectiveness of these vaccines, the immunity they induce is not long lasting [105]. This means that a booster vaccine must periodically be administered to maintain the antibody titres above the protective threshold in the immunized subjects. As an alternative, recombinant NDV vectoring BEFV G protein was constructed and used as a vaccine in cattle. After the initial and booster vaccination of 1-year-old cattle using the recombinant vaccine, a good neutralizing antibody titre was obtained [53], suggesting the potential of the recombinant virus to be used in the control of the disease. More importantly, the use of the vaccine will be highly relevant in disease eradication by making it possible to differentiate vaccinated from infected animals (DIVA).

Bovine herpes virus-1 is the causative agent of many diseases in cattle such as infectious pustular vulvovaginitis, balanoposthitis and infectious bovine rhinotracheitis. It is therefore an important pathogen that militates against cattle production in regions where it is endemic [106,107]. A number of vaccines are available in different preparations to control this virus. However, those vaccines are not without their shortcomings. The live vaccines, although highly immunogenic, may cause immunosuppression in the vaccinated host. On the other hand, the inactivated vaccines are very safe but poorly immunogenic [108]. The quest for more effective vaccines has led to the generation of recombinant NDV expressing the BHV-1 gD protein. This protein is among the three enveloped proteins of the virus that are involved in the initial infection and the induction of neutralizing immunity in the infected or vaccinated host. Interestingly, of the three envelope proteins, the gD protein has been shown to contain the most neutralizing T and B cell epitopes [109]. Little wonder, several recombinant vaccines against BHV-1 utilized the protein as an immunogen [110,111]. When the recombinant NDV expressing BHV-1 gD was used to immunize 3-month-old calves via the combined intranasal and intratracheal routes, BHV-1 specific IgA and IgG immune response were observed. Although the protection observed following challenge with the virulent BHV-1 was only partial [52], it was reasoned that a second dose of the vaccine might significantly improve the protective efficacy of this vaccine in cattle.

Rift valley fever (RVF) is another important disease of ruminants with high zoonotic potential. It is characterized by abortion, foetal abnormalities and neonatal mortality especially in lambs [112]. To control the disease, formalin killed or live attenuated vaccines are often used. The killed vaccines are generally less immunogenic and so, adjuvants and booster vaccinations are required to enhance their effectiveness [113]. The live vaccines on the other hand are effective following a single administration but their safety in young animals is questionable. Hence, the current trend in the control of RVF is the development of cheap and effective vaccines that can be used across all age groups of animals [114]. In line with this requirement, many recombinant vaccines in the form of epitope-based synthetic peptides, plasmid DNA encoding the protein of interest, or viral vectors-based vaccines have been generated. Of these vaccines, the viral vectored vaccines seem to be the most useful tools for efficient delivery of RFV immunogens. A recombinant Venezuelan equine encephalitis virus expressing the RFV G2 protein was found to induce strong protective immunity in vaccinated mice [114]. However, this chimeric virus failed to grow to high titres in cell cultures, a problem that diminishes its potential as a vaccine. Interestingly, the recovery of a recombinant NDV expressing Gn protein of RVF as a potential vaccine was reported. When this vaccine was used to intramuscularly immunize calves at a dose of 2×10^7 TCID$_{50}$ per animal, a good level of anti RFV Gn neutralizing antibodies was observed. Although the serum neutralizing titres were only moderate, they were thought to be enough to protect against the virulent form of the disease [51]. With this level of immunogenicity added to the growth in high titres following inoculation in chicken embryonated eggs, NDV vectored RFV vaccine stands to be useful in the control of the disease.

Recombinant NDV is also useful in the delivery of vaccine antigens in dogs and cats. Canine distemper, a severe life threatening viral infection of carnivores, is currently controlled using modified live vaccines whose safety is still questionable. Immunization with those vaccines has been reported to cause clinical disease due to reversion back to virulence [115,116]. Furthermore, pre-existing maternal immunity strongly interferes with the efficacy of these vaccines. Thus, a safer and more efficacious vaccine platform is needed to improve the current control strategy. One attractive strategy that could potentially overcome these challenges is the use of a recombinant viral vector system. Recently, a recombinant NDV expressing the H protein of CDV (rLaSota-CDV HN) was shown to elicit solid neutralizing immunity that protected against virulent CDV challenge in minks [67]. In view of the numerous advantages of the NDV vector, particularly its host restriction and non-pathogenicity in mammals, rLaSota-CDV stands to be a good alternative to the not-very-safe conventional live attenuated CDV vaccines.

Rabies is a highly fatal zoonotic viral disease of all warm-blooded animals associated with severe pathology in the nervous system. Humans predominantly get infected via bites from a rabid dog [117,118]. Thus, controlling the disease in dogs is crucial to its prevention in humans. Interestingly, both live attenuated and recombinant vectored vaccines are available for the control of rabies [119]. However, those vaccines possess some limitations that make the search for improved vaccines a continuous priority. For instance, most of the live attenuated rabies vaccines are not very safe and only induce suboptimal immune response in the vaccinated animals [119]. Furthermore, recombinant vaccinia virus vectored rabies vaccine has been linked to severe integumentary inflammation, and the vector may cause persistent infection in humans [120]. As a result, effective rabies disease control in both man and animals demands the development of an improved vaccine that overcomes those challenges. Using the LaSota strain as a backbone, a recombinant NDV expressing rabies virus glycoprotein (rLaSota-RVG) was constructed and recovered as a potential vaccine [59]. When used to vaccinate dogs and cats at a dose of $10^{9.8}$ EID50, the rLaSota-RVG induced a robust long lasting neutralizing immunity that effectively protected against challenge with a wild type rabies virus street strain.

Collectively, the NDV vector is not only an efficient vaccine vector in poultry, it has also been shown to be a promising vector in other animals such as cattle, camels, dogs, etc. Since NDV is apathogenic in non-avian species, efforts should be intensified towards its use as a vaccine vector in different animal species particularly against those diseases whose control strategies are confronted with a lot shortcomings and challenges.

4.4. Engineered NDV as a Vaccine Vector in Man

Genetic manipulation of NDV has also been shown to be applicable in the delivery of human vaccines [111]. This is of particular relevance in the case of pathogens with serious biosafety concerns. For instance, Ebola virus, which causes a highly fatal disease in humans, can only be handled under BSL-4 facility because of its potential for human-to-human spread [121]. Therefore, conventional live attenuated Ebola vaccines are not likely to be safe due to the possibility of their reversion back to virulence. Moreover, inactivated Ebola vaccines that have been shown to be safe can only induce a suboptimal immune response that incompletely protects against lethal Ebola virus challenge [122]. Thus, there is the need for a safer and more effective vaccine platform. Among all the next generation Ebola vaccines, recombinant virus vectored vaccines appear to be the most promising. Rabies virus and Venezuelan equine encephalitis virus expressing Ebola GP are some of the recent vaccines that are currently under clinical trials in humans [123,124]. Recombinant NDV expressing Ebola GP has also been shown to stimulate strong neutralizing systemic and mucosal immunity following booster vaccination in monkeys [68,69]. This signifies the ability of NDV to effectively deliver Ebola virus immunogens as vaccines.

Successful delivery of HIV antigens has also been achieved using recombinant NDV vehicle [125]. Several live and inactivated vaccine candidates have been unsuccessful in the fight against the HIV pandemic, which continuously ravages the human population in different countries [126]. An ideal HIV vaccine should not only stimulate a long-lasting neutralizing antibody response, it should also elicit strong cell-mediated immunity, particularly a CD8$^+$ response, which is believed to play a crucial role in virus clearance [127]. With the gag and pol proteins being the most important elicitors of robust immune responses against HIV [128], attempts have been made to construct a recombinant vectored vaccine expressing these proteins. In one study, genetically engineered NDV expressing HIV gp140 or gp160 have been shown to induce a strong anti-HIV humoral immune response following parenteral or intranasal administration in mice [61]. Similarly, recombinant NDV expressing HIV gag protein induced a strong cell-mediated immunity specifically directed against the gag antigen in mouse models. Indeed, co-immunisation of mice with recombinant NDV expressing HIV gag and another recombinant NDV expressing HIV envelope protein was shown to induce strong neutralizing antibody and cell-mediated immunity especially following intranasal vaccination [61].

The potential of NDV as a vaccine vector has also been evaluated against emerging human diseases such as Nipah virus encephalitis and severe acute respiratory syndrome (SARS). In one instance, NDV was engineered to express Nipah virus surface glycoproteins (F and G). When the engineered NDV was used to immunize mice at the dose of 10^8EID50, a strong and long lasting Nipah virus-specific neutralizing immunity was generated [70]. Furthermore, recombinant mesogenic NDV engineered to express SARS-CoV spike S glycoprotein was found to not only induce a strongly protective immune response in mice, it also drastically reduced the viral load in the lungs, trachea and nasal turbinate following challenge with 10^6TCID50 of SARS coronavirus [49]. Other human vaccines vectored by NDV have been extensively reviewed elsewhere [111].

5. Engineered NDV as an Improved Cancer Vaccine

5.1. Molecular Mechanisms of NDV Induced Oncolysis

As far back as seven decades ago, NDV has been recognized to replicate more efficiently in mammalian cancerous cells than in normal cells. This natural oncolytic tendency of NDV has been demonstrated both in cell culture systems and different animal models [129]. In fact, many oncolytic NDV strains are currently at various stages of clinical trials for use in humans. One of the possible mechanisms of this selective replication of NDV in cancerous cells is the differential activation of interferon signaling pathways in normal and cancerous cells. Normal human cells are equipped with the RIG-I receptor that efficiently senses the presence of NDV and elicits a measurable interferon response [130]. On the other hand, most human cancers have defective interferon pathways, and so they cannot easily antagonize the invasion and replication of NDV. Nevertheless, some NDV strains have been shown to efficiently replicate in cancers with an intact interferon antiviral system, suggesting the involvement of other mechanisms of NDV-induced oncolysis [131]. Previously, NDV has been shown to stimulate the upregulated expression of TNF-related apoptotic-inducing ligands (TRAIL) on monocytes leading to tremendous oncolytic effects in tumor cells bearing the TRAIL-R2 receptor [132]. Similarly, NDV has been shown to induce apoptosis in many cancer cell lines by triggering certain endoplasmic reticulum stress in a manner independent of P53 gene expression [133]. Thus, the oncolytic efficacy of NDV could also be related to its ability to stimulate both intrinsic and extrinsic apoptosis pathways. Other mechanisms of NDV-induced oncolysis include the upregulated expression of major histocompatibility molecules and cell adhesion receptors as well as the increased secretion of proinflammatory cytokines around the tumor site [134]. In summary, the strategies of NDV-mediated oncolysis can be broadly classified into those involving the manipulation of the cellular antiviral and anti-apoptotic pathways as well as other indirect mechanisms that involve the activation of innate and adaptive immune responses.

5.2. Recombinant NDV as an Improved Oncolytic Agent

Despite the widely known oncolytic efficacy of wild type NDV, certain types of cancer are still resistant to the activity of the virus [134]. With the advent of reverse genetics, however, strategies of improving the oncolytic efficacy of the virus have emerged [135]. One of those strategies involves the manipulation of the F protein cleavage site of the virus. In one study, avirulent NDV was reported to have been engineered to encode polybasic instead of monobasic F cleavage site. When the recombinant virus was evaluated for anticancer activity, it was found to have a dramatically increased cell-to-cell fusion activity, replication kinetics, and tumoricidal effects in many human cancerous cell lines compared to the wild type virus [136]. This suggests an improved therapeutic potential of the virus. However, since NDV with a polybasic F cleavage site is potentially pathogenic in chicken, its usage as a cancer vaccine in humans is nowadays discouraged because of the fear of accidental spillage into the environment. Hence, the current focus is the generation of recombinant NDV that demonstrates excellent antitumor effects in humans while retaining its safety profile in chicken [137].

In line with the current research focus in NDV oncolytics, [138] generated a type-1 interferon sensitive avirulent NDV by abrogating its V protein expression. Although the recombinant virus failed to grow efficiently in normal cells due to interferon antagonism, it grew to a much higher titre and indeed induced apoptosis in many human cancerous cells with defective interferon signaling systems. Unfortunately, the oncolytic prospects of this recombinant virus are limited to only cancers with a non-functional interferon pathway since the virus is interferon sensitive. Therefore, in order to overcome this limitation, [135] created two major alterations in the genome of an avirulent NDV and recovered the recombinant virus using reverse genetics. These alterations are the modification of the F cleavage site from monobasic to polybasic and the insertion of influenza NS-1 gene into a lentogenic NDV backbone. While the creation of the polybasic cleavage site stands to improve cell-to-cell spread of the virus in cancerous cells, the influenza NS-1 protein through its potent anti-interferon and anti-apoptotic properties, allows the virus to evade the innate immune response following infection [136]. Thus, when the recombinant NDV expressing this protein was used to treat various cancerous cell lines, efficient replication, better syncytia formation, and enhanced oncolysis were noted especially among malignant melanoma cell lines.

Other approaches used to enhance the oncolytic efficacy of NDV include reprogramming the virus to constitutively express certain interferons and proinflammatory cytokines [134]. A recombinant NDV expressing soluble IL-2 has been shown to effectively regress hepatocellular carcinoma in mice and lead to the establishment of strong immunity that completely protected the cured mice from future challenge with cancerous cells [139]. Furthermore, co-expression of IL-2 and IL-12 on the NDV backbone led to enhanced anti-hepatoma activity in mice. In another study, NDV was engineered to express IL-12 and or IL-2, and both recombinant viruses proved to be better anticancer agents than the wild type virus [140]. Furthermore, recombinant NDV strain AF2240 engineered to express IL-12 was found to be highly efficacious against human breast cancer [141]. In addition, [142] in a proof-of-concept study showed that recombinant NDV expressing GM-CSF demonstrated a superb oncolytic activity and an enhanced immunostimulation of innate immune cells compared to the wild type virus. Thus, although NDV is naturally oncolytic, reverse genetics technology can be used to improve its properties to overcome the potential limitations associated with the use of the wild type virus. Recently, recombinant NDV engineered to express IL-24 has been shown to demonstrate improved oncolytic efficacy in murine melanoma models [143]. A summary of the approaches used in enhancing the oncolytic potential of NDV using reverse genetics is shown in Table 2.

Table 2. Some strategies used to enhance the oncolytic efficacy of Newcastle disease virus.

	NDV Strain	Genetic Modification	In Vitro Effects	In Vivo Effects	Reference
1.	NDV Lasota	Change of F cleavage site from monobasic to polybasic	Enhanced oncolysis of neuroblastoma cells via intrinsic and extrinsic caspase independent pathways	Not done	[144]
2.	NDV Lasota	Expression of GM-CSF	Induction of strong interferon response in PBMC; Substantial tumor growth inhibition caused by vaccine cells modified with the virus	Not done	[142]
3.	NDV Hitchner B1	Modification of F cleavage site and insertion of influenza NS1 gene	Profound cytotoxicity on human myeloma cell line SKMel-2 and mouse melanoma cell line B16-F10	i. Infiltration of CD4 and CD8 positive cells ii. suppression of mouse footpad melanoma growth	[136]
4.	NDV Hitchner B1	Modification of F cleavage site and insertion of IL2	Not done	Complete colon cancer regression characterized by marked T cell infiltration in mice	[135]
5.	NDV-HUJ	Change of F cleavage site from polybasic to monobasic	Enhanced apoptosis of chemoresistant primary melanoma cells	Not done	[131]
6.	NDV Beaudette C	Truncation of V protein expression	Not done	Complete regression of duodenum adenocarcinoma in Balb/c mice	[138]
7.	Clone 30	Expression of IL2 and IL12	Enhanced tumor cell death on U251, HepG2, Hela, and A549 cells	Enhanced oncolytic effect on hepatocarcinoma in mouse	[140]
8.	NDV 73-T	Change of cleavage site from monobasic to polybasic; Insertion of 198 nucleotides at the HN-L junction	Enhanced oncolytic effect on CCD1125 and HT1080 cells	Inhibition of tumor growth in HT1080 xenograft mouse tumor model	[145]
9.	NDV MTH68	Expression of heavy and light chains of monoclonal antibody directed against Edb fibronectin antigen	Enhanced tumor selective cytotoxicity on HT 29 colon cancer cells	Not done	[146]
10.	FMW	Expression of chicken infectious anaemia virus proapoptotic protein	Enhanced killing of adenocarcinomic human alveolar cells	Significant regression of treated tumor	[147]

6. Concluding Remarks

For several decades, NDV was only known as a poultry pathogen [148,149] with some potential to treat human cancer [150]. However, with the discovery of reverse genetics, so many other prospects of the virus have been unearthed. Today, the virus is easily programmable into a protective bivalent vaccine against highly virulent NDV and other economically important poultry diseases. The virus can also be manipulated to generate rationally designed vaccines against several emerging infectious diseases of various domestic animals. Furthermore, the virus has demonstrated the potential to not only deliver vaccine antigens against fatal human diseases, but also serve as an improved oncolytic agent against a variety of human cancers. Thus, the impacts of engineered NDV in modern vaccinology are enormous. Given its simple genome, efficient replication, host restriction, and non-pathogenicity in most mammals, NDV is likely to be the vector of choice against many other emerging diseases of man and domestic animals.

Author Contributions: M.B.B., A.R.O., K.Y. and A.H.J. provided the needed literature. M.B.B. wrote the manuscript. A.R.O., B.P.H.P., A.I., and M.H.-B. critically reviewed the manuscript. All authors have read and agreed to the published version of the manuscript.

Funding: The article processing charge (APC) was funded by Higher Institution Center of Excellence (HICOE), grant number 6369101.

Conflicts of Interest: The authors declare no conflict of interest

References

1. Greenwood, B. The contribution of vaccination to global health: Past, present and future. *Philos. Trans. R. Soc. B Biol. Sci.* **2014**, *369*, 20130433. [CrossRef] [PubMed]
2. Taylor, W.P.; Bhat, P.N.; Nanda, Y.P. The principles and practice of rinderpest eradication. *Vet. Microbiol.* **1995**, *44*, 359–367. [CrossRef]
3. Henderson, D.A. The eradication of smallpox-An overview of the past, present, and future. *Vaccine* **2011**, *29*, D7–D9. [CrossRef] [PubMed]
4. Burnett, M.S.; Wang, N.; Hofmann, M.; Barrie Kitto, G. Potential live vaccines for HIV. *Vaccine* **2000**, *19*, 735–742. [CrossRef]
5. Chen, G.L.; Subbarao, K. Live attenuated vaccines for pandemic influenza. *Curr. Top. Microbiol. Immunol.* **2009**, *333*, 109–132.
6. Gendon Iu, Z. Advantages and disadvantages of inactivated and live influenza vaccine. *Vopr. Virusol.* **2004**, *49*, 4–12.
7. He, Y.; Rappuoli, R.; De Groot, A.S.; Chen, R.T. Emerging vaccine informatics. *J. Biomed. Biotechnol.* **2010**, *2010*. [CrossRef]
8. Bello, M.B.; Yusoff, K.; Ideris, A.; Hair-Bejo, M.; Peeters, B.P.H.; Omar, A.R. Diagnostic and Vaccination Approaches for Newcastle Disease Virus in Poultry: The Current and Emerging Perspectives. *Biomed. Res. Int.* **2018**. [CrossRef]
9. Alexander, D.J. Newcastle disease. *Br. Poult. Sci.* **2001**, *42*, 5–22. [CrossRef]
10. Dimitrov, K.M.; Abolnik, C.; Afonso, C.L.; Albina, E.; Bahl, J.; Berg, M.; Briand, F.X.; Brown, I.H.; Choi, K.S.; Chvala, I.; et al. Updated unified phylogenetic classification system and revised nomenclature for Newcastle disease virus. *Infect. Genet. Evol.* **2019**, *74*, 103917. [CrossRef]
11. Peeters, B.P.; de Leeuw, O.S.; Koch, G.; Gielkens, A.L. Rescue of Newcastle disease virus from cloned cDNA: Evidence that cleavability of the fusion protein is a major determinant for virulence. *J. Virol.* **1999**, *73*, 5001–5009. [CrossRef] [PubMed]
12. Nagai, Y.; Hamaguchi, M.; Toyoda, T. Molecular Biology of Newcastle Disease Virus. *Prog. Vet. Microbiol. Immunol.* **1989**, *5*, 16–64. [PubMed]
13. Yusoff, K.; Tan, W.S. Newcastle disease virus: Macromolecules and opportunities. *Avian Pathol.* **2001**, *30*, 439–455. [CrossRef] [PubMed]
14. Czeglédi, A.; Ujvári, D.; Somogyi, E.; Wehmann, E.; Werner, O.; Lomniczi, B. Third genome size category of avian paramyxovirus serotype 1 (Newcastle disease virus) and evolutionary implications. *Virus Res.* **2006**, *120*, 36–48. [CrossRef] [PubMed]
15. Murulitharan, K.; Yusoff, K.; Omar, A.R.; Molouki, A. Characterization of Malaysian velogenic NDV strain AF2240-I genomic sequence: A comparative study. *Virus Genes* **2013**, *46*, 431–440. [CrossRef] [PubMed]
16. Curran, J.; Kolakofsky, D. Replication of paramyxoviruses. *Adv. Virus Res.* **1999**, *54*, 403–422.
17. Munir, M.; Zohari, S.; Abbas, M.; Berg, M. Sequencing and analysis of the complete genome of Newcastle disease virus isolated from a commercial poultry farm in 2010. *Arch. Virol.* **2012**, *157*, 765–768. [CrossRef]
18. Dortmans, J.C.F.M.; Rottier, P.J.M.; Koch, G.; Peeters, B.P.H. The viral replication complex is associated with the virulence of Newcastle disease virus. *J. Virol.* **2010**, *84*, 10113–10120. [CrossRef]
19. Ganar, K.; Das, M.; Sinha, S.; Kumar, S. Newcastle disease virus: Current status and our understanding. *Virus Res.* **2014**, *184*, 71–81. [CrossRef]

20. Rout, S.N.; Samal, S.K. The Large Polymerase Protein Is Associated with the Virulence of Newcastle Disease Virus. *J. Virol.* **2008**, *82*, 7828–7836. [CrossRef]
21. Collins, M.S.; Strong, I.; Alexander, D.J. Evaluation of the molecular basis of pathogenicity of the variant Newcastle disease viruses termed? pigeon "MV-1 viruses". *Arch. Virol.* **1994**, *134*, 403–411. [CrossRef] [PubMed]
22. Panda, A.; Huang, Z.; Elankumaran, S.; Rockemann, D.D.; Samal, S.K. Role of fusion protein cleavage site in the virulence of Newcastle disease virus. *Microb. Pathog.* **2004**, *36*, 1–10. [CrossRef] [PubMed]
23. Samal, S.; Khattar, S.K.; Kumar, S.; Collins, P.L.; Samal, S.K. Coordinate deletion of N-glycans from the heptad repeats of the fusion F protein of Newcastle disease virus yields a hyperfusogenic virus with increased replication, virulence, and immunogenicity. *J. Virol.* **2012**, *86*, 2501–2511. [CrossRef] [PubMed]
24. Heiden, S.; Grund, C.; Röder, A.; Granzow, H.; Kühnel, D.; Mettenleiter, T.C.; Römer-Oberdörfer, A. Different regions of the newcastle disease virus fusion protein modulate pathogenicity. *PLoS ONE* **2014**, *9*, e113344. [CrossRef]
25. Zhao, W.; Zhang, Z.; Zsak, L.; Yu, Q. Effects of the HN gene C-terminal extensions on the Newcastle disease virus virulence. *Virus Genes* **2013**, *47*, 498–504. [CrossRef]
26. Jin, J.; Zhao, J.; Ren, Y.; Zhong, Q.; Zhang, G. Contribution of HN protein length diversity to Newcastle disease virus virulence, replication and biological activities. *Sci. Rep.* **2016**, *6*, 36890. [CrossRef]
27. Huang, Z.; Krishnamurthy, S.; Panda, A.; Samal, S.K. Newcastle disease virus V protein is associated with viral pathogenesis and functions as an alpha interferon antagonist. *J. Virol.* **2003**, *77*, 8676–8685. [CrossRef]
28. Nagai, Y. Paramyxovirus replication and pathogenesis. Reverse genetics transforms understanding. *Rev. Med. Virol.* **1999**, *9*, 83–99. [CrossRef]
29. Noton, S.L.; Fearns, R. Initiation and regulation of paramyxovirus transcription and replication. *Virology* **2015**, *479–480*, 545–554. [CrossRef]
30. Peeters, B.P.H.; Gruijthuijsen, Y.K.; De Leeuw, O.S.; Gielkens, A.L.J. Genome replication of Newcastle disease virus: Involvement of the rule-of-six. *Arch. Virol.* **2000**, *145*, 1829–1845. [CrossRef]
31. Whelan, S.P.J.; Barr, J.N.; Wertz, G.W. Transcription and replication of nonsegmented negative-strand RNA viruses. *Curr. Top. Microbiol. Immunol.* **2004**, *283*, 61–119. [PubMed]
32. Conzelmann, K.K. Reverse genetics of mononegavirales. *Curr. Top. Microbiol. Immunol.* **2004**, *283*, 1–41. [PubMed]
33. Vulliémoz, D.; Roux, L. "Rule of six": How does the Sendai virus RNA polymerase keep count? *J. Virol.* **2001**, *75*, 4506–4518. [CrossRef] [PubMed]
34. Gururaj, K.; Kirubaharan, J.J.; Gupta, V.K.; Pawaiya, R.S. Review Article Past and Present of Reverse Genetics in Animal Virology with Special Reference to Non–Segmented Negative Stranded RNA Viruses: A Review. *Adv. Anim. Vet. Sci.* **2014**, *2*, 40–48. [CrossRef]
35. Taniguchi, T.; Palmieri, M.; Weissmann, C. Qβ DNA-containing hybrid plasmids giving rise to QB phage formation in the bacterial host. *Nature* **1978**, *274*, 223–228. [CrossRef] [PubMed]
36. Conzelmann, K.K.; Schnell, M. Rescue of synthetic genomic RNA analogs of rabies virus by plasmid-encoded proteins. *J. Virol.* **1994**, *68*, 713–719. [CrossRef]
37. Jiang, Y.; Liu, H.; Liu, P.; Kong, X. Plasmids driven minigenome rescue system for Newcastle disease virus V4 strain. *Mol. Biol. Rep.* **2009**, *36*, 1909–1914. [CrossRef]
38. Su, J.; Dou, Y.; You, Y.; Cai, X. Application of minigenome technology in virology research of the Paramyxoviridae family. *J. Microbiol. Immunol. Infect.* **2015**, *48*, 123–129. [CrossRef]
39. Liu, H.; Albina, E.; Gil, P.; Minet, C.; de Almeida, R.S. Two-plasmid system to increase the rescue efficiency of paramyxoviruses by reverse genetics: The example of rescuing Newcastle Disease Virus. *Virology* **2017**, *509*, 42–51. [CrossRef]
40. Peeters, B.; de Leeuw, O. A single-plasmid reverse genetics system for the rescue of non-segmented negative-strand RNA viruses from cloned full-length cDNA. *J. Virol. Methods* **2017**, *248*, 187–190. [CrossRef]
41. Schirrmacher, V.; Fournier, P. Newcastle disease virus: A promising vector for viral therapy, immune therapy, and gene therapy of cancer. *Methods Mol. Biol.* **2009**, *542*, 565–605. [PubMed]
42. Shirvani, E.; Paldurai, A.; Manoharan, V.K.; Varghese, B.P.; Samal, S.K. A Recombinant Newcastle Disease Virus (NDV) Expressing S Protein of Infectious Bronchitis Virus (IBV) Protects Chickens against IBV and NDV. *Sci. Rep.* **2018**, *8*, 1–14. [CrossRef] [PubMed]

43. Zhao, W.; Zhang, Z.; Zsak, L.; Yu, Q. P and M gene junction is the optimal insertion site in Newcastle disease virus vaccine vector for foreign gene expression. *J. Gen. Virol.* **2015**, *96*, 40–45. [CrossRef] [PubMed]

44. Huang, Z.; Krishnamurthy, S.; Panda, A.; Samal, S.K. High-level expression of a foreign gene from the most 3′-proximal locus of a recombinant Newcastle disease virus. *J. Gen. Virol.* **2001**, *82*, 1729–1736. [CrossRef]

45. Zhang, Z.; Zhao, W.; Li, D.; Yang, J.; Zsak, L.; Yu, Q. Development of a Newcastle disease virus vector expressing a foreign gene through an internal ribosomal entry site provides direct proof for a sequential transcription mechanism. *J. Gen. Virol.* **2015**, *96*, 2028–2035. [CrossRef]

46. Kieft, J.S. Viral IRES RNA structures and ribosome interactions. *Trends Biochem. Sci.* **2008**, *33*, 274–283. [CrossRef]

47. Susta, L.; Cornax, I.; Diel, D.G.; Garcia, S.C.; Miller, P.J.; Liu, X.; Hu, S.; Brown, C.C.; Afonso, C.L. Expression of interferon gamma by a highly virulent strain of Newcastle disease virus decreases its pathogenicity in chickens. *Microb. Pathog.* **2013**, *61–62*, 73–83. [CrossRef]

48. Hu, H.; Roth, J.P.; Zsak, L.; Yu, Q. Engineered Newcastle disease virus expressing the F and G proteins of AMPV-C confers protection against challenges in turkeys. *Sci. Rep.* **2017**, *7*, 1–8. [CrossRef]

49. DiNapoli, J.M.; Kotelkin, A.; Yang, L.; Elankumaran, S.; Murphy, B.R.; Samal, S.K.; Collins, P.L.; Bukreyev, A. Newcastle disease virus, a host range-restricted virus, as a vaccine vector for intranasal immunization against emerging pathogens. *Proc. Natl. Acad. Sci. USA* **2007**, *104*, 9788–9793. [CrossRef]

50. Liu, R.; Ge, J.; Wang, J.; Shao, Y.; Zhang, H.; Wang, J.; Wen, Z.; Bu, Z. Newcastle disease virus-based MERS-CoV candidate vaccine elicits high-level and lasting neutralizing antibodies in Bactrian camels. *J. Integr. Agric.* **2017**, *16*, 2264–2273. [CrossRef]

51. Kortekaas, J.; Dekker, A.; de Boer, S.M.; Weerdmeester, K.; Vloet, R.P.M.; De Wit, A.A.C.; Peeters, B.P.H.; Moormann, R.J.M. Intramuscular inoculation of calves with an experimental Newcastle disease virus-based vector vaccine elicits neutralizing antibodies against Rift Valley fever virus. *Vaccine* **2010**, *28*, 2271–2276. [CrossRef]

52. Khattar, S.K.; Collins, P.L.; Samal, S.K. Immunization of cattle with recombinant Newcastle disease virus expressing bovine herpesvirus-1 (BHV-1) glycoprotein D induces mucosal and serum antibody responses and provides partial protection against BHV-1. *Vaccine* **2010**, *28*, 3159–3170. [CrossRef] [PubMed]

53. Zhang, M.; Ge, J.; Wen, Z.; Chen, W.; Wang, X.; Liu, R.; Bu, Z. Characterization of a recombinant Newcastle disease virus expressing the glycoprotein of bovine ephemeral fever virus. *Arch. Virol.* **2017**, *162*, 359–367. [CrossRef] [PubMed]

54. Zhao, R.; Sun, J.; Qi, T.; Zhao, W.; Han, Z.; Yang, X.; Liu, S. Recombinant Newcastle disease virus expressing the infectious bronchitis virus S1 gene protects chickens against Newcastle disease virus and infectious bronchitis virus challenge. *Vaccine* **2017**, *35*, 2435–2442. [CrossRef] [PubMed]

55. Zhao, W.; Spatz, S.; Zhang, Z.; Wen, G.; Garcia, M.; Zsak, L.; Yu, Q. Newcastle disease virus (NDV) recombinants expressing infectious laryngotracheitis virus (ILTV) gB and gD glycoproteins protect chickens against ILTV and NDV challenge. *J. Virol.* **2014**, *88*, 8397–8406. [CrossRef] [PubMed]

56. Schröer, D.; Veits, J.; Keil, G.; Römer-Oberdörfer, A.; Weber, S.; Mettenleiter, T.C. Efficacy of Newcastle disease virus recombinant expressing avian influenza virus H6 hemagglutinin against Newcastle disease and low pathogenic avian influenza in chickens and turkeys. *Avian Dis.* **2011**, *55*, 201–211. [CrossRef]

57. Kim, S.H.; Samal, S.K. Innovation in newcastle disease virus vectored avian influenza vaccines. *Viruses* **2019**, *11*, 300. [CrossRef]

58. Huang, Z.; Elankumaran, S.; Yunus, A.S.; Samal, S.K. A recombinant Newcastle disease virus (NDV) expressing VP2 protein of infectious bursal disease virus (IBDV) protects against NDV and IBDV. *J. Virol.* **2004**, *78*, 10054–10063. [CrossRef]

59. Ge, J.; Wang, X.; Tao, L.; Wen, Z.; Feng, N.; Yang, S.; Xia, X.; Yang, C.; Chen, H.; Bu, Z. Newcastle Disease Virus-Vectored Rabies Vaccine Is Safe, Highly Immunogenic, and Provides Long-Lasting Protection in Dogs and Cats. *J. Virol.* **2011**, *85*, 8241–8252. [CrossRef]

60. Wang, J.; Cong, Y.; Yin, R.; Feng, N.; Yang, S.; Xia, X.; Xiao, Y.; Wang, W.; Liu, X.; Hu, S.; et al. Generation and evaluation of a recombinant genotype VII Newcastle disease virus expressing VP3 protein of Goose parvovirus as a bivalent vaccine in goslings. *Virus Res.* **2015**, *203*, 77–83. [CrossRef]

61. Khattar, S.K.; Samal, S.; Devico, A.L.; Collins, P.L.; Samal, S.K. Newcastle disease virus expressing human immunodeficiency virus type 1 envelope glycoprotein induces strong mucosal and serum antibody responses in Guinea pigs. *J. Virol.* **2011**, *85*, 10529–10541. [CrossRef] [PubMed]

62. Wang, J.; Yang, J.; Ge, J.; Hua, R.; Liu, R.; Li, X.; Wang, X.; Shao, Y. Newcastle disease virus-vectored West Nile fever vaccine is immunogenic in mammals and poultry. *Virol. J.* **2016**, *13*, 1–11. [CrossRef] [PubMed]

63. Martinez-Sobrido, L.; Gitiban, N.; Fernandez-Sesma, A.; Cros, J.; Mertz, S.E.; Jewell, N.A.; Hammond, S.; Flano, E.; Durbin, R.K.; García-Sastre, A.; et al. Protection against respiratory syncytial virus by a recombinant Newcastle disease virus vector. *J. Virol.* **2006**, *80*, 1130–1139. [CrossRef] [PubMed]

64. Kong, D.; Wen, Z.; Su, H.; Ge, J.; Chen, W.; Wang, X.; Wu, C.; Yang, C.; Chen, H.; Bu, Z. Newcastle disease virus-vectored Nipah encephalitis vaccines induce B and T cell responses in mice and long-lasting neutralizing antibodies in pigs. *Virology* **2012**, *432*, 327–335. [CrossRef]

65. Zhang, M.; Ge, J.; Li, X.; Chen, W.; Wang, X.; Wen, Z.; Bu, Z. Protective efficacy of a recombinant Newcastle disease virus expressing glycoprotein of vesicular stomatitis virus in mice. *Virol. J.* **2016**, *13*, 31. [CrossRef]

66. Kim, S.-H.; Chen, S.; Jiang, X.; Green, K.Y.; Samal, S.K. Newcastle Disease Virus Vector Producing Human Norovirus-Like Particles Induces Serum, Cellular, and Mucosal Immune Responses in Mice. *J. Virol.* **2014**, *88*, 9718–9727. [CrossRef]

67. Ge, J.; Wang, X.; Tian, M.; Gao, Y.; Wen, Z.; Yu, G.; Zhou, W.; Zu, S.; Bu, Z. Recombinant Newcastle disease viral vector expressing hemagglutinin or fusion of canine distemper virus is safe and immunogenic in minks. *Vaccine* **2015**, *33*, 2457–2462. [CrossRef]

68. Wen, Z.; Zhao, B.; Song, K.; Hu, X.; Chen, W.; Kong, D.; Ge, J.; Bu, Z. Recombinant lentogenic Newcastle disease virus expressing Ebola virus GP infects cells independently of exogenous trypsin and uses macropinocytosis as the major pathway for cell entry. *Virol. J.* **2013**, *10*, 331. [CrossRef]

69. DiNapoli, J.M.; Yang, L.; Samal, S.K.; Murphy, B.R.; Collins, P.L.; Bukreyev, A. Respiratory tract immunization of non-human primates with a Newcastle disease virus-vectored vaccine candidate against Ebola virus elicits a neutralizing antibody response. *Vaccine* **2010**, *29*, 17–25. [CrossRef]

70. Bukreyev, A.; Huang, Z.; Yang, L.; Elankumaran, S.; St. Claire, M.; Murphy, B.R.; Samal, S.K.; Collins, P.L. Recombinant Newcastle Disease Virus Expressing a Foreign Viral Antigen is Attenuated and Highly Immunogenic in Primates. *J. Virol.* **2005**, *79*, 13275–13284. [CrossRef]

71. Yu, Q.; Roth, J.P.; Hu, H.; Estevez, C.N.; Zhao, W.; Zsak, L. Protection by Recombinant Newcastle Disease Viruses (NDV) Expressing the Glycoprotein (G) of Avian Metapneumovirus (aMPV) Subtype A or B against Challenge with Virulent NDV and aMPV. *World J. Vaccines* **2013**, *3*, 130–139. [CrossRef]

72. Wen, G.; Chen, C.; Guo, J.; Zhang, Z.; Shang, Y.; Shao, H.; Luo, Q.; Yang, J.; Wang, H.; Wang, H.; et al. Development of a novel thermostable Newcastle disease virus vaccine vector for expression of a heterologous gene. *J. Gen. Virol.* **2015**, *96*, 1219–1228. [CrossRef] [PubMed]

73. Ryan, M.D.; Drew, J. Foot-and-mouth disease virus 2A oligopeptide mediated cleavage of an artificial polyprotein. *EMBO J.* **1994**, *13*, 928–933. [CrossRef] [PubMed]

74. Stobart, C.C.; Moore, M.L. RNA virus reverse genetics and vaccine design. *Viruses* **2014**, *6*, 2531–2550. [CrossRef] [PubMed]

75. Hu, Z.; Hu, S.; Meng, C.; Wang, X.; Zhu, J.; Liu, X. Generation of a Genotype VII Newcastle Disease Virus Vaccine Candidate with High Yield in Embryonated Chicken Eggs. *Avian Dis. Dig.* **2011**, *6*, e7–e8. [CrossRef]

76. Xiao, S.; Nayak, B.; Samuel, A.; Paldurai, A.; Kanabagattebasavarajappa, M.; Prajitno, T.Y.; Bharoto, E.E.; Collins, P.L.; Samal, S.K. Generation by Reverse Genetics of an Effective, Stable, Live-Attenuated Newcastle Disease Virus Vaccine Based on a Currently Circulating, Highly Virulent Indonesian Strain. *PLoS ONE* **2012**, *7*, e52751. [CrossRef]

77. Roohani, K.; Tan, S.W.; Yeap, S.K.; Ideris, A.; Bejo, M.H.; Omar, A.R. Characterisation of genotype VII Newcastle disease virus (NDV) isolated from NDV vaccinated chickens, and the efficacy of LaSota and recombinant genotype VII vaccines against challenge with velogenic NDV. *J. Vet. Sci.* **2015**, *16*, 447–457. [CrossRef]

78. Hutchinson, E.C. Influenza Virus. *Trends Microbiol.* **2018**, *26*, 809–810. [CrossRef]

79. Lee, C.W.; Saif, Y.M. Avian influenza virus. *Comp. Immunol. Microbiol. Infect. Dis.* **2009**, *32*, 301–310. [CrossRef]

80. Swayne, D.E. Avian influenza vaccines and therapies for poultry. *Comp. Immunol. Microbiol. Infect. Dis.* **2009**, *32*, 351–363. [CrossRef]

81. Suarez, D.L.; Pantin-Jackwood, M.J. Recombinant viral-vectored vaccines for the control of avian influenza in poultry. *Vet. Microbiol.* **2017**, *206*, 144–151. [CrossRef] [PubMed]

82. Park, M.S.; Steel, J.; García-Sastre, A.; Swayne, D.; Palese, P. Engineered viral vaccine constructs with dual specificity: Avian influenza and Newcastle disease. *Proc. Natl. Acad. Sci. USA* **2006**, *103*, 8203–8208. [CrossRef] [PubMed]

83. Liu, I.; Ma, J.; Bawa, B.; Krammer, F.; Lyoo, Y.S.; Lang, Y.; Morozov, I.; Mahardika, G.N.; Ma, W.; Garcia-Sastre, A.; et al. Newcastle Disease Virus-Vectored H7 and H5 Live Vaccines Protect Chickens from Challenge with H7N9 or H5N1 Avian Influenza Viruses. *J. Virol.* **2015**, *89*, 7401–7408. [CrossRef] [PubMed]

84. Colvero, L.P.; Villarreal, L.Y.B.; Torres, C.A.; Brañdo, P.E. Assessing the economic burden of avian infectious bronchitis on poultry farms in Brazil. *Rev. Sci. Tech.* **2015**, *34*, 993–999. [CrossRef] [PubMed]

85. Bande, F.; Arshad, S.S.; Hair Bejo, M.; Moeini, H.; Omar, A.R. Progress and challenges toward the development of vaccines against avian infectious bronchitis. *J. Immunol. Res.* **2015**, *2015*. [CrossRef]

86. McKinley, E.T.; Hilt, D.A.; Jackwood, M.W. Avian coronavirus infectious bronchitis attenuated live vaccines undergo selection of subpopulations and mutations following vaccination. *Vaccine* **2008**, *26*, 1274–1284. [CrossRef]

87. Bickerton, E.; Keep, S.M.; Britton, P. Reverse genetics system for the avian coronavirus infectious bronchitis virus. In *Methods in Molecular Biology*; Humana Press: New York, NY, USA, 2017; Volume 1602, pp. 83–102. ISBN 0022-538X.

88. Abozeid, H.H.; Paldurai, A.; Varghese, B.P.; Khattar, S.K.; Afifi, M.A.; Zouelfakkar, S.; El-Deeb, A.H.; El-Kady, M.F.; Samal, S.K. Development of a recombinant Newcastle disease virus-vectored vaccine for infectious bronchitis virus variant strains circulating in Egypt. *Vet. Res.* **2019**, *50*, 12. [CrossRef]

89. Sharma, J.M.; Kim, I.J.; Rautenschlein, S.; Yeh, H.Y. Infectious bursal disease virus of chickens: Pathogenesis and immunosuppression. *Dev. Comp. Immunol.* **2000**, *24*, 223–235. [CrossRef]

90. Morla, S.; Deka, P.; Kumar, S. Isolation of novel variants of infectious bursal disease virus from different outbreaks in Northeast India. *Microb. Pathog.* **2016**, *93*, 131–136. [CrossRef]

91. Adamu, J.; Owoade, A.A.; Abdu, P.A.; Kazeem, H.M.; Fatihu, M.Y. Characterization of field and vaccine infectious bursal disease viruses from Nigeria revealing possible virulence and regional markers in the VP2 minor hydrophilic peaks. *Avian Pathol.* **2013**, *42*, 420–433. [CrossRef]

92. Wang, X.-N.; Zhang, G.-P.; Zhou, J.-Y.; Feng, C.-H.; Yang, Y.-Y.; Li, Q.-M.; Guo, J.-Q.; Qiao, H.-X.; Xi, J.; Zhao, D.; et al. Identification of neutralizing epitopes on the VP2 protein of infectious bursal disease virus by phage-displayed heptapeptide library screening and synthetic peptide mapping. *Viral Immunol.* **2005**, *18*, 549–557. [CrossRef]

93. Broor, S.; Bharaj, P. Avian and human metapneumovirus. *Proc. Ann. N. Y. Acad. Sci.* **2007**, *1102*, 66–85. [CrossRef]

94. Cecchinato, M.; Catelli, E.; Lupini, C.; Ricchizzi, E.; Prosperi, S.; Naylor, C.J. Reversion to virulence of a subtype B avian metapneumovirus vaccine: Is it time for regulators to require availability of vaccine progenitors? *Vaccine* **2014**, *32*, 4660–4664. [CrossRef]

95. Irvine, R.; Holmes, P. Diagnosis and control of goose parvovirus. *Practice* **2010**, *32*, 382–386. [CrossRef]

96. Ou, S.-C. Infectious laryngotracheitis virus in chickens. *World J. Virol.* **2012**, *1*, 142. [CrossRef]

97. Fulton, R.M.; Schrader, D.L.; Will, M. Effect of route of vaccination on the prevention of infectious laryngotracheitis in commercial egg-laying chickens. *Avian Dis.* **2009**, *44*, 8–16. [CrossRef]

98. Coppo, M.J.C.; Noormohammadi, A.H.; Browning, G.F.; Devlin, J.M. Challenges and recent advancements in infectious laryngotracheitis virus vaccines. *Avian Pathol.* **2013**, *42*, 195–205. [CrossRef]

99. Johnson, D.I.; Vagnozzi, A.; Dorea, F.; Riblet, S.M.; Mundt, A.; Zavala, G.; García, M. Protection Against Infectious Laryngotracheitis by In Ovo Vaccination with Commercially Available Viral Vector Recombinant Vaccines. *Avian Dis. Dig.* **2010**, *54*, 1251–1259. [CrossRef]

100. Kanabagatte Basavarajappa, M.; Kumar, S.; Khattar, S.K.; Gebreluul, G.T.; Paldurai, A.; Samal, S.K. A recombinant Newcastle disease virus (NDV) expressing infectious laryngotracheitis virus (ILTV) surface glycoprotein D protects against highly virulent ILTV and NDV challenges in chickens. *Vaccine* **2014**, *32*, 3555–3563. [CrossRef]

101. Lu, L.; Liu, Q.; Du, L.; Jiang, S. Middle East respiratory syndrome coronavirus (MERS-CoV): Challenges in identifying its source and controlling its spread. *Microbes Infect.* **2013**, *15*, 625–629. [CrossRef]

102. Du, L.; Han, G.Z. Deciphering MERS-CoV Evolution in Dromedary Camels. *Trends Microbiol.* **2016**, *24*, 87–89. [CrossRef]

103. Chen, X.; Chughtai, A.A.; Dyda, A.; MacIntyre, C.R. Comparative epidemiology of Middle East respiratory syndrome coronavirus (MERS-CoV) in Saudi Arabia and South Korea. *Emerg. Microbes Infect.* **2017**, *6*, e51. [CrossRef]

104. Nandi, S.; Negi, B.S. Bovine ephemeral fever: A review. *Comp. Immunol. Microbiol. Infect. Dis.* **1999**, *22*, 81–91. [CrossRef]

105. Walker, P.J.; Klement, E. Epidemiology and control of bovine ephemeral fever. *Vet. Res.* **2015**, *46*, 124. [CrossRef]

106. Graham, D.A. Bovine herpes virus-1 (BoHV-1) in cattle–a review with emphasis on reproductive impacts and the emergence of infection in Ireland and the United Kingdom. *Ir. Vet. J.* **2013**, *66*, 15. [CrossRef]

107. Nandi, S.; Kumar, M.; Manohar, M.; Chauhan, R.S. Bovine herpes virus infections in cattle. *Anim. Heal. Res. Rev.* **2009**, *10*, 85–98. [CrossRef]

108. Patel, J.R. Characteristics of live bovine herpesvirus-1 vaccines. *Vet. J.* **2005**, *169*, 404–416. [CrossRef]

109. Peralta, A.; Molinari, P.; Conte-Grand, D.; Calamante, G.; Taboga, O. A chimeric baculovirus displaying bovine herpesvirus-1 (BHV-1) glycoprotein D on its surface and their immunological properties. *Appl. Microbiol. Biotechnol.* **2007**, *75*, 407–414. [CrossRef]

110. Pérez Filgueira, D.M.; Zamorano, P.I.; Domínguez, M.G.; Taboga, O.; Del Médico Zajac, M.P.; Puntel, M.; Romera, S.A.; Morris, T.J.; Borca, M.V.; Sadir, A.M. Bovine herpes virus gD protein produced in plants using a recombinant tobacco mosaic virus (TMV) vector possesses authentic antigenicity. *Vaccine* **2003**, *21*, 4201–4209. [CrossRef]

111. Kim, S.H.; Samal, S.K. Newcastle disease virus as a vaccine vector for development of human and veterinary vaccines. *Viruses* **2016**, *8*, 183. [CrossRef]

112. Flick, R.; Bouloy, M. Rift Valley fever virus. *Curr. Mol. Med.* **2005**, *5*, 827–834. [CrossRef]

113. Ikegami, T.; Makino, S. Rift Valley fever vaccines. *Vaccine* **2009**, *27*, D69–D72. [CrossRef]

114. Faburay, B.; LaBeaud, A.D.; McVey, D.S.; Wilson, W.C.; Richt, J.A. Current status of rift valley fever vaccine development. *Vaccines* **2017**, *5*, 29. [CrossRef]

115. Rikula, U.; Pänkälä, L.; Jalkanen, L.; Sihvonen, L. Distemper vaccination of farmed fur animals in Finland. *Prev. Vet. Med.* **2001**, *49*, 125–133. [CrossRef]

116. Van Heerden, J.; Bingham, J.; van Vuuren, M.; Burroughs, R.E.J.; Stylianides, E. Clinical and serological response of wild dogs (*Lycaon pictus*) to vaccination against canine distemper, canine parvovirus infection and rabies. *J. S. Afr. Vet. Assoc.* **2002**, *73*, 8–12. [CrossRef]

117. Udow, S.J.; Marrie, R.A.; Jackson, A.C. Clinical features of dog- and bat-acquired rabies in humans. *Clin. Infect. Dis.* **2013**, *57*, 689–696. [CrossRef]

118. Susilawathi, N.M.; Darwinata, A.E.; Dwija, I.B.N.P.; Budayanti, N.S.; Wirasandhi, G.A.K.; Subrata, K.; Susilarini, N.K.; Sudewi, R.A.A.; Wignall, F.S.; Mahardika, G.N.K. Epidemiological and clinical features of human rabies cases in Bali 2008–2010. *BMC Infect. Dis.* **2012**, *12*, 81. [CrossRef]

119. Cleaveland, S.; Kaare, M.; Knobel, D.; Laurenson, M.K. Canine vaccination-Providing broader benefits for disease control. *Vet. Microbiol.* **2006**, *117*, 43–50. [CrossRef]

120. Brochier, B.; Blancou, J.; Thomas, I.; Languet, B.; Artois, M.; Kieny, M.P.; Lecocq, J.P.; Costy, F.; Desmettre, P.; Chappuis, G.; et al. Use of recombinant vaccinia-rabies glycoprotein virus for oral vaccination of wildlife against rabies: Innocuity to several non-target bait consuming species. *J.Wildl.Dis.* **1989**, *25*, 540–547. [CrossRef]

121. Qureshi, A.I. Ebola Virus Disease. In *Ebola Virus Disease*; Academic Press: Cambridge, MA, USA, 2016; pp. 139–157. ISBN 9780128042304.

122. Venkatraman, N.; Silman, D.; Folegatti, P.M.; Hill, A.V.S. Vaccines against Ebola virus. *Vaccine* **2017**, *36*, 5454–5459. [CrossRef]

123. Papaneri, A.B.; Wirblich, C.; Cann, J.A.; Cooper, K.; Jahrling, P.B.; Schnell, M.J.; Blaney, J.E. A replication-deficient rabies virus vaccine expressing Ebola virus glycoprotein is highly attenuated for neurovirulence. *Virology* **2012**, *434*, 18–26. [CrossRef]

124. Pushko, P.; Bray, M.; Ludwig, G.V.; Parker, M.; Schmaljohn, A.; Sanchez, A.; Jahrling, P.B.; Smith, J.F. Recombinant RNA replicons derived from attenuated Venezuelan equine encephalitis virus protect guinea pigs and mice from Ebola hemorrhagic fever virus. *Vaccine* **2000**, *19*, 142–153. [CrossRef]

125. Duan, Z.; Xu, H.; Ji, X.; Zhao, J. Recombinant Newcastle disease virus- vectored vaccines against human and animal infectious diseases. *Future Microbiol.* **2015**, *10*, 1307–1323. [CrossRef]

126. Sheppard, H. Inactivated- or Killed-Virus HIV/AIDS Vaccines. *Curr. Drug Target Infect. Disord.* **2005**, *5*, 131–141. [CrossRef]

127. Nanjundappa, R.H.; Wang, R.; Xie, Y.; Umeshappa, C.S.; Xiang, J. Novel CD8 + T cell-based vaccine stimulates Gp120-specific CTL responses leading to therapeutic and long-term immunity in transgenic HLA-A2 mice. *Vaccine* **2012**, *30*, 3519–3525. [CrossRef]

128. Khattar, S.K.; Palaniyandi, S.; Samal, S.; Labranche, C.C.; Montefiori, D.C.; Zhu, X.; Samal, S.K. Evaluation of humoral, mucosal, and cellular immune responses following co-immunization of HIV-1 gag and Env proteins expressed by newcastle disease virus. *Hum. Vaccines Immunother.* **2015**, *11*, 504–515. [CrossRef]

129. Tayeb, S.; Zakay-Rones, Z.; Panet, A. Therapeutic potential of oncolytic Newcastle disease virus: A critical review. *Oncolytic Virother.* **2015**, *4*, 49–62.

130. Ravindra, P.V.; Tiwari, A.K.; Ratta, B.; Chaturvedi, U.; Palia, S.K.; Chauhan, R.S. Newcastle disease virus-induced cytopathic effect in infected cells is caused by apoptosis. *Virus Res.* **2009**, *141*, 13–20. [CrossRef]

131. Lazar, I.; Yaacov, B.; Shiloach, T.; Eliahoo, E.; Kadouri, L.; Lotem, M.; Perlman, R.; Zakay-Rones, Z.; Panet, A.; Ben-Yehuda, D. The Oncolytic Activity of Newcastle Disease Virus NDV-HUJ on Chemoresistant Primary Melanoma Cells Is Dependent on the Proapoptotic Activity of the Inhibitor of Apoptosis Protein Livin. *J. Virol.* **2010**, *84*, 639–646. [CrossRef]

132. Washburn, B.; Weigand, M.A.; Grosse-Wilde, A.; Janke, M.; Stahl, H.; Rieser, E.; Sprick, M.R.; Schirrmacher, V.; Walczak, H. TNF-Related Apoptosis-Inducing Ligand Mediates Tumoricidal Activity of Human Monocytes Stimulated by Newcastle Disease Virus. *J. Immunol.* **2003**, *170*, 1814–1821. [CrossRef]

133. Fabian, Z.; Csatary, C.M.; Szeberenyi, J.; Csatary, L.K. p53-independent endoplasmic reticulum stress-mediated cytotoxicity of a Newcastle disease virus strain in tumor cell lines. *J. Virol.* **2007**, *81*, 2817–2830. [CrossRef]

134. Zamarin, D.; Palese, P. Oncolytic Newcastle disease virus for cancer therapy: old challenges and new directions. *Future Microbiol.* **2012**, *7*, 347–3675. [CrossRef] [PubMed]

135. Vigil, A.; Park, M.S.; Martinez, O.; Chua, M.A.; Xiao, S.; Cros, J.F.; Martínez-Sobrido, L.; Woo, S.L.C.; García-Sastre, A. Use of reverse genetics to enhance the oncolytic properties of newcastle disease virus. *Cancer Res.* **2007**, *67*, 8285–8292. [CrossRef] [PubMed]

136. Zamarin, D.; Martínez-Sobrido, L.; Kelly, K.; Mansour, M.; Sheng, G.; Vigil, A.; García-Sastre, A.; Palese, P.; Fong, Y. Enhancement of oncolytic properties of recombinant newcastle disease virus through antagonism of cellular innate immune responses. *Mol. Ther.* **2009**, *17*, 697–706. [CrossRef] [PubMed]

137. Kalyanasundram, J.; Hamid, A.; Yusoff, K.; Chia, S.L. Newcastle disease virus strain AF2240 as an oncolytic virus: A review. *Acta Trop.* **2018**, *183*, 126–133. [CrossRef]

138. Elankumaran, S.; Chavan, V.; Qiao, D.; Shobana, R.; Moorkanat, G.; Biswas, M.; Samal, S.K. Type I Interferon-Sensitive Recombinant Newcastle Disease Virus for Oncolytic Virotherapy. *J. Virol.* **2010**, *84*, 3835–3844. [CrossRef]

139. Wu, Y.; He, J.; An, Y.; Wang, X.; Liu, Y.; Yan, S.; Ye, X.; Qi, J.; Zhu, S.; Yu, Q.; et al. Recombinant Newcastle disease virus (NDV/Anh-IL-2) expressing human IL-2 as a potential candidate for suppressing growth of hepatoma therapy. *J. Pharmacol. Sci.* **2016**, *132*, 24–30. [CrossRef]

140. Ren, G.; Tian, G.; Liu, Y.; He, J.; Gao, X.; Yu, Y.; Liu, X.; Zhang, X.; Sun, T.; Liu, S.; et al. Recombinant Newcastle Disease Virus Encoding IL-12 and/or IL-2 as Potential Candidate for Hepatoma Carcinoma Therapy. *Technol. Cancer Res. Treat.* **2016**, *15*, NP83–NP94. [CrossRef]

141. Mohamed Amin, Z.; Che Ani, M.A.; Tan, S.W.; Yeap, S.K.; Alitheen, N.B.; Syed Najmuddin, S.U.F.; Kalyanasundram, J.; Chan, S.C.; Veerakumarasivam, A.; Chia, S.L.; et al. Evaluation of a Recombinant Newcastle Disease Virus Expressing Human IL12 against Human Breast Cancer. *Sci. Rep.* **2019**, *9*, 13999.

142. Janke, M.; Peeters, B.; de Leeuw, O.; Moorman, R.; Arnold, A.; Fournier, P.; Schirrmacher, V. Recombinant Newcastle disease virus (NDV) with inserted gene coding for GM-CSF as a new vector for cancer immunogene therapy. *Gene Ther.* **2007**, *14*, 1639–1649. [CrossRef]

143. Xu, X.; Yi, C.; Yang, X.; Xu, J.; Sun, Q.; Liu, Y.; Zhao, L. Tumor Cells Modified with Newcastle Disease Virus Expressing IL-24 as a Cancer Vaccine. *Mol. Ther. Oncolytics* **2019**, *14*, 213–221. [CrossRef]

144. Elankumaran, S.; Rockemann, D.; Samal, S.K. Newcastle Disease Virus Exerts Oncolysis by both Intrinsic and Extrinsic Caspase-Dependent Pathways of Cell Death. *J. Virol.* **2006**, *80*, 7522–7534. [CrossRef]

145. Cheng, X.; Wang, W.; Xu, Q.; Harper, J.; Carroll, D. Genetic modification of oncolytic Newcastle disease virus for cancer therapy. *J. Virol.* **2016**, *90*, 5343–5352. [CrossRef]

146. Pühler, F.; Willuda, J.; Puhlmann, J.; Mumberg, D.; Römer-Oberdörfer, A.; Beier, R. Generation of a recombinant oncolytic Newcastle disease virus and expression of a full IgG antibody from two transgenes. *Gene Ther.* **2008**, *15*, 371–383. [CrossRef]
147. Wu, Y.; Zhang, X.; Wang, X.; Wang, L.; Hu, S.; Liu, X.; Meng, S. Apoptin enhances the oncolytic properties of newcastle disease virus. *Intervirology* **2012**, *55*, 276–286. [CrossRef]
148. Bello, M.B.; Yusoff, K.M.; Ideris, A.; Hair-Bejo, M.; Peeters, B.P.H.; Jibril, A.H.; Tambuwal, F.M.; Omar, A.R. Genotype Diversity of Newcastle Disease Virus in Nigeria: Disease Control Challenges and Future Outlook. *Adv. Virol.* **2018**. [CrossRef]
149. Rasoli, M.; Yeap, S.K.; Tan, S.W.; Moeini, H.; Ideris, A.; Bejo, M.H.; Alitheen, N.B.M.; Kaiser, P.; Omar, A.R. Alteration in lymphocyte responses, cytokine and chemokine profiles in chickens infected with genotype VII and VIII velogenic Newcastle disease virus. *Comp. Immunol. Microbiol. Infect. Dis.* **2014**, *37*, 11–21. [CrossRef]
150. Ravindra, P.V.; Tiwari, A.K.; Sharma, B.; Chauhan, R.S. Newcastle disease virus as an oncolytic agent. *Indian J. Med. Res.* **2009**, *130*, 507–513.

Article

Virus-Induced Flowering by Apple Latent Spherical Virus Vector: Effective Use to Accelerate Breeding of Grapevine

Kiyoaki Maeda [1], Teppei Kikuchi [1], Ichiro Kasajima [2], Chungjiang Li [1], Noriko Yamagishi [2], Hiroyuki Yamashita [3] and Nobuyuki Yoshikawa [1,2,*]

[1] Faculty of Agriculture, Iwate University, Morioka 020-8550, Japan; g0418033@iwate-u.ac.jp (K.M.); a0116014@iwate-u.ac.jp (T.K.); cjli_xm@aliyun.com (C.L.)

[2] Agri-Innovation Center, Iwate University, Morioka 020-8550, Japan; kasajima@iwate-u.ac.jp (I.K.); nyamagi@iwate-u.ac.jp (N.Y.)

[3] Experimental Farm, Faculty of Life and Environmental Science, University of Yamanashi, Kofu 400-8510, Japan; hyamashita@yamanashi.ac.jp

* Correspondence: yoshikawa@iwate-u.ac.jp; Tel./Fax: +81-19-621-6150

Received: 11 December 2019; Accepted: 3 January 2020; Published: 7 January 2020

Abstract: Apple latent spherical virus (ALSV) was successfully used in promoting flowering (virus-induced flowering, VIF) in apple and pear seedlings. In this paper, we report the use of ALSV vectors for VIF in seedlings and in vitro cultures of grapevine. After adjusting experimental conditions for biolistic inoculation of virus RNA, ALSV efficiently infected not only progeny seedlings of *Vitis* spp. 'Koshu,' but also in vitro cultures of *V. vinifera* 'Neo Muscat' without inducing viral symptoms. The grapevine seedlings and in vitro cultures inoculated with an ALSV vector expressing the 'florigen' gene (Arabidopsis *Flowering locus T, AtFT*) started to set floral buds 20–30 days after inoculation. This VIF technology was successfully used to promote flowering and produce grapes with viable seeds in in vitro cultures of F_1 hybrids from crosses between *V. ficifolia* and *V. vinifera* and made it possible to analyze the quality of fruits within a year after germination. High-temperature (37 °C) treatment of ALSV-infected grapevine disabled virus movement to newly growing tissue to obtain ALSV-free shoots. Thus, the VIF using ALSV vectors can be used to shorten the generation time of grapevine seedlings and accelerate breeding of grapevines with desired traits.

Keywords: grapevine; apple latent spherical virus vector; virus-induced flowering; reduced generation time; breeding of grapevine; virus elimination

1. Introduction

Most plant viruses consist of small genomic RNA and capsid proteins that autonomously propagate upon infection of host plant cells and then spread throughout the plant by cell-to-cell and long-distance movement [1]. Virus vector technologies use the ability of viruses to suppress plant genes to engineer gene silencing or expression of exogenous genes such as green fluorescent protein (GFP) through relatively short experimental procedures [2]. Suppression of plant gene expression with virus vectors is called virus-induced gene silencing (VIGS) [3–5], which is becoming a popular technique in model plants such as *Arabidopsis thaliana* and *Nicotiana benthamiana*. Unlike genetic transformation, virus vectors can be applied to a wide range of plant species [5]. In recent years, virus vectors used in major crops were reported for gene silencing or expression of exogenous genes; tobacco rattle virus vector in solanaceous crops, pea early browning virus and bean pod mottle virus vectors in legumes, barley stripe mosaic virus vector in barley and wheat, brome mosaic virus vector in rice, maize, and barley, and apple latent spherical virus and tobacco ringspot virus vectors in cucurbits, etc. [6–16].

Compared with virus vectors for herbaceous crops, virus vectors available to woody crops like fruit trees are limited because an efficient method for inoculation in woody plants has not been fully developed. Virus vectors stably maintained in host plants would also be needed for efficient gene silencing and gene expression in woody crops because of the long growth period of fruit trees. Attempts have also been made to develop viral vectors for citrus and grapevine trees, which are globally important crops, and several vectors have been reported, including the citrus tristeza virus (CTV) and citrus leaf blotch virus vectors to promote gene expression and gene silencing in citrus plants, respectively [17–19]. In grapes, silencing the phytoene desaturase (PDS) gene and GFP expression have been reported using vectors based on grapevine virus A and grapevine leafroll-associated virus-2 [20,21]. The future possibilities of viral vectors in woody crops are summarized in a review by Dawson and Folimonova [22], although the CTV vector in citrus is the main focus.

We have used the apple latent spherical virus (ALSV) vector for driving gene silencing and gene expression in various plant species including fruit trees [6,23–26]. ALSV is a spherical virus consisting of a bipartite single-stranded RNA genome (RNA1 and RNA2) and three capsid proteins (Vp25, Vp20, and Vp24) that is in the *Cheravirus* genus and the *Secoviridae* family [27]. RNA1 encodes a polyprotein for a protease cofactor, an NTP-binding helicase, a viral protein genome-linked, a cysteine protease, and an RNA polymerase. RNA2 also encodes a polyprotein for a movement protein (MP) and three capsid proteins [28,29]. The ALSV vector has several advantages as a genetic tool; for example, ALSV has a relatively broad spectrum of host plants (*Caryophyllaceae, Chenopodiaceae, Cryptomeria, Fabaceae, Cucurbitaceae, Gentianaceae, Pinus, Rosaceae, Rutaceae, Solanaceae,* and Arabidopsis) [6,23,24,30–33]. ALSV does not cause viral symptoms in most hosts, making it possible to evaluate the functions of silenced genes. ALSV also invades to shoot meristems and induces uniform gene silencing throughout the plant [6,24,25,34,35]. We also succeeded in promoting flowering (virus-induced flowering, VIF) in apple and pear seedlings by simultaneous expression of the Arabidopsis *Flowering locus T* gene (*AtFT*) and suppression of apple *Terminal flower 1* gene (*MdoTFL1*) [25,26,36]. Apple seedlings generally take at least 6–7 years from germination to flowering in the wild. However, using VIF, apple seedlings blossom 1–2 months after virus inoculation and the generation time was shortened to within one year. ALSV is not transferred to most next-generation seedlings [37] and the virus can be removed from infected apple and pear trees with a simple heat treatment [36].

Grapevine is one of the most popular fruit crops worldwide and global grape production currently amounts to more than 75 million tons per year [38]. In grapevines, virus-based vectors could be important biotechnological tools to improve plant disease protection and support traditional varieties used in wine making [22]. If the ALSV vector could be applied to grapevine, the technology would be very useful for basic research including gene function analysis by VIGS and in breeding of new grape varieties using VIF.

In this research, we develop experimental procedures to efficiently infect grapevine seedlings and in vitro cultures with ALSV vectors. These vectors were successfully used for VIGS and VIF in grapevine plants. The VIF using ALSV vectors shortened the generation time of grapevine seedlings to within a year, indicating its use as a technique for accelerating grapevine breeding in combination with marker-assisted selection.

2. Materials and Methods

2.1. Plants

Progeny seeds of grapevine, *Vitis* spp. cultivar 'Koshu' and *V. ficifolia* var. *ganebu* (Ganebu) [39–41] generated by either self-pollination and F_1 seeds from Ganebu × *V. vinifera* cv. 'Nehelescol' were preserved at 4 °C until use. Seeds were germinated at 25 °C on wet paper in petri dishes under illumination with fluorescent light. For preparation of in vitro cultures, F_1 seeds of Ganebu × 'Nehelescol', *V. vinifera* 'Cabernet Sauvignon' × Ganebu, and Ganebu × *V. vinifera* 'Shine Muscat' crosses were germinated aseptically and grown in plant-boxes on solid medium (half-strength Murashige-Skoog salts and vitamins, 2% glucose, 0.8% agar, pH 5.8; 40 mL per each plant-box). For propagation of cultured plants, extended stems were cut every two to three nodes. In vitro cultures of cultivar 'Neo Muscat' kindly supplied by Dr. Ikuko Nakajima (Institute of Fruit Tree and Tea Science, NARO) were maintained on solid media as above and used for virus inoculation.

2.2. Construction of ALSV Vectors

The pCALSR1SM and pCLASR2XSB plasmids were used for construction of RNA1 and RNA2 ALSV vectors (Figure 1). In these vectors, RNA1 and RNA2 sequences were driven by cauliflower mosaic virus 35S RNA promoter and ended with the nopaline synthase terminator. These vectors were based on pCAMBIA1300 for introduction into *Agrobacterium tumefaciens* [36,39]. Vector structures are summarized in Figure 1. A 201-base fragment (nucleotide positions 104–304) of the *VvPDS* gene (accession No. EU816356) was synthesized and used as a template to amplify a fragment with 'VvPDS-XhoI (+)' and 'VvPDS-BamHI (−)' primers (Table S1) to attach *Xho*I and *Bam*HI sites. Amplified DNA was digested with *Xho*I and *Bam*HI and introduced into the X/S/B site of pCALSR2XSB in frame with virus polypeptide to generate pCALSR2-VvPDS. The pCALSR2-AtFT plasmid possessing full-length coding sequence excluding the stop codon of the *AtFT* gene (AB027504, 525 bp) was reported previously [36]. The full-length coding sequence of the *VvFT* gene (525 bp, EF157728) was amplified with 'VvFT-XhoI (+)' and 'VvFT-SmaI (−)' primers (Table S1) and introduced into the X/S/B site of pCALSR2XSB to generate pCALSR2-VvFT. The 201-base fragments of the *VvTFL1A* (nucleotide positions 71–271, DQ871591), *VvTFL1B* (nt positions 181–381, DQ871592), and *VvTFL1* (nt positions 223–413, DQ871593) genes were synthesized and used as a template to amplify fragments with primer pairs 'VvTFL1A-XhoI (+)' and 'VvTFL1A- SmaI (−)', 'VvTFL1B- XhoI (+)' and 'VvTFL1B- SmaI (−)', or 'VvTFL1C-XhoI (+)' and 'VvTFL1C- SmaI (−)' as described above (Table S1). The fragments were each inserted into the X/S/B site of pCALSR2XSB to generate pCALSR2-VvTFL1A, B, C (Figure 1). A fragment of the *VvTFL1A* (201 bp, nt positions 71–271) and *VvTFL1B* (nt positions 429–611) genes were also amplified with primer pairs, 'VvTFL1A-SalI' and 'VvTFL1A-MluI', and VvTFL1B-SalI' and 'VvTFL1B-MluI', respectively; excised at *Sal*I and *Mlu*I sites, and then introduced into the S-M site of pCALSR1SM to generate pCALSR1-VvTFL1A, B (Figure 1).

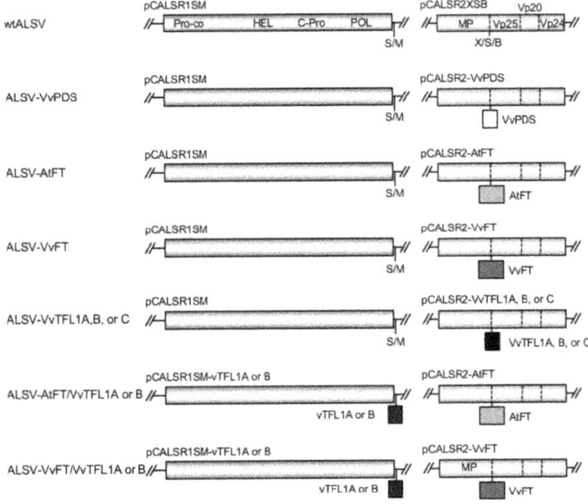

Figure 1. Schematic representation of infectious apple latent spherical virus (ALSV)-RNA clones (pCALSR1SM and pCALSR2XSB) and ALSV vectors containing a part of the phytoene desaturase (PDS) gene from grapevine (ALSV-VvPDS), full-length *FT* gene (*AtFT* from *A. thaliana* or *VvFT* from grapevine), and part of the *VvTFL A*, *B*, or *C* gene from grapevine. S/M and X/S/B indicate cloning sites of *Sal* I/*Mlu* I and *Xho* I/*Sam* I/*Bam* HI, respectively.

2.3. Agro-Inoculation

Competent *A. tumefaciens* strain GV3101::pMP90 cells were transformed with RNA1 or RNA2 vector plasmids by electroporation with a Gene Pulser (Bio-Rad Laboratories, Hercules, CA, USA). Agrobacterium strain EHA101 expressing pBE2113::HC-Pro [38] was also used. After confirming the insertion sequences in ALSV vectors by colony PCR, agrobacterium strains were cultured in standard Lysogeny broth (LB) media overnight at 28 °C. Agrobacterium pellets were recovered after brief centrifugation and diffused in agro-inoculation buffer (10 mM $MgCl_2$, 10 mM MES-KOH pH 5.7, 150 µM acetosyringone) at OD_{600} = 1.0. This suspension was kept at 22 °C in the dark for 2 h before inoculation. *N. benthamiana* (1–1.5 mo after sowing) was infiltrated with a mixture of three Agrobacterium strains expressing ALSV RNA1 and RNA2 clones, and an expression vector for the silencing suppressor gene *HC-Pro* as previously described [42]. Agro-inoculated *N. benthamiana* was kept in the dark overnight and grown for an additional two weeks. Virus infection was tested by RT-PCR analysis using upper non-inoculated leaves.

2.4. Preparation of RNAs from Infected Leaves

RNAs for inoculation by particle bombardment were prepared as follows. Upper leaves of infected *N. benthamiana* plants were harvested and stored at −80 °C, or immediately crushed with a mortar and pestle in liquid nitrogen. Infected leaves were homogenized in twice (*v/w*) the volume of extraction buffer (0.1 M Tris pH 7.8, 0.1 M NaCl, 5 mM $MgCl_2$), and the extracts were used to inoculate *Chenopodium quinoa* leaves. Two to three weeks after inoculation, leaves with mosaic symptoms were collected and stored at −80 °C. In an ice-cold blender, infected *C. quinoa* leaves (10 g) were ground in 30 mL extraction buffer supplemented with 1% β-mercaptoethanol. Crude sap was roughly strained through two layers of cotton mesh and centrifuged at 10,233× *g* for 10 min at 4 °C to recover the supernatant. Extracts were clarified with bentonite and concentrated with PEG-6000, followed by RNA extraction with phenol-chloroform as reported previously [29,43]. The RNA was resuspended in water and the concentration was measured at 260 nm.

2.5. ALSV Inoculation by Particle Bombardment

After coating gold particles with RNA from infected leaves, grapevine seedlings (10–15 plants/each inoculation) were inoculated using the Helios Gene Gun system (Bio-Rad Laboratories, München, Germany) following a previously reported method [43]. The air pressure used was 1379 kilopascal (kPa) and each cotyledon was shot once with gold particles. Grapevine in vitro cultures were inoculated using the PDS-1000/He Particle Delivery System (Bio-Rad Laboratories, München, Germany) at 1379 kPa, with two shots per petri dish. The GDS-80 gene gun system (Nepa Gene Co., Ltd., Ichikawa, Japan) was also used for inoculation of in vitro cultures at 207 kPa.

2.6. RT-PCR for Detection of Virus Infection

Total RNA was extracted from grapevine leaves three weeks after inoculation. Approximately 50 mg grapevine leaf tissue were frozen at −80 °C, crushed in a Micro Smash MS-100R (TOMY, Tokyo, Japan), and mixed with 600 µL extraction buffer (2% cetyltrimethylammonium bromide, 2% polyvinylpyrrolidone, 100 mM tris (hydroxymethyl) aminomethane, pH 8.0, 25 mM ethylenediaminetetraacetic acid, 2% β-mercaptoethanol). The homogenates were incubated at 65 °C for 20 min and mixed with 600 µL chloroform for 2 min. After centrifugation at 17,860× *g* for 10 min at 4 °C, the aqueous phase was recovered. One-third volume of 7.5 M LiCl was added to this solution, mixed well, and incubated at −80 °C for 30 min, −20 °C for 1 h, or 4 °C overnight. After centrifugation at 17,860× *g* for 30 min at 4 °C, RNA pellets were rinsed with 80% ethanol and dissolved in sterilized deionized water to a concentration of 1 µg $µL^{-1}$. One microgram of RNA was reverse transcribed with ReverTra Ace (Toyobo, Osaka, Japan) following the manufacturer instructions, and PCR was performed with Ex Taq (Takara) using a Thermal Cycler Dice Version III (Takara, Kusatsu, Japan).

The annealing temperature was 55 °C for the RT-PCR primers used (Table S1). ALR2-999 (+) and ALR2-1437 (−) primers were used to detect wild-type ALSV (wtALSV). ALR2-1418 (+) and ALR2-1511 (−) were used to detect VvPDS, AtFT, and VvFT inserts. ALR1-6598 (+) and ALR1-6691 (−) were used to detect VvTFL1A insert.

2.7. Quantitative RT-PCR

Total RNA was extracted from grapevine cultures (three independent plants inoculated with ALSV-VvPDS) with the Plant/Fungi Total RNA Purification Kit (NORGEN, Thorold, ON, Canada). Extracted RNA was further treated with DNase I. A total of 500 ng RNA per 20 μL reaction was reverse transcribed with ReverTra Ace. Relative cDNA concentrations were quantified with the ECO Real-Time PCR System (Illumina, San Diego, CA, USA) using SYBR Premix Ex Taq II (Takara). Amplification of the target sequences was confirmed by melting curve analysis. The VvPDS gene was amplified with 'VvPDS (+)' and 'VvPDS (−)' primers (Table S1), which targeted a 125-bp portion of VvPDS mRNA (JQ319631). A 128-bp portion of grapevine *Elongation factor 1α* gene (EF1α) was amplified with primers 'VvEF1α (+)' and 'VvEF1α (−)' (Table S1) as an expression control. Expression of VvPDS was standardized with *EF1α*. The fluorescence signal of *EF1α* reached the threshold level within 20–24 cycles, indicating it is a good internal control.

2.8. Tissue-Blot Hybridization

Detection of ALSV infection in in vitro plant cultures was conducted with tissue-blot hybridization as described previously [33]. Briefly, grapevine plants were placed on positively charged nylon membranes (Hybond-N+; GE Healthcare, Tokyo, Japan), covered with plastic wrap, and frozen by pouring liquid nitrogen over the membranes. Plant tissues were crushed with a pastry pin to transfer plant exudates to the membranes. Membranes were soaked in 0.05 N NaOH for 30 min, washed in 20× SSC (saline-sodium citrate) buffer for 30 min, then RNA was fixed to the membranes by UV illumination. An RNA probe was hybridized to viral RNA and detected with an Image Quant LAS4000 (GE Healthcare) after reaction with CDP-Star (GE Healthcare) for 5 min.

2.9. In Situ Hybridization

In situ hybridization analysis was conducted as described previously [37]. Shoots, tendrils, and flowers were excised from infected grapevine and fixed in formalin:ethanol:acetic acid:water (10:50:5:35 *v/v*), dehydrated with an ethanol/lemozol concentration series, and embedded in Paraplast Plus (Sigma-Aldrich Corp., St. Louis, MO, USA). Tissue sections of 12-μm thickness were prepared, extended on an APS-coated glass slide (Matsunami Glass, Osaka, Japan), deparaffinized, and hydrated. The sections were then treated with proteinase K and fixed once again. A DIG (digoxigenin)-labeled probe complementary to the Vp24 region of the ALSV-RNA2 [44] was used to hybridize slides. As a negative control, a DIG-labeled probe complementary to the P1 region of the soybean mosaic virus (SMV) genome was used [44]. The hybridized probes were labeled with sheep anti-DIG conjugated alkaline phosphatase (F. Hoffmann-La Roche SG, Basel, Switzerland) and stained with 5-bromo-4-chloro-3-indolyl phosphate (BICIP)/nitro blue tetrazolium (NBT) (F. Hoffmann-La Roche SG) to generate dark blue indigo dyes. The samples were dehydrated with an ethanol concentration series, dried, mounted in Entellan New (Merck KGaA, Darmstadt, Germany), and observed under a Leica MMLB optical microscope.

2.10. Analysis of Sugar, Acid, and Anthocyanin Content in Grape Berries

Mature and colored berries were collected from early-flowered lines; Ganebu × 'Nehelescol' (240-1), 'Cabernet Sauvignon' × Ganebu (4-23), and Ganebu × 'Shine Muscat' (264-T31). Three individuals from each line were used for analysis. Berries of Ganebu, Yamabudo, and 'Merlot' were collected from field-grown grapevines. Commercial grape varieties used as control ('Delaware', 'Kyoho', and 'Steuben') were purchased from a fruit shop. The sugar and acid content of berries were measured with a sugar acidity meter (ATAGO Co., Ltd., Tokyo, Japan) following the manufacturer protocol. Total anthocyanin was analyzed as described by Shiozaki and Murakami [40]. Berry skin (2–5 mg) was incubated in 10 mL 50% (*v/v*) aqueous acetic acid at 4 °C overnight. Absorbance of the extracts at 530 nm was measured using a spectrophotometer (Novaspec Plus, Biochrom Ltd., Cambridge, UK). The amount of total anthocyanin was determined by comparing the sample absorbance to a standard curve from Malvidin 3-glucoside chloride (Funakoshi Co., Ltd., Tokyo, Japan).

2.11. High-Temperature Treatment for Virus Elimination

According to the procedure established in apple [36], infected F1 hybrid (G × N) seedlings grown to the 10 true-leaf stage at 25 °C were then incubated at 27 °C for 3 d, 30 °C for 5 d, 35 °C for 5 d, and then 37 °C for 30 d. After high-temperature treatment, the plants were grown at 25 °C until RT-PCR analysis. Five infected seedlings were subjected to the same high-temperature treatment and assayed for virus infection.

3. Results

3.1. Construction of ALSV Vectors for Grapevine

ALSV-RNA1 and -RNA2 sequences were introduced into binary expression vectors (accession ViralMultiSegProj15367) [36,42]. For cloning purposes, the RNA1 vector included a S/M site behind the polyprotein, while the RNA2 vector included a X/S/B site in the middle of the polyprotein (between MP and Vp25). These vectors were designated pCALSR1SM and pCALSR2XSB (Figure 1).

The infectious clones of ALSV vectors were first co-transformed into *N. benthamiana* by agro-inoculation [39] and infection was assayed with reverse transcription-polymerase chain reaction (RT-PCR). Virus was further propagated through rub-inoculation of *C. quinoa* (a propagation host) plants, and viral RNA was extracted from infected leaves. ALSV prepared from a combination of the empty vectors (pCALSR1SM and pCALSR2XSB) was designated as wild-type ALSV (wtALSV) for this report (Figure 1). To silence the *VvPDS* gene, a 201-bp fragment of *VvPDS* was introduced into the RNA2 vector (pCALSR2-VvPDS) and ALSV-VvPDS virus was prepared from pCALSR1SM and pCALSR2-VvPDS (Figure 1).

To express *AtFT* and grapevine *FT* (*VvFT*) in the ALSV vector, the full-length gene coding sequences were introduced into the RNA2 vector (pCALSR2-AtFT and pCALSR2-VvFT). The ALSV-AtFT and ALSV-VvFT vectors were prepared as a mixture of pCALSR1SM and pCLASR2-AtFT, and pCALSR1SM and pCLASR2-VvFT, respectively. To silence *VvTFL1A*, *VvTFL1B*, or *VvTFL1C*, 201-bp fragments of these genes were introduced separately into the RNA2 vector (pCALSR2-VvTFL1A, B, or C). ALSV-VvTFL1A, ALSV-VvTFL1B, or ALSV-VvTFL1C virus was prepared from pCALSR1SM and pCALSR2-VvTFL1A, B, or C (Figure 1). For simultaneous expression of *FT* and suppression of *VvTFL1A* or *B*, *A. tumefaciens* clones carrying pCALSR1-VvTFLA or B were mixed with those carrying pCALSR2-AtFT or pCALSR2-VvFT to generate ALSV-AtFT/VvTFL1A and ALSV-AtFT/VvTFL1B or ALSV-VvFT/VvTFL1A and ALSV-VvFT/VvTFL1B viruses (Figure 1).

All inserts in these viruses (e.g., VvPDS, VvTFL1A, B, and C, AtFT, and VvFT) were stably maintained after infection in *N. benthamiana*, *C. quinoa*, or grapevine plants, as determined by RT-PCR. The AtFT and VvFT inserts were sometimes deleted from virus RNA and the plants with deletions were eliminated from analysis.

3.2. ALSV Vector Inoculation Conditions for Grapevine Seedlings and In Vitro Cultures

C. quinoa and *N. benthamiana* are susceptible to ALSV and can be systemically infected at young or even mature growth stages [25]. However, for ALSV infection of fruit trees such as apple and pear, cotyledons of seedlings immediately after germination should be inoculated by bombardment with gold particles coated with concentrated viral RNA for efficient infection [36,43]. We first tested whether ALSV can infect grapevine seedling by particle bombardment. Progeny seedlings derived from

Vitis spp. cv. 'Koshu' at three different growth stages were used in this analysis; seedlings immediately after germination with folded cotyledons, seedlings with expanded cotyledons, and seedlings with three true leaves (Figure 2A), were assessed in this experiment. The wtALSV or ALSV-VvPDS was used to inoculate cotyledons at the 'folded cotyledon' and 'expanded cotyledon' stages, or the first and second true leaves at the 'three true leaves' stage by particle bombardment. Local infection within inoculated leaves and systemic infection to upper leaves were assessed by RT-PCR. Table 1 and Figure 2B show that systemic infection was found in approximately 90% of seedlings when inoculated at the 'folded cotyledons' stage, whereas only 10% was infected when inoculated at the 'expanded cotyledons' stage, and no seedling was systemically infected when inoculated at the 'three true leaves' stage. Although systemic infection did not occur, all seedlings were locally infected in inoculated leaves at the 'expanded cotyledons' and the 'three true leaves' stages (Table 1), indicating that systemic movement of the virus was severely restricted in plants at these developmental stages.

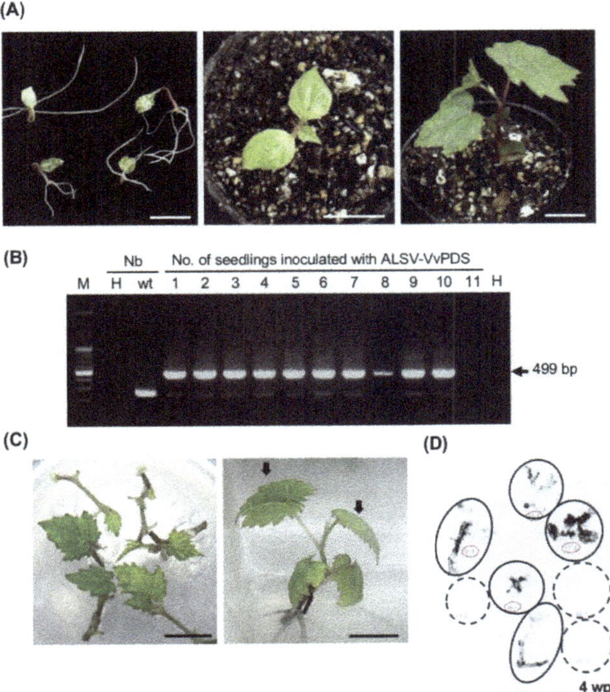

Figure 2. Inoculation of grapevine seedlings and in vitro cultures with ALSV. (**A**) Three growth stages of grapevine seedlings ('Koshu') used for ALSV vector inoculation. Left: 'Folded cotyledons', center: 'Expanded cotyledons', and right: 'Three true leaves' stages. (**B**) RT-PCR detection of ALSV-VvPDS in grapevine seedlings inoculated at the 'folded cotyledons' stage. The third true leaves were assayed. Nb, *N. benthamiana*; H, non-inoculated healthy plant; wt; wtALSV-infected plant. (**C**) Three true leaf stages of in vitro cultures ('Neo Muscat') used for ALSV vector inoculation in non-rooted (**left**) and rooted (**right**) cultures. (**D**) Tissue blot hybridization of cultures inoculated with wtALSV in the non-rooted culture by PDS-1000/He™ system. Circles and dotted circles indicate infected and uninfected plantlets, respectively.

Table 1. Infection of wild-type ALSV (wtALSV) and ALSV-VvPDS to the seedlings and in vitro cultures of grapevine by particle bombardment.

Material (Cultivar)	Growth Stage	System for Particle Bombardment *	ALSV Vector	No. of Infected/Inoculated Plants (%)	
				Local **	Systemic ***
Seedlings ('Koshu')	Folded cotyledons	Helios Gene Gun	wtALSV	nt	11/12 (92)
	Expanded cotyledons	Helios Gene Gun	wtALSV	10/10 (100)	1/10 (10)
	Three true leaves	Helios Gene Gun	wtALSV	10/10 (100)	0/15 (0)
	Folded cotyledons	Helios Gene Gun	ALSV-VvPDS	nt	10/11 (91)
	Expanded cotyledons	Helios Gene Gun	ALSV-VvPDS	10/10 (100)	1/12 (8)
	Three true leaves	Helios Gene Gun	ALSV-VvPDS	15/15 (100)	0/15 (0)
Plants cultured in vitro ('Neo Muscat')	True leaf, non-rooted	PDS-1000/He™	wtALSV	nt	17/58 (29)
	True leaf, rooted	GDS-80	wtALSV	nt	7/10 (70)
	True leaf, non-rooted	PDS-1000/He™	ALSV-VvPDS	nt	17/98 (17)

* Helios Gene Gun and PDS-1000/He™ systems are by Bio-Rad Laboratories, Inc. and GDS-80 is a gene gun system by Nepa Gene Co., Ltd. ** Infection was found on inoculated leaves. nt; not tested. *** Infection was found on upper leaves.

Fruit trees usually have heterozygous genomic compositions and are propagated as clones through cutting or grafting. This means that genotypes and phenotypes of seedlings are usually different from the original variety from which they were derived. Therefore, ALSV infection of mature plants or in vitro cultured plantlets is desirable for genetic analysis and breeding of a specific variety. We prepared an in vitro culture of 'Neo Muscat', a table grape popular in Japan. After extension on media, vines of the 'Neo Muscat' culture were cut every two to three nodes to propagate. These cultures were inoculated by particle bombardment with wtALSV or ALSV-VvPDS at two different growth stages with a PDS-1000/He™ system or a GDS-80 gene gun system. The first stage was within one to two weeks after cutting in which vines had not formed roots (Figure 2C, left). Vines were rooted at the second stage three to four weeks after cutting (Figure 2C, right). Whole plants, including vines and small expanding leaves, were inoculated in the first stage with a PDS-1000/He™ system. In the second stage, the two uppermost leaves were inoculated with a GDS-80 gene gun system (Figure 2C, right). After assaying infection by tissue-blot hybridization (Figure 2D) or RT-PCR, we found that ALSV systemically infected 17–29% and 70% of in vitro cultures that were inoculated at the first stage (non-rooted) with a PDS-1000/He™ system and the second stage (rooted) with a GDS-80 gene gun system, respectively. From these results, in vitro cultures at the true leaf stage had a high infection rate with a GDS-80 gene gun system (Table 1).

Figure 3A shows a grapevine seedling from 'Koshu' infected with wtALSV three months post inoculation (mpi). The plant was asymptomatic but ALSV was detected in all (1st to 12th) true leaves by RT-PCR (Figure 3B). Grapevine seedlings or in vitro cultures infected with wtALSV did not show any discernible difference in appearance (i.e., shape and color) from non-infected plants. In situ hybridization analysis of shoot apical tissues showed that ALSV was distributed in leaf primordia of infected grapevine seedlings (Figure 3C). ALSV was also distributed throughout all tissues including the apical meristem cells of tendrils in infected plants (Figure 3D).

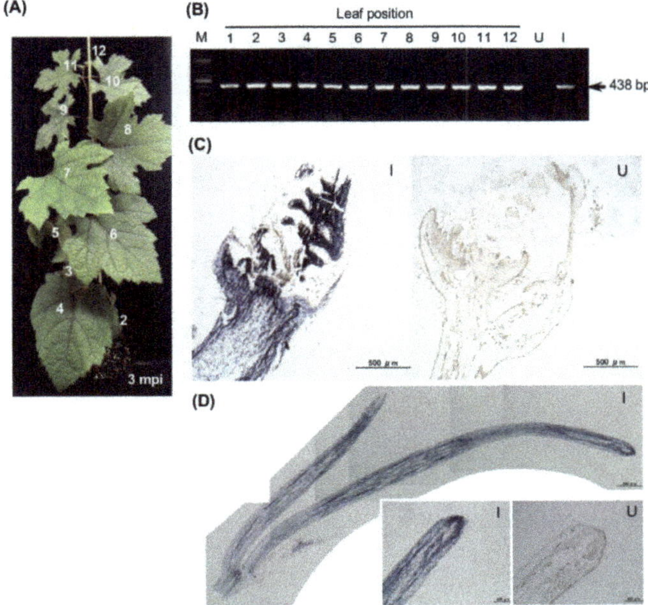

Figure 3. Systemic distribution of wtALSV in a grapevine seedling inoculated at the 'folded cotyledons' stage. (**A**) An infected grapevine seedling without visible symptoms three months post inoculation (mpi). Numbers indicate leaf positions used for detection of infection in (**B**). (**B**) Detection of wtALSV by RT-PCR from true leaves from the grapevine seedling shown in (**A**). M, DNA size maker; U, uninfected sample; I, infected sample. (**C**) In situ hybridization analysis of shoot tips from infected (I, left) and uninfected (U, right) grapevine seedlings using an ALSV-Vp24 (−) probe. (**D**) In situ hybridization analysis of tendrils of infected (I) and uninfected (U) grapevine seedlings probed with ALSV-Vp24 (−). Blue colour indicates ALSV distribution.

3.3. VIGS in Grapevine Using the ALSV Vector

ALSV-VvPDS (Figure 1) was used to inoculate grapevine seedlings of 'Koshu' and Ganebu, and in vitro cultures of 'Neo Muscat' using particle bombardment as described above. Infection rates were similar to those inoculated with wtALSV (Table 1). All of the 'Koshu' seedlings infected with ALSV-VvPDS had a photo-bleaching phenotype caused by the loss of function of the *PDS* gene in portions of the first true leaf, and in the whole second or third true leaves as well as those that developed above these true leaves (Figure 4A), after which plant growth stopped. When in vitro cultures of 'Neo Muscat' were inoculated with ALSV-VvPDS, photo-bleaching first appeared along the veins of upper leaves in inoculated plants, spread to the upper leaves (Figure 4B), and then the leaves of the plants stopped growing. The average *VvPDS* mRNA accumulation decreased to 5% in white leaves, compared with green leaves and non-infected leaves, although there was variation in the calculated values (Figure 4C).

Figure 4. Photo-bleaching phenotype of grapevine seedlings and in vitro cultures inoculated with ALSV-VvPDS. (**A**) 'Kousyu' and Ganebu seedlings infected with ALSV-VvPDS at the 'folded cotyledons' stage. Dpi; days post inoculation. (**B**) A plantlet from tissue culture ('Neo Muscat') inoculated with ALSV-VvPDS at the true leaf stage as shown in Figure 2C (rooted culture) two months after inoculation (mpi). (**C**) A quantitative analysis of VvPDS-mRNA in white (W) and green (G) leaves of plants infected with ALSV-VvPDS as shown in (**B**). Reference gene: *VvEF1α*.

3.4. VIF in Grapevine Using the ALSV Vector

Natural flowering of grapevine seedlings generally requires a long period (several years) after germination. In this study, early flowering in grapevine was tested by infection of ALSV-AtFT, ALSV-VvFT, ALSV-VvTFL1A, B, or C, ALSV-AtFT/VvTFL1A or B and ALSV-VvFT/VvTFL1A or B viruses (Figure 1 and Table 2). Three out of eight 'Koshu' seedlings infected by ALSV-AtFT set flowers at the shoot apex 20–37 days post inoculation (dpi), with four to seven true leaves at flowering (Figure 5A). Floral buds also formed at the apices of axillary buds. In contrast, ALSV-VvFT and ALSV solely expressing *VvTFL1A*, *VvTFL1B*, or *VvTFL1C* inserts without *AtFT* (ALSV-VvTFL1A, ALSV-VvTFL1B, and ALSV-VvTFL1C) never induced precocious flowering (Table 2). Flowering rate was judged by the number of plants that set floral buds within three months of the experiments. Any seedlings or in vitro cultures infected with wtALSV alone did not set floral buds during the experiments (Table 2). Flowering was observed in seedlings infected with ALSV-AtFT, ALSV-AtFT/VvTFL1A, and ALSV-AtFT/VvTFL1B virus in 38–89% of 'Koshu' and 67–100% of Ganebu (Table 2). There seemed to be no difference in flowering rate and phenotypes among plants infected with ALSV-AtFT, ALSV-AtFT/VvTFL1A, or ALSV-AtFT/VvTFL1B. In vitro cultures of 'Neo Muscat' inoculated with ALSV-AtFT/VvTFL1A or ALSV-AtFT/VvTFL1B showed 69–71% flowering (Table 2).

Table 2. Precocious flowering of the seedlings and in vitro cultures of grapevine infected with ALSV vectors possessing *FT* and/or *VvTFL1* sequences *.

Materials (Cultivar)	ALSV Vectors	No. of Flowered/Infected Plants (%)
Seedlings ('Koshu')	wtALSV	0/10 (0)
	ALSV-AtFT	3/8 (38)
	ALSV-VvFT	0/6 (0)
	ALSV-VvTFL1A	0/9 (0)
	ALSV-VvTFL1B	0/6 (0)
	ALSV-VvTFL1C	0/5 (0)
	ALSV-AtFT/VvTFL1A	8/9 (89)
	ALSV-AtFT/VvTFL1B	5/9 (56)
	ALSV-VvFT/VvTFL1A	0/6 (0)
	ALSV-VvFT/VvTFL1B	0/3 (0)
Seedlings (Ganebu)	wtALSV	0/10 (0)
	ALSV-AtFT	6/6 (100)
	ALSV-VvFT	0/6 (0)
	ALSV-VvTFL1A	0/10 (0)
	ALSV-VvTFL1B	0/10 (0)
	ALSV-VvTFL1C	0/5 (0)
	ALSV-AtFT/VvTFL1A	8/12 (67)
	ALSV-AtFT/VvTFL1B	17/21 (81)
	ALSV-VvFT/VvTFL1A	0/5 (0)
	ALSV-VvFT/VvTFL1B	0/5 (0)
In vitro cultures ('Neo Muscut')	wtALSV	0/20 (0)
	ALSV-AtFT/VvTFL1A	11/16 (69)
	ALSV-AtFT/VvTFL1B	5/7 (71)

* Seedlings were inoculated at folded cotyledon stage by a Helios gene gun system (Bio-Rad). The GDS-80 gene gun system was used for inoculation of in vitro cultures.

Figure 5. Early flowering of grapevine seedlings and in vitro cultures infected with ALSV-AtFT or ALSV-AtFT/VvTFL1A. (**A**) 'Koshu' seedlings inoculated at the 'folded cotyledons' stage. Dpi, days post inoculation. (**B**) A Ganebu seedling infected with ALSV-AtFT/VvTFL1A (left). Male (center) and female (right) flowers from Ganebu seedlings infected with ALSV-AtFT/VvTFL1A. Scale bars represent 2 mm. (**C**) Precocious flowering of in vitro cultures of 'Neo Muscat' (left) and Ganebu × 'Nehelescol' (G × N, line 240-1) (center) inoculated with ALSV-AtFT/VvTFL1A at the true leaf stage. Bisexual flowers from G × N are shown in the right-hand picture. A scale bar represents 5 mm.

3.5. Use of VIF for Evaluation of F_1 Hybrids from Crossing between the Two Vitis Species

V. ficifolia var. ganebu (Ganebu) is expected to be a novel genetic source for breeding new varieties of grapevine because it has no endodormancy and a high anthocyanin content in the grape skins [39–41]. However, Ganebu is not suitable as a cultivated grapevine because it is a dioecious grapevine variety (Figure 5B). Early-flowering Ganebu seedlings in Table 2 separated into 17 male vs. 14 female plants out of 31 plants in total. In contrast, all seedlings from 'Koshu' and in vitro cultures of 'Neo Muscut' in Table 2 produced bisexual flowers.

At first, we applied VIF to the F_1 hybrid progeny from crossing between the two *Vitis* species; Ganebu and 'Nehelescol' (G × N) for selection of seedlings with bisexual flowers. Flowering was observed in 28/36 (flowering/infected, 78%) seedlings infected with ALSV-AtFT/VvTFL1A in G × N seedlings. Among the flowered seedlings, 46% (13/28) of plants produced bisexual flowers (Figure 5C right).

Subsequently, we inoculated ALSV vectors to F_1 hybrid lines of G × N (line 240-1), 'Cabernet Sauvignon' × Ganebu (CS × G, line 4-23), and Ganebu × 'Shine Muscat' (G × SM, line 264-T3), which were germinated aseptically and grown in plant-boxes on solid medium. These F_1 hybrids infected with ALSV-AtFT/VvTFL1A or ALSV-AtFT/VvTFL1B set first floral buds with bisexual flowers 20–30 dpi and formed fruits successively in growth chamber conditions. Infected plants continued precocious flowering for several months after inoculation, then the plants stopped flowering. The flowers of infected plants were self-pollinated, and berries were formed as shown in Figure 6A,B. The grape berries ripened 6–10 months post inoculation and the diameter of the grape berries from G × N, CS × G, and G × SM were 0.9–1.0 cm, 0.7–0.9 cm, and 0.95–1.15 cm, respectively (Figure 6C), which was similar to that of Ganebu grown in field conditions (0.7–1.0 cm).

Figure 6. Fruiting in the F_1 hybrid progeny from crossings between the *Vitis* species; Ganebu × 'Nehelescol' (G × N), 'Cabernet Sauvignon' × Ganebu (CS × G) and Ganebu × 'Shine Muscat' (G × SM) inoculated with the ALSV vector; and analysis of their grape berries. (**A**) Fruiting in a G × N (line 240–1) plant inoculated with ALSV-AtFT/VvTFL1A. (**B**) Fruiting in a CS × G (line 4–23) plant inoculated with ALSV-AtFT/VvTFL1B. (**C**) Grapevine berries produced by F_1 hybrids of G × N and G × SM. (**D**) Comparison of total anthocyanin content in grape berries from G × N and the commercial varieties 'Delaware' and 'Kyoho'.

To evaluate the quality of fruits on early-flowering F_1 hybrids, we measured the sugar and acid contents of berries. The sugar content of F_1 hybrids, G × N (240-1) and G × SM (264-T3) were almost the same as those of Ganebu and commercial grapevine varieties (i.e., 'Kyoho' and 'Steuben') (Table S2). In contrast, the acid content of G × N (240-1), CS × G (4-23), and G × SM (264-T3) was higher than that of eating varieties 'Delaware', 'Kyoho', and 'Steuben', and similar to that of Ganebu (Table S2). Total anthocyanin levels in the skins of F_1 hybrids, G × N (240-1) and CS × G (264-T3), were slightly lower than that of Ganebu, but higher than those in the skins of the varieties 'Delaware', 'Merlot', 'Kyoho', and 'Steuben' (Table S2, Figure 6D).

As described above, it was possible to select the seedlings with bisexual flowers in a short period of time. The VIF using ALSV vectors also made it possible to analyze the compounds of fruits on F_1 grapevine seedlings within a year after germination. Most grapes had viable seeds, which germinated and grew as F_2 seedlings.

3.6. No Seed Transmission to Progeny Seedlings and Virus Elimination from Infected Grapevine Plants

In infected apple plants, ALSV is distributed in pollen grains, ovaries, and ovules of flowers, and infected apple can transfer ALSV to their progeny at a seed transmission rate of 4.5% [37]. In contrast, ALSV was not present in pollen and ovules in gentian plants and not transmitted to the gentian progeny plants [41]. In the present study, we investigated ALSV distribution in the flower organs of grapevine (Ganebu and F_1 hybrid of G × N) that was infected with ALSV-AtFT/VvTFL1B. The in situ hybridization analysis indicated that ALSV was present in the anther wall, pollen grain, and ovule of flowers of infected plants (Figure 7A,B). This result is consistent with those reported for infected apple plants [37]. As described above, mature fruits from F_1 plants contained viable seeds that germinated and grew into F2 seedlings. Therefore, we tested whether ALSV was transmitted to the F2 progeny seedlings from early-flowering F_1 grapevine plants infected with ALSV-AtFT/VvTFL1B using RT-PCR and qRT-PCR [41]. ALSV was not detected from a total of 60 progeny F2 seedlings tested (data not shown), suggesting that ALSV was not transmitted to progeny seedlings from infected grapevine plants. This indicates that the F2 seedlings could be used for subsequent breeding plan as ALSV-free stocks.

The incubation of ALSV-infected apple seedlings at high-temperature (37 °C) for four weeks could disable virus movement to newly growing tissue to obtain ALSV-free shoots from infected trees [36]. We investigated whether heat treatment could eliminate ALSV from newly growing tissue in grapevines. Infected F_1 hybrid (G × N) seedlings were grown until the 10 true-leaf stage at 25 °C (Figure 8A) and then incubated at 27 °C for 3 days, 30 °C for 5 days, 35 °C for 5 days, and 37 °C for 30 days (Figure 8B). The plants developed newly about 10 true leaves for incubation at 27–37 °C (Figure 8B). Subsequently, the plants were further grown at 25 °C for 30 days and the presence of ALSV was assessed in all expanded leaves. Clear PCR product bands were detected until the 4th to 10th leaves, which had developed before high-temperature treatment, and faint bands were detected in the 11th to 15th leaves (Figure 8B). In contrast, no ALSV bands were detected in the 16th to 24th leaves even after incubation at 25 °C for 30 days following the high-temperature treatment (Figure 8B). The same results were obtained in all five plants treated with high temperatures. Detection of RT-PCR ALSV products was repeated after two, four, and six months, but no virus was detected in any upper leaves of all plants. From these results, high-temperature incubation of ALSV-infected grapevine seedlings inhibited the systemic movement of ALSV from infected tissues to newly developed leaves, and once long-distance movement of the virus was inhibited, ALSV could not infect the upper leaves.

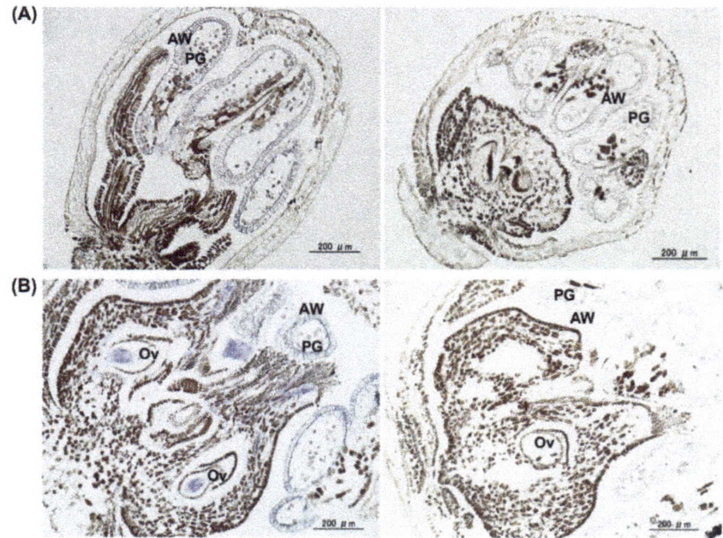

Figure 7. In situ hybridization analysis of ALSV distribution in the flower organs of Ganebu and F_1 hybrid (G × N, line 240-1) plants infected with ALSV-AtFT/VvTFL1A. (**A**) Male flower of Ganebu treated with the ALSV-Vp24 (−) probe (left) and SMV-P1 (−) probe (right) as a control. (**B**) ALSV distribution in bisexual flowers in the F_1 hybrid (G × N) detected with an ALSV-Vp24 (−) probe (left) and SMV-P1 (−) probe (right) as a control. Blue colour indicates ALSV distribution. AW; anther wall, PG; pollen grain, Ov; ovary.

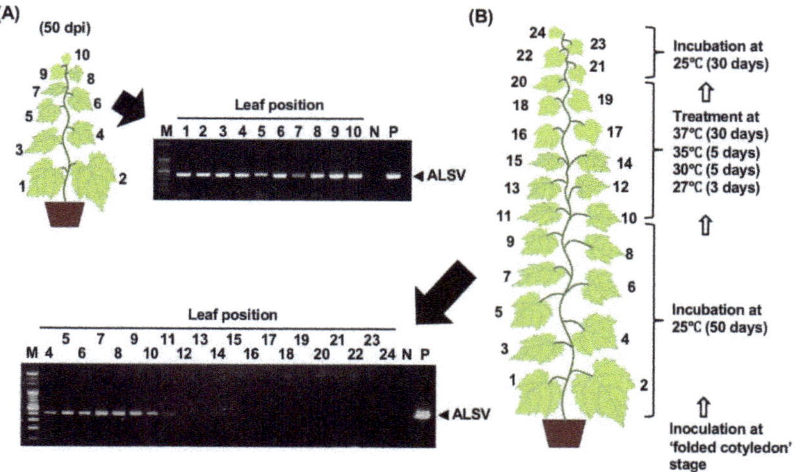

Figure 8. RT-PCR detection of ALSV in F_1 hybrid G × N seedlings before and after high-temperature treatment. (**A**) ALSV infection in inoculated seedlings was assayed by RT-PCR at 50 dpi. (**B**) Infected plants were incubated at 27 °C for 3 days, 30 °C for 5 days, 35 °C for 5 days, and then 37 °C for 30 days. The seedlings were then grown at 25 °C for 30 days and all expanded leaves were analyzed by RT-PCR to assess ALSV infection. Numbers indicate the leaf position on a grapevine seedling. M, size maker; N, negative control (non-infected tissue); P, positive control (infected tissue).

4. Discussion

The VIF and VIGS technologies using the ALSV vector are powerful tools for inducing early flowering and reducing generation time, as well as functional gene analysis in fruit trees [24,26,45]. If the technology is applicable in grapevine plants, it would be very useful for efficient breeding of new grapevine varieties.

To use ALSV vectors, it is important to establish an efficient inoculation method in the target plant species because it is generally difficult to efficiently infect fruit trees with viruses. Our results suggest that there is a significant difference in ALSV infection rates between three developmental stages of juvenile grapevine seedlings (Table 1 and Figure 2A). Only the 'folded cotyledons' stage was sensitive to systemic ALSV infection (Table 1). Although viral infection was detected in inoculated leaves at the 'expanded cotyledons' and 'three true leaves' stages, there was no evidence of systemic infection. These data indicated that ALSV is swiftly transferred from cotyledons to true leaves at the 'folded cotyledons' stage but does not undergo long-distance movement once cotyledons have expanded. Similar phenomena were observed in apple and pear; ALSV can infect inoculated leaves with particle bombardment but does not move into upper uninoculated leaves [43]. This limitation of viral movement may be a common feature in woody plants; however, once systemic infection was established, ALSV distributed throughout the plants without any symptoms after three months (Figure 3A,B). In situ hybridization analysis indicated ALSV invaded the shoot apical meristem and leaf primordia (Figure 3C,D), which was similar to ALSV distribution reported in apple plants [24,35].

It was worth investigating whether grapevine plants cultured in vitro could be infected with ALSV similarly to seedlings because this system is beneficial for the genetic analysis of preexisting varieties and for breeding new varieties. We found that in vitro cultures of 'Neo Muscat' with expanded true leaves (Figure 2C) were systemically infected with ALSV (Table 1 and Figure 2D). General differences in physiological states between soil-cultivated plants and tissue-cultured plants may exist that renders in vitro plants more susceptible to viral infection (Figure 2A,C). Indeed, there are differences in the expression of genes associated with natural immunity and/or disease resistance between tissue- and soil-cultured plants, such as those observed between ALSV and anthracnose infection resistance in soil-cultivated and tissue-cultured strawberry plants [23,46].

VIGS of the *VvPDS* gene in grapevine was success in this study (Figure 4). The upper-most leaves of infected grapevines were photo-bleached and mRNA levels of *VvPDS* were reduced to 5% of green leaves (Figure 4C). This suppression level was comparable to a previous study performed in apple, where the apple rubisco small subunit gene was silenced using ALSV [36]. Thus, ALSV can drive gene silencing in grapevine as well as in its original host plant. Photo-bleaching was observed not only in leaf blades, but also in petioles and vines of many grapevine plants infected with ALSV-VvPDS. Asymptomatic infection with ALSV can also facilitate analysis of functionally unknown genes in grapevine. So far, two virus vectors from GVA and GLRaV-2 have been reported to induce gene silencing in grapevine [20,21]. Both vectors were used to silence *PDS* or magnesium chelatase genes, and successfully induced photobleaching in grapevine leaves.

We have already reported early flowering in apple and pear with the expression of *AtFT*, which is known as 'florigen' [25,26,36,45]. Simultaneous expression of *AtFT* and suppression of *MdTFL1* improves early-flowering rates in apple seedlings [26]. Flowering in grapevine was induced approximately one month after inoculation of germinating 'Koshu', Ganebu, and Ganebu x 'Nehelescol' seedlings (Table 2 and Figure 5). Although accumulation of the AtFT protein was not examined in this study, AtFT was most likely expressed with these vectors based on the amplification of the *AtFT* insert from viral RNA isolated from infected plants, and the early-flowering phenotype observed in infected grapevine. Additionally, no flowering was observed in uninoculated plants or plants infected with wtALSV. Introduction of *VvFT* or *VvTFL1 A, B,* or *C* [47] alone with ALSV did not induce flowering in 'Koshu' and Genebu seedlings (Table 2). Although the reason that expression of *VvFT* does not induce flowering in grapevine is unknown, the same phenomenon occurs with a homologous combination of *MdFT1* or 2 from apple [26]. There is probably a complex mechanism in which flowering in grapevine cannot be induced only by expression of *VvFT* or suppression of *VvTFLs* expression, for example, other genetic factors are needed to be supplemented.

Flowering was induced one month after inoculation with ALSV-AtFT, ALSV-AtFT/VvTFL1A, and ALSV-AtFT/VvTFL1B of germinating seedlings and in vitro cultures. We expected that the concurrent expression of *AtFT* and suppression of grapevine *VvTFL1A* or *B* would improve flowering rates in grapevine seedlings, similar to apple infected with ALSV-AtFT/MdTFL1 [26]. However, there was no difference in flowering rates among plants expressing ALSV-AtFT, ALSV-AtFT/VvTFL1A, and ALSV-AtFT/VvTFL1B (Table 2), suggesting that early flowering is due to *AtFT* gene expression, not concurrent expression of *AtFT* and suppression of grapevine *VvTFL1A* or *B*.

When germinating seedlings were used for inoculation experiments, the plants flowered after four to five true leaves developed (Figure 5A) and continued flowering but could not grow enough to bear fruits with seed. Only small fruits set on parts of early-flowering plants, without viable seeds (Figure 5A). This could be due to the activity of *AtFT*, which induces grapevine flowering but changes all buds to flowers and terminates vegetative growth. However, in vitro cultures inoculated with ALSV vectors continued to grow and flower (Figure 5C), and consequently grew large enough to bear fruits (Figure 6 A,B).

Ganebu is a wild grape endemic to the subtropical southwestern islands of Japan [39,41]. It has no endodormancy and has high anthocyanin content in its grape skins and could be a novel source of flavonoids [39,40]. Therefore, a breeding program to develop new grapevine varieties with ever-bearing phenotypes could utilize crosses between Ganebu and other grapevine varieties. In this paper, the ALSV vector was successfully used to promote early flowering in seedlings of F_1 hybrids from Ganebu × 'Nehelescol', 'Cabernet Sauvignon' × Ganebu, and Ganebu × 'Shine Muscat'. This technique would be readily available for the rapid selection of seedlings with bisexual or unisexual flowers (Figure 5B,C). The early flowering F_1 hybrid plants (G × N, CS × G, and G × SM) with bisexual flowers produced fruits through self-pollination and the grapes were used for quality analysis. Comparison of the sugar, acid, and total anthocyanin content in Ganebu, F_1 hybrid berries (G × N, CS × G, and G × SM), a wild grapevine (*Vitis coignetiae*) and commercial varieties of grapevine showed that both G × N (240-1) and CS × G (4-23) F_1 hybrids had high anthocyanin and acid content comparable to those measured in Ganebu. Conversely, G × SM (264-T31) had low anthocyanin content compared with those of G × N (240-1) and CS × G (4-23) (Table S2). Thus, the VIF technology using ALSV vectors made it possible to analyze the quality of fruits on grapevine seedlings within a year after germination.

Mature fruits from F_1 plants contained viable seeds that germinated and grew into seedling plants. About 60 self-pollinated F_2 seedlings from G × N crosses were checked for ALSV seed transmission. Although ALSV invaded pollen and ovules of infected parent plants (Figure 7), all F_1 seedlings were not infected with ALSV. These results suggest that grapevine plants can block or minimize ALSV seed transmission by inhibiting viral propagation in embryos [44]. Currently, the F_2 seedlings from G × N crosses being cultivated to select plants with ever-bearing phenotypes.

ALSV elimination from infected grapevine plants by high-temperature treatment was investigated as previously reported in apple [36]. High-temperature treatment of infected seedlings at 37 °C for 30 days inhibited ALSV systemic movement into newly developed leaves, even after incubation at 25 °C for 30 days (Figure 8). Once the multiplication and movement of the virus was arrested, ALSV was not longer able to infect upper leaves, which were virus-free (Figure 8). Thus, ALSV-free shoots were easily obtained from infected grapevines and used as breeding-stocks.

5. Conclusions

The availability of ALSV vectors for VIF was demonstrated through this research and will be useful for accelerating the breeding of new varieties of grapevine, combined with marker-assisted selection [48]. Recently, the new plant breeding technique for gene modification using the CRISPR-Cas9 system was also applied in plants including grapevine [49–52]. Our ALSV system will provide a helpful tool to shorten the time necessary for evaluating fruit qualities in gene-edited grapevine and eliminate exogenous genes from transformed plants by genetic crosses.

Supplementary Materials: The following are available online at http://www.mdpi.com/1999-4915/12/1/70/s1, Table S1: Primers used for construction and detection of ALSV vectors, Table S2: Contents of sugar, acid, and total anthocyanin in fruits of F1 hybrids, Ganebu, Crimson glory vine, and commercial cultivars of grapevine

Author Contributions: K.M. and T.K. constructed ALSV vectors, I.K. and C.L. performed VIGS, N.Y. (Noriko Yamagishi) conducted VIF, H.Y. prepared plant materials, and N.Y. (Nobuyuki Yoshikawa) designed the study and wrote the article. All authors have read and agreed to the published version of the manuscript.

Funding: This research was supported by a grant from Cross-Ministerial Strategic Innovation Promotion Program (SIP), and "Technologies for creating next-generation agriculture, forestry and fisheries" (grant numbers: 26012A and 26012AB, funding agency: Bio-oriented Technology Research Advancement Institution, NARO).

Acknowledgments: We are grateful to Ikuko Nakajima (NARO Institute of Fruit Tree Science, Fujimoto 2-1, Tsukuba, Ibaraki 305-8605 Japan) for kindly supplying the in vitro culture of 'Neo Masucat'; and Takayuki Momma, Kohei Yagi, Miyoko Kawasaki, and Motoko Sato for their excellent technical support.

Conflicts of Interest: The authors declare no conflict of interest. The funders had no role in the design of the study; in the collection, analyses, or interpretation of data; in the writing of the manuscript, or in the decision to publish the results.

References

1. Carrington, J.C.; Kasschau, K.D.; Mahajan, S.K.; Schaad, M.C. Cell-to-cell and long-distance transport of viruses in plants. *Plant Cell* **1996**, *8*, 1669–1681. [CrossRef]
2. Scholthof, H.B.; Scholthof, K.B.G.; Jackson, A.O. Plant virus gene vectors for transient expression of foreign proteins in plants. *Annu. Rev. Phytopathol.* **1996**, *34*, 299–323. [CrossRef] [PubMed]
3. Burch-Smith, T.M.; Andersen, J.C.; Martin, G.B.; Dinesh-Kumar, S.P. Applications and advantages of virus-induced gene silencing for gene function studies in plants. *Plant J.* **2004**, *39*, 734–746. [CrossRef] [PubMed]
4. Kumagai, M.H.; Donson, J.; Della-Cioppa, G.; Harvey, D.; Hanley, K.; Grill, L.K. Cytoplasmic inhibition of carotenoid biosynthesis with virus-derived RNA. *Proc. Natl. Acad. Sci. USA* **1995**, *92*, 1679–1683. [CrossRef] [PubMed]
5. Robertson, D. VIGS vectors for gene silencing: Many targets, many tools. *Annu. Rev. Plant Biol.* **2004**, *55*, 495–519. [CrossRef]
6. Igarashi, A.; Yamagata, K.; Sugai, T.; Takahashi, Y.; Sugawara, E.; Tamura, A.; Yaegashi, H.; Yamagishi, N.; Takahashi, T.; Isogai, M.; et al. *Apple latent spherical virus* vectors for reliable and effective virus-induced gene silencing among a broad range of plants including tobacco, tomato, *Arabidopsis thaliana*, cucurbits, and legumes. *Virology* **2009**, *386*, 407–416. [CrossRef]
7. Constantin, G.D.; Krath, B.N.; MacFarlane, S.A.; Nicolaisen, M.; Johansen, I.E.; Lund, O.S. Virus-induced gene silencing as a tool for functional genomics in a legume species. *Plant J.* **2004**, *40*, 622–631. [CrossRef]
8. Dommes, A.B.; Gross, T.; Herbert, D.B.; Kivivirta, K.I.; Becker, A. Virus-induced gene silencing: Empowering genetics in non-model organism. *J. Exp. Bot.* **2019**, *70*, 757–779. [CrossRef]
9. Grønlund, M.; Constantin, G.; Piendnoir, E.; Kovacev, J.; Johansen, I.E.; Lund, O.S. Virus-induced gene silencing in *Medicago truncatula* and *Lathyrus odorata*. *Virus Res.* **2008**, *135*, 345–349. [CrossRef]
10. Holzberg, S.; Brosio, P.; Gross, C.; Pogue, G.P. Barley stripe mosaic virus-induced gene silencing in a monocot plant. *Plant J.* **2002**, *30*, 315–327. [CrossRef]
11. Liu, Y.; Schiff, M.; Dinesh-Kumar, S.P. Virus-induced gene silencing in tomato. *Plant J.* **2002**, *31*, 777–786. [CrossRef] [PubMed]
12. Purkayastha, A.; Dasgupta, I. Virus-induced gene silencing: A versatile tool for discovery of gene functions in plants. *Plant Physiol. Biochem.* **2009**, *47*, 967–976. [CrossRef] [PubMed]
13. Scofield, S.R.; Nelson, R.S. Resources for virus-induced gene silencing in the grasses. *Plant Physiol.* **2009**, *149*, 152–157. [CrossRef] [PubMed]
14. Valentine, T.; Shaw, J.; Blok, V.C.; Phillips, M.S.; Oparka, K.J.; Lacomme, C. Efficient virus-induced gene silencing in roots using a modified Tobacco rattle virus vector. *Plant Physiol.* **2004**, *136*, 3999–4009. [CrossRef] [PubMed]

15. Zhang, C.; Bradshaw, J.D.; Whitham, S.A.; Hill, J.H. The development of efficient multipurpose Bean pod mottle virus viral vector set for foreign gene expression and RNA silencing. *Plant Physiol.* **2010**, *153*, 52–65. [CrossRef]

16. Zhao, F.; Lim, S.; Igori, D.; Too, R.H.; Kwon, S.-Y.; Kwon, S.-Y. Development of tobacco ringspot virus-based vectors for foreign gene expression and virus-induced gene silencing in a variety of plants. *Virology* **2016**, *492*, 166–178. [CrossRef]

17. Folimonov, A.S.; Folimonova, S.Y.; Bar-Joseph, M.; Dawson, W.O. A stable RNA virus-based vector for citrus trees. *Virology* **2007**, *368*, 205–216. [CrossRef]

18. Agüero, J.; Ruiz-Ruiz, S.; Vives, M.D.C.; Velázquez, K.; Navarro, L.; Peña, L.; Moreno, P.; Guerri, J. Development of viral vectors based on *Citrus leaf blotch virus* to express foreign proteins or analyze gene function in citrus plants. *Mol. Plant Microbe Interact.* **2012**, *25*, 1326–1337. [CrossRef]

19. Agüero, J.; Vives, M.D.C.; Velázquez, K.; Pina, J.A.; Navarro, L.; Moreno, P.; Guerri, J. Effectiveness of gene silencing induced by viral vectors based on *Citrus leaf blotch virus* is different in *Nicotiana benthamiana* and citrus plants. *Virology* **2014**, *460–461*, 154–164. [CrossRef]

20. Muruganantham, M.; Moskovitz, Y.; Haviv, S.; Horesh, T.; Fenigstein, A.; du Preez, J.; Stephan, D.; Burger, J.T.; Mawassi, M. Grapevine virus A-mediated gene silencing in *Nicotiana benthamiana* and *Vitis vinifera*. *J. Virol. Methods* **2009**, *155*, 167–174. [CrossRef]

21. Kurth, E.G.; Peremyslov, V.V.; Prokhnevsky, A.I.; Kasschau, K.D.; Miller, M.; Carrington, J.C.; Dolja, V.V. Virus-derived gene expression and RNA interference vector for grapevine. *J. Virol.* **2012**, *86*, 6002–6009. [PubMed]

22. Dawson, W.O.; Folimonova, S.Y. Virus-based transient expression vectors for woody crops: A new frontier for vector design and use. *Annu. Rev. Phytopathol.* **2013**, *51*, 321–337. [CrossRef] [PubMed]

23. Li, C.; Yamagishi, N.; Kasajima, I.; Yoshikawa, N. Virus-induced gene silencing and virus induced flowering in strawberry (*Fragaria* × *ananassa*) using apple latent spherical virus vector. *Hortic. Res.* **2019**, *6*, 18. [CrossRef] [PubMed]

24. Sasaki, S.; Yamagishi, N.; Yoshikawa, N. Efficient virus-induced gene silencing in apple, pear and Japanese pear using *Apple latent spherical virus* vectors. *Plant Methods* **2011**, *7*, 15. [CrossRef]

25. Yamagishi, N.; Sasaki, S.; Yamagata, K.; Komori, S.; Nagase, M.; Wada, M.; Yamamoto, T.; Yoshikawa, N. Promotion of flowering and reduction of a generation time in apple seedlings by ectopical expression of the *Arabidopsis thaliana FT* gene using the *Apple latent spherical virus* vector. *Plant Mol. Biol.* **2011**, *75*, 193–204. [CrossRef]

26. Yamagishi, N.; Kishigami, R.; Yoshikawa, N. Reduced generation time of apple seedlings to within a year by means of a plant virus vector: A new plant-breeding technique with no transmission of genetic modification to the next generation. *Plant Biotechnol. J.* **2014**, *12*, 60–68. [CrossRef]

27. Thompson, J.R.; Dasgupta, I.; Fuchs, M.; Iwanami, T.; Karasev, A.V.; Petrzik, K.; Sanfacon, H.; Tzanetakis, I.; van der Vlugt, R.; Wetzel, T.; et al. ICTV virus taxonomy profile: *Secoviridae*. *J. Gen. Virol.* **2017**, *98*, 529–531. [CrossRef]

28. Li, C.; Sasaki, N.; Isogai, M.; Yoshikawa, N. Stable expression of foreign proteins in herbaceous and apple plants using Apple latent spherical virus RNA2 vectors. *Arch. Virol.* **2004**, *149*, 1541–1558. [CrossRef]

29. Li, C.; Yoshikawa, N.; Takahashi, T.; Ito, T.; Yoshida, K.; Koganezawa, H. Nucleotide sequence and genome organization of Apple latent spherical virus: A new virus classified into the family *Comoviridae*. *J. Gen. Virol.* **2000**, *81*, 541–547. [CrossRef]

30. Fekih, R.; Yamagishi, N.; Yoshikawa, N. Apple latent spherical virus vector-induced flowering for shortening the juvenile phase in Japanese gentian and lisianthus plants. *Planta* **2016**, *244*, 203–214. [CrossRef]

31. Fujita, N.; Kazama, Y.; Yamagishi, N.; Watanabe, K.; Ando, S.; Tsuji, H.; Kawano, S.; Yoshikawa, N.; Komatsu, K. Development of the VIGS system in the dioecious plant *Silene latifolia*. *Int. J. Mol. Sci.* **2019**, *20*, 1031. [CrossRef] [PubMed]

32. Takahashi, R.; Yamagishi, N.; Yoshikawa, N. A MYB transcription factor controls flower color in soybean. *J. Hered.* **2013**, *104*, 149–153. [CrossRef] [PubMed]

33. Yamagishi, N.; Yoshikawa, N. Virus-induced gene silencing in soybean seeds and emergence stage of soybean plants with *Apple latent spherical virus* vectors. *Plant Mol. Biol.* **2009**, *71*, 15–24. [CrossRef] [PubMed]

34. Tamura, A.; Kato, T.; Taki, A.; Sone, A.; Satoh, N.; Yamagishi, N.; Takahashi, T.; Ryo, B.-S.; Natsuaki, T.; Yoshikawa, N. Preventive and curative effects of Apple latent spherical virus vectors harboring part of the target virus genome against potyvirus and cucumovirus infections. *Virology* **2013**, *446*, 314–324. [CrossRef]

35. Yamagishi, N.; Yoshikawa, N. Expression of *FLOWERING LOCUS T* from Arabidopsis thaliana induces precocious flowering in soybean irrespective of maturity group and stem growth habit. *Planta* **2011**, *233*, 561–568. [CrossRef] [PubMed]

36. Yamagishi, N.; Li, C.; Yoshikawa, N. Promotion of flowering by *Apple latent spherical virus* vector and virus elimination at high temperature allow accelerated breeding of apple and pear. *Front. Plant Sci.* **2016**, *7*, 171. [CrossRef] [PubMed]

37. Nakamura, K.; Yamagishi, N.; Isogai, M.; Komori, S.; Ito, T.; Yoshikawa, N. Seed and pollen transmission of *Apple latent spherical virus* in apple. *J. Gen. Plant Pathol.* **2011**, *77*, 48–53. [CrossRef]

38. Where Are Grapes Grown? Worldatlas. Available online: https://www.worldatlas.com/articles/top-grape-growing-countries.html (accessed on 6 January 2020).

39. Matsui, H. Physiological and Ecological Characteristics of Wild Grapes Originating from Japan. *J. Brew. Soc. Jpn.* **1989**, *84*, 687–693. (In Japanese) [CrossRef]

40. Shiozaki, S.; Murakami, K. Flavonoid profiles of wild grapes native to Japan: *Vitis coignetiae* Pulliat and *Vitis ficifolia* Bunge var. ganebu Hatusima. *Agric. Sci.* **2017**, *8*, 239–252.

41. Yamashita, H.; Mochioka, R. Wild grape germplasms in Japan. *Adv. Hortic. Sci.* **2014**, *28*, 214–224.

42. Kon, T.; Yoshikawa, N. Induction and maintenance of DNA methylation in plant promoter sequences by apple latent spherical virus-induced transcriptional gene silencing. *Front. Microbiol.* **2014**, *5*, 595. [CrossRef] [PubMed]

43. Yamagishi, N.; Sasaki, S.; Yoshikawa, N. Highly efficient inoculation method of apple viruses to apple seedlings. *Julius-Kühn-Archiv* **2010**, *427*, 226–229.

44. Kamada, K.; Omata, S.; Yamagishi, N.; Kasajima, I.; Yoshikawa, N. Gentian (*Gentiana triflora*) prevents transmission of apple latent spherical virus (ALSV) vector to progeny seeds. *Planta* **2018**, *248*, 1431–1441. [CrossRef]

45. Kasajima, I.; Ito, M.; Yamagishi, N.; Yoshikawa, N. Apple latent spherical virus (ALSV) vector as a tool for reverse genetic studies and non-transgenic breeding of a variety of crops. In *Plant Epigenetics*; Rajewsky, N., Jurga, S., Barciszewski, J., Eds.; RNA Technologies' Series; Springer International Publishing AG: Cham, Switzerland, 2017; pp. 513–536.

46. Namai, K.; Matsushima, Y.; Morishima, M.; Amagai, M.; Natusaki, T. Resistance to anthracnose is decreased by tissue culture but increased with longer acclimation in the resistant strawberry cultivar. *J. Gen. Plant Pathol.* **2013**, *79*, 402–411. [CrossRef]

47. Carmona, M.J.; Calonje, M.; Martinez-Zapater, J.M. The *FT/TFL1* gene family in grapevine. *Plant Mol. Biol.* **2007**, *63*, 637–650. [CrossRef]

48. Yang, S.; Fresnedo-Ramirez, J.; Wang, M.; Cote, L.; Schweitzer, P.; Barba, P.; Takacs, E.M.; Clark, M.; Luby, J.; Manns, D.C.; et al. A next-generation marker genotyping platform (AmpSeq) in heterozygous crops: A case study for marker-assisted selection in grapevine. *Hortic. Res.* **2016**, *3*, 16002.

49. Costa, L.D.; Malony, M.; Gribaudo, I. Breeding next generation tree fruit: Technical and legal challenges. *Hortic. Res.* **2017**, *4*, 17067.

50. Malnoy, M.; Viola, R.; Jung, M.-H.; Koo, O.-J.; Kim, S.; Kim, J.-S.; Velasco, R.; Kanchiswarmy, C.N. DNA-free genetically edited grapevine and apple protoplast using CRISPER/Cas9 ribonucleoproteins. *Front. Plant Sci.* **2016**, *7*, 1904.

51. Nakajima, I.; Ban, Y.; Azuma, A.; Onoue, N.; Moriguchi, T.; Yamamoto, T.; Toki, S.; Endo, M. CRISPER/Cas9-mediated targeted mutagenesis in grape. *PLoS ONE* **2017**, *12*, e0177966.

52. Ren, C.; Liu, X.; Zhang, Z.; Wang, Y.; Duan, W.; Li, S.; Liang, Z. CRISPR/Cas9-mediated efficient targeted mutagenesis in Chardonnay (*Vitis vinifera* L.). *Sci. Rep.* **2016**, *6*, 32289. [CrossRef]

 *viruses*

Article

Diversity and Host Specificity Revealed by Biological Characterization and Whole Genome Sequencing of Bacteriophages Infecting *Salmonella enterica*

Karen Fong [1], Denise M. Tremblay [2,3], Pascal Delaquis [4], Lawrence Goodridge [5], Roger C. Levesque [6], Sylvain Moineau [2,3,7], Curtis A. Suttle [8] and Siyun Wang [1,*]

[1] Food, Nutrition and Health, The University of British Columbia, Vancouver, BC V6T 1Z4, Canada; karen.fong@ubc.ca

[2] Félix d'Hérelle Reference Center for Bacterial Viruses, Faculté de Médecine Dentaire, Université Laval, Québec City, QC G1V 0A6, Canada; denise.tremblay@greb.ulaval.ca (D.M.T.); sylvain.moineau@bcm.ulaval.ca (S.M.)

[3] Groupe de Recherche en Écologie Buccale, Faculté de Médecine Dentaire, Université Laval, Québec City, QC G1V 0A6, Canada

[4] Agriculture and Agri-Food Canada, Summerland, BC V0H 1Z0, Canada; pascal.delaquis@canada.ca

[5] Food Science Department, University of Guelph, Guelph, ON N1G 2W1, Canada; lawrence.goodridge@mcgill.ca

[6] Institut de Biologie Intégrative et des Systèmes (IBIS), Université Laval, Quebec City, QC G1V 0A6, Canada; rclevesq@ibis.ulaval.ca

[7] Département de Biochimie, de Microbiologie, et de Bio-Informatique, Faculté des Sciences et de Génie, Université Laval, Québec City, QC G1V 0A6, Canada

[8] Departments of Earth, Ocean and Atmospheric Sciences, Microbiology and Immunology, and Botany, and the Institute for Oceans and Fisheries, The University of British Columbia, Vancouver, BC V6T 1Z4, Canada; suttle@science.ubc.ca

[*] Correspondence: siyun.wang@ubc.ca

Received: 2 August 2019; Accepted: 5 September 2019; Published: 14 September 2019

Abstract: Phages infecting members of the opportunistic human pathogen, *Salmonella enterica*, are widespread in natural environments and offer a potential source of agents that could be used for controlling populations of this bacterium; yet, relatively little is known about these phages. Here we describe the isolation and characterization of 45 phages of *Salmonella enterica* from disparate geographic locations within British Columbia, Canada. Host-range profiling revealed host-specific patterns of susceptibility and resistance, with several phages identified that have a broad-host range (i.e., able to lyse >40% of bacterial hosts tested). One phage in particular, SE13, is able to lyse 51 out of the 61 *Salmonella* strains tested. Comparative genomic analyses also revealed an abundance of sequence diversity in the sequenced phages. Alignment of the genomes grouped the phages into 12 clusters with three singletons. Phages within certain clusters exhibited extraordinarily high genome homology (>98% nucleotide identity), yet between clusters, genomes exhibited a span of diversity (<50% nucleotide identity). Alignment of the major capsid protein also supported the clustering pattern observed with alignment of the whole genomes. We further observed associations between genomic relatedness and the site of isolation, as well as genetic elements related to DNA metabolism and host virulence. Our data support the knowledge framework for phage diversity and phage–host interactions that are required for developing phage-based applications for various sectors, including biocontrol, detection and typing.

Keywords: Bacteriophage; *Salmonella*; biocontrol; comparative genomics; phage diversity

1. Introduction

Bacteriophages are the most abundant biological entity on Earth and have been estimated to kill 20% to 25% of microbes daily [1–4]. Moreover, phages are key contributors to bacterial ecology and evolution through obligate parasitism, using either lytic or temperate life cycles thereby resulting in direct or delayed lysis of bacterial hosts, respectively [5]. Phage–host interactions have contributed vastly to genetic flux through horizontal gene transfer that is responsible for the dissemination and acquisition of important bacterial phenotypes, such as enhanced colonization of the human gut epithelium, antimicrobial resistance and toxin production [2,6].

Phage diversity is immense and the global phage gene pool likely represents the greatest biodiversity and largest potential source of novel genes, providing new insights on phage diversity and evolutionary relationships in disparate environments [2,7,8]. Moreover, this vast diversity is a potential reservoir of antibacterial agents for developing "phage therapies" or "biocontrol" strategies to control bacterial pathogens. Phage-based biocontrol of bacterial pathogens in foods and food processing environments is an attractive alternative to using synthetic antimicrobial agents or physical disinfection treatments that can have harmful effects on humans, animals and plants [9–11]. At least, phages most suited to this purpose should exhibit a broad host range and are free of genes encoding for lysogeny and resistance to antimicrobial agents and/or virulence [10].

Non-typhoidal *Salmonella enterica* is a foodborne pathogen causing high rates of mortality and morbidity worldwide [12,13]. Globally, bacteria in the genus *Salmonella* cause 93 million enteric infections and 155,000 diarrheal deaths each year [12], and although there are animal reservoirs including poultry and swine [8], its presence in other food products such as nuts, produce and ready-to-eat products [14] confirms that it can adapt to diverse environments [15].

Comparative genomics approaches have been used to aid in the development of phage-based products targeting several genera including *Acinetobacter*, *Pseudomonas*, *Mycobacterium*, *Lactococcus*, *Vibrio*, and *Salmonella* [2]. These analyses provided insights at genomic and phylogenetic levels (e.g., phage relatedness and the elucidation of novel genetic elements), associations among phage communities across disparate environments, and elucidation of novel phage–host interactions [2,16–20]. Nevertheless, an in-depth understanding of *Salmonella* phage diversity and phenotype-genotype characteristics is lacking. Here, we present comparative phenotypic, genomic and phylogenetic analyses of 45 new phage isolates from British Columbia, Canada, that infect non-typhoidal strains of *Salmonella*.

2. Materials and Methods

2.1. Bacterial Strains and Growth Conditions

Salmonella strains ($n = 61$) were obtained from various sources, including the International Life Sciences Institute, the *Salmonella* foodborne syst-OMICS database, or were isolated from the Lower Mainland of British Columbia (Table S1). Strains were maintained at −80 °C in Brain–Heart-Infusion broth (BD/Difco, East Rutherford, NJ, United States) supplemented with 20% glycerol. Working stocks were prepared and maintained on tryptic soy agar (TSA; BD/Difco) at 4 °C for a maximum of one month. Fresh overnight liquid cultures were prepared prior to each experiment by inoculating an isolated colony into 10 mL tryptic soy broth (TSB; BD/Difco). Cultures were incubated for 16 h at 37 °C with shaking at 170 rpm.

2.2. Bacteriophage Isolation and Purification

Bacteriophages were isolated from sediment (S), cattle feces (F), sewage effluent (E), irrigation water (I), and water tanks from an aquaculture facility (W) in British Columbia, Canada, and as specified in the phage name (Figure 1 and Figure S1). Four broad-host range phages SI1, SF1, SS1, and SS4 were isolated previously [21].

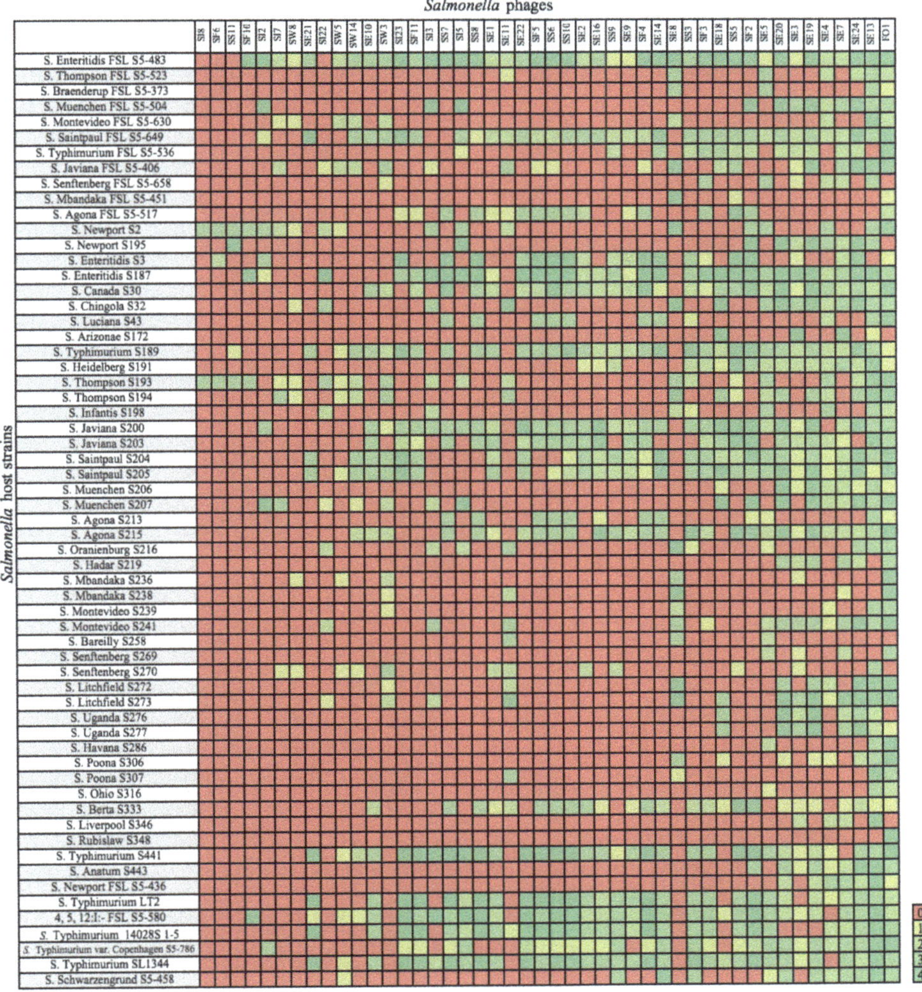

Figure 1. Host range of isolated phages. *Salmonella* strains susceptible to phage infection are indicated by a clearing of 1 to 4; 0 = no lysis.

Sample enrichment, phage isolation and purification were carried out as described elsewhere [21]. Briefly, an effort was made to enrich the abundance of phage by mixing 10 g of sample, 90 mL TSB and 1 mL of a cocktail of 7 indicator strains of *Salmonella* (Table S1) that were grown for 16 h as described above. Serotypes represented by these strains include high-risk types important to food safety and have exhibited high prevalence in human cases of salmonellosis [22,23]. Suspensions were then incubated at 37 °C for 22 ± 2 h. The phage-enriched samples were spun at 4000× g to remove bacteria, and the supernatant passed through a 0.45-µm pore-size polyethersulfone filter membrane (Pall Corporation, Port Washington, NY, United States). Afterwards, 100 µL of filtrate was mixed with 300 µL of each of the indicator *Salmonella* strains (diluted 10-fold after growing for 16 h) and 4 mL of 0.7% TSA top agar, according to the double-agar overlay method [24]. Plates were then incubated at 37 °C for 18 ± 2 h for plaque visualization. Plaques were lifted from the agar surface using a truncated sterile pipette tip and re-suspended in 200 µL salt–magnesium (SM) buffer (0.05 M Tris-HCl; 0.1 M NaCl and 0.01 M MgSO$_4$; adjusted to pH 7.5). Suspensions were allowed to rest for at least 6 h at room temperature. Double agar overlays were then prepared with the suspension as described previously [21]. A minimum of 3 single plaque isolations was performed in series to obtain a clonal phage isolate. Phages were subsequently concentrated by centrifugation and stored at 4 °C until further analyses.

2.3. Host Range Determination

Purified phage lysates were standardized to a concentration of 10^9 PFU/mL in SM buffer as recommended previously [24]. Felix-O1, a strictly lytic phage, was used as a control as it infects most members of the Salmonellae [25]. Host range was determined by spotting 5 μL of lysate, in duplicate, on a lawn of *Salmonella* cells (representing rare and common strains and serotypes) [26] grown for 16 h at 37 °C (n = 61 strains; Figure 1). The drops were allowed to dry at room temperature prior to incubation at 37 °C for 18 ± 2 h. Zones of cell lysis were assessed with a scaling system [21,22,27], where 0 indicated a zone with complete turbidity (no infection) and +4 indicated a completely clear zone with no turbidity. These values were converted into a heat map as shown in Figure 1.

2.4. Phage DNA isolation

Prior to DNA isolation, 1 mL of the lysates was filtered with a 0.45-μm pore-size cellulose–acetate membrane (Pall Corporation). Subsequently, 5 μL of both 10X RNAse A (Invitrogen, Carlsbad, CA, United States) and 1X DNAse (Invitrogen) were added to the filtered lysates for removal of contaminating host nucleic acid. $MgSO_4$ was also added to a final concentration of 10 mM and the suspension was incubated at 37 °C for 30 min. Then, 100 μL of lysis solution (2.5% sodium dodecyl–sulfate, 0.25 M EDTA and 0.50 M Tris-HCl (pH 9.0)) was added to the mixture, followed by incubation at 65 °C for 30 min. Subsequently, 125 μL of 8 M potassium acetate (Amresco, Solon, OH, United States) was added and the suspensions were placed on ice for 30 min, then spun at 17,000 rpm ($27,141\times g$) for ten min at 4 °C. Afterwards, 500 μL of phenol–chloroform (Amresco) was added, the contents were mixed on a vortex for 1 min, and spun at 14,000 rpm ($18,407\times g$) for 10 min at room temperature. The upper phase containing the DNA was carefully transferred to a clean microcentrifuge tube and an equal volume of isopropanol (Amresco) was added, followed by centrifugation at 17,000 rpm ($27,141\times g$) for 10 min at 4 °C. The supernatant was discarded, and the pellet washed 3 times with 70% ethanol and allowed to dry for 15 min. Finally, the pellet was re-suspended in 20 μL Tris-HCl (pH 8.0) and stored at −80 °C until analyzed [28].

2.5. DNA Sequencing and Annotation Workflow

The DNA library was prepared using the Nextera XT DNA Library Preparation Kit (Illumina, Hayward, CA, United States) according to the manufacturer's instructions and shotgun-sequenced using the Illumina MiSeq platform with the MiSeq Reagent Kit v2 (Illumina). Contigs were assembled de novo from the paired-end reads with the Ray assembler version 2.2.0 [29]. Depth of sequencing in the newly isolated phages ranged from 40- to 2564-fold coverage (Table S2).

Open reading frames (ORFs) were identified and annotated with the Rapid Annotation using Subsystems Technology (RAST) pipeline [30]. Annotations were also subsequently verified using the BLASTp algorithm (NCBI), employing an E-value cut-off of 0.01 [31].

2.6. Genomic Analysis

Genomic analysis was conducted on the phage genomes to identify phages with desirable characteristics for biocontrol purposes and also to probe the biodiversity of our novel isolates. An in silico approach was taken to predict phage morphotypes by comparison with closely related phage genomes using the BLASTp algorithm (NCBI). Genomes that were most closely related (i.e., possessing the highest E-value and >50% query coverage) were chosen to aid in assigning newly sequenced phages to putative families. ARAGORN was used to identify genes encoding putative tRNAs, which employs heuristics and homology comparisons with tRNA consensus sequences for prediction of the tRNA secondary structure [32].

Phylogenetic trees were constructed in MEGA X [33]. Nucleotide sequences were aligned using the ClustalW algorithm and the phylogenetic tree constructed using the Maximum-Likelihood method employing 1000 bootstrap replicates. Clusters in the phylogenetic trees were identified using ClusterPicker [34] with an inter-cluster threshold of 50% nucleotide identity, as has been demonstrated for other phage genomes [2]. Amino-acid and nucleotide comparisons were conducted for individual phages within clusters using alignment tools such as MEGA X. Visual representations and comparisons of whole genomes were produced using EasyFig [35].

Phages were annotated using a combination of automatic (i.e., RAST) and manual (NCBI BLASTp) approaches. Phages were classified as putatively temperate when a gene encoding integrase (for integration into the bacterial host chromosome) could be identified; whereas, phages without this gene were classified as putatively lytic [20,35].

3. Results

The genus *Salmonella* comprises a diverse group of microorganisms with well-characterized pan genomes, routes of transmission, pathogenesis, and epidemiology [20,36]. Despite their potential relevance to the genomic and biological attributes of the genus, there is comparatively little genomic information on *Salmonella* phages [20].

3.1. Host Range of Phages Infecting Salmonella

Given the diversity of strains in the Salmonellae in terms of disease attribution and pathogenicity, we chose strains representing serotypes that (i) cause the most illnesses (e.g., Enteritidis, Typhimurium); (ii) are rarer (compared to Enteritidis and Typhimurium), yet emerging in North America as subtypes associated with foodborne disease (e.g., Heidelberg, Saintpaul); and (iii) demonstrate multi-drug resistance (e.g., Typhimurium SL1344, Schwarzengrund S5-458) [13,21,26].

Overall, the *Salmonella* strains representing serotypes Enteritidis and Typhimurium were lysed by most of the phages (Figure 1), with S. Enteritidis FSL S5-483 exhibiting the greatest phage susceptibility. All of the antibiotic-resistant strains were susceptible to at least one phage isolate. Moreover, the phages infected some rarer, yet emerging strains [21,26], although a subset of bacterial strains (both within and between serotypes) were resistance to phage infection (Figure 1).

The phages exhibited a variety of host ranges. Some isolates (e.g., SE21, SE10, SI23) had relatively narrow host ranges, infecting 10, 15 and 16 *Salmonella* strains, respectively; others (e.g., SE13, SE7, SE20) were broader and infected 51, 38, and 35 strains, respectively. Felix-O1 had the broadest host range, infecting 54 of 61 tested strains. Originally isolated in England [37], Felix-O1 is a virulent phage that infects 98.2% of all *Salmonella* strains and is commonly used in diagnostics and typing [38,39].

Of the newly isolated phages, SE13, from sewage, had the broadest host range, infecting 51 of 61 strains, including some that were weakly resistant (e.g., S. Liverpool S346) or fully resistant (e.g., S. Arizonae S172) to Felix-O1 infection (Figure 1). All *Salmonella* strains were infected by at least one phage except for S. Rubislaw S348 which was only infected by Felix-O1.

3.2. General Genomic Characterization

Genomic characterization was performed to identify phages that possessed particular genomic features that would be pertinent for biocontrol (i.e, rendering them either unable to be used for food biocontrol and/or possessing genes (e.g., tRNA genes, DNA-replication elements) that could potentially constitute an infection/replication advantage in the host) (Table S3). The complete genomes have been deposited into Genbank with accession numbers: MK761195—MK761199, MK770409—MK770415, MK972685—MK972699, MK972700—MK972717 (Table S4). Additionally, we describe here distinct patterns of diversity that were revealed through our genomic analysis.

Phages were organized into 12 distinct clusters on the basis of 50% nucleotide similarity (Figure 2). There were four genomic singletons (including Felix-O1), which did not cluster into any sub-groups, although they exhibited similarity to previously sequenced phages. For instance, singleton phage

SE13 showed ~93% nucleotide sequence identity to *Salmonella* phage BP63 (NC_031250), while SE5, interestingly, was 98% similar to Erwinia phage phiEa21-4 (NC_015292) (Table 1). Interestingly, phages of Cluster 6 did not show any significant matches to previously isolated phages, indicating the presence of a novel genus or cluster.

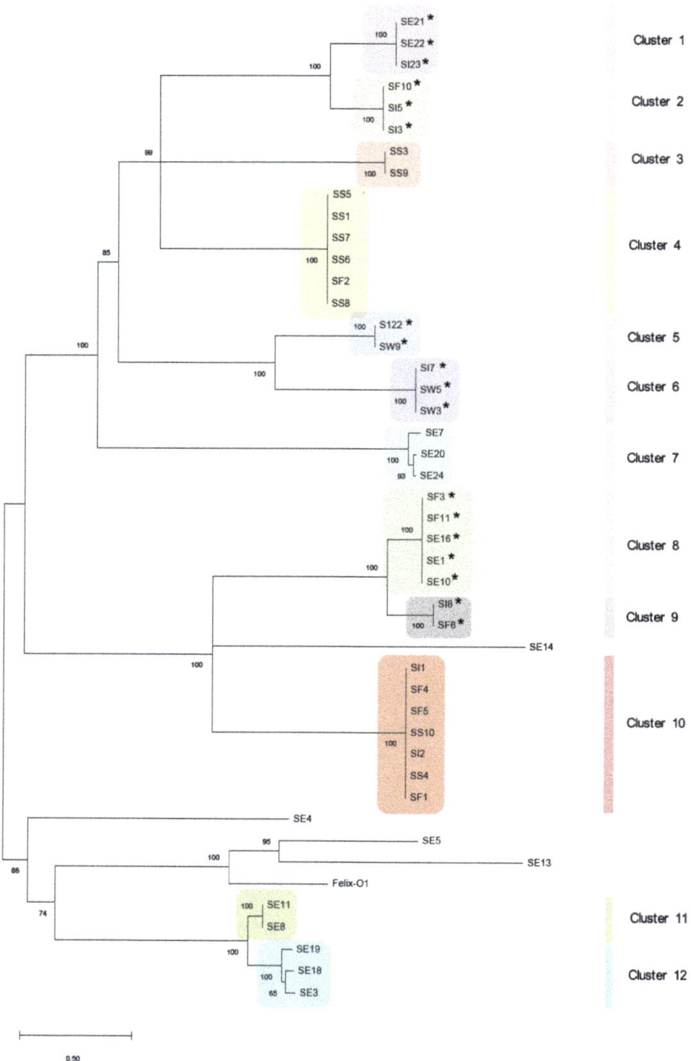

Figure 2. Dendrogram of whole genome nucleotide alignment. Tree was constructed using the ClustalW alignment and the Maximum-Likelihood method in MEGA X with 1000 bootstrap replicates. Bootstrap percentages are shown next to each node. Scale represents the number of nucleotide substitutions per site. Putatively temperate phages are indicated with an asterisk (*).

Table 1. Closest related sequenced phages to newly isolated phages. The closest related phages (possessing highest E-value and >50% query coverage) and their respective genera were determined through nucleotide homology using NCBI BLASTn.

Newly Isolated Phage	Cluster	Closest Related Phage (NCBI Best Match)	Genus
SE21	1	*Salmonella* phage 103203_sal5	*Lederbergvirus*
SE22	1	*Salmonella* phage 103203_sal5	*Lederbergvirus*
SI23	1	*Salmonella* phage 103203_sal5	*Lederbergvirus*
SF10	2	*Salmonella* phage ST160	*Lederbergvirus*
SI5	2	*Salmonella* phage ST160	*Lederbergvirus*
SI3	2	*Salmonella* phage ST160	*Lederbergvirus*
SS3	3	*Salmonella* phage GG32	*Cvivirinae*
SS9	3	*Salmonella* phage GG32	*Cvivirinae*
SS5	4	*Salmonella* phage vB_SenS_SB3	*Guernseyvirinae*
SS1	4	*Salmonella* phage vB_SenS_SB3	*Guernseyvirinae*
SS7	4	*Salmonella* phage vB_SenS_SB3	*Guernseyvirinae*
SS6	4	*Salmonella* phage vB_SenS_SB3	*Guernseyvirinae*
SF2	4	*Salmonella* phage vB_SenS_SB3	*Guernseyvirinae*
SS8	4	*Salmonella* phage ST3	*Guernseyvirinae*
SI22	5	*Salmonella* phage FSL SP-004	*Peduovirinae*
SW9	5	*Salmonella* phage FSL SP-004	*Peduovirinae*
SI7	6	N/A	N/A
SW5	6	N/A	N/A
SW3	6	N/A	N/A
SE7	7	*Salmonella* phage S147	*Tequintavirus*
SE20	7	*Salmonella* phage Seabear	*Tequintavirus*
SE24	7	*Salmonella* phage S126	*Tequintavirus*
SF3	8	*Salmonella* phage 103203_sal5	*Lederbergvirus*
SF11	8	*Salmonella* phage 103203_sal5	*Lederbergvirus*
SE16	8	*Salmonella* phage 103203_sal5	*Lederbergvirus*
SE1	8	*Salmonella* phage 103203_sal5	*Lederbergvirus*
SE10	8	*Salmonella* phage 103203_sal5	*Lederbergvirus*
SI8	9	*Salmonella* phage ST160	*Lederbergvirus*
SF6	9	*Salmonella* phage SE1	*Lederbergvirus*
SE14	Singleton	*Salmonella* phage S115	*Cvivirinae*
SI1	10	*Salmonella* phage vB_SenS_SB3	*Guernseyvirinae*
SF4	10	*Salmonella* phage vB_SenS_SB3	*Guernseyvirinae*
SF5	10	*Salmonella* phage vB_SenS_SB3	*Guernseyvirinae*
SS10	10	*Salmonella* phage vB_SenS_SB3	*Guernseyvirinae*
SI2	10	*Salmonella* phage vB_SenS_SB3	*Guernseyvirinae*
SS4	10	*Salmonella* phage vB_SenS_SB3	*Guernseyvirinae*
SF1	10	*Salmonella* phage vB_SenS_SB3	*Guernseyvirinae*
SE4	Singleton	*Salmonella* phage ZCSE2	N/A
SE5	Singleton	*Erwinia* phage phiEa21-4	*Ounavirinae*
SE13	Singleton	*Salmonella* phage BP63	N/A
SE11	11	*Salmonella* phage SP01	*Tequintavirus*
SE8	11	*Salmonella* phage SP01	*Tequintavirus*
SE19	12	*Salmonella* phage SP01	*Tequintavirus*
SE18	12	*Salmonella* phage BSP22A	*Tequintavirus*
SE3	12	*Salmonella* phage S147	*Tequintavirus*

Overall, clusters of phages with a high degree of genetic similarity could be isolated from disparate sampling sites. For instance, phages in Cluster 8 were isolated from sewage effluent and cattle feces, while Cluster 10 contained phages from sediment, irrigation water and cattle feces (Figure 2). However, phages from sewage also occurred across different clusters (e.g., Clusters 1, 7, 8, 11, and 12) or could not be assigned a specific cluster, indicating sewage is a rich source of phage diversity.

A lysogeny module containing a gene encoding for integrase (i.e., responsible for integration into the host chromosome) was identified in 18 of the 45 phages, indicating a possible temperate lifestyle (Figure 2 and Table S3). Furthermore, these phages were also isolated from diverse environments (Table S5). The putative temperate phages belonged to seven clusters: Clusters 1, 2, 5, 6, 8, and 9 (Figure 2). Accessory proteins associated with recombination (e.g., C protein and Cox proteins) were also identified in the lysogeny modules of Cluster 6 phages.

Interestingly, we also noted similar clustering patterns upon comparison of the whole genome and major capsid protein (MCP) phylogenetic trees (Figure S1). For instance, 5/6 of the phages in Cluster 4 in the whole genome dendrogram (Figure 2) are also clustered together in Cluster 4 of the MCP dendrogram (Figure S1). All phages within Clusters 5 and 6 on both figures are also grouped together similarly. Clusters 8 and 9 of the whole genome dendrogram are also grouped together to form Cluster 7 in the MCP dendrogram. On the basis of the MCP, singletons SE13, SE4, and SE14, as identified in Figure 2, are either grouped together or with other phages (Cluster 4, Figure S1), indicating the conservation of this shared core gene. Given the patterns of clustering, these results suggest the MCP may be suitable as a phylogenetic marker for whole genome clustering.

3.3. Phage Classifications

The taxonomic assignments of 45 newly isolated *Salmonella* phages were predicted by in silico analysis (Table 2) as has been performed by others [40]. For those phages that could be assigned a taxonomic rank, 46.7% (21/45) were classified in the family Siphoviridae, 28.9% (13/45) and 17.8% (8/45) were assigned to the families Podoviridae and Myoviridae, respectively, and 6.7% (3/45) could not be assigned to a family. Previously, phages SI1, SF1, SS1, and SS4 were classified as Siphoviridae, based on morphology determined by transmission electron microscopy [21]. The predicted morphotypes also correlated with cluster analysis of the whole genome and the MCP. For example, Cluster 3 of Figure S1 solely comprised phages predicted as Myoviruses according to the in silico analysis. Additionally, Cluster 4 contained predicted Siphoviruses (Figure S1).

Table 2. Genotypes and taxonomic assignments predicted from in silico analysis of 45 *Salmonella* phage genomes. Asterisks indicate morphotypes which have been confirmed by transmission electron microscopy [21]. Sources of phages are denoted as follows: sediment (S), cattle feces (F), sewage effluent (E), irrigation water (I), and water tanks from an aquaculture facility (W).

Morphotype Classification

Myoviridae	Siphoviridae	Podoviridae	Unclassified
SE4	SE3	SE1	SI7
SE5	SE7	SE10	SW3
SE13	SE8	SE16	SW5
SE14	SE11	SE21	
SI22	SE18	SE22	
SS3	SE19	SF3	
SS9	SE20	SF6	
SW9	SE24	SF10	
	SF1*	SF11	
	SF2	SI3	
	SF4	SI5	
	SF5	SI8	
	SI1*	SI23	
	SI2		
	SS1*		
	SS4*		
	SS5		
	SS6		
	SS7		
	SS8		
	SS10		

tRNA Genes

Phage	Number	Phage	Number	Phage	Number	Phage	Number	Phage	Number
SE1	0	SF1	0	SI1	0	SS1	0	SW3	0
SE3	29	SF2	0	SI2	0	SS3	4	SW5	0
SE4	0	SF3	0	SI3	1	SS4	0	SW9	0
SE5	27	SF4	0	SI5	1	SS5	0		
SE7	29	SF5	0	SI7	0	SS6	0		
SE8	22	SF6	1	SI8	1	SS7	0		
SE10	0	SF10	1	SI22	0	SS8	0		
SE11	22	SF11	0	SI23	0	SS9	4		
SE13	0					SS10	0		
SE14	4								
SE16	0								
SE18	28								
SE19	29								
SE20	29								
SE21	0								
SE22	0								
SE24	29								

arnC

Phage	Cluster	Source
SE21	1	E
SE22	1	I
SI23	1	E
SE16	10	E
SF11	10	F
SE10	10	E
SF3	10	F
SE1	10	E

vriC

Phage	Cluster	Source
SS3	3	S
SS9	3	S
SE14	Singleton	E

exoZ

Phage	Cluster	Source
SI3	2	I
SI5	2	I
SF10	2	F
SI8	11	I
SF6	11	F

3.4. Identification of Putative Phage tRNAs

A range of putative tRNA genes were identified in our phage collection (Table 2), with at least one tRNA identified in 36% (16/45) of the phages. The distribution of tRNA-containing phages varied based on the isolation source, with 59% (10/17) of phages from sewage, 25% (2/8) from cattle feces, 38% (3/8) from irrigation water; 22% (2/9) from sediment; and none (0/3) from aquaculture possessing at least one tRNA-encoding gene. The findings suggest that tRNA genes are not uncommon within *Salmonella* phages and may vary geographically.

3.5. Genomic Analysis of SE13

Phage SE13 isolated from sewage had the broadest host range of the 45 isolates, lysing nearly all of the antibiotic-resistant strains tested (except *S.* Agona S5-517) and some rare strains (e.g., *S.* Poona S306, S307; uncommonly seen in outbreaks) that were resistant to infection from other phages in our collection, including Felix-O1 (e.g., *S.* Arizonae S172) (Figure 1). SE13 also lysed the serotypes responsible for the highest rates of infection worldwide, *S.* Enteritidis and *S.* Typhimurium [39].

Based on BLASTn, the 52,438 bp genome (G+C = 45.8%) of SE13 revealed 93% identity to *Salmonella* phage BP63, with putative genes involved in structure, host recognition, and metabolism/replication. RAST identified 73 ORFs, suggesting that approximately 9% of the genome is non-coding (Figure 3). RAST assigned functions to 13 of 73 ORFs and subsequent verification with NCBI BLASTp further assigned functions to five additional ORFs, including the major capsid protein. The remaining 55 ORFs were classified as hypothetical (Table S3). No lysogeny-related modules encoding integrase, nor antibiotic-resistance and/or virulence factors were identified.

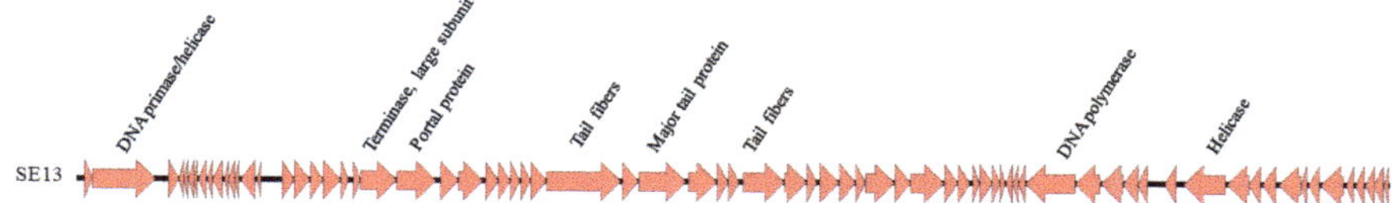

Figure 3. Linear whole-genome representation of phage SE13. Large open reading frames (ORFs) greater than 1000 bp are indicated. Arrows indicate the direction of transcription.

3.6. Genome Size, G+C content and Identification of DNA Metabolism-Related Genes

Cluster analysis separated the phages into 12 groups (Figure 2). ORF prediction using a combination of RAST and BLASTp revealed genetic elements shared among phages in a cluster, as well as distinct genotypes exhibited among clusters. Genome sizes ranged from 30,037 bp to 158,539 bp, and G+C values from 39.2% to 54.4%, in concordance with other phages of *Salmonella* [5,20].

Phages with genomes >100,000 bp are represented in Clusters 3, 7, 11, and 12; additionally, SE14, a genomic singleton, possesses a genome of 152,926 bp and 198 ORFs. These large genome phages possess accessory genes encoding for proteins involved in phage replication (e.g., DNA polymerase, DNA helicase, DNA primase, replication factor C, sliding clamp loader subunit) that were adjacent (i.e., modular in its arrangement), or separated by a non-related ORF. Further, these genes are positioned on the same strand, implying they are likely to be transcribed together as part of a module [41]. Phages of Cluster 3 are represented in Figure 4.

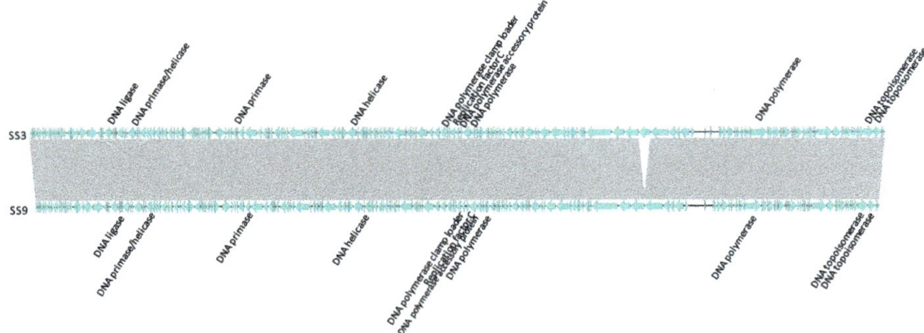

Figure 4. Linear whole genome comparison of phages in Cluster 3. ORFs encoding for DNA replication elements are indicated. Grey regions indicate nucleotide homology of >96%. Directions indicate the direction of transcription.

These modules were not identified in clusters of phages with smaller genomes (<50,000 bp), nor in genomic singletons SE13 (52,438 bp), SE4 (53, 494 bp), and SE5 (84,567 bp). However, in some of these phages, a small number of replication-related elements were dispersed throughout the genome as well as duplicate copies of select genes encoding for replication-related functions (e.g., DNA ligase, DNA topoisomerase, DNA helicase). Concordantly, duplicate copies of some of these genes were identified in Clusters 3, 7, 11, and 12, representing phages with larger genomes (Figure 4).

3.7. Identification of Genes Encoding for Putative Virulence Factors

Putative virulence factors were identified in three phages (Table 2). Three genes putatively encode virulence-related functions, including a polymyxin resistance protein ArnC (311 amino acids). Additionally, a gene encoding for exopolysaccharide production protein ExoZ (380 amino acids) was identified in phages in Clusters 2 and 11 (Table 2). Lastly, a large virulence protein VriC (1613 amino acids) was identified in phages of Cluster 3 (SS3 and SS9), as well as the genomic singleton, SE14. A nucleotide alignment using ClustalW revealed that vriC in SS3 and SS9 was nearly identical with one nucleotide mismatch at position 4374, while vriC in SE14 was slightly more divergent, possessing 96% identity to that of SS3 and SS9.

4. Discussion

Here we describe 45 newly isolated phages that have important phenotypic and genomic properties for biocontrol and their patterns of diversity. Overall, we observed very diverse host ranges within our collection of phages; with phage SE13 possessing the broadest host range. Screening of putative genes revealed the carriage of genes encoding integrase, virulence and antimicrobial resistance in some phages, although SE13 was devoid of these genetic elements. A cluster analysis of the phage genomes revealed an abundance of diversity and also novel associations between genes encoding for replication-related functions, phage genome size, and the G+C content association between phage and *Salmonella* host. Generally, a larger genome size and discordant G+C content was correlated with an increased number of genes encoding for replication-related functions, which may be advantageous if using these phages in biocontrol agents. Lastly, cluster analysis of the major capsid protein revealed a similar grouping pattern with that of the whole genomes, potentiating its use as a marker for genetic relatedness.

4.1. Host Range of Phages Infecting Salmonella

We screened the phage isolates against a panel of 61 *Salmonella* strains to determine their host ranges (Figure 1), as lysis of a broad range of *Salmonella* strains is critical for biocontrol applications [8]. Ideally, phage cocktails are formulated from several phages with broad host ranges [42].

Strains representing serotypes Enteritidis and Typhimurium were lysed by most of the phages (Figure 1), which is important considering the global association of these serotypes with salmonellosis [43]. Further, all of the antibiotic-resistant strains were susceptible to at least one phage, which is noteworthy given the need for new strategies to tackle antibiotic resistance. One *Salmonella* strain, *S.* Rubislaw S348 could not be lysed by the newly isolated phages, which may be due to the lack of a suitable receptor, or other phage-resistance mechanisms.

Some groups of phages exhibited unique host ranges. Phages isolated from irrigation water displayed comparatively limited host ranges, which suggests that the available host spectrum is narrower than in other sampled environments. For example, phages from sewage exhibited broader host ranges (Figure 1), especially SE13, which exhibited the broadest host range of the newly isolated phages. This is consistent with sewage containing an abundance of potential hosts and phages [42]. However, phage abundance tends to oscillate with host abundance. The "kill the winner" hypothesis posits that the host that grows most quickly (i.e., the "winner") will be the most susceptible to phage infection, leading to greater diversity of hosts and phage. Phages with broader host range are better adapted to survive environments where host diversity is high, and the abundance of any given genotype is relatively low, or where environmental conditions and bacterial populations change often, for example sewage [44]. Sewage water and sludge are good sources of phages with broad host ranges, as are some non-sewage sources [42].

It is important to note that the phages isolated here may not represent the proportions of phages in the environment. Laboratory isolation by plaque formation and infection of indicator strains likely sample a subset of phages. However, because the indicator strains used for isolation are common North American foodborne outbreak isolates, the phages described herein may prove highly useful for the control of *Salmonella* in this jurisdiction.

4.2. General Genomic Characterization

Cluster analysis assigned phages into 12 clusters with four genomic singletons (including Felix-O1) (Figure 2). Overall, some clusters contained phages with high sequence identity that were isolated from disparate sites (e.g., clusters 8 and 10). These observations, coupled with the fact that these phages were isolated from proximate regions within British Columbia, suggest that they may be genetically endemic [20] and/or transmissible via unknown vectors (e.g., wildlife moving among sites). It has also been suggested that phages within a cluster may share common hosts [19], which implies that host populations within a cluster may also be genetically similar [19]; future metagenomic analyses may verify this hypothesis. We also saw evidence of rich diversity across clusters; phages from sewage were assigned to different clusters (e.g., Clusters 1, 7, 8, 11, and 12) or could not be assigned a specific cluster (Figure 2). As wastewater is abundant in nutrients, bacterial hosts, and high rates of horizontal gene transfer, phages isolated from these sites represent a reservoir of novel and diverse genetic materials [45].

A lysogeny module containing a gene encoding integrase was identified in 18 of the 45 phages (Figure 2). Although the use of temperate phages is counter-indicated for some biocontrol applications (e.g., in food systems) [9], phage modification by deletion of lysogeny modules may be considered for other applications, particularly if the phage has a broad host range. It should also be cautioned that in addition to genomic analysis, in vitro transduction assays should be carried out to confirm a virulent lifestyle [20,21].

The MCP was identified and functionally annotated in all phages (Figure S1), suggesting that significant divergence of this protein is constrained due to its important conserved role in capsid assembly [46] and maintenance of the viral capsid structure [38]. Because it represents a core gene and is thus not known to be horizontally transferred, multiple studies have assessed the role of MCP as a phylogenetic marker [16,46–49], further exaggerated by the fact that there is no benchmark gene used to study phage diversity. Indeed, comparisons of the whole genome and the MCP in our subset of phages show similar clustering patterns (Figure 2 and Figure S1), potentiating its use as an inference gene for genetic relatedness. Furthermore, determination of the MCP sequence may be useful for pre-assignment of novel phages into groups or clusters which share desirable characteristics for biocontrol (e.g., broad host range and infection efficiency) [48–50].

4.3. Phage Classifications

Electron microscopy and nucleic acid content have largely provided the basis for taxonomic classification [51], with the currently classified viruses exhibiting genomic relatedness in concordance with their morphotypes [52]. This correlation was further corroborated with our cluster analysis (Figure 2). The largest proportion (46.7%) of our phages were assigned to the family *Siphoviridae*, while 28.9% and 17.8% were assigned to the families *Podoviridae* and *Myoviridae*, respectively, and 6.7% could not be assigned to a family (Table 2). The diversity of morphotypes in this collection is important when informing optimal cocktail design. The sensitivity of phage to external factors (e.g., storage, temperature, and pH) varies between morphological families [53], thus cocktails comprising different phage morphotypes should be considered. Additionally, it has been previously reported that *Salmonella* phages of different morphotypes use different host receptors [54], which, when incorporated into a cocktail, diminishes the occurrence of host resistance [55].

4.4. Identification of Putative Phage tRNAs

The presence of tRNA genes is relatively common and has been observed in many phage genomes [56]. It has been proposed that particular tRNA genes benefit phage replication by corresponding to codons used by the phage genome rather than the host [57]. Accordingly, phages with similar codon usages to that of their hosts will not benefit from retention of tRNA genes and would use that of their host [58]. It has also been hypothesized that temperate phages integrate at the position of a host tRNA gene, with the phage tRNA compensating for the interruption in the host tRNA gene [59].

At least one tRNA was identified in 36% of the newly isolated phages, consistent with a large-scale comparative analysis of 827 mycobacteriophages, which revealed that 41.4% contained at least one tRNA gene, and that these displayed cluster specificity [56]. In our study, most phages from sewage had at least one tRNA. As sewage represents a rich source of host diversity, having different tRNAs might enhance phage genome replication in multiple hosts [58]. However, different numbers of phages were recovered from each site; thus, these distribution patterns are preliminary.

4.5. Genomic Analysis of SE13

Of the newly isolated phages, phage SE13 had the broadest host range (Figure 1) and no evidence of an integrase, and thus was likely lytic. Additionally, as there were no antibiotic-resistance or virulence factors identified, it is a good candidate for biocontrol of *Salmonella*.

Genetically, SE13 possesses synteny, with predicted ORFs (Table S3) encoding genes for structure (major capsid protein, scaffold protein, tail proteins), packaging (large terminase subunit, portal protein), a lysozyme, and a likely cognate holin in close proximity, as has been identified in other phage genomes [60,61]. Given the broad host range of SE13, the tail fibers are of specific interest. ORFs 31 and 45 encode for putative tail fibers of 986 and 445 amino acids, respectively, and are 93 and 97% similar, respectively, to those of *Salmonella* phage BP63, another broad host range phage that is a component of SalmoPro® (Phagelux, Inc.), a GRAS-certified antimicrobial processing aid for controlling *Salmonella* on foods [61]. A variety of tail-associated accessory proteins were also clustered together at this locus, including tail-fiber assembly proteins, ORFs 46 and 47 (Table S3). This locus also encoded for tail-associated protein products located on the same strand of the genome, indicating they are likely transcribed together as a module [62].

SE13 also possesses a variety of DNA metabolism-related genes (e.g., thymidylate synthase, deoxycytidylate deaminase, guanylate kinase, and nicotinate phosphoribosyltransferase) (Table S3). Due to their disparate positions in the genome, it suggests that these genes were acquired by separate horizontal gene transfer events with hosts, prophages or other lytic phages during co-infection, particularly since sewage, from which SE13 was isolated, is an environment facilitating a high frequency of horizontal gene transfer and rearrangements [63].

4.6. Genome size, G+C Content and Identification of DNA Metabolism-Related Genes

The cluster analysis revealed that genes encoding for self-replication-related functions were common within and among clusters, and that the number of these elements were associated with the phage genome size. We also observed duplicate copies of genes encoding for a variety of replication-related functions in Clusters 3, 7, 11, and 12, which also comprised phages with larger genomes. Duplicate copies may enhance synthesis of proteins involved in replication, and hence increase phage production and evolutionary fitness.

Although the role of these replication modules in large phages remains unclear, it is possible that larger genomes carry accessory genes that are not essential, but which enable more efficient phage replication. Efficient replication may lead to an increased burst size and/or reduced latent period, both of which are desirable when selecting phages for biocontrol purposes [64]. The fact that Clusters 3, 7, 11, and 12 have a substantially different G+C content than that of their hosts suggests that having more genes for self-replication may be particularly advantageous [20,65,66]. For instance, *Salmonella* has a G+C content of 50% to 52% [20]; whereas, the G+C content of phages in Cluster 3 is ~44%, suggesting the eleven DNA replication elements in these phages may be advantageous (Figure 4). Clusters 7, 11, and 12 also have G+C contents ranging from 39.2% to 44.7%. Moreover, some clusters with a G+C content similar to that of *Salmonella* (e.g., Clusters 4, 5, 6, 8, 10) have fewer self-replication elements (i.e., ranging from zero to three). Concordantly, *Salmonella* phages isolated from dairy farms in rural New York State with G+C contents differing from that of their hosts also harbored anywhere from 1–12 DNA replication elements [20].

4.7. Identification of Genes Encoding for Putative Virulence Factors

The selection of phages devoid of genetic elements that could pose a risk to human health is critical to biological control applications [9]. Phages can transfer DNA between hosts via transduction [65], which may result in the insertion or deletion of cryptic and/or functional genetic elements, and alter host phenotype [67]. These genetic elements may reside in the phage genome for extended durations until a susceptible host is encountered [68].

Some phages harbored one or more copies of a polymyxin resistance protein ArnC (Table 2), which has also been identified in P22-like viruses 103203_sal5, 146851_sal4, 103203_sal4 and 101962B_sal5, albeit shorter by an amino acid [69]. Naturally synthesized by the bacterium *Bacillus polymyxa*, the polymyxins are a family of last-resort oligopeptide antibiotics used in human medicine that bind to the lipopolysaccharide (LPS) of Gram-negative bacteria, increasing membrane permeability and leakage of intracellular material [70]. Alterations in the moieties comprising the LPS may confer resistance to polymyxins. For instance, the synthesis and transfer of 4-amino-L-arabinose to the LPS is carried out by multiple genes in the *arn* operon [70,71], therefore it is unclear if alterations in one gene in this locus would confer resistance to polymyxin. Phage genomes possessing *arnC* occurred in Clusters 1 and 8, which also comprise putatively temperate phages, suggesting a specialized transduction mechanism (Figure 2). Further, most phages possessing *arnC* were sourced from sewage (SE21, SE22, SE16, SE10, and SE1), a known reservoir of antibiotic resistance genes, and "hotspots" of horizontal gene transfer [64]. Gene *arnC* has also been identified in *Salmonella* phages 22 and 34 isolated in India [72]. However, the absence of antibiotic resistance elements in putatively lytic phages highlights their relatively low frequency of generalized transduction, and suitability for biocontrol.

We also identified virulence factors in a small subset of phages (Table 2). Virulence factors are naturally found in a broad variety of foodborne pathogens and contribute to enhanced host invasion and environmental fitness [15]. Cluster 3, comprising phages SS3 and SS9 from sediment, carried a gene encoding an identical large virulence protein VriC of 1,613 amino acids (Table 2). SE14, which could not be assigned to a specific cluster, also possessed *VriC* which possessed 99% amino acid identity to that of the Cluster 3 phages. Homologs of VriC have been found to occur elsewhere, for instance, in *Salmonella* phages SFP10 (99.32% amino acid identity), Sh19 (99.01% amino acid identity) [73], and *Escherichia coli* phage PhaxI (99.13% amino acid identity) [74]. Interestingly, *Salmonella* phage 38 appears to possess a truncated form of VriC of 465 amino acids [72]. Although phage-encoded, the origin and function of this protein is unclear, therefore it is unknown if homologs possessing near-identical amino acid sequences would possess the same function. Of the newly isolated phages, SS3, SS9 and SE14 are not classified as temperate, suggesting a generalized transduction mechanism. Although the frequency of generalized transduction is quite rare [67], it has been shown to transfer large genome cassettes and pathogenicity islands [68]. However, it is unclear if these phages possess a high-transducing frequency, and if *vriC* was transduced into a bacterial host would result in a functional virulence factor.

An exopolysaccharide production protein ExoZ was defined in phages in Clusters 2 and 9. *ExoZ* has been identified in a limited set of phages, including PhWands-1 and PhWands-2 [75]. Functionally, this locus encodes virulent effector proteins, as found in clinically-relevant strains of *Pseudomonas aeruginosa* found in cystic fibrosis patients [66]. Additionally, some species of *Rhizobium* encode a homolog of ExoZ involved in the acetyl modification of succinoglycan, an exopolymer [76]. However, other genes in the *exo* locus are involved in the production of exopolysaccharide [77]; therefore, if introduced alone it is unlikely to cause phenotypic conversion. Nonetheless, as exopolysaccharide production is involved in biofilm formation [77], phages encoding these genes should not be used to control bacterial pathogens, particularly since the rate of transduction is not known.

5. Conclusions

Here we described and compared 45 newly isolated *Salmonella* phage isolates on both their basis for biocontrol and biodiversity. Overall, patterns of diversity of the *Salmonella* phages isolated from British Columbia, Canada are complex, although some similarities in the whole genome, MCP sequences and morphotypes occurred among phages isolated from different sites. A novel broad host range phage (SE13) that is genetically distinct from other phage, and which shows no evidence of virulence-associated genes, represents a promising biocontrol agent against *Salmonella*.

We found several putative virulence genes (e.g., *arnC, vriC, exoZ*) in our phages which have only been reported in a few studies [20,72,75]. We also saw evidence for a novel association between genome size and DNA metabolism and GC content in our phages, suggesting links between these genes and enhanced phage replication. The carriage of these genetic elements provides insight into the phage–host interactions and, provided they are appropriately assessed on a genetic level, suggests that our phages may be good candidates for future pathogen mitigation strategies.

The advent of high-throughput sequencing has led to an explosion of insight into microbial genomes; although, sequencing of phages has lagged behind that of bacteria, despite their critical roles in bacterial evolution. The characterization of this collection of phages contribute to the limited knowledge surrounding phage diversity and phage–host interactions, and will aid with the development of biocontrol strategies against *Salmonella*.

Supplementary Materials: The following are available online at http://www.mdpi.com/1999-4915/11/9/854/s1, Figure S1: Dendrogram of the major capsid protein nucleotide alignment, Table S1: *Salmonella* strains used for phage isolation and host range determination, Table S2: Sequencing coverage of newly isolated phages, Table S3: Annotated ORFs of the sequenced phages, Table S4: Accession numbers of newly isolated phages, Table S5: Source, indicator strain and plaque morphologies of sequenced phages.

Author Contributions: Conceptualization, P.D.; L.G.; R.C.L.; and S.W.; methodology, K.F.; D.M.T.; and S.M.; formal analysis, K.F.; investigation, K.F.; data curation, K.F.; and S.W.; writing—original draft preparation, K.F.; writing—review and editing, K.F.; D.M.T.; P.D.; L.G.; R.C.L.; S.M.; C.A.S.; and S.W.; funding acquisition, S.W.; L.G.; and R.C.L.

Funding: This research was funded by Genome Canada, grant number 8505 and the National Sciences and Engineering Research Council of Canada, NSERC Discovery Grant RGPIN-2015-04871. S.M. holds the Tier 1 Canada Research Chair in Bacteriophages.

Acknowledgments: The authors gratefully acknowledge Stéphanie Loignon from the University of Laval for helpful discussions and Yvonne Ma and Abner Bogan from the University of British Columbia for technical assistance. We also thank researchers who contributed to and/or maintained The *Salmonella* Foodborne Syst-OMICS Database, including Sandeep Tamber at Health Canada, Martin Wiedmann at Cornell University, Michelle Danyluk at University of Florida, and researchers at the Institute for Integrative and Systems Biology, Université Laval.

Conflicts of Interest: The authors declare no conflict of interest. The funders had no role in the design of the study; in the collection, analyses, or interpretation of data; in the writing of the manuscript, or in the decision to publish the results.

References

1. Hendrix, R.W. Bacteriophages: Evolution of the majority. *Theor. Popul. Biol.* **2002**, *61*, 471–480. [CrossRef] [PubMed]
2. Turner, D.; Ackermann, H.W.; Kropinski, A.M.; Lavigne, R.; Sutton, J.M.; Reynolds, D.M. Comparative analysis of 37 *Acinetobacter* bacteriophages. *Viruses* **2018**, *10*, 5. [CrossRef] [PubMed]
3. Gilmore, B.F. Bacteriophages as anti-infective agents: Recent developments and regulatory challenges. *Exp. Rev. Anti-Infect. Ther.* **2012**, *10*, 533–535. [CrossRef] [PubMed]
4. Suttle, C.A. Marine viruses—Major players in the global ecosystem. *Nat. Rev. Microbiol.* **2017**, *5*, 801–812. [CrossRef] [PubMed]
5. Mikalová, L.; Bosák, J.; Hříbková, H.; Dědičová, D.; Benada, O.; Šmarda, J.; Šmajs, D. Novel temperate phages of *Salmonella enterica* subsp. *salamae* and subsp. *diarizonae* and their activity against pathogenic *S. enterica* subsp. *enterica* isolates. *PLoS ONE* **2017**, *12*, e0170734. [CrossRef]
6. Seed, K.D. Battling phages: How bacteria defend against viral attack. *PLoS Pathog.* **2015**, *11*, e1004847. [CrossRef] [PubMed]
7. Clokie, M.R.J.; Millard, A.D.; Letarov, A.V.; Heaphy, S. Phages in nature. *Bacteriophage* **2011**, *1*, 31–45. [CrossRef]
8. Cortés, P.; Spricigo, D.A.; Bardina, C.; Llagostera, M. Remarkable diversity of *Salmonella* bacteriophages in swine and poultry. *FEMS Microbiol. Lett.* **2015**, *362*, 1–7. [CrossRef] [PubMed]
9. Goodridge, L.; Fong, K.; Wang, S.; Delaquis, P. Bacteriophage-based weapons for the war against foodborne pathogens. *Curr. Opin. Food Sci.* **2018**, *20*, 69–75. [CrossRef]
10. Hagens, S.; Loessner, M.J. Application of bacteriophages for detection and control of foodborne pathogens. *Appl. Microbiol. Biotechnol.* **2007**, *76*, 513–519. [CrossRef]
11. Abedon, S.T. Lysis from without. *Bacteriophage* **2011**, *1*, 46–49. [CrossRef] [PubMed]
12. Majowicz, S.; Musto, J.; Scallan, E.; Angulo, F.; Kirk, M.; O'Brien, S.J.; Jones, T.F.; Fazil, A.; Hoekstra, R.M. International Collaboration on Enteric Disease 'Burden of Illness' Studies. The global burden of nontyphoidal *Salmonella* gastroenteritis. *Clin. Infect. Dis.* **2010**, *50*, 882–889. [CrossRef] [PubMed]
13. Scallan, E.; Hoekstra, R.M.; Angulo, F.J.; Tauxe, R.V.; Widdowson, M.A.; Roy, S.L.; Jones, J.L.; Griffin, P.M. Foodborne illness acquired in the United States—Major pathogens. *Emerg. Infect. Dis.* **2011**, *17*, 7–15. [CrossRef] [PubMed]
14. Reports of Selected *Salmonella* Outbreak Investigations. Available online: https://www.cdc.gov/salmonella/outbreaks.html (accessed on 2 November 2018).
15. Fong, K.; Wang, S. Strain-specific survival of *Salmonella enterica* in peanut oil, peanut shell, and chia seeds. *J. Food Prot.* **2016**, *79*, 361–368. [CrossRef] [PubMed]
16. Finke, J.F.; Winget, D.M.; Chan, A.M.; Suttle, C.A. Variation in the genetic repertoire of viruses infecting *Micromonas pusilla* reflects horizontal gene transfer and links to their environmental distribution. *Viruses* **2017**, *9*, 116. [CrossRef] [PubMed]
17. Liu, W.; Lin, Y.R.; Lu, M.W.; Sung, P.J.; Wang, W.H.; Lin, C.S. Genome sequences characterizing five mutations in RNA polymerase and major capsid of phages φA318 and φAs51 of *Vibrio alginolyticus* with different burst efficiencies. *BMC Genom.* **2014**, *15*, 505. [CrossRef] [PubMed]
18. Marinelli, L.J.; Fitz-Gibbon, S.; Hayes, C.; Bowman, C.; Inkeles, M.; Loncaric, A.; Russell, D.A.; Jacobs-Sera, D.; Cokus, S.; Pellegrini, M.; et al. *Propionibacterium acnes* bacteriophages display limited genetic diversity and broad killing activity against bacterial skin isolates. *mBio* **2012**, *3*, e00279-12. [CrossRef] [PubMed]

19. Pope, W.H.; Bowman, C.A.; Russell, D.A.; Jacobs-Sera, D.; Asai, D.J.; Cresawn, S.G.; Jacobs, W.R.; Hendrix, R.W.; Lawrence, J.G.; Hatfull, G.F. Whole genome comparison of a large collection of mycobacteriophages reveals a continuum of phage genetic diversity. *eLIFE* **2015**, *4*, e06416. [CrossRef] [PubMed]

20. Switt, A.I.M.; Orsi, R.H.; Den Bakker, H.C.; Vongkamjan, K.; Altie, C.; Wiedmann, M. Genomic characterization provides new insight into *Salmonella* phage diversity. *BMC Genom.* **2013**, *14*, 481.

21. Fong, K.; LaBossiere, B.; Switt, A.I.M.; Delaquis, P.; Goodridge, L.; Levesque, R.C.; Danyluk, M.D.; Wang, S. Characterization of four novel bacteriophages isolated from British Columbia for control of non-typhoidal *Salmonella* in vitro and on sprouting alfalfa seeds. *Front. Microbiol.* **2017**, *8*, 2193. [CrossRef]

22. Kutter, E. Phage host range and efficiency ofplating. In *Bacteriophage*; Clokie, M.R.J., Kropinski, A.M., Eds.; Humana Press: New York, NY, USA, 2009.

23. Ferrari, R.G.; Rosario, D.K.A.; Cunha-Neto, A.; Mano, S.B.; Figueiredo, S.F.; Conte-Jubior, C.A. Worldwide epidemiology of *Salmonella* serovars in animal-based foods: A meta-analysis. *Appl. Environ. Microbiol.* **2019**, *85*, e00591-19. [CrossRef] [PubMed]

24. Snyder, T.R.; Boktor, S.W.; M'Ikanatha, N.M. Salmonellosis outbreaks by food vehicle, serotype, season, and geographical location, United States, 1998 to 2015. *J. Food Prot.* **2019**, *82*, 1191–1199. [CrossRef] [PubMed]

25. Adams, M.H. *Bacteriophages*; Interscience Publishers: New York, NY, USA, 1959.

26. Ellis, A.; Preston, M.; Borczyk, A.; Miller, B.; Stone, P.; Hatton, B.; Shagla, A.; Hockin, J. A community outbreak of *Salmonella* Berta associated with a soft cheese product. *Epidemiol. Infect.* **1998**, *120*, 29–35. [CrossRef] [PubMed]

27. Khan Mirzaei, M.; Nilsson, A.S. Isolation of phages for phage therapy: A comparison of spot tests and efficiency of plating analyses for determination of host range and efficacy. *PLoS ONE* **2015**, *10*, e0118557. [CrossRef] [PubMed]

28. Moineau, S.; Pandian, S.; Klaenhammer, T.R. Evolution of a lytic bacteriophage via DNA acquisition from the *Lactococcus lactis* chromosome. *Appl. Environ. Microbiol.* **1994**, *60*, 1832–1841. [PubMed]

29. Boisvert, S.; Laviolette, F.; Corbeil, J. Ray: Simultaneous assembly of reads from a mix of high-throughput sequencing technologies. *J. Comput. Biol.* **2010**, *17*, 1519–1533. [CrossRef] [PubMed]

30. Aziz, R.K.; Bartels, D.; Best, A.A.; DeJongh, M.; Disz, T.; Edwards, R.A.; Formsma, K.; Gerdes, S.; Glass, E.M.; Kubal, M.; et al. The RAST server: Rapid annotations using subsystems technology. *BMC Genom.* **2008**, *9*, 75. [CrossRef]

31. Dutilh, B.E.; Cassman, N.; McNair, K.; Sanchez, S.E.; Silva, G.G.Z.; Boling, L.; Barr, J.J.; Speth, D.R.; Seguritan, V.; Aziz, R.K.; et al. A highly abundant bacteriophage discovered in the unknown sequences of human faecal metagenomes. *Nat. Commun.* **2014**, *5*, 4498. [CrossRef]

32. Laslett, D.; Canback, B. ARAGORN, a program to detect tRNA genes and tmRNA genes in nucleotide sequences. *Nucleic Acids Res.* **2004**, *32*, 11–16. [CrossRef]

33. Kumar, S.; Stecher, G.; Li, M.; Knyaz, C.; Tamura, K. MEGA X: Molecular evolutionary genetics analysis across computing platforms. *Mol. Biol. Evol.* **2018**, *35*, 1547–1549. [CrossRef]

34. Ragonnet-Cronin, M.; Lycett, S.J.; Hodcroft, E.; Hue, S.; Fearnhill, E.; Dunn, D.; Delpech, V.; Leigh Brown, A.J.; Lycett, S. Automated analysis of phylogenetic clusters. *BMC Bioinform.* **2013**, *14*, 317. [CrossRef] [PubMed]

35. Sullivan, M.J.; Petty, N.K.; Beatson, S.A. Easyfig: A genome comparison visualizer. *Bioinformatics* **2011**, *27*, 1009–1010. [CrossRef] [PubMed]

36. Colavecchio, A.; D'Souza, Y.; Tompkins, E.; Jeukens, J.; Freschi, L.; Emond-Rheault, J.G.; Kukavica-Ibrulj, I.; Boyle, B.; Bekal, S.; Tamber, S.; et al. Prophage integrase typing is a useful indicator of genomic diversity in *Salmonella enterica*. *Front. Microbiol.* **2017**, *8*, 1283. [CrossRef] [PubMed]

37. Felix, A.; Callow, B.R. Typing of paratyphoid B bacilli by means of Vi bacteriophage. *Br. Med. J.* **1943**, *2*, 127–130. [CrossRef] [PubMed]

38. Kuhn, J.; Suissa, M.; Wyse, J.; Cohen, I.; Weiser, I.; Reznick, S.; Lubinsky-Mink, S.; Stewart, G.; Ulitzur, S. Detection of bacteria using foreign DNA: The development of a bacteriophage reagent for *Salmonella*. *Int. J. Food Microbiol.* **2002**, *74*, 229–238. [CrossRef]

39. Welkos, S.; Schreiber, M.; Baer, H. Identification of *Salmonella* with the O-1 bacteriophage. *Appl. Microbiol.* **1974**, *28*, 618–622. [PubMed]

40. Wichels, A.; Biel, S.S.; Gelderblom, H.R.; Brinkhoff, T.; Muyzer, G.; Schütt, C. Bacteriophage diversity in the North Sea. *Appl. Environ. Microbiol.* **1998**, *64*, 4128–4133.

41. Murphy, J.; Bottacini, F.; Mohony, J.; Kelleher, P.; Neve, H.; Zomer, A.; Nauta, A.; van Sinderen, D. Comparative genomics and functional analysis of the 936 group of lactococcal *Siphoviridae* phages. *Sci. Rep.* **2016**, *6*, 21345. [CrossRef]

42. Weber-Dabrowska, B.; Jończyk-Matysiak, E.; Żaczek, M.; Łobocka, M.; Łusiak-Szelachowska, M.; Górski, A. Bacteriophage procurement for therapeutic purposes. *Front. Microbiol.* **2016**, *7*, 1177. [CrossRef]

43. Galanis, E.; Lo Fo Wong, D.M.; Patrick, M.E.; Binsztein, N.; Cieslik, A.; Chalermchikit, T.; Aidara-Kane, A.; Ellis, A.; Angulo, F.J.; Wegener, H.C.; et al. Web-based surveillance and global *Salmonella* distribution, 2000–2002. *Emerg. Infect. Dis.* **2006**, *12*, 381–388. [CrossRef]

44. Díaz-Muñoz, S.L.; Koskella, B. Bacteria-phage interactions in natural environments. *Adv. Appl. Microbiol.* **2014**, *89*, 135–183. [PubMed]

45. Parmar, K.; Dafale, N.; Pal, R.; Tikariha, H.; Purohit, H. An insight into phage diversity at environmental habitats using comparative metagenomics approach. *Curr. Microbiol.* **2018**, *75*, 132–141. [CrossRef] [PubMed]

46. Deveau, H.; Labrie, S.J.; Chopin, M.C.; Moineau, S. Biodiversity and classification of Lactococcal phages. *Appl. Environ. Microbiol.* **2006**, *72*, 4338–4346. [CrossRef] [PubMed]

47. Adriaenssens, E.M.; Cowan, D.A. Using signature genes as tools to assess environmental viral ecology and diversity. *Appl. Environ. Microbiol.* **2014**, *80*, 4470–4480. [CrossRef] [PubMed]

48. Grose, J.H.; Casjens, S.R. Understanding the enormous diversity of bacteriophages: The tailed phages that infect the bacterial family *Enterobacteriaceae*. *Virology* **2014**, *468–480*, 421–443. [CrossRef] [PubMed]

49. Casjens, S.R.; Grose, J.H. Contributions of P2-and P22-like prophages to understanding the enormous diversity and abundance of tailed bacteriophages. *Virology* **2016**, *496*, 255–276. [CrossRef]

50. Born, Y.; Knecht, L.E.; Eigenmann, M.; Bolliger, M.; Klumpp, J.; Fieseler, L. A major-capsid-protein-based multiplex PCR assay for rapid identification of selected virulent bacteriophage types. *Arch. Virol.* **2019**, *164*, 819–830. [CrossRef]

51. Adriaenssens, E.M.; Brister, J.R. How to name and classify your phage: An informal guide. *Viruses* **2017**, *9*, 70. [CrossRef]

52. Simmonds, P.; Aiewasakun, P. Virus classification—Where do you draw the line? *Arch. Virol.* **2018**, *163*, 2037–2046. [CrossRef]

53. Bodier-Montagutelli, E.; Morello, E.; l'Hostis, G.; Guillon, A.; Dalloneau, E.; Respaud, R.; Pallaoro, N.; Blois, H.; Vecellio, L.; Gabard, J.; et al. Inhaled phage therapy: A promising and challenging approach to treat bacterial respiratory infections. *Expert Opin. Drug Deliv.* **2017**, *14*, 959–972. [CrossRef]

54. Shin, H.; Lee, J.H.; Kim, H.; Choi, Y.; Heu, S.; Ryu, S. Receptor diversity and host interaction of bacteriophages infecting *Salmonella enterica* serovar Typhimurium. *PLoS ONE* **2012**, *7*, e43392. [CrossRef] [PubMed]

55. Rohde, C.; Resch, G.; Pirnay, J.P.; Blasdel, B.G.; Debarbieux, L.; Gelman, D.; Górski, A.; Hazan, R.; Huys, I.; Kakabadze, E.; et al. Expert opinion on three phage therapy related topics: Bacterial phage resistance, phage training and prophages in bacterial production strains. *Viruses* **2018**, *10*, 178. [CrossRef] [PubMed]

56. Delesalle, V.A.; Tanke, N.T.; Vill, A.C.; Krukonis, G.P. Testing hypotheses for the presence of tRNA genes in mycobacteriophage genomes. *Bacteriophage* **2016**, *6*, e1219441. [CrossRef] [PubMed]

57. Bailly-Bechet, M.; Vergassola, M.; Rocha, E. Causes for the intriguing presence of tRNAs in phages. *Genome Res.* **2007**, *17*, 1486–1495. [CrossRef] [PubMed]

58. Kunisawa, T. Functional role of mycobacteriophage transfer RNAs. *J. Theor. Biol.* **2000**, *205*, 167–170. [CrossRef] [PubMed]

59. Cheetham, B.F.; Katz, M.E. A role for bacteriophages in the evolution and transfer of bacterial virulence determinants. *Mol. Microbiol.* **1995**, *18*, 201–208. [CrossRef] [PubMed]

60. Bardina, C.; Colom, J.; Spricigo, D.; Otero, J.; Sánchez-Osuna, M.; Cortés, P.; Llagostera, M. Genomics of three new bacteriophages useful in the biocontrol of *Salmonella*. *Front. Microbiol.* **2016**, *7*, 545. [CrossRef] [PubMed]

61. Kang, H.W.; Kim, J.W.; Jung, T.S.; Woo, G.J. wksl3, a new biocontrol agent for *Salmonella enterica* serovars Enteritidis and Typhimurium in foods: Characterization, application, sequence analysis, and oral acute toxicity study. *Appl. Environ. Microbiol.* **2013**, *79*, 1956–1968. [CrossRef] [PubMed]

62. Weigel, C.; Seitz, H. Bacteriophage replication modules. *FEMS Microbiol. Rev.* **2006**, *30*, 321–381. [CrossRef] [PubMed]

63. Karkman, A.; Do, T.T.; Walsh, F.; Virta, M.P.J. Antibiotic-resistance genes in waste water. *Trends Microbiol.* **2018**, *26*, 220–228. [CrossRef] [PubMed]

64. Pereira, C.; Moreirinha, C.; Lewicka, M.; Almeida, P.; Clemente, C.; Cunha, A.; Delgadillo, I.; Romalde, J.L.; Nunes, M.L.; Almeida, A. Bacteriophages with potential to inactivate *Salmonella* Typhimurium: Use of single phage suspensions and phage cocktails. *Virus Res.* **2016**, *220*, 179–192. [CrossRef] [PubMed]

65. Dupuis, M.E.; Moineau, S. Genome organization and characterization of the virulent lactococcal phage 1358 and its similarities to *Listeria* phages. *Appl. Environ. Microbiol.* **2010**, *76*, 1623–1632. [CrossRef] [PubMed]

66. Miller, E.S.; Kutter, E.; Mosig, G.; Arisaka, F.; Kunisawa, T.; Rüger, W. Bacteriophage T4 genome. *Microbiol. Mol. Biol. Rev.* **2003**, *67*, 86–156. [CrossRef] [PubMed]

67. Howard-Varona, C.; Hargreaves, K.R.; Abedon, S.T.; Sullivan, M.B. Lysogeny in nature: Mechanisms, impact and ecology of temperate phages. *ISME J.* **2017**, *11*, 1511–1520. [CrossRef] [PubMed]

68. Goh, S. Phage transduction. *Methods Mol. Biol.* **2016**, *476*, 177–185.

69. Paradiso, R.; Riccardi, M.G.; Orsini, M.; Galiero, G.; Borriello, G. Complete genome sequences of three bacteriophages infecting *Salmonella enterica* serovar Enteritidis. *Genome Announc.* **2016**, *4*, e00939-16. [CrossRef] [PubMed]

70. Moskowitz, S.M.; Brannon, M.K.; Dasgupta, N.; Pier, M.; Sgambati, N.; Miller, A.K.; Selgrade, S.E.; Miller, S.I.; Denton, M.; Conway, S.P.; et al. PmrB mutations promote polymyxin resistance of *Pseudomonas aeruginosa* isolated from colistin-treated cystic fibrosis patients. *Antimicrob. Agents Chemother.* **2012**, *56*, 1019–1030. [CrossRef]

71. Abraham, N.; Kwon, D.H. A single amino acid substitution in PmrB is associated with polymyxin B resistance in clinical isolates of *Pseudomonas aeruginosa*. *FEMS Microbiol. Lett.* **2009**, *298*, 249–254. [CrossRef] [PubMed]

72. Karpe, Y.A.; Kanade, G.D.; Pingale, K.D.; Arankalle, V.A.; Banerjee, K. Genomic characterization of *Salmonella* bacteriophages isolated from India. *Virus Genes* **2016**, *52*, 117–126. [CrossRef] [PubMed]

73. Oh, B.; Moyer, C.L.; Hendrix, R.W.; Duda, R.L. The delta domain of the HK97 major capsid protein is essential for assembly. *Virology* **2014**, *456–457*, 171–178. [CrossRef]

74. Shahrbabak, S.S.; Khodabandehlou, Z.; Shahverdi, A.R.; Skurnik, M.; Ackermann, H.W.; Varjosalo, M.; Yazdi, M.T.; Sepehizadeh, Z. Isolation, characterization and complete genome sequence of PhaxI: A phage of *Escherichia coli* O157:H7. *Microbiology* **2013**, *159*, 1629–1638. [CrossRef] [PubMed]

75. Moreno Switt, A.I.; den Bakker, H.C.; Cummings, C.A.; Rodriguez-Rivera, L.D.; Govoni, G.; Raneiri, M.L.; Degoricija, L.; Brown, S.; Hoelzer, K.; Peters, J.E.; et al. Identification and characterization of novel *Salmonella* mobile elements involved in the dissemination of genes linked to virulence and transmission. *PLoS ONE* **2012**, *7*, e41247. [CrossRef]

76. York, G.M.; Walker, G.C. The succinyl and acetyl modifications of succinoglycan influence susceptibility of succinoglycan to cleavage by the *Rhizobium meliloti* glycanases ExoK and ExsH. *J. Bacteriol.* **1998**, *95*, 4912–4917.

77. Aird, E.L.H.; Brightwell, G.; Jones, M.A.; Johnston, A.W.B. Identification of the *exo* loci required for exopolysaccharide synthesis in *Agrobacterium radiobacter* NCIB11883. *Microbiology* **1991**, *137*, 2287–2297.

MDPI

St. Alban-Anlage 66

4052 Basel

Switzerland

Tel. +41 61 683 77 34

Fax +41 61 302 89 18

www.mdpi.com

Viruses Editorial Office

E-mail: viruses@mdpi.com

www.mdpi.com/journal/viruses